Biological Events Probed
by Ultrafast Laser Spectroscopy

Contributors

R. R. Alfano
Robert R. Birge
Jacques Breton
J. Buchert
Robert Callender
Edward Dolan
A. G. Doukas
Thomas G. Ebrey
Laura Eisenstein
Hans Frauenfelder
Nicholas E. Geacintov
B. I. Greene
J. C. Hindman
Barry Honig

Joseph J. Katz
Bacon Ke
V. V. Klimov
N. Leontis
David Mauzerall
Thomas L. Netzel
Lewis J. Noe
F. Pellegrino
K. S. Peters
C. V. Shank
Stanley L. Shapiro
V. A. Shuvalov
Charles E. Swenberg
Daniel Wong

WITHDRAWN
University of
Illinois Library
at Urbana-Champaign

The person charging this material is responsible for its return to the library from which it was withdrawn on or before the **Latest Date** stamped below.

Theft, mutilation, and underlining of books are reasons for disciplinary action and may result in dismissal from the University.

To renew call Telephone Center, 333-8400

UNIVERSITY OF ILLINOIS LIBRARY AT URBANA-CHAMPAIGN

NOV 16 1983
NOV 30 1983
JAN 2 1986
JAN 09 REC'D
MAY 14 1993
JAN 08 REC'D
DEC 13 1983
OCT 19 1988
Jan 3
NOV 22 REC'D
JAN 20 1984
APR 21 1989
JAN 25 REC'D
JUN 9 REC'D
FEB 23 1984
DEC 22 1989
FEB 13 REC'D
JUN 22 1984
MAR 29 REC'D
July 5
JUL 3 REC'D
OCT 19 1990
NOV 5 1990
APR 05 1985
OCT 31 REC'D
APR 30 REC'D
DEC 26 1991
DEC 31 REC'D
MAY 29 1986
JUN 11 1992
JUN 2 REC'D
AUG 27
JUN 02

Biological Events Probed by Ultrafast Laser Spectroscopy

Edited by

R. R. ALFANO

Picosecond Laser and Spectroscopy Laboratory
Department of Physics
The City College of The City University of New York
New York, New York

1982

ACADEMIC PRESS
A Subsidiary of Harcourt Brace Jovanovich, Publishers
New York   London
Paris   San Diego   San Francisco   São Paulo   Sydney   Tokyo   Toronto

COPYRIGHT © 1982, BY ACADEMIC PRESS, INC.
ALL RIGHTS RESERVED.
NO PART OF THIS PUBLICATION MAY BE REPRODUCED OR
TRANSMITTED IN ANY FORM OR BY ANY MEANS, ELECTRONIC
OR MECHANICAL, INCLUDING PHOTOCOPY, RECORDING, OR ANY
INFORMATION STORAGE AND RETRIEVAL SYSTEM, WITHOUT
PERMISSION IN WRITING FROM THE PUBLISHER.

ACADEMIC PRESS, INC.
111 Fifth Avenue, New York, New York 10003

United Kingdom Edition published by
ACADEMIC PRESS, INC. (LONDON) LTD.
24/28 Oval Road, London NW1 7DX

Library of Congress Cataloging in Publication Data
Main entry under title:

Biological events probed by ultrafast laser spectroscopy.

   Includes bibliographical references and index.
   1. Biology--Technique. 2. Laser spectroscopy.
3. Laser pulses, Ultrashort. 4. Photosynthesis.
5. Vision. 6. Deoxyribonucleic acid. 7. Hemoproteins.
I. Alfano, R. R.
QH324.9.L37B56   574   82-1613
ISBN 0-12-049950-9   AACR2

PRINTED IN THE UNITED STATES OF AMERICA

82 83 84 85   9 8 7 6 5 4 3 2 1

Contents

LIST OF CONTRIBUTORS ... xi
PREFACE ... xiii

Part I PHOTOSYNTHESIS

Chapter 1 Primary Processes of Oxygen-Evolving Photosynthesis
Daniel Wong

I. Introduction ... 3
II. The Photosynthetic Apparatus ... 5
III. Fundamental Concepts of Photosynthesis ... 11
IV. The Primary Events ... 15
V. Conclusion ... 22
References ... 23

Chapter 2 Time-Resolved Fluorescence Spectroscopy
F. Pellegrino and R. R. Alfano

I. Introduction ... 27
II. Time-Resolved Fluorescence Spectroscopy Techniques ... 31
III. Fluorescence Kinetic Measurements in Photosynthesis ... 34
IV. Future Directions ... 49
References ... 50

Chapter 3 Early Photochemical Events in Green Plant Photosynthesis: Absorption and EPR Spectroscopic Studies
Bacon Ke, Edward Dolan, V. A. Shuvalov, and V. V. Klimov

I. Introduction ... 55
II. The Primary Electron Donors of the Two Photosystems ... 57
III. Early Photochemical Events in the Electron Acceptor Systems of Photosystems I and II ... 59
References ... 76

Chapter 4 Electron Transfer Reactions in Reaction Centers of Photosynthetic Bacteria and in Reaction Center Models

Thomas L. Netzel

I. *In Vivo* Electron Transfer Reactions	79
II. *In Vitro* Electron Transfer Reactions	97
References	114

Chapter 5 Photoprocesses in Chlorophyll Model Systems

Joseph J. Katz and J. C. Hindman

I. Introduction	119
II. Chlorophyll and the Photosynthetic Unit	121
III. Chlorophyll Properties Relevant to Models	123
IV. Covalently Linked Pairs as Photoreaction Center Models	138
V. Energy Transfer in Self-Assembled Systems	144
VI. Energy Transfer in the Chl_{sp} Antenna Model	150
VII. A Model for Electron Transfer from the Primary Donor	152
References	152

Chapter 6 Exciton Annihilation and Other Nonlinear High-Intensity Excitation Effects

Nicholas E. Geacintov and Jacques Breton

I. Introduction	158
II. Photophysical Phenomena	158
III. Laser Characteristics in Relation to the Different Photophysical Processes	162
IV. Single Picosecond Pulse Excitation	163
V. Nanosecond Pulse Excitation	174
VI. Microsecond Pulse Excitation	183
VII. Picosecond Pulse Excitation Studies on Absorption Changes in Bacterial Reaction Centers	187
References	189

Chapter 7 Fluorescence Decay Kinetics and Bimolecular Processes in Photosynthetic Membranes

Charles E. Swenberg

I. Introduction	193
II. Fluorescence Decay Profiles: Low-Intensity Regime	194
III. Exciton–Exciton Annihilation: High Excitation Intensities	201
IV. Conclusion	212
References	213

Chapter 8 Statistical Theory of the Effect of Multiple Excitation in Photosynthetic Systems

David Mauzerall

I. Introduction	215
II. Statistics	216
III. Kinetics	225
IV. Application	228
V. Comparison with Other Treatments	230
VI. Conclusions	232
Appendix I: Calculation of Yield with Escape at Traps	233
Appendix II	234
References	235

Part II VISION

Chapter 9 An Introduction to Visual Pigments and Purple Membranes and Their Primary Processes

Robert Callender

I. Introduction	239
II. The Free Chromophore	241
III. Chromophore Binding and Color	243
IV. Light and Dark Reactions	246
V. The Primary Photochemical Event	250
References	254

Chapter 10 Dynamics of the Primary Events in Vision

K. S. Peters and N. Leontis

I. Rhodopsin	259
II. Isorhodopsin	266
III. Hypsorhodopsin	267
IV. Concluding Remarks	267
References	268

Chapter 11 Primary Events in Bacteriorhodopsin

Thomas G. Ebrey

I. Introduction	271
II. Absorption Measurements	273
III. Fluorescence Measurements	277
References	280

Chapter 12 Theoretical Aspects of Photoisomerization in Visual Pigments and Bacteriorhodopsin

Barry Honig

I. Introduction	281
II. Absorption Spectra	283
III. The Primary Photochemical Event	286
IV. Excited State Processes	290
V. Concluding Remarks	295
References	296

Chapter 13 Simulation of the Primary Event in Rhodopsin Photochemistry Using Semiempirical Molecular Dynamics Theory

Robert R. Birge

I. Introduction	299
II. Theoretical Approaches in the Simulation of Intramolecular Dynamics	300
III. Semiempirical Molecular Dynamics Theory and Rhodopsin Photochemistry	302
IV. Comments and Conclusions	315
References	316

Part III HEMOPROTEINS

Chapter 14 Introduction to Hemoproteins

Laura Eisenstein and Hans Frauenfelder

I. The Importance of Hemoprotein Studies	321
II. Structure and States	322
III. Fast Elementary Processes	326
IV. The Individual Steps	327
References	335

Chapter 15 The Study of the Primary Events in the Photolysis of Hemoglobin and Myoglobin Using Picosecond Spectroscopy

Lewis J. Noe

I. Introduction	339
II. Review of Binding and Photodissociation of Molecular Oxygen and Carbon Monoxide in Heme Compounds	340
III. Experimental Results	346

IV. Picosecond Photodissociation Experiments, Related Experiments, and Results	348
V. Conclusions	354
References	355

Part IV DNA

Chapter 16 Ultrafast Techniques Applied to DNA Studies

Stanley L. Shapiro

I. Introduction	361
II. Nonlinear Optical Effects and Selective Action—Proposed Infrared Schemes	363
III. Photochemical Reactions in Nucleic Acid Components Induced by Visible and Ultraviolet Pulses	365
IV. Selectivity in a Mixture of Bases and in More Complex Nucleic Acid Components	369
V. Selectivity in a RNA Component and Multiphoton Experiments with Viruses and Plasmids	371
VI. Theory of Selectivity on Nucleic Acid Components	372
VII. Selective Photodamage of Dye–Biomolecule Complexes	375
VIII. Picosecond Depolarization Measurements and Torsional Motions in DNA	378
References	382

Part V ULTRAFAST LASER TECHNIQUES

Chapter 17 Picosecond Laser Techniques and Design

A. G. Doukas, J. Buchert, and R. R. Alfano

I. Introduction	387
II. General Principles of Operation of Picosecond Lasers	388
III. Laser Components and Designs	392
IV. Single-Pulse Selection and Amplification	397
V. Obtaining the Right Wavelength	400
VI. Time Measurements and Applications	405
References	414

Chapter 18 Subpicosecond Ultrafast Laser Technique—Application and Design

C. V. Shank and B. I. Greene

I. Introduction	417
II. Subpicosecond Pulse Generation	417

III.	Measurement Techniques	420
IV.	Future Directions	426
	References	427

INDEX   429

List of Contributors

Numbers in parentheses indicate the pages on which the authors' contributions begin.

R. R. ALFANO (27, 387), *Picosecond Laser and Spectroscopy Laboratory, Department of Physics, The City College of The City University of New York, New York, New York 10031*

ROBERT R. BIRGE (299), *Department of Chemistry, University of California, Riverside, California 92521*

JACQUES BRETON (157), *Departement de Biologie, Service de Biophysique, Centre d'Etudes Nucleaires de Saclay, Gif sur Yvette, France*

J. BUCHERT (387), *Picosecond Laser and Spectroscopy Laboratory, Department of Physics, The City College of The City University of New York, New York, New York 10031*

ROBERT CALLENDER (239), *Department of Physics, The City College of The City University of New York, New York, New York 10031*

EDWARD DOLAN (55), *Charles F. Kettering Research Laboratory, Yellow Springs, Ohio 45387*

A. G. DOUKAS (387), *Picosecond Laser and Spectroscopy Laboratory, Department of Physics, The City College of The City University of New York, New York, New York 10031*

THOMAS G. EBREY (271), *Department of Physiology and Biophysics, University of Illinois at Urbana-Champaign, Urbana, Illinois 61801*

LAURA EISENSTEIN (321), *Department of Physics, University of Illinois at Urbana-Champaign, Urbana, Illinois 61801*

HANS FRAUENFELDER (321), *Department of Physics, University of Illinois at Urbana-Champaign, Urbana, Illinois 61801*

NICHOLAS E. GEACINTOV (157), *Chemistry Department, New York University, New York, New York 10003*

B. I. GREENE (417), *Bell Telephone Laboratories, Incorporated, Holmdel, New Jersey 07733*

J. C. HINDMAN (119), *Chemistry Division, Argonne National Laboratory, Argonne, Illinois 60439*

LIST OF CONTRIBUTORS

BARRY HONIG* (281), *Department of Physiology and Biophysics, University of Illinois at Urbana-Champaign, Urbana, Illinois 61801*

JOSEPH J. KATZ (119), *Chemistry Division, Argonne National Laboratory, Argonne, Illinois 60439*

BACON KE (55), *Charles F. Kettering Research Laboratory, Yellow Springs, Ohio 45387*

V. V. KLIMOV (55), *Institute of Photosynthesis, USSR Academy of Sciences, Pushchino, USSR*

N. LEONTIS (259), *Department of Chemistry, Harvard University, Cambridge, Massachusetts 02138*

DAVID MAUZERALL (215), *The Rockefeller University, New York, New York 10021*

THOMAS L. NETZEL (79), *Department of Chemistry, Brookhaven National Laboratory, Upton, New York 11973*

LEWIS J. NOE (339), *Department of Chemistry, University of Wyoming, Laramie, Wyoming 82071*

F. PELLEGRINO (27), *Picosecond Laser and Spectroscopy Laboratory, Department of Physics, The City College of The City University of New York, New York, New York 10031*

K. S. PETERS (259), *Department of Chemistry, Harvard University, Cambridge, Massachusetts 02138*

C. V. SHANK (417), *Bell Telephone Laboratories, Incorporated, Holmdel, New Jersey 07733*

STANLEY L. SHAPIRO (361), *Molecular Spectroscopy Division, National Bureau of Standards, Washington, D.C. 20234*

V. A. SHUVALOV (55), *Institute of Photosynthesis, USSR Academy of Sciences, Pushchino, USSR*

CHARLES E. SWENBERG (193), *Laboratory of Preclinical Studies, National Institute of Alcohol Abuse and Alcoholism, Rockville, Maryland 20852*

DANIEL WONG† (3), *Picosecond Laser and Spectroscopy Laboratory, Department of Physics, The City College of The City University of New York, New York, New York 10031*

---

* Present address: Department of Biochemistry, Columbia University, New York, New York 10032.

† Present address: Ames Division, Miles Laboratories, Inc., Elkhart, Indiana 46515.

Preface

The development of picosecond laser technology over the past ten years has enabled scientists to investigate some of the most important primary processes occurring in nature. Fundamental information on the structure and dynamics of these photobiological processes has been obtained from direct temporal measurements on the ultrafast time scale ranging from 10^{-13} to 10^{-9} sec. In particular, the primary processes that take place in photosynthesis, vision, hemoglobin, and DNA have been investigated with picosecond time-resolved laser spectroscopy. Indeed a great wealth of information has been obtained that has enhanced our perception of nature. Knowledge about these developments may be gathered from original research contributions scattered in scientific journals. Review articles have been published from time to time, but they have usually dealt with one particular topic or phenomenon. Available textbooks on biological processes do not cover primary events in any detail. There is a definite need for a book that covers the introductory as well as the current research work on the various aspects of ultrafast phenomena in biology.

This book reviews current progress in the experimental and theoretical understanding of primary phenomena that occur in biology on a picosecond and nanosecond time scale. It is possible for the interested reader to gain a complete overview of all the important breakthroughs and developments in understanding primary events during the past ten years (prior to 1982). In this volume, the reader will find a review of basic principles, an up-to-date survey of research results, and the current thinking of experts in the fields of photosynthesis, vision, hemoglobin, and DNA.

This book should prove a useful source both for the newcomer to the field of ultrafast phenomena and for the experienced. The book was written to attract biologists, chemists, physicists, and engineers who are interested in biological processes, and it will give them the opportunity to find all the necessary and relevant material in one presentation. The high quality of the research reviewed and covered by the contributors to this volume will enlighten the beginner as well as the expert. It is my goal that this book will stimulate future research in understanding primary events in biology.

The book is organized into five parts—the primary events in the various areas of biology are reviewed in the first four parts, and picosecond and subpicosecond laser techniques are covered in Part V.

The primary events in photosynthesis are reviewed in Part I. Wong presents an introduction to the primary processes in photosynthesis. Pellegrino and Alfano review fluorescence kinetic measurements, while Ke and co-workers review absorption kinetic measurements in higher plants. Netzel reviews kinetic measurements in bacteria photosynthesis. Work on model systems is described by Katz and Hindman. The theories of primary energy transfer appropriate to photosynthesis are covered by Swenberg. Multiexcitation processes are discussed in the articles by Geacintov and Breton and by Mauzerall.

The primary visual processes are discussed in Part II. Callender presents the introductory material. Peters and Leontis review the kinetic experimental work on rhodopsin and Ebrey reviews the kinetic work on bacteriorhodopsin. The theoretical concepts and current thinking on the primary events in vision are reviewed in the chapters by Birge and by Honig.

In Part III, Eisenstein and Frauenfelder present the introductory material on the kinetic models of hemoglobin and myoglobin. Interesting picosecond measurements on hemoglobin and myoglobin are reviewed by Noe.

Measurements on DNA are reviewed by Shapiro in Part IV. Part V covers the subject of picosecond and subpicosecond pulse generation and measurement techniques.

I wish to thank the contributors for their cooperation in this endeavor and Drs. R. Callender, A. G. Doukas, N. E. Geacintov, B. I. Greene, R. Knox, A. Lamola, B. Parson, F. Pellegrino, and S. L. Shapiro for their reviews, advice, and comments. Special thanks are given to Mrs. M. Gibbs for secretarial assistance, and to Academic Press for its continued interest and cooperation. I gratefully acknowledge NSF, NIH, AFOSR, NASA, and Hamamatsu for their support over the years.

PART I

Photosynthesis

CHAPTER 1

Primary Processes of Oxygen-Evolving Photosynthesis

Daniel Wong[*]

Picosecond Laser and Spectroscopy Laboratory
Department of Physics
The City College of The City University of New York
New York, New York

I.	Introduction	3
II.	The Photosynthetic Apparatus	5
	A. The Pigments	5
	B. Photosynthetic Membranes	8
	C. Chlorophyll–Protein Complexes	9
	D. Reaction Centers	10
III.	Fundamental Concepts of Photosynthesis	11
	A. The Photosynthetic Unit	11
	B. Two Light Reactions and the Z Scheme	13
IV.	The Primary Events	15
	A. Absorption of Light	15
	B. Fluorescence	17
	C. Energy Transfer and Migration	18
	D. Primary Photochemistry	20
V.	Conclusion	22
	References	23

I. Introduction

Energy flow into the biosphere (the thin layer of living materials on the surface of the earth and in the upper layer of the ocean) occurs through the conversion of solar energy to chemical energy by the process of photosynthesis in higher plants, algae, and certain bacteria. In the case of green plants, the process consists of the oxidation of water to oxygen and the reduction of carbon dioxide to carbohydrate in the presence of chlorophyll and light:

$$CO_2 + H_2O \xrightarrow[\text{light}]{\text{chlorophyll}} (CH_2O) + O_2 + 112 \text{ kcal/mole} \tag{1}$$

[*] Present address: Ames Division, Miles Laboratories, Inc., Elkhart, Indiana 46515.

Historically, the progress leading to this understanding was made in the late eighteenth and early nineteenth centuries. The capacity of plants to produce free oxygen was discovered by Joseph Priestly in 1772, the need for light and chlorophyll was observed by Jan Ingen-Housz in 1779, the necessity for carbon dioxide was recognized by Jean Senebier in 1782, and the involvement of water was found by Nicolas Theodore de Saussure in 1804. The realization that carbohydrate was the organic product came about 1850. The understanding that the fundamental physical function of photosynthesis was the conversion of light energy to chemical energy was due to Julius Robert von Mayer (discoverer of the principle of conservation of energy) in 1845. Over the next century progress in the general understanding of the process was slow and often dependent on technological progress. During that time, a number of discoveries were made and concepts developed which were later to provide important foundations for what can be called the modern era of photosynthesis, which began in 1960. The reader is referred to a paper by Myers (1974), titled "Conceptual Developments in Photosynthesis, 1924–1974," for a brief account of events during that exciting period of photosynthesis research.

Since 1905 it has been known that the reactions in oxygen-evolving photosynthetic systems are of two types—those activated by light (light reactions) and those that are not directly dependent on the presence of light (dark reactions). The light reactions involve two light-induced charge separation steps connected in series by a sequence of oxidation–reduction components, forming what is called the electron transport chain. The photochemical steps are known simply as photoreactions I and II, and the sites are the reaction centers. The net result of photoreaction I is the formation of a weak oxidant, the oxidized reaction center I, and a strong reductant, the reduced primary electron acceptor of photoreaction I, while that of photoreaction II is the formation of a strong oxidant, the oxidized reaction center II, and a weak reductant, the reduced primary electron acceptor of photoreaction II. Enhancing the efficiency of light collection for the reaction centers are a battery of light-harvesting chlorophylls (known also as bulk or antenna chlorophylls) and other accessory pigments, the chemical nature of which is dependent on the biological species of the organism.

The antenna, the reaction center, the electron transport intermediates, and the structural protein and lipid components make up the photosystem (PS), the functional apparatus for the photoreactions—hence PS I and PS II. Distinguished on the basis of the reaction rates, the light reactions are divided into two groups: primary events [half-times ($t_{1/2}$), 10^{-15}–10^{-9} sec] consisting of the absorption of light, transportation of excitation energy, and induction of charge separation by the excitation energy, and secondary events ($t_{1/2}$, 10^{-8}–10^{-1} sec) consisting of events such as the movement of electrons through the electron transport chain, production of adenosine triphosphate (ATP), evolution of oxygen, and movement of ions across membranes. Finally, the dark reactions, which occur on a time scale in excess of 1 sec, are coupled enzymatic reactions by which carbon dioxide is reduced to carbohydrate using the chemical energy previously stored in ATP and reduced nicotinamide adenine dinucleotide phosphate (NADPH) by the light reactions.

The dark reactions occur by one of two pathways. Most green plants incorporate CO_2 by combining it with ribulose 1,5-bisphosphate to produce two molecules of phosphoglyceric acid, a three-carbon compound. Hence the process is known as the three-carbon (C_3) pathway (also called the pentose phosphate pathway or the Calvin cycle). In subsequent reactions phosphoglyceric acid is further reduced to phosphoglyceraldehyde, from which various carbohydrates of the plant are synthesized, along with another molecule of ribulose diphosphate to regenerate the cycle. The alternative pathway is the C_4 pathway (also known as the dicarboxylic acid cycle or the Hatch–Slack pathway). Here CO_2 is combined with phosphoenolpyruvic acid in the mesophyll cells to yield oxalacetic acid and then malate and aspartate. The latter compounds are then transported to specialized bundle sheath cells where they are converted to carbohydrate by the Calvin cycle enzymes. In certain desert plants the C_4 pathway occurs in the same cells, but the reactions are separated in time (crassulacean acid metabolism).

The purpose of this chapter is to outline for the unfamiliar reader the basic concepts which are of importance in the study of the primary processes of photosynthesis. No attempt is made to provide a comprehensive picture of photosynthesis. For this, the reader is referred to the following monographs: *Bioenergetics of Photosynthesis* (1975) and *Photosynthesis: Energy Conversion by Plants and Bacteria* (1982), edited by Govindjee and published by Academic Press, and *The Intact Chloroplast* (1976) and *Primary Processes of Photosynthesis* (1977), volumes 1 and 2 of *Topics in Photosynthesis*, edited by J. Barber and published by Elsevier.

II. The Photosynthetic Apparatus

A. THE PIGMENTS

A number of pigments are associated with the process of light absorption in photosynthetic systems. These pigments can be classified into three main groups: chlorophylls, carotenoids, and phycobilins.

1. *Chlorophylls*

Chlorophylls constitute a small group of closely related, intensely green pigments long recognized as the primary light absorbers of photosynthetic organisms. Chlorophyll *a* (Chl *a*) is the most abundant member and is found in all organisms carrying out photosynthesis with the evolution of *molecular oxygen*. Chlorophyll *b* is found in higher plants and green algae, except for a few individual species and mutant forms. Chlorophylls c_1 and c_2 (Chl c_1 and Chl c_2), found in diatoms and marine brown algae, and chlorophyll *d* (Chl *d*), reported in some marine red algae are minor components.

The chemical structure of chlorophylls (Fig. 1) is related to that of porphyrins, particularly, protoporphyrin IX. The skeleton of chlorophyll consists of a square, planar tetrapyrrole macrocycle, the center of which is chelated to a Mg^{2+} ion, which if displaced by two hydrogen ions gives the corresponding pheophytin. In Chl *a*, Chl *b*, and Chl *d*, the pyrrole constituting ring IV is hydrogenated at positions 7 and 8. An important characteristic of the chlorophyll molecule is the presence of ring V, the cyclopentanone ring. Each chlorophyll molecule contains two carbonyl groups, both of which are esterified, one with methanol and the other with an isoprenoid alcohol, phytol. The latter divides the chlorophyll molecule into a polar "head" portion and a nonpolar lipophilic "tail" portion. The tetrapyrrole macrocycle provides a system of conjugated double bonds over which the π electrons can be delocalized. Also participating in this delocalization are the carbonyl groups, namely, the keto group in ring V and the aldehyde group in ring II in the case of Chl *b*.

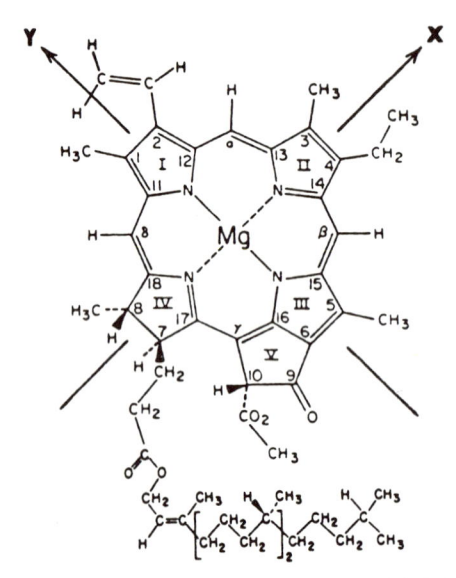

FIG. 1. Molecular structure of chlorophyll *a*. Replacement of CH_3 by CHO in ring II gives chlorophyll *b*, and replacement of $CHCH_2$ by O—CHO in ring I gives chlorophyll *d*. The two axes for electronic transitions are also shown.

The visible–near ultraviolet absorption spectrum of a chlorophyll molecule has two main absorption regions, the $S_0 \to S_1$ transition, giving the Q bands in the 550–700 nm region, and the $S_0 \to S_2$ transition, giving the Soret or B bands in the 350–500 nm region. In metal porphyrins both the excited singlet states S_1 and S_2 are orbitally degenerate, and the Q (0–0) and B (0–0) transitions appear as single bands. In chlorophyll, however, this degeneracy is lifted and the absorption bands are split, the x-polarized transition defined by the ring II–ring IV direction and the y-polarized transition by the ring I–ring III direction.

An important estimation routinely made in the laboratory is that of the chlorophyll concentration in a suspension of isolated photosynthetic membranes. Perhaps the most widely used method for this purpose is that of Mackinney (1941) in which the chlorophylls are extracted in an 80:20 mixture of acetone–water, the optical densities at 645 and 663 nm measured, and the chlorophyll concentrations calculated using the equations

$$D_{663} = 82.04 C_a + 9.27 C_b, \tag{2}$$

$$D_{645} = 16.75 C_a + 45.6 C_b, \tag{3}$$

where D is the optical density and C_a and C_b are the concentrations of Chl a and Chl b, respectively (in mg/ml). Generally, only the total chlorophyll concentration is needed, and it is more convenient to rewrite the above equation as

$$C = C_a + C_b = 8.04 D_{663} + 20.3 D_{645}, \tag{4}$$

where the units of C are μg/ml. However, when Eq. (4) is used, it is often referred to in the literature as Arnon's method (Arnon, 1949).

2. Carotenoids

Carotenoids are yellow or orange pigments found in nearly all photosynthetic organisms. They can be divided into two groups: carotenes and carotenols, which are oxygenated carotenes also known as xanthophylls. In higher plant carotenes, β-carotene is the major component and α-carotene is the minor component. Xanthophylls are numerous, but in higher plants the principal components are lutein, violaxanthin, and neoxanthin. *In vivo*, β-carotene absorbs in the region of 460–500 nm.

3. Phycobiliproteins

Phycobiliproteins are high-molecular-weight globular proteins from three groups of algae, the prokaryotic blue-green algae (Cyanophyta), the eukaryotic red algae (Rhodophyta), and the eukaryotic cryptomonads (Cryptophyceae). Organisms containing these accessory pigments do not contain

Chl *b*. There are three major categories of phycobiliproteins: blue phycocyanins (PCs), allophycocyanins (APCs), and red phycoerythrins (PEs). All blue-green algae contain APC and C-phycocyanin, while all red algae contain APC and R- or C-phycocyanin (Gantt, 1975; Glazer, 1976). Many blue-green algae and some red algae lack PE. All phycobilins are made up of two distinct polypeptide chains, forming the α and β subunits, which are present in equal amounts. The molecular masses for the small (α) subunits range from 10,000 to 20,000 daltons, and for the larger (β) subunit range from about 14,000 to 22,000 daltons (Bogorad, 1975). The actual absorbing species in the biliproteins are open-chain tetrapyrrole bile pigments which are covalently linked to the subunits. The number and chemical nature of these chromophores (phycocyanobilin, phycoerythrobilin, and phycourobilin) differ depending on the source of the biliprotein and the subunit in question. In general, the α subunit of PC contains one chromophore and the β subunit contains two. The subunits of APC contain one chromophore each (Glazer and Fang, 1973).

In vivo, phycobiliproteins are organized into macromolecular complexes known as phycobilisomes [phycobiliprotein bodies (Gantt, 1975) attached in a regular array to the outside of the photosynthetic membranes (thylakoids, discussed below)]. Studies on phycobilisomes and on intact cells indicate the following pathway of energy transfer (Gantt and Lipschultz, 1973; Ley and Butler, 1976; Porter *et al.*, 1978):

PE ⟶ PC ⟶ APC ⟶ APC-B ⟶ Chl *a*.

(~560 nm) (~620 nm) (~650 nm) (~670 nm) (~680 nm)

The approximate absorption maximum for each pigment is indicated in parentheses. The approximate peak locations of biliprotein fluorescence are 575 nm for PE, 636 nm for PC, and 660 and 675 nm for APC.

B. PHOTOSYNTHETIC MEMBRANES

The light reactions of oxygen-evolving photosynthetic organisms are localized in chlorophyll-containing membranes (or lamellae) of flattened sacs called thylakoids. In primitive blue-green algae the thylakoids are situated in the general cytoplasm of the cell, mainly toward the periphery. In red algae the photosynthetic membranes are segregated within the intracellular body (organelle) known as the chloroplast. In both types of organisms the main accessory pigments are the water-soluble biliproteins—PE, PC, and APC—localized in the phycobilisomes. Higher plants show still greater complexity in chloroplast structure.

A typical higher plant chloroplast (Fig. 2) is composed of two outer limiting envelope membranes enclosing an internal lamellae structure containing areas of stacked or appressed lamellae known as grana, connected

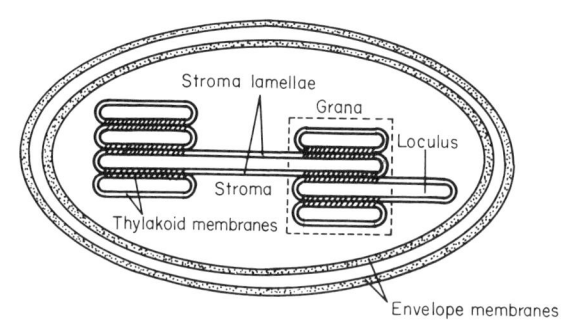

FIG. 2. Schematic diagram of a green plant chloroplast, ▰, stacked membrane regions; ▱, unstacked membrane regions.

by regions of unappressed membranes called stroma lamellae. The matrix between the inner lamellae and the outer envelopes form the stroma where the dark reactions occur. The region enclosed by the thylakoid is the loculus.

Chloroplasts are relatively large organelles that exist in various shapes and sizes depending on the species. The green, unicellular alga *Chlorella*, for example, contains a single cup- or bell-shaped chloroplast that fills most of the cytoplasm. The leaf cells of higher plants contain 10–100 ellipsoidal chloroplasts, each measuring $2-4 \times 5-10$ μm. The chemical composition of the chloroplast in terms of its dry weight is: 40–55% protein, 20–35% lipid, 5–10% chlorophyll, 1% pigments other than chlorophyll, and small amounts of genetic material. The thylakoid membrane is currently conceptualized in terms of the fluid mosaic model (Singer and Nicolson, 1972). The membrane is thus pictured as consisting of a phospholipid bilayer, ~ 70 Å thick, in which protein molecules are partially or wholly embedded, thus possessing much lateral mobility within the membrane plane.

C. CHLOROPHYLL–PROTEIN COMPLEXES

The estimated concentration of chlorophyll in the thylakoid membranes of higher plants is ~ 0.1 M (Rabinowitch, 1945), comprising about 10% of the dry mass or 20% of the lipid mass of the mature thylakoid membrane. Most of these pigment molecules function in a light-harvesting capacity with only a small fraction ($\sim 1/300$) of Chl a directly involved in photochemistry. Because chlorophylls play a key role in photosynthesis, a question of prime importance is their location in the thylakoid membrane. Biochemical studies on chlorophyll–protein fractions isolated from detergent-solubilized thylakoid membranes have revealed that virtually all chlorophylls are associated with proteins (Boardman *et al.*, 1978; Thornber *et al.*, 1979). The increased pace of research activity in this area over the past decade has uncovered numerous chlorophyll–polypeptide entities, but as yet little information on their structural and intermolecular interaction aspects (Arntzen, 1978). For the present, however, it appears that these chlorophyll–protein complexes fall into three main groups: (1) chlorophyll–protein complexes associated with PS I activity (commonly referred to as CP I) and containing about 30% of the total chlorophyll, (2) chlorophyll-protein complexes associated with PS II activity and containing about 20% of the total chlorophyll, and (3) light-harvesting chlorophyll–pigment complexes (also referred to as LHCs or CP II) with a Chl a/Chl b ratio approaching 1, as opposed to a value of about 3 in thylakoids.

D. Reaction Centers

Oxygen-evolving photosynthetic systems utilize two sites of photochemical activity known as traps or reaction centers. These consist of chlorophyll molecules designated P (pigment) followed by the number indicating the peak of the long-wavelength light-induced absorbance change—P700 for reaction center I and P680 (or P690) for reaction center II. P700, discovered by Kok in 1956 (Kok, 1956), is the primary electron donor in the photochemical act of PS I. The photooxidation of P700 is reversible and can be induced by far-red light. The light-minus-dark difference absorption spectrum is characterized by troughs at 430, 682, and 793 nm, the last two forming a double band suggestive of a dimer. The differential molar extinction coefficient of P700 has been found to be 67 ± 7 mM^{-1} cm^{-1} at 703 nm (Ke, 1978). The light-minus-dark difference circular dichroism (CD) spectrum shows a positive peak at about 697 nm and a negative one at 688 nm (Philipson et al., 1972), which can be interpreted as indicating exciton interaction in a chlorophyll dimer. Light also induces the appearance of an electron paramagnetic resonance (EPR) signal (Bearden and Malkin, 1975; Katz et al., 1978), the most prominent component (signal I) of which is rapidly reversible, has a g value of 2.0025 indicative of a free electron, has a Gaussian line shape, and has a peak-to-peak linewidth of ~ 7.5 G—about $1/\sqrt{2}$ narrower than that observed for a chemically oxidized monomeric Chl a cation radical. In addition, electron nuclear double resonance (ENDOR) spectroscopy shows photooxidized P700 with a proton hyperfine coupling constant about one-half that of the monomeric Chl a radical. All four lines of evidence support the PS I reaction center being a Chl a dimer. The models proposed for the dimer have generated much controversy (Katz et al., 1978). Currently, two models with C_2 symmetry appear to be favored.

In the model by Fong (1974) the two Chl a molecules are cross-linked through water molecules between the magnesium atom of one Chl a and the carbomethoxy ester C=O function of the other Chl *a*. The Chl *a* macrocycles are thus rotated by 60° with respect to each other, and are ~6 Å apart. In the model of Shipman *et al.* (1976), however, the cross-linking agent is a bifunctional ligand with hydrogen-bonding properties, e.g., R'XH, where X is O, NH, or S, and R' is H or an alkyl. Cross-linking is achieved by coordination of X to the magnesium atom of one Chl *a* molecule and hydrogen bonding of the hydrogen to the keto C=O function in ring V of the other Chl *a* molecule, placing the macrocycles at their van der Waals radii, 3.6 Å.

The reversible photooxidation of P680 is characterized by a light-minus-dark difference absorption spectrum showing bleaching at 430 and 682 nm (Doring *et al.*, 1969). Its light-induced EPR signal, known as signal II, has a g value of 2.0046, a linewidth of ~20 G, and a complicated hyperfine structure (see Knaff and Malkin, 1978). A reversible, light-induced absorption increase at $-170°$C with a peak at 825 nm and a half-time for relaxation of 3 msec has also been attributed to the oxidation of P680 (Mathis and Vermeglio, 1975). *In vitro* studies on the Chl *a* π cation radical gave a molar extinction coefficient of about 7000 M^{-1} cm^{-1} at about 830 nm (Borg *et al.*, 1970; Seki *et al.*, 1973).

In practice, however, it must be pointed out that, unlike for bacterial systems (see Chapter 4), biochemically well-defined preparations of reaction centers from green plant photosynthetic systems do not yet exist. Consequently, the properties of these reaction centers have been studied using subchloroplast fractions, known as particles, enriched in one of the reaction centers. Generally, of the two fractions enriched in reaction centers, the lighter one contains P700 and has a Chl *a*/Chl *b* ratio of >3, while the bulkier one is enriched in P680, contains LHC, and has a Chl *a*/Chl *b* ratio of <3 (Thornber *et al.*, 1979). More purified preparations containing the PS I reaction center include the chlorophyll-P700–protein complex (CP I) and a single polypeptide component of it having a molecular weight of 70,000 (Bengis and Nelson, 1975). Fewer attempts have been made to obtain preparations highly enriched in P680. A recently reported preparation contains ~1 PS II reaction center per 30–40 Chl molecules and is free of PS I reaction centers (Kilmov *et al.*, 1980).

III. Fundamental Concepts of Photosynthesis

A. THE PHOTOSYNTHETIC UNIT

In 1932, Emerson and Arnold (1932a,b) studied the effect of the flash repetition rate on the yield of oxygen evolution by a suspension of the green alga *Chlorella* using bright 10-μ sec flashes. They found the maximum yield per flash to be 1 oxygen molecule per 2000–2500 chlorophyll molecules. This finding led to the proposal by Gaffron and Wohl (1936a,b) that there was a cooperation of ~2500 chlorophyll molecules in a "unit" capable of producing one molecule of oxygen. Thus was born the idea of the photosynthetic unit (PSU) for oxygen evolution. With progress in our understanding of the photosynthetic process another term appeared in the literature—the PSU for a primary photoreaction. This can be looked upon in one of two ways. Our current definition of photosynthesis is that it is an oxidation–reduction process in which two molecules of water are oxidized to one molecule of oxygen by the transfer of four electrons (or four hydrogen atoms) to carbon dioxide, reducing it to carbohydrate. This implies that the transfer of one electron requires a minimum of ~600 chlorophyll molecules. Since the transfer of one electron to carbon dioxide requires two photoreactions, each photoact must be associated with 300 chlorophyll molecules. Alternatively, it is known that the minimum quantum requirement for the evolution of one oxygen molecule is eight quanta (Emerson, 1958), so that for maximum efficiency a minimum of 300 chlorophyll molecules must be associated with the absorption of each proton. The following definition of the PSU for a photoreaction was given by Duysens (1966): "The unit consists of a reaction center and those chlorophyll molecules that have the highest probability of transferring excitation energy to this reaction center when all reaction centers are in the trapping state." It must be noted that the size of the PSU is not necessarily uniform in all plants and that variations from 300 to 5000 chlorophyll molecules per CO_2 molecule reduced have been found (Schmid and Gaffron 1971).

The idea of a PSU was important to the field in that it suggested for the first time the role of electronic excitation energy transfer in photosynthesis. The concept also stimulated much research aimed at elucidating the structural and functional organization of the light-harvesting molecules in the photosynthetic apparatus. Although the PSU as initially proposed was strictly a statistical entity, it also created some interest in the search for a molecular moiety representing such a unit. In this regard, it must be cautioned that particles of dimensions 155 × 180 × 100 Å observed in electron micrographs, and called "quantasomes" by Park and Biggins (1964), not be identified with the PSU. In a critical evaluation of the quantasome by Arntzen and Briantais (1975) the following was noted: "There is, therefore, no direct biochemical evidence for the existence of an autonomous functional 'quantasome' unit based on fractionation evidence." Since it is now known that all chlorophylls are associated with proteins in the thylakoid membrane, it is conceivable that structural entities do exist which satisfy the definition of a PSU. The chlorophyll-P700-protein (CP I) complex could be such an entity for photoreaction I.

B. Two Light Reactions and the Z Scheme

A unique feature of our present understanding of oxygen-evolving photosynthesis is the concept that the process is driven by two light reactions. Associated with each photoreaction is a reaction center which, in the course of photochemical activity, utilizes the energy of a captured photon to to promote an electron to a state of higher reduction potential. In order to sustain photosynthesis a source and a sink for electrons must be available. This is achieved if the two light reactions are connected in series, with water as the ultimate electron donor and carbon dioxide the ultimate electron acceptor. Electron flow is facilitated by a series of oxidation–reduction components acting as intermediates. The first model representing such a sequence of events was the zig-zag scheme of Hill and Bendall (1960), which has been elaborated upon in the modern Z scheme of photosynthetic electron transport. In this scheme, the electron transport intermediates are generally positioned according to their midpoint potentials at pH 7 on a redox potential scale, so that a spontaneous electron flow is represented by a downward arrow and a light-induced step is indicated by an upward arrow. Figure 3 shows a detailed version of the Z scheme, with the reported

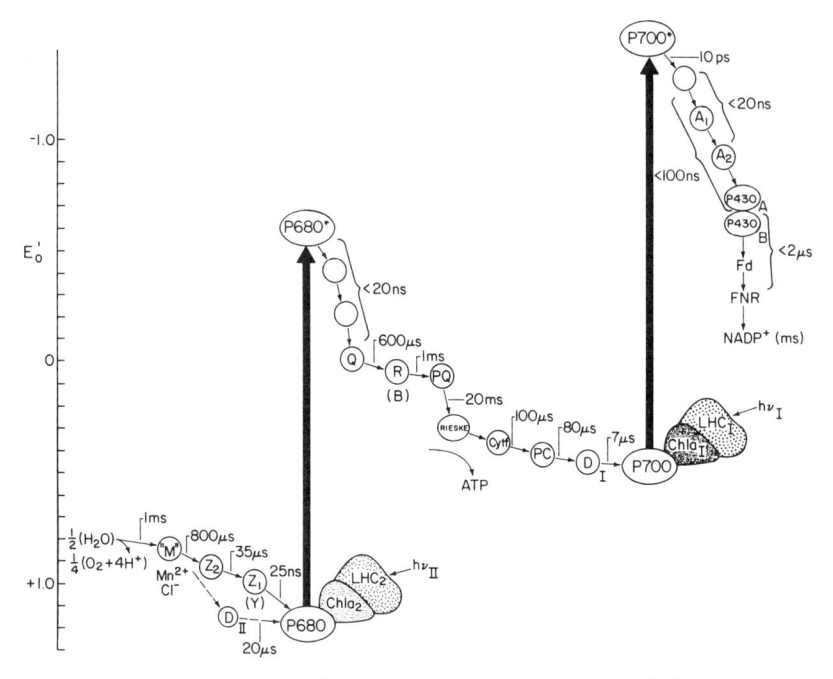

Fig. 3. Z scheme of photosynthesis. (From Govindjee, 1980.)

half-times or relaxation times for each electron transfer step included (Govindjee, 1980). Briefly, excitation (left vertical arrow) of the PS II reaction center (P680) results in the removal of an electron from P680, giving P680$^+$, and reduction of the first stable electron acceptor Q in $\lesssim 20$ nsec (instrument-limited; Doring et al., 1969), forming Q$^-$. Some experimental evidence supporting the participation of an intermediate between P680 and Q exists (Fajer et al., 1980). P680$^+$ is reduced by electrons from water oxidation, mediated by the charge accumulator, M, of the oxygen-evolving apparatus, and electron transfer intermediates Z_2 and Z_1 (or Y), with the approximate half-times indicated with the arrow for each step. Excitation (right vertical arrow in Fig. 3) of the PS I reaction center (P700) is followed by electron removal from P700, giving P700$^+$ in <10 psec (Fenton et al., 1979). The primary acceptor, A_1, has been suggested to be a monomeric Chl a (Fujita et al., 1978). However, the electron, after primary charge separation, is transferred to intermediates A_2, P430, ferredoxin, ferredoxin–NADP reductase (FNR), and finally NADP$^+$. The P700$^+$ is re-reduced by electron transfer from PS II through the intersystem electron transport chain as follows: The electron on Q$^-$ passes in sequence through a two-electron acceptor R (or B), which is a special form of plastoquinone (PQ), the PQ pool, a Rieske iron–sulfur protein, cytochrome f, plastocyanin (PC), an intermediate, D (most likely, bound plastocyanin), and finally P700$^+$.

The Z scheme is now well supported by experimental findings (see below). below). Although the model does not explain all the experimental results reported to date, it has shown ready adaptability in explaining these results. In fact, the scheme has shown greater consistency in explaining experimental results and greater versatility in accommodating new data than any other model thus far suggested (Hoch and Owens, 1963; Knaff and Arnon, 1969; Franck and Rosenberg, 1964).

Evidence Supporting the Two-Photoreaction Scheme

1. *The red drop phenomena*: In 1943 Emerson and Lewis (1943) found that the quantum yield of oxygen evolution as a function of excitation wavelength declined at the red end of the absorption spectrum, well within the absorption band of Chl a. This phenomenon is known as the red drop effect. A similar observation on Chl a fluorescence was made by Duysens (1952). The oxygen evolution results can readily be understood in terms of two light reactions if it is assumed that the two pigment systems associated with their respective light reactions have different absorptions and that far-red light is absorbed predominantly by pigment system I. The fluorescence results could readily be explained if the Chl a fluorescence yield is lower in pigment system I than in pigment system II.

2. *The Emerson enhancement effect*: The series two-light reaction model implies that maximum efficiency of electron transport occurs when the excitations of the two photosystems are balanced. The red drop effect demonstrates that far-red light causes unequal excitation of the photosystems. Thus it is readily seen that the red drop effect can be avoided by the addition of shorter-wavelength light absorbed predominantly by PS II (e.g., Chl *b* in green algae and phycobilins in red algae) to the far-red light. This improvement in the efficiency of oxygen evolution is known as the Emerson enhancement effect (see Myers, 1971).

3. *The chromatic transients*: In 1957, Blinks (1960) discovered that, when the red alga *Porphyra perforata* was alternately exposed to red (absorbed by Chl *a*, sensitizing PS I) and green (absorbed by PE, sensitizing PS II) light of equal steady state effective intensity, transient jumps or drops in the oxygen evolution rates occurred. In the transition from red to green illumination the oxygen output showed a transient increase, followed by a decrease, and then a recovery to the steady state. The reverse transition from green to red illumination only gave the decrease and recovery phases of the previous transient. This effect is readily understood in terms of two light reactions in which a chemical intermediate accumulates by the operation of one light reaction and is removed by the other.

4. *Oxidation–reduction changes in the electron transport components*: The redox components P700 (Kok, 1961), cytochrome *f* (Duysens *et al.*, 1961; Duysens and Amesz, 1962), and PQ (Amesz, 1964; Rumberg *et al.*, 1964) are all photooxidized in light predominantly absorbed by PS I and photoreduced by light predominantly absorbed by PS II.

5. *Partial reactions of photosynthesis*: The two light reactions can be studied separately by using appropriate substitute electron donors and electron acceptors (see Trebst, 1974).

6. *Physical separation of reaction centers*: The isolation of subchloroplast fractions enriched in P700 and P680 have provided evidence that two physically distinct light reactions exist in oxygen-evolving photosynthesis (cf. Boardman *et al.*, 1978; Thornber *et al.*, 1979).

IV. The Primary Events

A. Absorption of Light

The red absorption band of Chl *a in vivo* is broad (half-bandwidth, ~30 nm) compared to that in nonpolar solvents (half-bandwidth, ~10 nm). The absorption maximum of Chl *a* in the thylakoids is ~678 nm, red-shifted by about 15 nm from Chl *a* in diethyl ether. The broadening and red shift of the absorption band are apparent effects, originating from a series of narrow overlapping components commonly referred to as "biological forms of chlorophyll" and denoted by Ca followed by the number corresponding to the respective red absorption maximum. There are at least six distinct spectral forms of Chl a (French et al., 1972), and as many as 10 have even been suggested (see Litvin and Sineshchekov, 1975), based on analyses of low-temperature ($-196 C°$) and derivative absorption spectra. French et al. (1972) found four major Chl a forms—Ca 662, Ca 670, Ca 677, and Ca 684—with half-bandwidths between 7.6 and 12 nm present in all the spectra. Two forms absorbing at wavelengths longer than 685 nm were Ca 692 (half-width, 13–18 nm) and Ca 700–706 (half-width, 14–25 nm). Chl b exists in two forms with maxima at 640–642 and 648–650 nm (French et al., 1972). Comparison of 20 and -196 C° spectra shows a sharpening of the chlorophyll red band with cooling, a consequence of the narrowing of the individual components by a factor of 1.2–2. The definitive cause of the Chl a forms is unknown, but they are most probably due to physically different states of chlorophylls associated with different microenvironments in the thylakoid membrane, an effect analogous to solvent effects on many organic dyes (see e.g., Mataga and Kubota, 1970). Light absorption by chlorophyll or the accessory pigments takes place in $\leq 10^{-15}$ sec, giving

$$\text{Car} + h\nu \longrightarrow \text{Car*} \tag{5}$$

$$\text{Chl} + h\nu \longrightarrow \text{Chl*}, \tag{6}$$

where Car and Car* denote the ground state and excited state carotenoid, respectively, and Chl* denotes the singlet excited chlorophyll. The excited pigment molecule can dissipate its energy by a number of pathways:

Energy transfer to another molecule,

$$\text{Car*} + \text{Chl} \longrightarrow \text{Car} + \text{Chl*} \tag{7}$$

$$\text{Chl*} + \text{Chl} \longrightarrow \text{Chl} + \text{Chl*}, \tag{8}$$

Fluorescence,

$$\text{Chl*} \longrightarrow \text{Chl} + h\nu', \quad \text{where } \nu' < \nu, \tag{9}$$

Nonradiative dissipation,

$$\text{Chl*} \longrightarrow \text{Chl} + \text{heat}, \tag{10}$$

or photochemistry (if the molecule is a reaction center),

$$\text{Chl*} \longrightarrow \text{Chl}^+ + e^-, \tag{11}$$

where $e^- \equiv$ electron.

B. Fluorescence

The emission spectrum of Chl *a in vivo* at room temperature shows a major band at 685 ± 2 nm and a vibrational satellite band at about 740 nm. Both bands originate from a $\pi^* \to \pi$ transition to the ground state. Although most of the fluorescence comes from PS II (see Govindjee *et al.*, 1973), the fraction of PS II to PS I fluorescence attains a minimum in the 710–720 nm region (Lavorel, 1962; Vredenberg and Duysens, 1965; Briantais, 1969). Emission from the second excited state has never been reported, even with excitation into the blue-violet (Soret) absorption band. However, it is conceivable that such an emission could be induced by high-intensity laser excitation, as reported for bacteriochlorophyll (Kung and DeVault, 1978). Analyses of emission spectra, similar to those performed on absorption spectra, also suggest the existence of a number of different fluorescing forms of Chl *a*. As many as 10–12 such forms have been suggested (Litvin and Sineshchekov, 1975). These Chl *a* forms can be roughly classified into two groups: fluorescent and weakly fluorescent. This classification was first suggested by Duysens (1952) to explain the finding that light absorbed by the phycobilins of red algae were more effective in exciting Chl *a* fluorescence than light absorbed by Chl *a* itself. Today our understanding is that both the phycobilins and the fluorescent Chl *a* forms are associated with PS II and that the weakly fluorescent forms are associated with PS I (see Govindjee *et al.*, 1973). In the literature, the fluorescence bands are often labeled F, followed by the wavelength of the emission peak in nanometers—thus F685 for the main band of the room temperature fluorescence.

Cooling of chlorophyll-containing cells, chloroplasts, or thylakoids to −196°C results in the appearance of two additional bands, one at ∼695 nm (F695) and the other at 715–740 nm (F730), the exact location being species-dependent. For example, in green algae F730 is generally located at ∼720 nm, and in higher plants it is at ∼735 nm. The nature and function of the Chl *a* forms which emit F695 and F730 are still subjects of current interest (see Satoh and Butler, 1978). The following has generally served well as a working hypothesis for interpreting low-temperature fluorescence results: F685 is emitted by the bulk of Chl *a*, F695 is emitted by a Chl *a* form closely associated with the reaction center of PS II, and F730 is emitted by a Chl *a* form distinct from P700 but in its vicinity (Satoh and Butler, 1978; Murata *et al.*, 1966; Govindjee and Yang, 1966). In interpreting experimental results, it is important to bear in mind that the room temperature F730 is emitted mainly by PS II and the low-temperature F730 is emitted mainly by PS I. Emission at temperatures between 77 and 4 K has also been studied (Cho and Govindjee, 1970a,b; Rijgersberg *et al.*, 1979; Avarmaa *et al.*, 1979); no additional band appears, but significant changes in the relative intensities of the bands are observed.

The absolute yield of Chl *a* fluorescence in photosynthetic systems is small, not exceeding 0.1. In *Chlorella*, a value of 0.04 has been reported (Murty et al., 1965). The yield is a function of three factors: (1) temperature, (2) excitation wavelength, because of differences in the efficiency of energy transfer from different accessory pigments, and (3) oxidation-reduction state of the reaction center and quenchers in the electron transport chain, particularly Q (Duysens and Sweers, 1963), being low when the PS II reaction center is open (i.e., ready for photochemistry, with P680 in the reduced state and Q in the oxidized state). The last point has been involved in the interpretation of microsecond and millisecond transients in Chl *a* fluorescence yields when dark-adapted algae and thylakoids are illuminated (Mauzerall, 1972; Jursinic and Govindjee, 1977). Other fluorescence parameters of interest are the lifetime and the degree of polarization. A detailed discussion of fluorescence lifetime is given in Chapter 2. Studies on the fluorescence polarization has so far provided only qualitative information on energy transfer and chlorophyll alignment in the thylakoids (Becker et al., 1976; Garab and Breton, 1976; Wong and Govindjee, 1979, 1981).

Unlike the situation in absorption studies fewer artifacts are encountered in fluorescence measurements. The most common sample artifact overlooked is self-reabsorption of the fluorescence leading to a distortion at the F685 end of the spectrum. This problem can normally be avoided if the average chlorophyll concentration in the sample is kept at about 5–10 μg/ml.

C. Energy Transfer and Migration

All oxygen-evolving photosynthetic organisms contain Chl *a* and various light-harvesting pigments known collectively as accessory pigments. This distinction is a misnomer, since with the exception of a small fraction of Chl *a* molecules which make up the reaction centers, the bulk of the Chl *a* functions essentially like the accessory pigments. Transfer of excitation energy in photosynthetic systems is of two types: that between like molecules—Chl *a* to Chl *a*—and that between unlike molecules—e.g., accessory pigments to Chl *a*. Energy transfer in photosynthesis has two principal functions: (1) increasing the probability of the capture of light energy, and (2) broadening the spectral range for light absorption. To appreciate the first role, consider the variations in light intensity experienced by the organism in its natural environment. The chlorophyll molecule absorbs an average of 0.1 quantum/sec in dim light and $\sim$ 10 quanta/sec in direct sunlight. The cooperative effects of a few hundred chlorophylls in directing energy to the reaction center in a PSU, in effect, increase the absorption cross section of the reaction by Förster's inductive resonance mechanism (Knox, 1975, 1977; Duysens, 1964), in which the pairwise transfer rate is dependent upon the intermolecular distance ($\propto r^{-6}$, where r is the distance), the orientation factor κ^2 ($\kappa^2 \leq 4$) between the donor and acceptor dipoles, and the overlap of the donor fluorescence with the acceptor absorption spectrum (Förster, 1967a,b). The Förster mechanism represents one extreme of dipole-allowed transfer between molecules, one in which the vibrational relaxation rate is greater than the pairwise transfer rate. The other extreme of a dipole–dipole interaction is a fast transfer in which the vibrational relaxation rate is less than the transfer rate, the latter being proportional to the inverse third power of the distance. Recent investigations by Kenkre and Knox (1974) have suggested that the Förster mechanism may be safely applied to most excitation transfer in photosynthesis.

Another important aspect of light harvesting is the high efficiency with which energy absorbed by the antenna chlorophyll arrives at the reaction center. This question has been addressed a number of times in the past, and it is now recognized that the efficiency of energy transfer to the trap depends on three factors: the dimensionality of the PSU (Pearlstein, 1964), the presence of accessory pigments (Pearlstein, 1964; Borisov and Il'ina, 1971; Seely, 1973), and the pairwise transfer rate between neighboring Chl a molecules, which is a function of their mutual orientation. Models often considered for the PSU are of two kinds: (1) those with an ordered inhomogeneous arrangement of pigments in which the shorter-wavelength-absorbing forms are arranged toward the outside of the longer-wavelength-absorbing forms with respect to the reaction center (Seely, 1973), and (2) those with a homogeneous mixture of Chl a and Chl b throughout the PSU (Swenberg *et al.*, 1976; Kopelman, 1976). However, it appears that meaningful progress with regard to understanding photosynthesis must now await detailed information pertaining to the architecture of the biological system.

Energy transfer or migration is also affected by the extent of interaction between PSUs. The dynamic nature of the biological photosynthetic system can accommodate the wide range of possible PSU interactions, from none (corresponding to a puddle or unicentral model) to very extensive (corresponding to a lake or multicentral model) (Robinson, 1967). It appears that a limited amount of interaction generally exists. The regulation of PSU interaction is a subject of current interest in the literature (Williams, 1977).

D. Primary Photochemistry

As pointed out earlier, the sites of primary photochemistry in the photosynthesis of oxygen-evolving organisms are the two reaction centers. In general terms, each reaction center (P) is associated with a primary electron acceptor center by more than two orders of magnitude, improving the conditions for photosynthesis. The importance of the accessory pigments in facilitating the other role of energy transfer is readily appreciated once it is realized that the absorption region of chlorophyll excludes a major portion of the visible solar radiation spectrum reaching the surface of the earth.

Evidence for the transfer of excitation energy from the accessory pigments to Chl *a* is provided by the fluorescence of the latter. In the absence of energy transfer, the excitation spectrum of Chl *a* fluorescence should correspond to the absorption spectrum of Chl *a* in the organism. Instead, the fluorescence yield is higher than expected at excitation wavelengths where Chl *a* absorption is low and accessory pigment absorption is significant. A comprehensive investigation of the transfer of excitation energy in a number of photosynthetic systems was presented by Duysens (1952). By studying absorption spectra and action spectra for photosynthesis and fluorescence, he estimated the efficiencies of the various energy transfer steps involving the accessory pigments and Chl *a* in green, brown, blue-green, and red algae, and in diatoms. Based on this and subsequent studies by others, the following efficiencies for energy transfer have been compiled: $\sim 50\%$ in green algae (Emerson and Lewis, 1943; Duysens, 1952; Cho and Govindjee, 1970a,b) and $10-15\%$ in blue-green algae for carotenoid $\rightarrow$ Chl *a* (Duysens, 1952; Emerson and Lewis, 1943; Papageorgiou and Govindjee, 1967; Cho and Govindjee, 1970a,b), except $\sim 70\%$ for fucoxanthol $\rightarrow$ Chl *a* (Duysens, 1952; Tanada, 1951); $96 \pm 3\%$ for PE $\rightarrow$ PC (Porter et al., 1978); $78 \pm 8\%$ for PC $\rightarrow$ APC $\rightarrow$ Chl *a* (Porter et al., 1978); and $\sim 100\%$ from Chl *b* $\rightarrow$ Chl *a* (Cho and Govindjee, 1970a,b).

Energy transfer between molecules which are alike in chemical structure plays a crucial role in photosynthesis. Presumably, this occurs in the final stages of light harvesting prior to trapping at the reaction center. Evidence for the existence of energy transfer between Chl *a* molecules (also referred to as energy migration) came from the finding that Chl *a* fluorescence *in vivo* had a low degree of polarization (p) (Arnold and Meek, 1956; Mar and Govindjee, 1972) (typically, $p = 0.03-0.08$). The latter is consistent with the occurrence of some energy transfer between Chl *a* molecules which are partially aligned in regard to their transition moments, but is inconsistent with a situation in which there is either extensive transfer in a well-oriented system or poor transfer in a randomly oriented system. Precise interpretation of fluorescence polarization values in partially oriented systems, such as Chl *a* in the thylakoid membranes, is difficult at this time (Knox, 1975).

The mechanism of electronic excitation energy transfer in photosynthesis has been a subject of much interest for the past 40 yr, and many excellent discussions on the subject are available (e.g., Knox, 1975, 1977). It is generally accepted that excitation energy transfer in the photosynthetic system occurs

(A) and donor (D). Excitation of the reaction center leads to the transfer of a charge from P to A, followed by a stabilization of the charge separation by the re-reduction of P by D:

$$\underset{\text{(Chl)}}{\text{D·}\dot{\text{P}}\text{·A}} \xrightarrow{hv} \underset{\text{(Chl*)}}{\underset{\text{excitation}}{\text{D·}\dot{\text{P}}\text{·A}}} \longrightarrow \underset{\text{(Chl)}}{\underset{\text{transfer}}{\text{D·P*·A}}} \longrightarrow \underset{\text{(Chl)}}{\underset{\substack{\text{charge}\\\text{separation}}}{\text{D·}\dot{\text{P}}^+\text{·A}^-}} \longrightarrow \underset{\text{(Chl)}}{\underset{\text{stabilization}}{\text{D}^+\text{·P·A}^-}} \quad (12)$$

Depending upon the photosystem involved, P denotes P700 (PS I) or P680 (PS II). Both photoreactions possess the following properties: (1) They are extremely rapid, occurring in a picosecond time domain; (2) they occur at cryogenic temperatures; and (3) they proceed with high quantum efficiency.

1. *Primary Photochemistry in PS I*

In this case, D in Eq. (12) denotes plastocyanin (a copper protein) and A denotes the primary electron acceptor (suggested to be a monomeric Chl *a* (Fajer *et al.*, 1980). The first stable electron acceptor is the component with an absorption change at 430 nm, labeled P430 (Ke, 1973, 1978), which probably is a bound ferredoxin. Improved time resolution for EPR studies in recent years has led to the identification of two components between P700 and P430, designated A_1 and A_2, the latter being the component identified, until recently, as X. Although this area is still in a transitory stage, the following scheme is currently being proposed in the literature:

$$\text{Chl } a \text{ P700 } A_1 A_2 \text{ P430} \xrightarrow{hv} \text{Chl } a^* \text{ P700 } A_1 A_2 \text{ P430} \quad (13)$$

$$\text{Chl } a^* \text{ P700 } A_1 A_2 \text{ P430} \xrightarrow[<10 \text{ p sec}]{\text{transfer}} \text{Chl } a \text{ P700}^* A_1 A_2 \text{ P430} \quad (14)$$

$$\text{P700}^* A_1 A_2 \text{ P430} \xrightleftharpoons[3 \mu \text{ sec}]{<10 \text{ p sec}} \text{P700}^+ A_1^- A_2 \text{ P430} \quad (15)$$

$$\text{P700}^+ A_1^- A_2 \text{ P430} \xrightleftharpoons[250 \mu \text{ sec}]{\sim 200 \text{ p sec}} \text{P700}^+ A_1 A_2^- \text{ P430} \quad (16)$$

$$\text{P700}^+ A_1 A_2^- \text{ P430} \longrightarrow \text{P700}^+ A_1 A_2 \text{ P430}^- \quad (17)$$

$$\text{P700}^+ A_1 A_2 \text{ P430} \xrightarrow[\text{PC} \quad \text{PC}^+]{\sim 20 \mu \text{ sec}} \text{P700 } A_1 A_2 \text{ P430}^- \quad (18)$$

$$\text{P700 } A_1 A_2 \text{ P430}^- \xrightarrow[\text{Fd} \quad \text{Fd}^-]{\sim 5 \text{ m sec}} \text{P700 } A_1 A_2 \text{ P430,} \quad (19)$$

where PC denotes plastocyanin and Fd denotes ferredoxin. The identity of A_1 is unclear, but it has been suggested by Shuvalov *et al.* (1979) to be a

Chl *a* dimer, by Fenton *et al.* (1979) and Baltimore and Malkin (1980) to be either a chlorophyll or a pheophytin monomer and by Fajer *et al.* (1980) to be a monomeric Chl *a*. The proposal that a chlorophyll anion is the first reduced chemical product of PS I photochemistry was made earlier by Fujita *et al.* (1978).

2. *Primary Photochemistry of PS II*

Unlike that for PS I, the primary photochemistry for PS II is less well understood. For PS II, D in Eq. (12) denotes the unidentified electron donor Z and, for lack of any information to the contrary, A denotes the stable electron acceptor Q. The primary photoreaction of PS II is analogous to that of PS I; photooxidation of P680 is visualized to involve the following sequence of reactions (Duysens and Sweers, 1963; Kautsky *et al.*, 1960; see also Butler, 1973; Govindjee and Jursinic, 1979):

$$Z \cdot \overset{\text{Chl }a}{P680} \cdot Q \xrightarrow{h\nu} Z \cdot \overset{\text{Chl }a^*}{P680} \cdot Q$$

$$Z \cdot \overset{\text{Chl }a^*}{P680} \cdot Q \xrightarrow{\text{transfer}} Z \cdot \overset{\text{Chl }a}{P680^*} \cdot Q$$

$$Z \cdot P680^* \cdot Q \xrightarrow{< 20 \text{ n sec}} Z \cdot P680^+ \cdot Q^-$$

$$Z \cdot P680^+ \cdot Q^- \xrightarrow{30 \text{ n sec}} Z^+ \cdot P680 \cdot Q^-$$

$$Z^+ \cdot P680 \cdot Q^- \xrightarrow[R \quad R^-]{100-600 \ \mu\text{sec}} Z^+ \cdot P680 \cdot Q$$

V. Conclusion

The field of photosynthesis encompasses a spectrum of traditional disciplines of study. The above outline focuses on some physical and chemical aspects of importance to the primary processes of photosynthesis. The chapter is presented in three parts: the first deals with the pigments, pigment–protein complexes, and the photosynthetic membrane and its organization; the second is a synopsis of the two basic concepts of the modern era of photosynthesis—the PSU and the two light reactions; and the third covers some areas of immediate interest to ultrafast laser spectroscopy. Numerous references to review articles are included to provide sources of information on areas not discussed.

Note added in proof: Since the original submission of this article two concepts have been challenged. The first is the idea that the PS I reaction center is a Chl *a* dimer (p. 10). Wasielewski and co-workers [Wasielewski,

M. R., Norris, J. R., Shipman, L. L., Lin, C. P., and Svec, W. A. (1981). *Proc. Natl. Acad. Sci. U.S.A.* **78**, 2957–2961.] have suggested that a Chl monomer in a special environment better explains the linewidth results obtained from magnetic resonance experiments. The second area of contention is the explanation for the different spectral forms of Chl *a* (p. 16). Evidence has now accumulated to suggest that these may be manifestations of the different chemical species of chlorophyll [Rabeiz, C. A., and Lascelles, J. (1982). *In* "Photosynthesis: Energy Conversion by Plants and Bacteria" (Govindjee, ed.), Academic Press, New York; and Bazzaz. M. B. (1981). *Photobiochem. Photobiophys.* **2**, 199–207.]

Acknowledgments

This work was supported in part from NSF grant PCM 77-14966 and the City University PSC-BHE Research Award Program. The author thanks Dr. Govindjee for valuable criticism of the manuscript and for kindly authorizing the inclusion of Fig. 3.

References

Amesz, J. (1964). *Biochim. Biophys. Acta* **79**, 257–265.
Arnold, W., and Meek, E. S. (1956). *Arch. Biochem. Biophys.* **60**, 82–90.
Arnon, D. I. (1949). *Plant Physiol.* **24**, 1–15.
Arntzen, C. J. (1978). *Curr. Top. Bioenerg.* **8**, 111–160.
Arntzen, C. J., and Briantais, J.-M. (1975). *In* "Bioenergetics of Photosynthesis" (Govindjee, ed.), pp. 51–113. Academic Press, New York.
Avarmaa, R. A., Kuchubey, S. M., and Tamkivi, R. P. (1979). *FEBS Lett.* **102**, 139–142.
Baltimore, B. G., and Malkin, R. (1980). *Photochem. Photobiol.* **31**, 485–490.
Bearden, A. J., and Malkin, R. (1975). *Q. Rev. Biophys.* **7**, 131–177.
Becker, J. F., Breton, J., Geacintov, N. E., and Trentacosti, F. (1976). *Biochim. Biophys. Acta* **440**, 531–544.
Bengis, C., and Nelson, N. (1975). *J. Biol. Chem.* **250**, 2783–2788.
Blinks, L. R. (1960). *Proc. Natl. Acad. Sci. U.S.A.* **46**, 327–333.
Boardman, N. K., Anderson, J. M., and Goodchild, D. J. (1978). *Curr. Top. Bioenerg.* **8**, 35–109.
Bogorad, L. (1975). *Annu. Rev. Plant Physiol.* **26**, 369–401.
Borg, D. C., Fajer, J., Felton, R. H., and Dolphin, D. (1970). *Proc. Natl. Acad. Sci. U.S.A.* **67**, 813–820.
Borisov, A. Y., and Il'ina, M. D. (1971). *Biochim. Biophys. Acta* **305**, 364–371.
Briantais, J.-M. (1969). Thesis, Univ. Paris-Sud (Orsay).
Butler, W. L. (1973). *Acc. Chem. Res.* **6**, 177–184.
Cho, F., and Govindjee (1970a). *Biochim. Biophys. Acta* **216**, 139–150.
Cho, F., and Govindjee (1970b). *Biochim. Biophys. Acta* **216**, 151–161.
Doring, G., Renger, G., Vater, J., and Witt, H. T. (1969). *Z. Naturforsch., Teil B* **24**, 1139–1143.
Duysens, L. N. M. (1952). Ph.D. Thesis, Univ. Utrecht, Utrecht.
Duysens, L. N. M. (1964). *Prog. Biophys. Mol. Biol.* **17**, 1–104.
Duysens, L. N. M. (1966). *Brookhaven Symp. Biol.* **19**, 72–80.
Duysens, L. N. M., and Amesz, J. (1962). *Biochim. Biophys. Acta* **64**, 243–260.

Duysens, L. N. M., and Sweers, H. E. (1963). *In* "Studies on Microalgae and Photosynthetic Bacteria" (Japanese Society of Plant Physiology, eds.), pp. 353–372. University of Tokyo Press, Tokyo.
Duysens, L. N. M., Amesz, J. and Kamp, B. M. (1961). *Nature (London)* **190**, 510–511.
Emerson, R. (1958). *Annu. Rev. Plant Physiol.* **9**, 1–24.
Emerson, R., and Arnold, W. (1932a). *J. Gen. Physiol.* **15**, 391–420.
Emerson, R., and Arnold, W. (1932b). *J. Gen. Physiol.* **16**, 191–205.
Emerson, R., and Lewis, C. M. (1943). *Am. J. Bot.* **30**, 165–178.
Fajer, J., Davis, M. S., Forman, A., Klimov, V. V., Dolan, E., and Ke, B. (1980). *J. Am. Chem. Soc.* **102**, 7143–7145.
Fenton, J. M., Pellin, M. J., Govindjee, and Kaufmann, K. J. (1979). *FEBS Lett.* **100**, 1–4.
Fong, F. K. (1974). *Proc. Natl. Acad. Sci. U.S.A.* **71**, 3692–3695.
Förster, T. (1967a). *In* "Modern Quantum Chemistry, Part III, Action of Light and Organic Crystals" (O. Sinanoglu, ed.), pp. 93–137. Academic Press, New York.
Förster, T. (1967b). *In* "Comprehensive Biochemistry: Bioenergetics" (M. Florkin and E. H. Stotz, eds.), Vol. 22, pp. 61–80. Elsevier, Amsterdam.
Franck, J., and Rosenberg, J. L. (1964). *J. Theor. Biol.* **7**, 276–301.
French, C. S., Brown, J. S., and Lawrence, M. C. (1972). *Plant Physiol.* **49**, 421–429.
Fujita, I., Davis, M. S., and Fajer, J. (1978). *J. Am. Chem. Soc.* **100**, 6280–6282.
Gaffron, H., and Wohl, K. (1936a). *Naturwissenschaften* **24**, 81–90.
Gaffron, H., and Wohl, K. (1936b). *Naturwissenschaften* **24**, 103–107.
Gantt, E. (1975). *BioScience* **25**, 781–788.
Gantt, E., and Lipschultz, C. A. (1973). *Biochem. Biophys. Acta* **292**, 858–861.
Garab, G. I., and Breton, J. (1976). *Biochem. Biophys. Res. Commun.* **71**, 1095–1102.
Glazer, A. N. (1976). *Photochem. Photobiol. Rev.* **1**, 71–115.
Glazer, A. N., and Fang, S. (1973). *J. Biol. Chem.* **248**, 659–662.
Govindjee (1980). *Plant Biochem. J., S. M. Sircar Meml. Vol.* pp. 7–30.
Govindjee, and Jursinic, P. (1979). *Photochem. Photobiol. Rev.* **4**, 125–205.
Govindjee, and Yang, L. (1966). *J. Gen. Physiol.* **49**, 763–780.
Govindjee, Papageorgiou, G., and Rabinowitch, E. (1973). *In* "Practical Fluorescence: Theory, Methods, and Techniques" (G. G. Guilbault, ed.), pp. 543–575. Dekker, New York.
Hill, R., and Bendall, F. (1960). *Nature (London)* **186**, 136–137.
Hoch, G., and Owens, O. V. H. (1963). *In* "Photosynthetic Mechanisms of Green Plants" (B. Kok and A. T. Jagendorf, eds.), Publ. No. 1145, pp. 409–420. Natl. Acad. Sci.–Natl. Res. Counc., Washington, D. C.
Jursinic, P., and Govindjee (1977). *Biochim. Biophys. Acta* **461**, 253–267.
Katz, J. J., Norris, J. R., Shipman, L. L., Thurnauer, M. C., and Waiselewski, M. R. (1978). *Annu. Rev. Biophys. Bioeng.* **7**, 393–434.
Kautsky, H., Appel, W., and Amann, H. (1960). *Biochem. Z.* **332**, 277–292.
Ke, B. (1973). *Biochim. Biophys. Acta* **301**, 1–33.
Ke, B. (1978). *Curr. Top. Bioenerg.* **7**, 75–138.
Kenkre, V. M., and Knox, R. S. (1974). *Phys. Rev. Lett.* **33**, 803–806.
Klimov, V. V., Dolan, E., and Ke, B. (1980). *FEBS Lett.* **112**, 97–100.
Knaff, D. B., and Arnon, D. I. (1969). *Proc. Natl. Acad. Sci. U.S.A.* **64**, 715–722.
Knaff, D. B., and Malkin, R. (1978). *Curr. Top. Bioenerg.* **7**, 139–172.
Knox, R. S. (1975). *In* "Bioenergetics of Photosynthesis" (Govindjee, ed.), pp. 183–221. Academic Press, New York.
Knox, R. S. (1977). *In* "Primary Processes of Photosynthesis" (J. Barber, ed.), pp. 55–94. Elsevier/North-Holland, Amsterdam.
Kok, B. (1956). *Biochim. Biophys. Acta* **22**, 399–401.

Kok, B. (1961). *Biochim. Biophys. Acta* **48**, 524–533.
Kopelman, R. (1976). *J. Lumin.* **12/13**, 775–780.
Kung, M. C, and DeVault, D. (1978). *Biochim. Biophys. Acta* **501**, 217–231.
Lavorel, J. (1962). *Biochim. Biophys. Acta* **60**, 510–523.
Ley, A. C., and Butler, W. L. (1977). *Proc. Natl. Acad. Sci. U.S.A.* **73**, 3957–3960.
Litvin, F. L., and Sineshchekov, V. A. (1975). *In* "Bioenergetics of Photosynthesis" (Govindjee, ed.), pp. 619–661. Academic Press, New York.
Mackinney, G. (1941). *J. Biol. Chem.* **140**, 315–322.
Mar, T., and Govindjee (1972). *Photosynth., Two Centuries Its Discovery Joseph Priestley, Proc. Int. Congr. Photosynth. Res., 2nd, Stresa, Italy, 1971* **1**, 271–281.
Mataga, N., and Kubota, T. (1970). "Molecular Interactions and Electronic Spectra." Dekker, New York.
Mathis, P., and Vermeglio, A. (1975). *Biochim. Biophys. Acta* **396**, 371–381.
Mauzerall, D. (1972). *Proc. Natl. Acad. Sci. U.S.A.* **69**, 1358–1362.
Murata, N., Nishimura, M., and Takamiya, A. (1966). *Biochim. Biophys. Acta* **126**, 234–243.
Murty, N. R., Cederstrand, C., and Rabinowitch, E. (1965). *Photochem. Photobiol.* **4**, 917–921.
Myers, J. (1971). *Annu. Rev. Plant Physiol.* **22**, 289–312.
Myers, J. (1974). *Plant Physiol.* **54**, 420–426.
Papageorgiou, G., and Govindjee (1967). *Biochim. Biophys. Acta* **131**, 173–178.
Park, R. B., and Biggins, J. (1964). *Science* **144**, 1009–1011.
Pearlstein, R. M. (1964). *Proc. Natl. Acad. Sci. U.S.A.* **52**, 824–830.
Philipson, K. D., Sato, V. L., and Sauer, K. (1972). *Biochemistry* **11**, 4591–4595.
Porter, G., Tredwell, C. J., Searle, G. F. W., and Barber, J. (1978). *Biochim. Biophys. Acta* **501**, 232–245.
Rabinowitch, E. I. (1945). "Photosynthesis and Related Processes," Vol. 1, pp. 411–412. Wiley (Interscience), New York.
Rijgersberg, C. P., Melis, A., Amesz, J., and Swager, J. A. (1979). *Chlorophyll Organ. Energy Transfer Photosynth. Ciba Found. Symp.* No. 61, 305–322.
Robinson, G. W. (1967). *Brookhaven Symp. Biol.* **19**, 16–48.
Rumberg, B., Schmidt-Mende, P., and Witt, H. T. (1964). *Nature (London)* **201**, 466–468.
Satoh, K., and Bulter, W. L. (1978). *Biochim. Biophys. Acta* **502**, 103–110.
Schmid, G. H., and Gaffron, H. (1971). *Photochem. Photobiol.* **14**, 451–464.
Seely, G. R. (1973). *J. Theor. Biol.* **40**, 173–187.
Seki, H., Arai, S., Shida, T., and Imamura, M. (1973). *J. Am. Chem. Soc.* **95**, 3404–3405.
Shipman, L. L., Cotton, T. M., Norris, J. R., and Katz, J. J. (1976). *J. Am. Chem. Soc.* **98**, 8222–8230.
Shuvalov, V. A., Ke, B., and Dolan, E. (1979). *FEBS Lett.* **100**, 5–8.
Singer, S. J., and Nicolson, G. (1972). *Science* **175**, 720–731.
Swenberg, C. E., Dominijanni, R., and Geacintov, N. E. (1976). *Photochem. Photobiol.* **24**, 601–604.
Tanada, T. (1951). *Am. J. Bot.* **38**, 276–283.
Thornber, J. P., Markwell, J. P., and Runman, S. (1979). *Photochem. Photobiol.* **29**, 1205–1216.
Trebst, A. (1974). *Annu. Rev. Plant Physiol.* **25**, 425–458.
Vredenberg, W. J., and Duysens, L. N. M. (1965). *Biochim. Biophys. Acta* **94**, 355–370.
Williams, W. P. (1977). *In* "Primary Processes of Photosynthesis" (J. Barber, ed.), pp. 99–147. Elsevier, Amsterdam.
Wong, D., and Govindjee (1979). *FEBS Lett.* **97**, 373–377.
Wong, D., and Govindjee (1981). *Photochem. Photobiol.* **33**, 103–108.

CHAPTER 2

Time-Resolved Fluorescence Spectroscopy

F. Pellegrino and R. R. Alfano

Picosecond Laser and Spectroscopy Laboratory
Department of Physics
The City College of The City University of New York
New York, New York

I.	INTRODUCTION	27
	A. Energy Transfer in Photosynthesis	28
	B. Time-Resolved Fluorescence Emission in Photosynthesis	29
II.	TIME-RESOLVED FLUORESCENCE SPECTROSCOPY TECHNIQUES	31
III.	FLUORESCENCE KINETIC MEASUREMENTS IN PHOTOSYNTHESIS	34
	A. Interpretation of Time Dependence of Fluorescence Decay in Photosynthesis	37
	B. Fluorescence Quenching and Exciton Annihilation Effects	38
	C. Environmental Effects on the Fluorescence from Photosynthetic Systems	43
	D. Picosecond Fluorescence Temperature Studies	45
	E. Energy Transfer in the Accessory Pigment Complexes of Red and Blue-Green Algae	46
IV.	FUTURE DIRECTIONS	49
	REFERENCES	50

I. Introduction

The history of time-resolved fluorescence spectroscopy in photosynthesis has been integrally related to the history of technological advances in the fields of light pulse generation and temporal resolution techniques. This intrinsic relationship between discovery and technological advancement arises from the fundamental nature of the scientific process, which demands that the experimental conditions of excitation and measurement be always more accurately defined. The development in the late 1960s of mode-locked lasers capable of generating ultrashort pulses of picosecond (10^{-12}-sec) duration at well-defined regions of the spectrum, and the subsequent introduction of comparable time resolution techniques, ushered in a new era of scientific discovery. At last a means for exciting and probing a system on a time scale of physical interest comparable to molecular vibrational and rotational deexcitation was available. The recent application of these techniques to ultrafast photobiological processes has opened up new avenues for scientific investigation. Some of the important aspects of this research will be presented here.

If scientific investigation were not limited by quantum mechanical restrictions, the quest for understanding nature's basic processes would someday be rewarded by the ability to observe physical processes in detail in their natural state. In photosynthesis the scenario for this event would begin with observation of the absorption of a single quantum of light as it strikes the outer surface of a leaf at a typical radiant flux density of 10^{16}–10^{17} visible quanta per second per square centimeter, the incident solar flux at the earth's surface at noontime. The initial absorption would most likely take place in the accessory pigment complex composed of carotenoids, chlorophyll b (Chl b), and chlorophyll a (Chl a) in green plants or phycobiliprotein pigment complexes in red and blue-green algae. The molecular structure of the excited pigment complex would subsequently search out modes of deexcitation leading to molecular stability consistent with the structural and energy gradients present in the immediate domain of exciton interaction.

A. Energy Transfer in Photosynthesis

Molecular deexcitation can either proceed through nonradiative internal decay in the vibrational rotational manifold or be transferred by intersystem crossing from the lowest excited singlet state to a triplet state. Other processes contributing to the relaxation might involve second-order processes of excited state annihilation between two singlet excitations or singlet–triplet exciton annihilation. Intermolecular energy transfer between the singlet states of like molecules (homogeneous transfer) or the singlet states of unlike molecules (heterogeneous transfer) could also contribute to this relaxation. In the crowded accessory pigment system the exciton might very well visit many molecular sites during its lifetime before becoming hopelessly trapped in a low-lying state of a reaction center molecule where the energy might be photochemically converted or diverted to a radiationless trapping site where it may undergo a biologically wasteful deexcitation. Throughout its migration the exciton is also subjected to a radiative decay probability in the form of fluorescence, again a wasteful process from the biological point of view.

Possible energy transfer mechanisms which consider interaction between two molecules involve either a mass motion mechanism, as in an electron transfer or collision, or electromagnetic coupling, as in the radiative or resonance transfer process. Studies on Chl a in solution at the concentrations present in the photosynthetic unit (PSU), where the interchromophore separation is estimated at 11–17 Å (Thomas et al., 1956; Colbow, 1973; Bay and Pearlstein, 1963), have shown that little energy transfer occurs at these high concentrations (Beddard and Porter, 1976). Chlorophyll excimer formations apparently act as strong quenchers of the excitation at these high concentrations, allowing the energy to be transferred among chromophore molecules only at lower interchromophore separations (Trosper *et al.*, 1968). The fate of the absorbed photon must therefore necessarily depend on such factors as the excited state population of the donor, the spacing between molecules, the relative angles between the absorption and emission dipole of the acceptor and donor molecules, respectively, and the overlap between the absorption spectra of the acceptor and the emission spectra of the donor (Forster, 1948). Exciton energy migration in the photosynthetic pigment complex has been described by a variety of physical models which have taken the theoretical constructs of random walk theory (Montroll, 1969; Robinson, 1967), diffusion (Bay and Pearlstein, 1963), and exciton percolation (Kopelman, 1976a,b,c). These models have recently been applied to the physical structures present in photosynthetic systems. Some of these models are discussed in Chapter 7.

From the very first experimental observation of the role of sunlight in photosynthesis by Jon Ingenhousz (Rabinowitch and Govindjee, 1969), the very component which forms the driving force of photosynthesis, light, has been of invaluable assistance in probing the physical nature of the process. Ingenhousz published his findings on "the influence of the light of the sun upon the plant" in 1779, three years after Joseph Priestley's treatise on the "improvement of air by plants," which formed the foundation for the understanding of the photosynthetic process. Since then countless experiments have sought to unravel the energy transfer kinetics which occur in this all-important process.

B. TIME-RESOLVED FLUORESCENCE EMISSION IN PHOTOSYNTHESIS

The overall temporal evolution of the fluorescence emission from photosynthetic systems possesses an interesting characteristic behavior. The range of the emission spans a time scale from the ultrafast absorption process (10^{-15} sec) to the slow decay of delayed emission which lasts for many minutes subsequent to excitation (Strehler and Arnold, 1951; Lavorel, 1975; Malkin, 1977). To place the fluorescence kinetics of the primary process in its proper time perspective relative to the total fluorescence emission from the photosynthetic system, it is important to keep in mind that the fluorescence emission is a measure of the total depletion of the first excited singlet state population and is influenced by the state of the reaction center traps and the aggregation and orientational relationship of the pigment molecules in the photosystem. A characteristic absorption spectra and fluorescence emission spectra from spinach chloroplasts is shown in Fig. 1.

The overall temporal behavior of the relative fluorescence intensity in green plants and algae subsequent to excitation displays an initial onset (O)

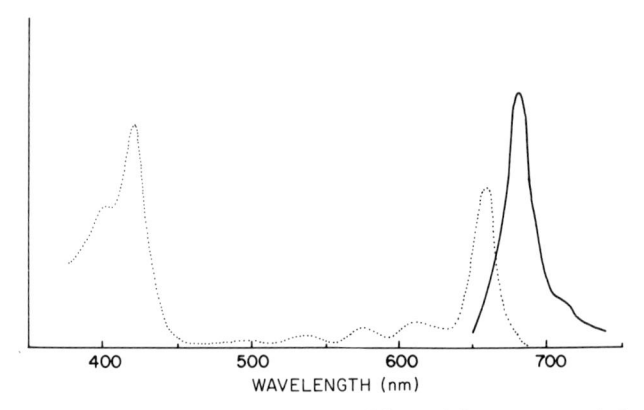

FIG. 1. Characteristic absorption spectrum of Chl *a* and fluorescence emission spectra of spinach chloroplasts at room temperature. Dotted line, absorption; solid line, fluorescence.

followed by a steep increase (I); then a slight dip (D) appears in the fluorescence emission and is subsequently followed by a slower increase to the peak (P) of the emission, which occurs approximately 0.4 sec subsequent to excitation. A gradual decline in the fluorescence to the steady state level (S) occurs, and the emission rises to a maximum (M) before decaying to a final terminal steady state level (T) (Murakami et al., 1975, p. 607; Papageorgiou, 1975). These events are depicted in Fig. 2.

Primary photochemistry is reflected only in the initial fluorescence emission (i.e., the O level in Fig. 2). The maximum limiting time for the primary photosynthetic process must necessarily be the time during which the excitation energy becomes completely dissipated in the structural framework of the photosystem, which we can take to be the time associated with the natural radiative lifetime that measures the unimolecular fluorescence decay in the absence of all other excitation decay mechanisms. For the primary pigment, Chl *a*, the natural radiative lifetime was calculated by Brody and

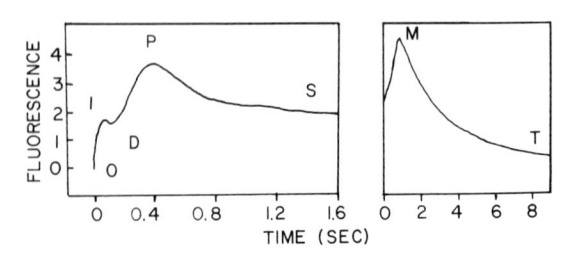

FIG. 2. Time dependence of *in vivo* fluorescence emission (fluorescence induction) of *C. pyrenoidosa*.

Rabinowitch (1957), from the area under the absorption curve, to be 15.2 nsec. The measured fluorescence lifetime of Chl *a*, however, is ~5 nsec, indicative of a fluorescence quantum yield of ~33% ($\phi = \tau/\tau_0$).

Thus in primary photosynthesis we shall be concerned only with the time regime from the initial absorption (10^{-15} sec) of the excitation energy to the time at which it has finally decayed by the slowest process available to it (~10^{-8} sec), the natural radiative lifetime. However, the fluorescence quantum yield of Chl *a in vivo* is only about 3%, which indicates an efficient transfer of the excitation energy for use in photosynthesis.

II. Time-Resolved Fluorescence Spectroscopy Techniques

The myriad of experiments that have been performed in the primary temporal regime offer a variety of information which, like the pieces of a giant jigsaw puzzle, must be carefully assembled in order to obtain an enlightening and cohesive picture of primary photosynthesis. However, the pieces that we must deal with do not lend themselves to sharp and well-defined interpretations which give rise to a precise solution of the puzzle. On the other hand, our pieces are complicated by the fact that we are often forced to compare information which differs in such basic characteristics as sample species investigated, sample homogeneity and cross-sectional thickness, wavelength of excitation, wavelength of observation, preparation and state of the sample, intensity of excitation, and temperature. Thus we must be very careful in deriving our conclusions from the apparent agreement, or lack thereof, of experimental observations.

In order to appreciate fully the progress of research on primary photosynthetic fluorescence kinetics a brief description of the various experimental techniques covering the excitation and detection methods used in these investigations is presented to acquaint the reader with some of the advantages and disadvantages of each technique. A detailed account of some picosecond experimental systems is presented in Chapter 17.

Various excitation sources are available for studying photosynthetic as well as other photobiological systems such as the visual process and DNA synthesis. These excitation sources include: the hydrogen flash lamp; gas lasers such as helium–neon, krypton ion, argon, and nitrogen lasers; solid state lasers such as the neodymium–glass, neodymium–YAG, and ruby lasers; and dye lasers which offer the advantage of wavelength tunability, high repetition rate, and short pulse duration. The excitation source can be characterized by a description of the wavelength and spectral bandwidth of the source, the energy, the pulse width (generally FWHM), and the pulse intensity profile. These characteristics are generally quite different for the excitation sources cited above and generally cannot be duplicated by any two such sources. Therefore each experiment is distinguished unequivocably by the excitation source used.

For purposes of biological investigation, it is important to consider the effective photobiological cross section that is active during the time period of observation. Some of the factors which must be considered include (1) the photorecovery of the system at the photon flux of the excitation source, (2) the effective local absorption of the fluorescing species at the wavelength of excitation, (3) the possibility of multiphoton processes occurring during absorption, (4) the formation of triplet excitations, (5) the formation of ionic species with localized conduction states, and (6) excitonic interactions among excited state molecules.

Fast photodetectors and large-bandwidth oscilloscopes form by far the simplest tools capable of temporally resolving the time course of fluorescence emission. However, the temporal resolution obtainable with such conventional devices is approximately 300 psec. Sampling oscilloscopes coupled with fast photodetectors (≤ 50 psec risetime) can give rise to a temporal resolution of a few tens of picoseconds in experiments with repetitive signals.

The earliest time-resolved fluorescence techniques made use of controlled flashing light excitation, phase fluorimetry, and photon counting. The flash excitation method makes use of conventional photodetection temporal resolution techniques (e.g., the photomultiplier and the oscilloscope) and is thus limited by the response time of this equipment, typically 0.3 nsec. The phase modulation technique (Bailey and Rollefson, 1953; Muller et al., 1969; Borisov and Godik, 1972a) involves exciting the sample with a light source that is modulated at a specific frequency and measuring the lag or phase delay between the excitation light and the emitted fluorescence light. The fluorescence signal is collected from the sample and compared with a reference beam from the excitation source. The phase difference between these two signals is determined by comparison with calibrated phase shifts that are electronically generated. The signal-to-noise ratio is enhanced by the fact that only the fluorescence signal at the modulated frequency is discriminated by the detection apparatus, thereby significantly reducing background light. Measurement of the phase angle determines the lifetime of the excited state, since the phase is related to the frequency of modulation and the lifetime by the relationship $\tan \phi = w\tau$.

Photon counting makes use of pulse amplitude discrimination and pulse amplification devices to detect weak photomultiplier signals. The discriminated pulses are counted electronically and thereby constitute a measure of the emission from the sample investigated. Since photon counting makes use of photomultiplier detection directly, it can record the high sensitivities available to photomultiplier detectors and thus forms one of the most sensitive detection techniques (Alfano and Ockman, 1968).

2. TIME-RESOLVED FLUORESCENCE SPECTROSCOPY

The direct measurement of fluorescence emission on a picosecond time scale can also be accomplished with a picosecond optical Kerr gate or streak camera. The picosecond optical Kerr gate (Duguay and Hansen, 1969) operates like an ultrafast shuttering device actuated by the intense electric field of the laser pulse. This technique is described in detail in Chapter 17. When activated by intense optical pulses of picosecond duration, such as from a neodymium–glass laser (~ 8 psec pulse width), a 10-psec resolution element can be selected from the fluorescent emission of the sample. The optical Kerr technique allows sequential sampling of the decay profile from a light-emitting event, which is achieved by merely delaying the arrival of the gate-opening pulse relative to the portion of the decay being sampled. Since the intensity of the transmitted light is proportional to the square of the intensity of the gate-activating pulse, and since in addition laser pulse stability is not easily obtainable in high-power, short-pulsed lasers, extensive signal averaging and normalization are required in order to obtain a complete temporal decay profile in a Kerr gate measurement.

A streak camera with picosecond resolution provides an ideal device for the measurement of subnanosecond fluorescence kinetics. The streak camera provides a continuous and direct measurement of the temporal profile of a light-emitting event ranging from a few picoseconds to several nanoseconds (Yu *et al.*, 1977; Pellegrino and Alfano, 1979; Shapiro *et al.*, 1975; Harris *et al.*, 1976). A typical apparatus employing a neodymium–glass laser excitation source and a streak camera detection system is shown in Fig. 3.

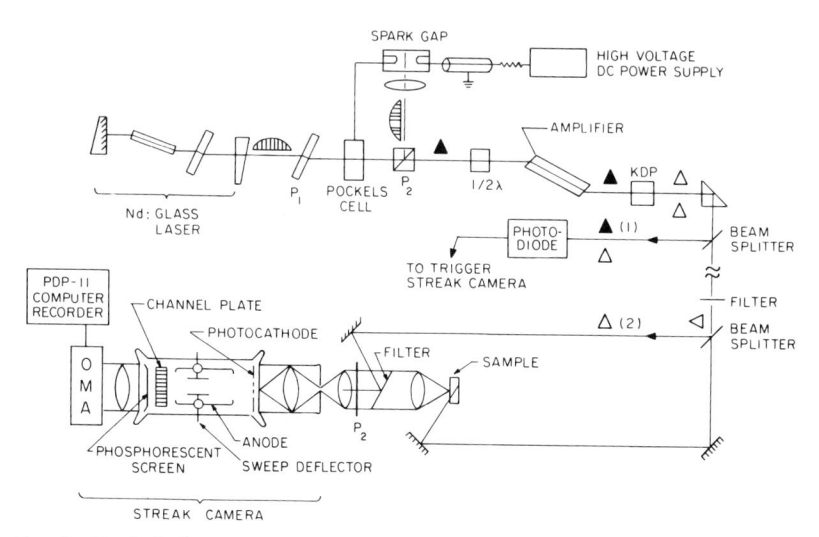

FIG. 3. Typical picosecond time-resolved fluorescence apparatus with neodymium–glass laser excitation and a streak camera detection system.

In the streak camera, photoelectrons emitted by light striking the photocathode are deflected by an applied voltage ramp, which causes the electrons to be transversely streaked across a phosphorescent screen. Photoelectrons released from the photocathode at a certain function of time will strike the phosphorescent screen at a corresponding function of position, causing a track representative of the incident fluorescence intensity temporal decay profile to be produced. The use of the streak camera in fluorescence kinetic measurements is also described in more detail in Chapter 17.

The streak camera provides certain distinct advantages over the Kerr gate. In particular, it is able to obtain the entire decay in a single shot, as opposed to the Kerr technique which provides only one point of the decay profile for a single laser shot. In addition, since the delay time in the Kerr gate is obtained by changing the optical path length of the gate-activating pulse by moving a prism over a linear translation stage, the time range which can be investigated with a Kerr gate is generally limited to a few nanoseconds.

Streak cameras are also generally plagued by "jitter" which causes the sweep deflection ramp to be initiated anywhere in a window which can be as wide as 100–200 psec, thus hindering signal averaging of the fluorescence signal from shot to shot. Mourou and Knox (1980) have greatly reduced jitter time by using direct optical triggering of a silicon semiconductor junction with the laser pulse itself.

III. Fluorescence Kinetic Measurements in Photosynthesis

The picosecond fluorescence emission from photosynthetic systems is generally interpreted as indicative of the energy migration and trapping that takes place in the PSU (Sauer, 1975). The parameters of fluorescence emission in primary photosynthesis are therefore precisely those variables which influence the fluorescence state, the first excited singlet, both prior to and during the trapping process.

The first measurement of the fluorescence lifetime from Chl *a* was obtained by Brody and Rabinowitch (1957) using nanosecond flash lamp excitation pulses and direct detection methods. Dimitrievsky *et al.* (1957) made indirect measurements through the use of phase fluorimetry, obtaining a value of ~1.5 nsec. Subsequent measurement of the *in vivo* fluorescence lifetime by Muller *et al.* (1969) indicated that the *in vivo* fluorescence lifetime was excitation intensity-dependent, and the fluorescence lifetime was even shown to be dependent on photosynthetic activity (Tumerman *et al.*, 1961). This intensity dependence of the fluorescence lifetime was later found to occur also in direct picosecond fluorescence measurements by Campillo *et al.* (1976b) and Mauzerall (1976).

Early measurements of fluorescence lifetimes using the phase technique developed by Bailey and Rollefson (1953) were undertaken by Murty and Rabinowitch (1965), Nicholson and Fortoul (1967), Muller et al. (1969), Tumerman and Sorokin (1967), Borisov and Godik (1972a,b), and recently by Moya et al. (1977), where the fluorescence lifetime, measured at low light intensities, in the range 0.001–1 J/m² sec, was found to vary from 0.35 to 0.8 nsec. In these measurements, the necessary assumptions of the experimental dependence of the fluorescence decay necessarily limit the interpretation of the fluorescence kinetics. Indeed, the assumption of a single exponential decay law for the *in vivo* fluorescence from photosynthetic systems cannot be justified in light of the possible effects from concentration quenching, different unimolecular decay rates for the various pigment components present in the photosynthetic system, and characteristic fluorescence emission from photosystems I and II (PSI and PSII) as well as from the light-harvesting complex (LHC). Fluorescence from nonphotosynthetically active species, as well as characteristic PSU structural influences on the fluorescence lifetime, must also be considered.

Merkelo et al. (1969) first introduced the use of mode-locked gas lasers to study fluorescence from photosynthetic systems by using a 0.8-nsec pulse from a helium–neon laser. Both direct (photomultiplier and oscilloscope) and indirect (phase) measurements of the fluorescence emission from *in vitro* and *in vivo* samples were made. *In vitro* measurements of Chl *b* (3.87 ± 0.05 nsec) and phycocyanin (1.14 ± 0.01 nsec) and *in vivo* measurements of Chl *a* fluorescence from *Chlorella pyrenoidosa* (1.4 ± 0.05 nsec) were obtained. Seibert et al. (1973) used a picosecond mode-locked solid state laser (neodymium–glass) to obtain a train of ~100 pulses at 1.06 μm, which were frequency doubled to 0.53 μm by the process of second-harmonic generation in a potassium dihydrogen phosphate (KDP) crystal to measure the decay kinetics from escarole chloroplasts. A Duguay Hansen optical Kerr gate was used for the time-resolved measurements, and a lifetime of 320 ± 50 psec for the *in vivo* Chl *a* fluorescence from escarole chloroplasts was obtained. The experiment was later repeated in spinach chloroplasts by Seibert and Alfano (1974), yielding a two-component decay with a characteristic dip in the fluorescence occurring at 50 psec into the emission. This dip was later attributed to statistical fluctuations of the laser output as well as a possible insufficient number of data points in that region and did not show up in later measurements. Two maxima were observed in the fluorescence emission, one at 15 psec and the other at 90 psec. From their measurements Seibert and Alfano (1974) attributed a lifetime of 10 psec to the PS I emission component and a lifetime of 210 psec to the PS II emission component. The 90-psec delay in the appearance of the second peak was attributed to possible energy transfer from carotenoids to PS II.

Subsequent experiments by the same group (Yu et al., 1975) on isolated PS I- and PS II-enriched preparations of spinach chloroplasts confirmed the dual exponential nature of the decay and in particular identified the fast-emission component (60 psec) with emission from PS I and the slow (200-psec) emission as arising from PS II. These results were repeated in intact chloroplast preparations from spinach (Yu et al., 1977) and again verified the presence of a two-component fluorescence decay at room temperature and 685 nm, of 56 and 220 psec, nearly identical in value to the previously reported results for enriched PS I and PS II preparations of spinach chloroplasts (Yu et al., 1975). At room temperature the 730-nm fluorescence was found to decay as a single exponential with a lifetime of 100 psec. The risetime of the 730-nm emission at low temperature (90 K) was measured to be 13 psec. These experiments were carried out at an intensity of 2×10^{14} photons/cm^2 per pulse, with 530-nm pulse train excitation and a Kerr gate time resolution apparatus. The spectral and temperature characteristics of the $t = 0$ emission from spinach chloroplasts was reported by Yu et al. (1977) and Pellegrino et al. (1978) and is discussed later in this chapter.

The streak camera detection technique, with its simplicity of operation and capability of obtaining the complete fluorescence decay measurement in a single shot, facilitated the investigation of fluorescence measurements from photosynthetic systems. Paschenko et al. (1975) measured the fluorescence emission from one pulse in a train of exciting pulses at an intensity of 3×10^{14} photons/cm^2 per pulse using a streak camera and ruby laser excitation. They reported the observation of a three-component decay for green pea chloroplasts with lifetimes of 80 psec (wavelength of observation greater than 730 nm) which was attributed to PS I fluorescence, 300 psec (wavelength of observation greater than 650 nm) which was attributed to PS II fluorescence, and 4500 psec attributed to fluorescence from chlorophyll pigments not involved in photosynthesis. Borisov and Il'ina (1973) had estimated the lifetime for PS I from pea subchloroplast particles to be ~ 30 psec by use of phase fluorimetry. For PS II fluorescence, Muller et al. (1969) had initially reported a lifetime of 380 psec through phase fluorimetry measurements, while a value of 700 ± 200 psec using the nanosecond flash lamp method had been reported by Singhal and Rabinowitch (1969).

Kollman et al. (1975) and Shapiro et al. (1975) have used streak camera detection and picosecond pulse train excitation from a neodymium–glass laser to measure Chl a fluorescence from *C. pyrenoidosa* and *Anacystis nidulans*, reporting lifetimes as short as 75 psec (for wavelength of observation greater than 640 nm) for *A. nidulans* and 41 psec (for wavelength of observation greater than 640 nm) for *C. pyrenoidosa*. These low values of fluorescence lifetimes were attributed to concentration quenching, since comparable lifetimes were obtained in concentrated solutions of Chl a and

Chl *b* in chloroform. A similar experiment performed by Beddard *et al.* (1975), again using picosecond pulse train excitation with a photon flux of ~5 × 10^{14} photons/pulse, reported short lifetimes of 134 psec (wavelength of observation greater than 580 nm) for Chl *a* fluorescence from spinach chloroplasts, 108 psec for *Chlorella*, and 92 psec for *Porphyridium cruentum*, in general agreement with the lifetimes reported by Kollman *et al.* (1975). However, even though Beddard *et al.* (1975) varied the excitation intensity by a factor of 10 in their measurement of the fluorescence lifetime from *P. cruentum*, they did not observe a change in the lifetime. Indeed most experiments using picosecond pulse train excitation performed during this period appeared to result in anomalously fast lifetimes, and the picosecond fluorescence lifetime measurements generally seemed to result in faster values than those obtained earlier with the use of conventional techniques. These fast lifetimes were later attributed to annihilation effects. (See also Chapters 6–8.)

A. INTERPRETATION OF TIME DEPENDENCE OF FLUORESCENCE DECAY IN PHOTOSYNTHESIS

For dark-adapted *Chlorella*, Harris *et al.* (1976) measured a double-component decay of 32 and 90 psec at room temperature and with a pulse train excitation of 10^{15} photons/cm^2 per pulse. At low intensity (~10^{14} photons/cm^2 per pulse) and also under various experimental conditions, it was found that the decay could be well described by a $t^{1/2}$ time dependence. Recent measurements by Beddard *et al.* (1979), however, have characterized the fluorescence decay from spinach chloroplasts at room temperature as a double exponential; a short component of 410 ps and a long component of 1.4 nsec were reported. These measurements were obtained with mode-locked dye laser excitation and the use of a photon-counting apparatus. Searle *et al.* (1979) have also recently analyzed their results in terms of a dual exponential component. It is interesting to compare the findings of Beddard *et al.* (1979) with the results of Moya (1974) and Moya *et al.* (1977) who reported a lifetime change from 0.4 to 1.4 nsec upon the addition of 3-(3',4'-dichlorophenyl)-1,1-dimethylurea (DCMU) to chloroplasts in order to close the reaction center traps of PS II. It appears by simple analogy that one could associate the short component measured by Beddard *et al.* (1979) with photochemical trapping, and the long component as representing general exciton diffusion in the photosynthetic membrane.

The question as to whether photosynthetic fluorescence emission can be interpreted in terms of a double exponential or in terms of a $t^{1/2}$ dependence is a subject of some controversy. Generally speaking, although an exponential decay and $t^{1/2}$ and a $t^{1/3}$ time dependences can be easily distinguished theoretically (Fig. 4), in practice fluorescence kinetic data obtained at a low

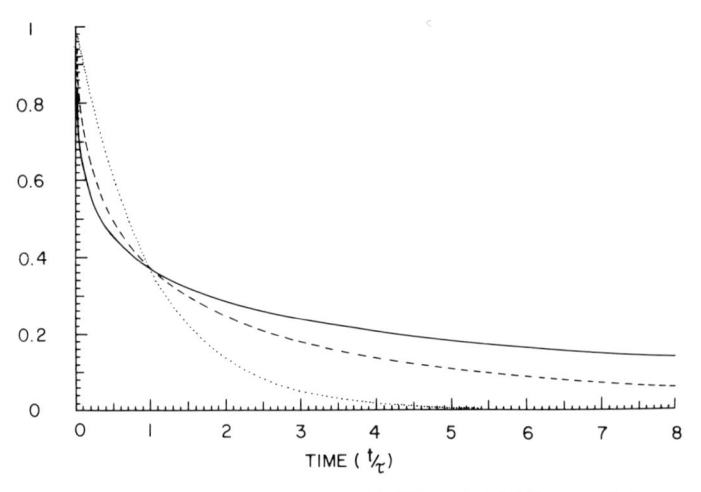

FIG. 4. Exponential time dependence: t/τ, $(t/\tau)^{1/2}$, and $(t/\tau)^{1/3}$ Dotted line, $\exp -(t/\tau)$, dashed line, $\exp -(t/\tau)^{1/2}$; solid line, $\exp -(t/\tau)^{1/3}$.

excitation intensity appear to be most readily described by a simple exponential decay. A random diffusion of the excitation energy in a three-dimensional system via Forster energy transfer mechanism does indeed predict a $t^{1/2}$ time behavior, however, it is generally agreed that the photosynthetic apparatus is best described by a two-dimensional model (Paillotin and Swenberg, 1979), and in this case a $t^{1/3}$ dependence holds for Forster energy transfer (Nakashima and Yoshihara, 1980).

B. Fluorescence Quenching and Exciton Annihilation Effects

The experiments of Mauzerall (1976) and Campillo et al. (1976b) gave a first clue to the possible reason for the observed short lifetimes by presenting a clear indication of the effect of excitation pulse intensity on the fluorescence yield and lifetime in photosynthetic systems. These observations led to a greater awareness of the multiple-excitation problem and exciton interactions arising from both picosecond pulse train excitation and high-intensity single picosecond pulse excitation. The possibility of creating long-lived triplet states or ionic species from laser pulse train excitation was clearly recognized as affecting the results of measurements of the prompt fluorescence. Detailed presentations of exciton annihilation affects may be found in Chapters 6–8.

Mauzerall's (1976) observed decrease in the fluorescence quantum yield of *Chlorella* with increasing excitation intensity from a 7-nsec pump pulse was interpreted in terms of a Poisson statistical theory governing the distribution of excitations in the PSU. However, for the case of closed reaction center traps, 20% of the fluorescence quantum yield appeared to be unaffected by quenching. These results were verified for single picosecond pulse excitations by Campillo *et al.* (1976a) who showed the onset of a similar decrease in the fluorescence quantum yield of *Chlorella* at an excitation intensity of approximately 10^{13} photons/cm^2 per pulse (Fig. 5). This quenching of the fluorescence from multiple excitations reflects the inability of the photosynthetic domain to make appropriate use of or efficiently transfer excitations above a certain threshold value.

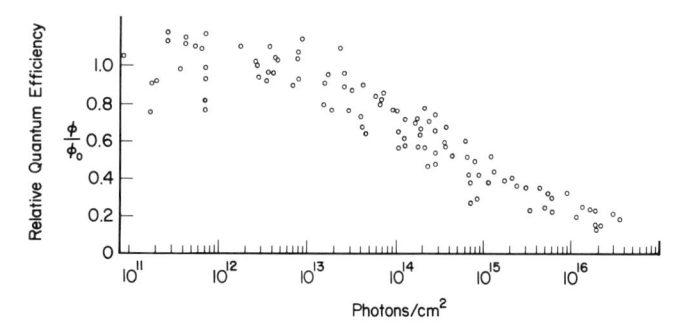

FIG. 5. Fluorescence quantum yield from the alga *C. pyrenoidosa* as a function of the intensity of a single picosecond excitation pulse. (From Campillo *et al.*, 1976a.)

The fluorescence quenching of the quantum yield under single-pulse excitation probing singlet–singlet interactions and microsecond pulse excitation probing singlet–triplet interactions was investigated by Geacintov *et al.* (1977a,b). In addition, Breton and Geacintov (1976) and Geacintov and Breton (1977) made a thorough investigation of the effect of excitation intensity on the fluorescence yield in spinach chloroplasts, at both room temperature and 100 K. A decrease in both the 685- and 735-nm fluorescence yield at 100 K was observed subsequent to exposure to an increasing number of 10-psec pulses spaced 5 nsec apart from a dye laser with an output wavelength of 610 nm. When a whole train of pulses (~300) was used for excitation, it was found that the low-temperature (100 K) 735-nm fluorescence began to quench at an intensity of 10^{15} photons/cm^2 for the whole pulse train, while the 685-nm fluorescence quenched at an intensity at least a factor of 10 higher. For single-pulse excitation, however, both wavelength component emissions were found to quench in a similar manner, and the quenching of the room temperature fluorescence at 685 nm subsequent to exposure to multiple excitation pulses at 610 nm was found to be less pronounced than at low temperature. Geacintov *et al.* (1977b) also observed quenching of the fluorescence lifetime from spinach chloroplasts at low temperature (77 K). Although the lifetime shortening for the 690-nm component qualitatively followed their observed quantum yield decrease, the lifetime of the 735-nm component appeared to be intensity independent (for the two excitation intensities used) even though the 735-nm yield curve showed a marked decrease over this range of intensities. Since annihilation was found to commence subsequent to excitation by the sixth pulse in their pulse train of 610-nm picosecond laser pulses, spaced 10 nsec apart, and since the fluorescence quenching was attributed to the presence of triplets, they additionally were able to deduce that the triplet diffusion time in PS I was on the order of 50 nsec. Porter et al. (1977) also have observed singlet–triplet quenching of fluorescence from *Chlorella* subsequent to excitation from multiple picosecond laser pulses.

Recently Breton et al. (1980) have shown that, for pulse train excitation, the creation of ionic species is negligible, and in addition the fluorescence quenching is sensitive to the presence of oxygen. From these observations they have attributed the quenching effects to the formation of triplet excitations from prior pulses in the pulse train. Mathis et al. (1979) have also recently shown that carotenoid triplets can act as fluorescence quenchers in both whole chloroplasts and PS I, PS II, and LHC preparations of spinach chloroplasts. Breton et al. (1980) also measured the onset of triplet fluorescence quenching in spinach chloroplasts and in a light-harvesting chlorophyll protein complex (LHCP), containing three Chl *a*, three Chl *b*, and one carotenoid molecule, isolated by sodium dodecyl sulfate (SDS) solubilization and polyacrylamide gel electrophoresis. A Poisson statistical behavior of the fluorescence intensity as a function of excitation intensity was observed consistent with the smaller domain size present in the complex particle. As expected from the smaller number of accessory pigment molecules present in the complex particle as opposed to whole chloroplasts, quenching in whole chloroplasts was found to occur at an excitation energy a factor of 100 lower than the complex particle.

Singlet–singlet annihilation and its effect on the fluorescence lifetime was directly demonstrated by Campillo et al. (1976b), who measured $1/e$ point lifetimes at a wavelength of 700 nm for *C. pyrenoidosa* of <50, 175, and 375 psec for single-pulse (20 psec) excitation intensities of 3×10^{15}, 3×10^{14}, and 10^{14} photons/cm^2 per pulse, respectively (Fig. 6).

Searle et al. (1977) have measured the fluorescence lifetime in PS I and PS II preparations of spinach chloroplasts with single picosecond pulse excitation. The PS I fluorescence was found to be intensity-independent over the range 5×10^{13}–10^{16} photons/cm^2 at room temperature, with a $1/e$ time of 100 psec ($\lambda \geq 650$ nm). The PS II fluorescence, however, was found to be intensity-dependent with a $1/e$ time of 500 psec ($\lambda \geq 650$ nm), for

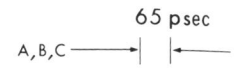

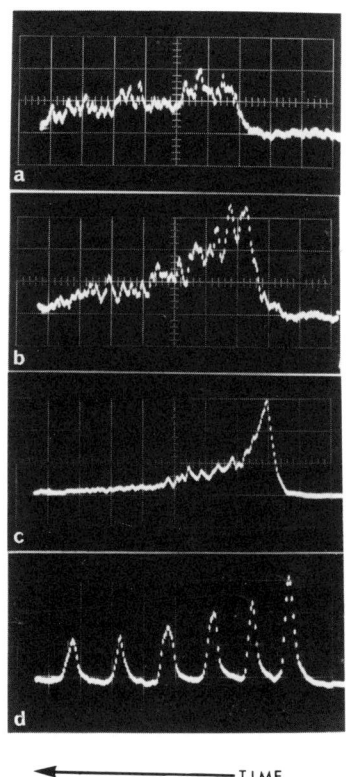

FIG. 6. Fluorescence lifetime of the alga *C. pyrenoidosa* measured as a function of the intensity of a single picosecond excitation pulse. From Campillo *et al.*, 1976b. Copyright 1976 by the American Association for the Advancement of Science.

measurements in the range 5×10^{13}–10^{14} photons/cm^2 per pulse, decreasing to 150 psec at the higher intensities of 5×10^{15}–10^{16} photons/cm^2 per pulse. An increase in the PS I fluorescence to 1.90 nsec at 77 K and 5×10^{13} photons/cm^2 excitation was measured for wavelengths greater than 700 nm. The PS II low-temperature lifetime was measured to be 2.47 nsec ($\lambda \geq 650$ nm).

The interpretation of the fluorescence quenching in terms of exciton annihilation introduces a mechanism which is independent of the photosynthetic activity state of the system but is dependent on the existence of closely spaced excitations. A theoretical study on singlet–singlet and singlet–triplet annihilation has been made by Breton *et al.* (1980). Swenberg *et al.* (1976) have obtained an expression for the fluorescence quantum yield dependence on the intensity:

$$\phi = (2k/\Gamma I)\log[1 + I(\Gamma/2k)],$$

where k is the inverse of the fluorescence lifetime, I is the intensity of excitation, and Γ is the singlet–singlet annihilation parameter related to the singlet–singlet annihilation coefficient γ_{ss}. This equation has been used to fit the fluorescence quantum yield measurements both in spinach chloroplasts (Geacintov et al., 1977b) and phycobilisomes (Pellegrino et al., 1981), as well as in isolated phycobiliproteins from Nostoc sp. (Wong et al., 1981). Paillotin et al. (1979) and Paillotin and Swenberg (1979) have considered a more general approach and have developed a model for exciton annihilation which encompasses the descriptions proposed by Mauzerall (1976) and Swenberg et al. (1976) as special cases. This model is found to be supportive of the "lake" description of energy migration in photosynthetic systems.

Exciton annihilation has also been investigated in organic and inorganic systems. Excitons in organic molecular crystals may undergo interactions similar to those of excitons in the PSU (Knox, 1975). An extraordinary variety of exciton interactions have been measured in aromatic hydrocarbons such as tetracene (Arnold et al., 1976; Swenberg and Stack, 1968; Geacintov et al., 1969). Indeed a theory describing the efficient directed migration or percolation of excitons among molecules in a mixed crystal has been proposed by Kopelman (1976a,b), and it may prove interesting to probe energy transfer in photosynthetic systems by modeling energy transfer in parallel artificial systems.

The exciton annihilation phenomenon can be divided into two basic categories when applied to photosynthetic systems, namely, singlet–singlet and singlet–triplet annihilation, which can contribute to a decrease in the excited state population and shorten the fluorescence lifetime and fluorescence quantum yield. Since the build-up of a sufficient triplet population usually requires several nanoseconds, singlet–triplet annihilation is expected to occur only during multiple-pulse excitation with pulse spacing of several nanoseconds. Singlet–singlet annihilation on the other hand depends solely on the singlet state population and is thus dependent on the intensity of the excitation pulse. The role of triplets in quenching the fluorescence was first noted by Breton and Roux (1975).

Although the exciton annihilation phenomenon can be a hindrance in measurements of true in vivo kinetics, it can prove useful in revealing the structural configuration of the system investigated as well as energy transfer within the PSU. Singlet–singlet annihilation and singlet–triplet annihilation processes reflect the density of both singlet and triplet state excitations. Systems with different structural arrangements, whether in terms of different interchromophore spacing or the presence or absence of traps, will affect the singlet and triplet excited state population numbers and density which subsequently are reflected in the fluorescence kinetics and quantum yield measurements. Campillo *et al.* (1977a) have used the singlet–singlet exciton annihilation process to measure topological differences in the PSUs of the mutants PM-8 *dpl*, wild-type strain 2.4.1, and a carotenoid-altered *Ga* mutant species of *Rhodopseudomonas spheroides*. These measurements are discussed in greater detail in Chapter 4. Measurement of the relative fluorescence quantum yield as a function of the intensity of excitation of a single 20-psec pulse at 530 nm revealed different quenching curves for each of these mutants. In this case the lack of reaction centers for the mutant PM-8 *dpl* was considered responsible for the earlier and stronger quenching observed for this mutant with respect to the reaction center containing wild-type strain 2.4.1 and *Ga* mutant species. This conclusion is also supported by their measurement of a longer lifetime for the PM-8 *dpl* mutant (1100 $\pm$ 200 psec) than for the *Ga* or wild-type 2.4.1 mutant (100 $\pm$ 25 psec) species, as well as a carotenoidless strain of *R. spheroides*, R-26. Most recently exciton annihilation effects in intact phycobilisomes and isolated phycobiliproteins from the blue-green alga *Nostoc* sp. have been studied by Pellegrino *et al.* (1981) and Wong *et al.* (1981), respectively.

C. ENVIRONMENTAL EFFECTS ON THE FLUORESCENCE FROM PHOTOSYNTHETIC SYSTEMS

Various physical factors which influence the photosynthetic activity of the system being measured can significantly affect the measurement of fluorescence kinetics. Among such factors are the presence of background illumination and the presence of poisons which act to block trap recovery by preventing electron transfer from the primary acceptor to the electron transport chain.

Phase fluorescence measurements (Muller *et al.*, 1969; Briantais *et al.*, 1972; Moya *et al.*, 1977) have shown the fluorescence lifetime from higher green plants to increase from 0.35 to 1.92 nsec as the reaction centers become closed by illumination from intensities ranging from 0.1 to 100 J/m^2 sec. An increase in PS I fluorescence from 80 to 200 psec and in PS II fluorescence from 300 to 600 psec, with the use of saturating background illumination, was reported by Paschenko *et al.*, (1975) along with a similar increase in the lifetime of PS II fluorescence with the addition of DCMU. It is known that *in vivo* chlorophyll fluorescence lifetimes increase by a factor of $\sim$3 when the reaction centers are closed (Harris *et al.*, 1976). However, since the *in vivo* chlorophyll fluorescence yield is approximately 1–3%, this corresponds to a total increase in the fluorescence yield to only about 9%, which is still about a factor of 3 lower than the measured *in vitro* fluorescence yield. Therefore either other decay mechanisms are present *in vivo* or equivalently not all the *in vivo* chlorophyll gives rise to fluorescence (Duysens, 1952). Sauer and Brewington (1978) have measured a lifetime change from 0.2 to 0.48 nsec in chloroplasts as the PS II reaction center traps are closed by the addition of DCMU. Hervo *et al.* (1975) measured a similar change from 750 psec to 1.15 nsec in the fluorescence lifetime with the addition of DCMU, while Beddard *et al.* (1979) have recently reported only minor changes in lifetime under similar conditions. Along with the increase in the fluorescence yield, a corresponding increase in the fluorescence lifetime under conditions which act to close the reaction center traps have also been observed by Briantais *et al.* (1972), Tumerman and Sorokin (1967), and Yu *et al.* (1977). When DCMU is added, a much longer fluorescence lifetime should be observed, since normal electron transfer to the traps is inhibited by the nonrecovery of closed traps. If indeed primary photochemistry is an efficient process, the closing of the reaction center traps should then make available to the fluorescent emission process all photons normally used for photoconversion, but this does not appear to be the case. Emerson has also shown an opposite effect on the fluorescence quantum yield when multiple excitation wavelengths are used on the sample. A significant drop in the fluorescence quantum yield, accompanied by an increase in the photosynthetic activity, is obtained upon providing additional excitation in the wavelength region lying in the active spectrum of PS I. This phenomenon is known as the Emerson enhancement effect and clearly demonstrates the presence of a second active photosystem (PS I) present in the overall process.

Interactions between the LHC and antenna components of PS I and PS II can affect the intensity dependence of the yield and lifetime. Moreover millimolar concentrations of monovalent cations (e.g., Na^+) can enhance the LHC–PS I interaction relative to the interaction between the LHC and PS II (Barber, 1976; Williams, 1977), while divalent cations (e.g., Mg^{2+}) can produce the reverse effect (Arntzen and Ditto, 1976; Butler and Strasser, 1978). Barber *et al.* (1978) have observed that Mg^{2+} affects the energy transfer rate from PS II to PS I in DCMU-treated chloroplasts. However, the effects of DCMU on the lengthening of the fluorescence lifetime and on the increase in the fluorescence quantum yield appear to be rather inconclusive. Beddard *et al.* (1979) and Searle *et al.* (1979) have observed only quantum yield changes, without changes in the lifetimes, in similar experiments with DCMU-treated chloroplasts. The relative amount of coupling between these antenna components may be investigated by using the tripartite model proposed by Butler and Kitajima (1975) and Butler (1978), which describes interactions within and among PS I, PS II, and the LHC. It appears that conformational changes in the stromal thylakoid membrane lie at the heart of this coupling dynamics (Butler, 1977).

D. Picosecond Fluorescence Temperature Studies

Time-resolved fluorescence measurements at low temperatures indicate that the fluorescence from green plants and algae at 77 K possesses two emission maxima, one at 735 nm and another at 685 nm, attributed to the emissions from PS I and PS II, respectively (Yu et al., 1977). Recently, Moya (1979) has observed four emission components from chloroplasts at low temperatures in the 680–760 nm region. A decrease in the temperature to near liquid nitrogen values acts to immobilize electron transport from the primary acceptor to the electron transport chain and thus affects an increase in the population density of closed traps. Yu et al. (1977) measured the fluorescence kinetics from spinach chloroplast preparations at both room temperature and 90 K. A sixfold change was observed in the lifetime of the 730-nm fluorescence from 100 to 600 psec as the temperature was lowered to 90 K, while the fluorescence kinetics of the 685-nm component was found to be temperature-independent. In addition, a risetime of 13 psec was measured for the 730-nm component at 90 K, although all other risetimes (685 and 695 nm), at both room temperature and 90 K, were resolution-limited (≤ 10 psec).

The measured temperature-dependent change in the fluorescence kinetics at 730 nm was also supported by measurement of the time-resolved spectra at room temperature and 90 K, measured relative to the 685-nm emission. At room temperature, the spectra (for wavelengths greater than 700 nm) was found to be larger within 10 psec of excitation than the time-integrated spectra. This was interpreted as indicative of the presence of a faster component at wavelengths greater than 700 nm relative to the lifetime of the component at 685 nm. In addition, as expected from the low-temperature kinetic measurement of the increased fluorescence lifetime of the 730-nm component at 90 K, the 90 K time-integrated spectra was found to be greater (for wavelengths greater than 700 nm) than the spectrum obtained within 10 psec of excitation, indicating the presence of a longer-lived decay at low temperature measured relative to the 685-nm emission. The ratio of the 730 nm fluorescence intensity measured relative to the 685-nm emission was found to be 1:9 at room temperature and 1:2 at 90 K. Govindjee and Yang (1966) placed this ratio at a value of 1:11 at room temperature and 1.4:1 at 77 K.

Geacintov et al. (1977b) and Campillo et al. (1976a) have shown that the 735-nm fluorescence kinetics at 77 K under single-pulse excitation is intensity-independent, while the emission at 685 nm is intensity-dependent. Campillo et al. (1977b) have measured the rise of the 735-nm fluorescence at 77 K and low intensity ($\sim 10^{14}$ photons/cm^2 per pulse) to be 140 psec. The slow rise of the PS I emission at low temperatures followed the decay of the

LHC fluorescence in this intensity region and is supportive of the idea of energy transfer from PS II or the LHC to PS I. The 730-nm risetime, however, is different from the measurements of Yu et al. (1977) and warrants further investigation. Butler et al. (1979) reported an increase in both the lifetime and quantum yield of the 735-nm emission as the temperature was lowered from 213 to 77 K. This emission component was attributed to a chlorophyll c-705 which was interpreted as being an efficient trap for PS I excitations at higher temperatures, channeling the energy to the photochemical trap of PS I, P700. The observed change in the fluorescence characteristics was attributed to the dynamic quenching of the fluorescence from singlet–singlet exciton annihilation in the Chl a LHC.

E. Energy Transfer in the Accessory Pigment Complexes of Red and Blue-Green Algae

The transfer of energy in the accessory pigment complex of photosynthetic systems is most easily followed in the phycobilisomes (Gantt, 1975) of red and blue-green algae. These phycobiliprotein bodies consist of an aggregation of the pigments, phycoerythrin (PE), phycocyanin (PC), and allophycocyanin (APC), which make up a regularly structured arrangement of chromophores in organized units in both red and blue-green algae. A representation of the arrangement of the phycobiliprotein units in phycobilisomes is shown in Fig. 7.

Energy transfer in phycobilisomes is generally believed to take place by the Forster inductive resonance energy transfer mechanism from the shorter to the longer-wavelength-absorbing elementary pigments: PE → PC → APC

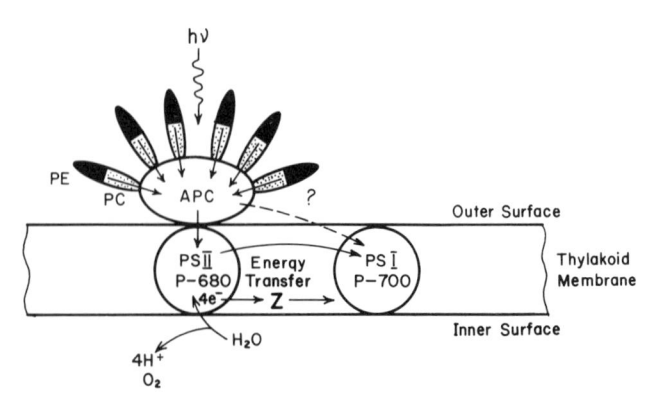

FIG. 7. Representation of the arrangement of the phycobiliprotein units—PE, PC, and APC—in phycobilisomes.

(Gantt and Lipschultz, 1973; Grabowski and Gantt, 1978; Duysens, 1952; Dale and Teale, 1970). The isolated phycobiliproteins, C-PE, C-PC, and APC, each contain different amounts of bilin chromophores. For example, C-PE contains 18 chromophores in the trimer form, while C-PC contains 18 chromophores in the hexamer form and APC contains 6 chromophores in the trimer form. The bilin chromophores are covalently attached to the apoproteins which are 120 Å in diameter and 30 Å thick in the basic trimer unit. Because of this chromophore packing arrangement, one should expect to observe fluorescence quenching due to singlet–singlet exciton annihilation within each phycobiliprotein. In intact phycobilisomes a lesser amount of quenching is expected in PE and PC as the PE → PC → APC energy transfer is also present in this case. However, there also exists a tendency for increased quenching in intact phycobilisomes because of the higher effective local absorption and the closer packing arrangement of the phycobiliprotein units.

Energy transfer times from B-phycoerythrin (B-PE) to R-phycocyanin of 300 ± 200 psec and from APC to Chl a of 500 ± 200 psec have been reported by Tomita and Rabinowitch (1962). Estimates of the Forster critical distances and the interchromophore distances by Dale and Teale (1970), when combined with the intrinsic fluorescence lifetime of the bilin prosthetic group, yield a Forster pairwise transfer time of 10 psec. Recently, Grabowski and Gantt (1978) have used a hemispherical model consisting of layered concentric shells for the phycobilisomes and have estimated the mean energy transfer time from the PE layer to the PC layer to be 280 ± 40 psec.

Searle et al. (1978) have shown that energy transfer in phycobilisomes isolated from the red alga P. cruentum can be followed through picosecond fluorescence kinetic measurements. Their measurement of the rise and decay times of B-PE and APC from the phycobilisomes of P. cruentum revealed that B-PE possessed a fast fluorescence risetime (rising within the duration of the excitation pulse) with a $1/e$ decay time of 70 psec. The decay time of B-PE was observed to be excitation intensity-independent from 10^{13} to 10^{15} photons/cm^2. The APC fluorescence risetime was measured to be 120 psec with a single exponential decay behavior for excitation pulse intensities within the range of 2×10^{13} to 4×10^{14} photons/cm^2. However, the relaxation time of APC decreased from 4 to 2 nsec over this intensity range, with a more complex exponential behavior occurring at higher intensities. In contrast, the decay time for APC in solution has been reported to be 2.6 ± 0.1 nsec by Grabowski and Gantt (1978) with use of the photon-counting technique and, as there are likely to be more decay mechanisms active for the APC fluorescence decay in PBS as opposed to in solution, an even shorter lifetime than 2.6 nsec is therefore expected for the APC fluorescence emission from PBS.

Wong et al. (1981) and Pellegrino et al. (1981) have recently measured the fluorescence rise and decay kinetics and fluorescence quantum yield from isolated phycobiliproteins C-PE, C-PC, and APC forms I, II, III, and B, as well as the C-PE and C-PC + APC emission from intact phycobilisomes of the blue-green alga *Nostoc* sp., as a function of the intensity of a single 530-nm, 6-psec excitation pulse over the intensity range 10^{13}–10^{15} photons/cm^2.

For isolated phycobiliproteins Wong et al. (1981) have found a single exponential decay with a lifetime of 1552 ± 31 psec for C-PE at single-pulse excitation intensities $<10^{14}$ photons/cm^2. For higher excitation intensities, the kinetics were found to decay as a double exponential with a fast component of 130 ± 23 psec and a slower component of 1477 ± 29 psec. The normalized fluorescence yield was found to decline significantly for intensities $>10^{14}$ photons/cm^2. The decrease arises from singlet–singlet annihilation. For isolated phycobiliprotein C-PC the kinetics were observed to be exponential from 10^{13} to 10^{15} photons/cm^2, and the lifetime declined from a value of 2111 ± 83 psec at an excitation intensity of $\sim 3 \times 10^{13}$ photons/cm^2 per pulse to a value of 1376 ± 24 at an intensity of 1.3×10^{15} photons/cm^2 per pulse. The decay times of the isolated phycobiliprotein APC forms I, II, III, and B were measured to be 1932 ± 165, 1870 ± 90, 1816 ± 88, and 2577 ± 121 psec, respectively, the average for APC forms I, II, and III being 1869 ± 62 psec. The APC lifetimes were found to be only slightly dependent on the excitation intensity, although the relative fluorescence quantum yield was constant throughout the intensity range investigated. All the risetimes in the isolated phycobiliproteins were found to be resolution-limited to ≤ 12 psec.

Measurement of the fluorescence rise and decay kinetics in the phycobilisomes of *Nostoc* sp. by Pellegrino et al. (1981) showed a clear indication of the presence of energy transfer in the phycobiliprotein complex. The decay time measurement of the C-PE component emission from phycobilisomes (560–600 nm) was found to be 31 ± 4 psec, which is considerably shorter than the measurement for isolated C-PE in solution (1552 ± 31 psec). This value is also shorter than the measurement reported by Searle et al. (1978) ($1/e$ time of 70 psec) in phycobilisomes of *P. cruentum* and is consistent with the suggestion of Grabowski and Gantt (1978) that energy transfer in the phycobilisomes isolated from blue-green algae is more efficient than energy transfer in the phycobilisomes isolated from red algae. While the risetime of the C-PE emission was resolution-limited, the risetime of the C-PC + APC emission component ($\lambda \geq 600$ nm) was found to be 34 ± 13 psec, reflecting the decay time measurement of the C-PE emission component (Fig. 8). Thus energy transfer from C-PE to C-PC + APC is clearly indicated to occur within this time. The decay kinetics of the C-PC + APC emission component were fit with a double exponential decay. The fast component was found to

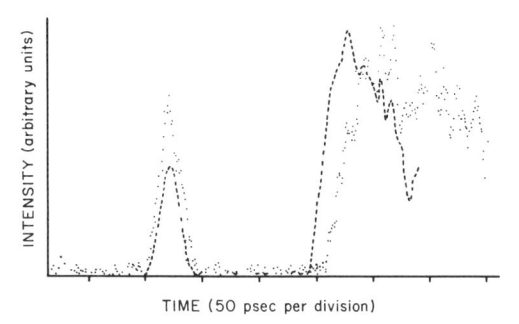

FIG. 8. Risetime of C-PC + APC fluorescence emission from the phycobilisomes of *Nostoc* sp. in PBS. The risetime of 34 ± 13 psec was found to be consistent with the measured decay time of 31 ± 4 psec for the C-PE emission component. The resolution-limited (≤ 12 psec) fluorescence risetime measurement of the fluorescence from erythrosin in water is shown for comparison. Dotted line, C-PC + APC; dashed line, erythrosin.

decrease from 212 psec at 2.7×10^{13} photons/cm^2 per pulse to 83 psec at 2.7×10^{15} photons/cm^2 per pulse, while the long component decreased from a value of 1174 psec to 716 psec over the same intensity region. The fluorescence quantum yield for this emission was found to decrease by a factor of 13 over this intensity range.

The decrease in lifetime and quantum yield for the C-PC + APC emission component, as compared to the case of isolated phycobiliproteins was interpreted in terms of the efficient transfer of excitation energy from C-PE to C-PC and APC, resulting in a higher effective absorption for APC in phycobilisomes. Singlet–singlet annihilation in APC accounts for the reduction in τ and ϕ.

IV. Future Directions

Time-resolved fluorescence spectroscopy with picosecond time resolution techniques has enabled researchers to probe energy transfer directly in the primary photosynthetic apparatus. The investigations of the last decade have provided a solid foundation upon which further studies can be based. Future directions appear to lie in the area of model systems made up of either naturally occurring photosynthetic components or artificial analogs. Indeed, in some recent studies, attempts have been made to duplicate the properties of Chl *a* in the photochemical reaction centers of green plants (Boxer and Closs, 1976; Shipman *et al.*, 1976; Wasielewski *et al.*, 1976; Fong and Koester, 1976; Galloway *et al.*, 1979). Such model systems are discussed in Chapter 5. Interest in the artificial duplication of reaction center chlorophylls arises from the fact that these forms possess the capacity to dissociate water (Fong and Galloway, 1978), which may prove instrumental in developing a means to generate hydrogen fuel from the dissociation of water in photosynthetic systems (Porter and Archer, 1976; Hall, 1977). Fluorescence studies on Chl a–water aggregate systems have recently been reported (Fong and Koester, 1976; Galloway et al., 1979). Galloway et al. (1979) have measured the fluorescence emission at 780 nm of the chlorophyll dihydrate polycrystal (Chl $a \cdot 2\,H_2O)_n$ with an emission peak at 760 nm. The fluorescence decay at 780 nm was found to be biphasic and temperature-dependent, with the fast-component lifetime decreasing to less than 50 psec at 294 K.

With the development of new laser excitation sources possessing wavelength tunability, subpicosecond pulse widths, and low-intensity excitation, combined with streak cameras with comparable time resolution and high repetition rates, a means of investigating the finer details of energy transfer will become available. These studies will eventually lead to a complete description of the mechanisms responsible for the primary process in photosynthesis.

Acknowledgments

The authors wish to acknowledge the support from the National Science Foundation Biophysics Division and City University Faculty Award Program.

References

Alfano, R. R., and Ockman, N. (1968). *J. Opt. Soc. Am.* **58**, 90–95.
Arnold, S., Alfano, R. R., Pope, M., Yu, W., Ho, P., Tharrats, T., and Swenberg, C. E. (1976). *J. Chem. Phys.* **64**, 5104.
Arntzen, C. J., and Ditto, C. L. (1976). *Biochim. Biophys. Acta* **449**, 259–274.
Bailey, E. A., and Rollefson, G. K. (1953). *J. Chem. Phys.* **21**, 1315.
Barber, J. (1976). *In* "Topics in Photosynthesis: The Intact Chloroplast", Vol. 1 (J. Barber, ed.), pp. 89–134. Elsevier, Amsterdam.
Barber, J., Searle, G. F. W., and Tredwell, C. J. (1978). *Biochim. Biophys. Acta* **501**, 174–182.
Bay, Z., and Pearlstein, R. M. (1963). *Proc. Natl. Acad. Sci. U.S.A.* **50**, 1071–1078.
Beddard, G. S., and Porter, G. (1976). *Nature (London)* **260**, 366–367.
Beddard, G. S., Porter, G., and Tredwell, C. J. (1975). *Nature (London)* **258**, 166–168.
Beddard, G. S., Fleming, G. R., Porter, G., Searle, G. F. W., and Synowiec, J. A. (1979). *Biochim. Biophys. Acta* **545**, 165–174.
Borisov, A. Y., and Godik, V. I. (1972a). *Bioenergetics* **3**, 211.
Borisov, A. Y., and Godik, V. I. (1972b). *Bioenergetics* **3**, 515.
Borisov, A. Y., and Il'ina, M. D. (1973). *Biochim. Biophys. Acta* **305**, 364–371.
Boxer, S. G., and Closs, G. L. (1976). *J. Am. Chem. Soc.* **98**, 5406–5408.
Breton, J., and Geacintov, N. (1976). *FEBS Lett.* **69**, 86.
Breton, J., and Roux, E. (1975). *In* "Lasers in Physical Chemistry and Biophysics" (J. Joussot-Dubien, ed.), pp. 379–388. Elsevier, Amsterdam.
Breton, J., Geacintov, N. E., and Swenberg, C. E. (1980). *Biochim. Biophys. Acta* **548**, 616–635.
Briantais, J. M., Govindjee, and Merkelo, H. (1972). *Photosynthetica* **6**, 133–141.

Brody, S. S., and Rabinowitch, E. (1957). *Science* **125**, 555.
Butler, W. L. (1977). *Brookhaven Symp. Biol.* **28**, 338–344.
Butler, W. L. (1978). *Annu. Rev. Plant Physiol.* **29**, 345–378.
Butler, W. L., and Kitajima, M. (1975). *Biochim. Biophys. Acta* **386**, 72–85.
Butler, W. L., and Strasser, R. J. (1978). *Proc. Int. Congr. Photosynth., 4th, Reading, Engl., 1977* pp. 11–20.
Butler, W. L., Tredwell, C. J., Malkin, R., and Barber, J. (1979). *Biochim. Biophys. Acta* **545**, 309–315.
Campillo, A. J., Shapiro, S. L., Kollman, V. H., and Hyer, R. C. (1976a). *Biophys. J.* **16**, 93–97.
Campillo, A. J., Kollman, V. H., and Shapiro, S. L. (1976b). *Science* **193**, 227–229.
Campillo, A. J., Hyer, R. C., Monger, T. G., Parsons, W. W., and Shapiro, S. L. (1977a). *Proc. Natl. Acad. Sci. U.S.A.* **74**, 1997–2001.
Campillo, A. J., Shapiro, S. L., Geacintov, N. E., and Swenberg, C. E. (1977b). *FEBS Lett.* **83**, 316–320.
Colbow, K. (1973). *Biochim. Biophys. Acta* **314**, 320–327.
Dale, R. E., and Teale, F. W. J. (1970). *Photochem. Photobiol.* **12**, 99–117.
Dimitrievsky, U., Ermolaev, V., and Terenin, A. (1957). *Dokl. Biol. Sci. (Engl. Transl.)* **114**, 468.
Duguay, M. A., and Hansen, J. W. (1969). *Appl. Phys. Lett.* **15**, 192.
Duysens, L. N. M. (1952). Ph.D Thesis, Univ. Utrecht, Utrecht.
Fong, F. K., and Galloway, L. (1978). *J. Am. Chem. Soc.* **100**, 3594–3596.
Fong, F. K., and Koester, V. J. (1976). *Biochim. Biophys. Acta* **423**, 52–64.
Forster, T. (1948). *Ann. Phys. (Leipzig)* **2**, 55.
Galloway, L., Matthews, T. G., Lytle, F. E., and Fong, F. K. (1979). *J. Am. Chem. Soc.* **101**, 229.
Gantt, E. (1975). *BioScience* **25**, 781–788.
Gantt, E., and Lipschultz, C. A. (1973). *Biochim. Biophys. Acta* **292**, 858–861.
Geacintov, N. E., and Breton, J. (1977). *Biophys. J.* **18**, 1–15.
Geacintov, N. E., Pope, M., and Vogel, F. (1969). *Phys. Rev. Lett.* **22**, 593.
Geacintov, N. E., Breton, J., Swenberg, C. E., and Paillotin, G. (1977a). *Photochem. Photobiol.* **26**, 619–638.
Geacintov, N. E., Breton, J., Swenberg, C., Campillo, A. J., Hyer, R. C., and Shapiro, S. L. (1977b). *Biochim. Biophys. Acta* **461**, 306–312.
Govindjee, and Yang, L. (1966). *J. Gen. Physiol.* **49**, 763–780.
Grabowski, J., and Gantt, E. (1978). *Photochem. Photobiol.* **28**, 39–45.
Hall, D. O. (1977). *Trends Biochem. Sci.* **2**, 99–101.
Harris, L., Porter, G., Synoweic, J. A., Tredwell, C. J., and Barber, J. (1976). *Biochim. Biophys. Acta* **449**, 329–339.
Hervo, G., Paillotin, G., and Thiery, J. (1975). *J. Chem. Phys.* **72**, 761–766.
Knox, R. S. (1975). *In* "Bioenergetics of Photosynthesis" (Govindjee, ed.), pp. 183–219. Academic Press, New York.
Kollman, V. H., Shapiro, S. L., and Campillo, A. J. (1975). *Biochem. Biophys. Res. Commun.* **63**, 917.
Kopelman, R. (1976a). *In* "Radiationless Processes" (F. K. Fong, ed.), pp. 297–347. Springer-Verlag, Berlin and New York.
Kopelman, R. (1976b). *J. Phys. Chem.* **80**, 2191–2195.
Kopelman, R. (1976c). *J. Lumin.* **12**, 775.
Lavorel, J. (1975). *In* "Bioenergetics of Photosynthesis" (Govindjee, ed.), pp. 223–317. Academic Press, New York.
Malkin, S. (1977). *In* "Primary Processes of Photosynthesis, Vol. 2, Topics in Photosynthesis" (J. Barber, ed.), pp. 349–431. Elsevier, Amsterdam.

Mathis, P., Butler, W. L., and Satoh, K. (1979). *Photochem. Photobiol.* **30**, 603–614.
Mauzerall, D. (1976). *J. Phys. Chem.* **80**, 2306–2309.
Merkelo, H., Hartmann, S. R., Mar, T., Singhal, G. S., and Govindjee (1969). *Science* **164**, 301.
Montroll, E. W. (1969). *J. Math. Phys.* **10**, 753–765.
Mourou, G., and Knox, R. (1980). *Appl. Phys. Lett.* **36**, 492.
Moya, I. (1974). *Biochim. Biophys. Acta* **358**, 214–227.
Moya, I. (1979). Ph.D Thesis, Univ. Paris-Sud (Orsay).
Moya, I., Govindjee, Vernotte, C., and Briantais, J. M. (1977). *FEBS Lett.* **75**, 13–18.
Muller, A., Lumry, R., and Walker, M. S. (1969). *Photochem. Photobiol.* **9**, 113–126.
Murakami, S., Torres-Pereira, J., and Packer, L. (1975). In "Bioenergetics of Photosynthesis" (Govindjee, ed.), pp. 555–618. Academic Press, New York.
Murty, N. R., and Rabinowitch, E. (1965). *Biophys. J.* **5**, 655–661.
Nakashima, N., and Yoshihara, K. (1980). *J. Chem. Phys.* **73**, 3553–3559.
Nicholson, W. J., and Fortoul, J. L. (1967). *Biochim. Biophys. Acta* **143**, 577–582.
Paillotin, G., and Swenberg, C. E. (1979). *Chlorophyll Organ. Energy Transfer Photosynth., Ciba Found. Symp.* No. 61, 201–209.
Paillotin, G., Swenberg, C. E., Breton, J., and Geacintov, N. E. (1979). *Biophys. J.* **25**, 513–534.
Papageorgiou, G. (1975). In "Bioenergetics of Photosynthesis" (Govindjee, ed.), pp. 319–371. Academic Press, New York.
Paschenko, V. Z., Protasov, S. P., Rubin, A. B., Timofeev, K. N., Zamazova, L. M., and Rubin, L. B. (1975). *Biochim. Biophys. Acta* **408**, 145.
Pellegrino, F., and Alfano, R. R. (1979). In "Multichannel Image Detectors" (Y. Talmi, ed.), ACS Symposium Series, No. 102, pp. 183–198. *Am. Chem. Soc.*, Washington, D.C.
Pellegrino, F., Yu, W., and Alfano, R. R. (1978). *Photochem. Photobiol.* **28**, 1007–1012.
Pellegrino, F., Wong, D., Alfano, R. R., and Zilinskas, B. A. (1981). *Photochem. Photobiol.* (To be published.)
Porter, G., and Archer, M. D. (1976). *Interdiscip. Sci. Rev.* **1**, 119–143.
Porter, G., Synowiec, J. A., and Tredwell, C. J. (1977). *Biochim. Biophys. Acta* **459**, 329–336.
Rabinowitch, E., and Govindjee, (1969). "Photosynthesis." Wiley, New York.
Robinson, G. W. (1967). *Brookhaven Symp. Biol.* **19**, 16.
Sauer, K. (1975). In "Bioenergetics of Photosynthesis" (Govindjee, ed.), pp. 115–181. Academic Press, New York.
Sauer, K., and Brewington, G. T. (1978). *Proc. Int. Congr. Photosynth., 4th, Reading, Engl.* pp. 409–421.
Searle, G. F. W., Barber, J., Harris, L., Porter, G., and Tredwell, C. J. (1977). *Biochim. Biophys. Acta* **459**, 390–401.
Searle, G. F. W., Barber, J., Porter, G., and Tredwell, C. J. (1978). *Biochim. Biophys. Acta* **501**, 246–256.
Searle, G. F. W., Tredwell, C. J., Barber, J., and Porter, G. (1979). *Biochim. Biophys. Acta* **545**, 496–507.
Seibert, M., and Alfano, R. R. (1974). *Biophys. J.* **14**, 269.
Seibert, M., Alfano, R. R., and Shapiro, S. L. (1973). *Biochim. Biophys. Acta* **292**, 493.
Shapiro, S. L., Kollman, V. H., and Campillo, A. J. (1975). *FEBS Lett.* **54**, 358.
Shipman, L. L., Cotton, T. M., Norris, J. R., and Katz, J. J. (1976). *Proc. Natl. Acad. Sci. U.S.A.* **73**, 1791–1794.
Singhal, S. S., and Rabinowitch, E. (1969). *Biophys. J.* **9**, 586–591.
Strehler, B., and Arnold, W. (1951). *J. Gen. Physiol.* **34**, 809–820.
Swenberg, C. E., and Stack, W. T. (1968). *Chem. Phys. Lett.* **2**, 327.
Swenberg, C. E., Geacintov, N. E., and Pope, M. (1976). *Biophys. J.* **16**, 1447–1452.
Thomas, J. B., Minnaert, K., and Elbers, P. F. (1956). *Acta Bot. Neerl.* **5**, 315–321.

Tomita, G., and Rabinowitch, E. (1962). *Biophys. J.* **2**, 483–499.
Trosper, T., Park, R. B., and Sauer, K. (1968). *Photochem. Photobiol.* **7**, 451–469.
Tumerman, L. A., and Sorokin, E. M. (1967). *Mol. Biol.* **1**, 628–638.
Tumerman, L. A., Borisova, O. F., and Rubin, A. B. (1961). *Biophysics (Engl. Transl.)* pp. 723–728.
Wasielewski, M. R., Studier, M. H., and Katz, J. J. (1976). *Proc. Natl. Acad. Sci. U.S.A.* **73**, 4282–4286.
Williams, W. P. (1977). *In* "Primary Processes of Photosynthesis" (J. Barber, ed.), pp. 99–147. Elsevier, Amsterdam.
Wong, D., Pellegrino, F., Alfano, R. R., and Zilinskas, B. A. (1981). *Photochem. Photobiol.* **33**, 651–662.
Yu, W., Ho, P., Alfano, R. R., and Seibert, M. (1975). *Biochim. Biophys. Acta* **387**, 159.
Yu, W., Pellegrino, F., and Alfano, R. R. (1977). *Biochim. Biophys. Acta* **460**, 171–181.

CHAPTER 3

Early Photochemical Events in Green Plant Photosynthesis: Absorption and EPR Spectroscopic Studies*

Bacon Ke and Edward Dolan

Charles F. Kettering Research Laboratory
Yellow Springs, Ohio

and

V. A. Shuvalov and V. V. Klimov

Institute of Photosynthesis
USSR Academy of Sciences
Pushchino, USSR

I.	Introduction	55
II.	The Primary Electron Donors of the Two Photosystems	57
III.	Early Photochemical Events in the Electron Acceptor Systems of Photosystems I and II	59
	References	76

I. Introduction

In green plants the overall photosynthetic reaction, namely, conversion of water and carbon dioxide to oxygen and carbohydrate, occurs in subcellular organelles called chloroplasts. The early stages, including light absorption, release of oxygen from water, reduction of $NADP^+$, and conversion of ADP to ATP, are confined to the thylakoid membranes in the chloroplast interior. In these extended, double-layer structures are embedded large arrays of light-harvesting chlorophyll molecules which serve to channel absorbed light energy to reaction centers which in turn are coupled to electron transport chains. The eventual reduction of carbon dioxide to carbohydrate, etc., with the help of NADPH and ATP, occurs subsequently in the surrounding stroma regions of the chloroplasts.

---

* Contribution No. 700 from the Charles F. Kettering Research Laboratory.

The initial photochemical charge separation and subsequent flow of electrons from water to $NADP^+$ is presented (Fig. 1) in the Z scheme with two photochemical systems, in which P designates the primary electron donor and A the primary acceptor of photosystems I and II (PS I and PS II) and the subscripts 1, 2, 3, etc., represent the sequential location of the electron carriers in the transport chain. Light energy absorbed by the chlorophyll arrays migrates to the two reaction centers to produce the primary charge transfer states, $P_I^+ \cdot A_{I,1}^-$ or $P_{II}^+ \cdot A_{II,1}^-$, with an efficiency very close to 100%. By way of chains of electron carriers in the thylakoid membrane, the positive charge on P_{II}^+ liberates oxygen from water and the negative charge on $A_{I,1}^-$ is transferred to $NADP^+$, while the free energy derived from charge recombination between $A_{II,1}^-$ and P_I^+ may be coupled to the conversion of ADP to ATP. The course of these electron transfers is determined by the various redox potentials, transition state energies, and structural relationships of the electron carriers bound within the lipid–protein matrix of the thylakoid. It should be noted, however, that while many electron carriers have been characterized to some degree, the precise identity of some components in the chain and, in some instances, their exact function still remain unclear. Moreover, investigation will probably reveal still other components, particularly in the water-splitting portion of the chain. (Not shown is the

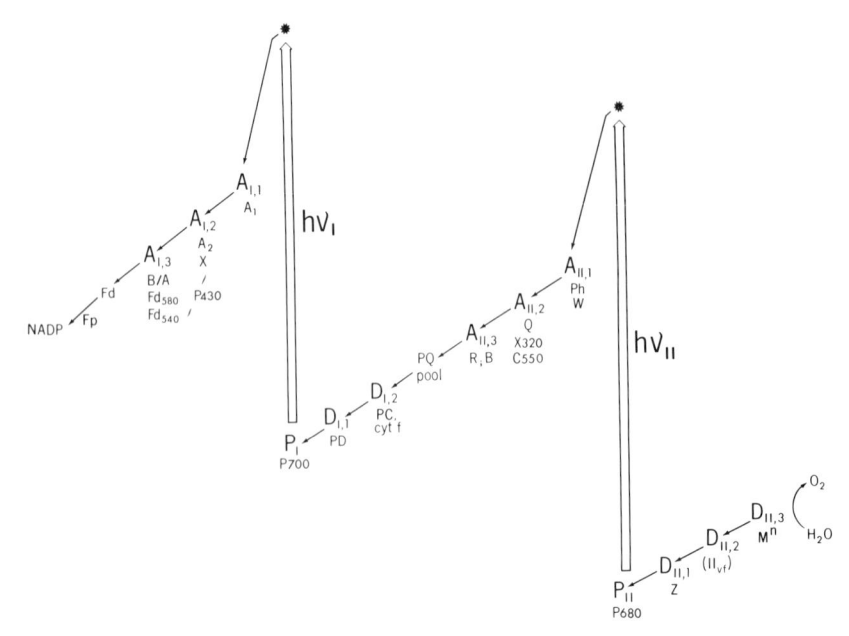

FIG. 1

light-harvesting chlorophyll antenna complex containing about 200–300 chlorophyll molecules associated with each reaction center.)

Evidence for the existence of these two series-coupled reaction centers is provided by the nonadditive effects of different wavelengths of excitation light on the rates of photosynthesis as measured by oxygen evolution, with the long-wavelength threshold in the action spectrum of PS II shifted about 15–20 nm to the blue relative to that for PS I. The Z scheme is also in accord with the requirement of about 10 light quanta per oxygen molecule formed, if some cyclic electron flow around PS I (coupled to the conversion of ADP to ATP) is allowed, as evidence has indicated.

Steady state illumination coupled with spectrophotometric measurements has been widely adapted to the study of photosynthesis since the early 1950s (Duysens, 1952). One of the most popular methods for studying the primary photochemical events in photosynthesis is to activate the system with a pulse of light and then monitor it by measuring absorbance changes specific to the components of interest. It was through the improvement in shortening the duration of actinic flashes that new insights into the reaction kinetics were obtained (Ke, 1972a) and, in recent years, has progressed from milli- or microseconds generated in flash lamps energized by capacitor discharge to nanoseconds in Q-switched (ruby) lasers to picoseconds in mode-locked lasers (Seibert, 1978; Holten and Windsor, 1978).

Picosecond spectroscopy has been applied to photosynthesis studies only recently. Its application has been limited by the signal-to-noise problem associated with the inherent wide bandwidth of the technique, as well as by consideration of spectral requirements, expense, and experimental expertise. It was initially employed for studying the photochemistry in photosynthetic bacteria by using purified reaction center preparations yielding relatively large signals (Kaufmann et al., 1975; Rockley et al., 1975) on a chlorophyll basis, but for green plant systems no such reaction center preparations have yet been obtained. However, the first attempts in 1978 to apply picosecond spectroscopy to subchloroplast fragments highly enriched in PS I reaction centers (Shuvalov et al., 1979b; Fenton et al., 1979) appeared successful. For PS II studies, only kinetic studies extending into the nanosecond range have been reported so far.

II. The Primary Electron Donors of the Two Photosystems

In 1956 a reversible, light-induced bleaching in the 700-nm region was observed in a variety of photosynthetic systems and attributed to photooxidation of the primary electron donor of PS I (Kok, 1956). The light-minus-dark difference spectrum of this species, designated P700, was attributed to oxidation of a dimer of chlorophyll a. The appearance of two bleaching bands at 700 and 682 nm could then be explained as a loss of excitation coupling in the reduced state of the *dimer*, consistent with circular dichroism changes of opposite sign observed in the two bands (Phillipson et al., 1972).

Features observed in a stoichiometrically related electron paramagnetic resonance (EPR) signal were also found to be consistent with a dimeric chlorophyll: a g value of 2.0035, a narrowing of the linewidth from that of monomeric $Chl^{\ddot{+}}$ by about a factor of $(2)^{1/2}$ (Katz and Norris, 1973), and a two fold reduction of hyperfine splitting values (derived from the structureless EPR signal by the electron nuclear double resonance (ENDOR) technique) (Norris et al., 1974; Feher et al., 1975) from those determined for monomeric $Chl^{\ddot{+}}$.

There has not been close agreement between redox potential values for the $P700^+$–$P700$ couple, but most reported values fall in the range $+400$ to $+500$ mV (Ke et al., 1975). It is now recognized that most of the variations in previously reported values were probably due to changes in the environment of P700 in the various types of particles as a result of different preparative procedures, particularly with respect to degree of exposure to detergent (Setif and Mathis, 1980).

The differential extinction coefficient of P700 in digitonin-fractionated subchloroplast particles is 64,000 M^{-1} cm^{-1} (Hiyama and Ke, 1972) at the maximum of the bleaching band near 700 nm. A similar value has also been obtained for intact chloroplasts by measuring the quantitative relationship between the number of electrons generated by light in PS II and those transferred to oxidized P700 (Haehnel, 1976).

Early estimates of the time required for P700 photooxidation, less than 25–100 nsec (Witt and Wolff, 1970; Ke, 1972b), were limited either by the duration of the actinic flash or by the time resolution of the detector circuitry. Recent measurements in the picosecond range have shown the P700 photooxidation time to be in fact less than 10–50 psec (Shuvalov et al., 1979b,c; Fenton et al., 1979).

Compared to P700, much less is known about the primary electron donor of PS II, P680, the investigation of which is hampered in two ways: (1) Because of the extremely positive redox potential (actual value not yet determined) of $P680^{\ddot{+}}/P680$ its oxidized-minus-reduced difference spectrum has not yet been determined directly by chemical means, as has been done for that of P700; and (2) oxidized $P680^+$, especially when formed photochemically, is usually very rapidly reduced by some secondary electron donor, thus making accumulation of $P680^+$ difficult.

The difference spectrum of P680 in the visible region, first constructed from a 200-μsec decay component in flash-induced absorption change transients in chloroplasts, strongly resembled the Chl $a^{\ddot{+}}$-minus-Chl a, difference spectrum, having a major bleaching band at 680 nm (Döring et al., 1967). However, the magnitude of the absorbance change, compared with an assumed chlorophyll extinction coefficient, suggested a P680/P700 ratio much less than 1:1. Only later on a faster time scale was a 35-μsec decay component observed and the expected stoichiometry satisfied (Gläser et al., 1974).

When chloroplasts were later examined under more physiological conditions, no absorbance change at all was observed, even with a 3-μsec resolution (Van Best and Mathis, 1978). It was concluded that P680$^+$ was probably being reduced by a secondary (physiological) electron donor in an appreciably shorter time, thus rendering the expected absorbance change unobservable under the experimental conditions used. This possibility was supported by the observation of a 1-μsec decay in luminescence following a flash, this decay time being consistent with the presumed rapid reduction of P680$^+$ by a secondary donor (Van Best and Duysens, 1977). The longer-lived component in absorption kinetics, as well as a kinetic counterpart observed in flash-induced luminescence, has since been ascribed to a charge recombination between P680$^+$ and the reduced primary electron acceptor, $Q^{\bar{\cdot}}$, after the chloroplasts have been modified by low pH, Tris washing, or exposure to detergents.

A narrow, irreversible EPR signal ($\Delta H_{pp} = 8$ G, $g = 2.003$), produced upon illuminating PS II subchloroplast particles at low temperatures, was tentatively assigned to P680$^{\ddot{+}}$ (Malkin and Bearden, 1973; Ke et al., 1973b). However, because optical spectroscopy indicated that photochemically formed P680$^{\ddot{+}}$ did not accumulate, because of rapid reduction by either the secondary donor or the photoreduced primary acceptor, it was alternately suggested that the free-radical EPR signal initially attributed to P680$^{\ddot{+}}$ may actually belong to an oxidized chlorophyll molecule whose role is that of an immediate secondary donor to P680$^{\ddot{+}}$. The rather narrow ΔH_{pp} subsequently measured and assumed to be representative of P680$^{\ddot{+}}$, suggests a dimer form (like P700$^{\ddot{+}}$) for this chlorophyll radical (Visser, 1975; Malkin and Bearden, 1975; however, see Davis et al., 1979).

III. Early Photochemical Events in the Electron Acceptor Systems of Photosystems I and II

It has been found during the past few years that the electron acceptor systems of both PS I and PS II, as well as that of the bacterial reaction center, consist of a chain of electron acceptors capable of rapid and efficient transfer of the electron away from the primary donor molecule, permitting its reduction by the donor system and thus avoiding wasteful back reactions.

The electron transfer among the earlier acceptors, probably because of suitable structural and spatial arrangements, can take place even at cryogenic temperatures. Each step of the electron transfer along this chain is accompanied by the loss of some fraction of the absorbed energy (perhaps between 0.05 and 0.5 eV) and by increasing charge stabilization caused by greater spatial separation of the transferred electron from the photooxidized donor, P^+.

Because of their specific role in generating strong reducing power for the reduction of ferredoxin and NADP, the PS I acceptors possess highly negative potentials: the three iron–sulfur centers representing $A_{I,2}$, $A_{I,3}$, etc., have E_m values of -0.73, -0.58, and -0.54 V, respectively (versus a standard hydrogen electrode), the last presumably being the most remote acceptor.

While the primary electron donor of PS II must have an extremely positive potential, undoubtedly greater than $+0.8$ V, as required for water oxidation, the acceptors are accordingly less reducing than those of PS I. Interestingly, recent findings on the acceptor system of PS II indicate that its chemical makeup is very similar to that of the bacterial reaction center, with $A_{II,1}$ and $A_{II,2}$ being a pheophytin and a quinone, respectively, the latter associated with Fe^{2+} in some not clearly defined manner.

An early electron acceptor of the PS I reaction center, discovered and characterized by both optical spectroscopy (Hiyama and Ke, 1971) (the spectral species being designated P430) and EPR spectroscopy (Malkin and Bearden, 1971), has been identified as iron–sulfur centers (A and B) (Evans et al., 1974). Center B has an E_m of ≤ -0.58 V and possesses characteristic features in its EPR spectrum at g values of 2.05, 1.92, and 1.89, while center A has an E_m of -0.54 V and characteristic g values of 2.05, 1.94, and 1.86 (Ke et al., 1973a; Evans et al., 1974). At redox potentials not too negative, low-temperature illumination of chloroplasts or PS I subchloroplast particles results in an appearance of the EPR signals of P700$^+$ and reduced center A$^-$ (Malkin and Bearden, 1971; Ke et al., 1973a). If center A has been reduced chemically in the dark beforehand, illumination of the sample is accompanied by an appearance of the EPR signals of P700$^+$ and reduced center B (Heathcote et al., 1978).

Redox blocking, as illustrated above, has recently been used successfully for the detection and characterization of earlier electron acceptor(s). Sauer et al. (1978) reported that, by poising the PS I subchloroplast particles at a potential sufficiently negative to reduce both centers A and B, P700$^+$ was still formed in a flash but decayed (by charge recombination) in 250 μsec rather than 40 msec, thus suggesting an earlier acceptor. If the subchloroplast particles were also illuminated by strong background light in an effort to reduce this assumed earlier acceptor, the light-induced P700 photo-

3. GREEN PLANT PHOTOSYNTHESIS: ABSORPTION AND EPR STUDIES 61

oxidation signal then decayed in about 3 μsec. These two newly observed decay times therefore suggested the presence of not just one but two other electron acceptors, $A_{I,1}$ and $A_{I,2}$, between P700 and the two bound iron–sulfur centers, A and B.

More detailed studies on the spectral and kinetic aspects of these earlier acceptors in PS I have been carried out using optic and EPR spectroscopy plus a combination of other approaches, namely, picosecond time resolution for the detection of extremely short-lived transient species, low temperatures to alter kinetic behavior, and highly reducing conditions to provide selective blocking of acceptor components along the electron transport chain.

As shown in Fig. 2a (inset), when bound iron–sulfur centers A and B are chemically reduced at -0.62 V, the light-induced absorption change decays in two phases at 5 K, with lifetimes of 1.3 and 130 msec, respectively (Shuvalov et al., 1979a). The light-minus-dark difference spectrum of the 130-msec component was interpreted as the sum of contributions from P700 photooxidation and photoreduction of the acceptor $A_{I,2}$, whose spectrum could be extracted by subtracting the known difference spectrum for P700 oxidation. This possibility was verified by measuring the absorption changes induced by continuous illumination at 4°C in PS I particles poised at -0.62 V (Fig. 2b) (Shuvalov et al., 1979a). Under these conditions, the rate of electron donation to P700^{+} from the low-potential mediators in the medium becomes competitive with the rate of recombination between P700^{+} and $A_{I,2}^{-}$ formed

FIG. 2. (a) Spectrum of light-induced absorbance changes with a lifetime of 130 msec at 5 K in TSF-I fragments poised at -0.62 V. This difference spectrum represents the changes P700^{+} $-$ P700 plus $A_{I,1}^{-} - A_{I,1}$. The inset shows the biphasic kinetics ($t_{1/2} = 1.3$ and 130 msec) of absorbance changes at 430 nm induced by 710-nm dye laser excitation. (b) Light-minus-dark difference absorption spectrum at 4°C of TSF-I fragments poised at -0.62 V. This difference spectrum, representing the change $A_{I,2}^{-} - A_{I,2}$, was measured in a dual-wavelength spectrophotometer using continuous background illumination.

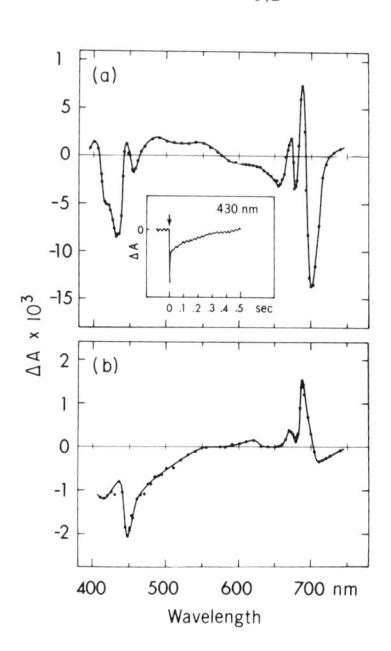

in the light, thus resulting in an accumulation of $A_{I,2}^-$ (the absence of any bleaching at 700 nm confirmed that P700 was maintained in its reduced state). The broad bleaching in the blue–green region of the difference spectrum was interpreted as due to the reduction of an iron–sulfur center, while absorbance changes in the 450-nm and 670–720 nm regions were attributed to electrochromic shifts induced in the absorption spectrum of some chlorophyll molecules by the electric field of the reduced acceptor, $A_{I,2}^-$. The similarity in the difference spectra for $A_{I,2}$ produced in the flash kinetics and photoaccumulation experiments supported the original assignment.

The iron–sulfur center nature of acceptor $A_{I,2}$ involved in the 130-msec decay was further supported by a kinetic correlation with the EPR signal of an iron–sulfur center designated X. Figure 3a shows the light-minus-dark EPR spectrum at 9 K of a PS I particle poised at -0.62 V in the $g = 1.78$ region characteristic of X (Shuvalov et al., 1979a). Although the risetime for formation of this EPR signal was limited by the time response of the spectrometer, the signal clearly decays with $t_{1/2} = 130$ msec, which agrees with the decay time of the optical signal attributed to reversal of the $P700^{\pm} \cdot A_{I,2}^-$ state. Thus these data are consistent with the assumption that $A_{I,2}$ and X represent the same acceptor, probably another iron–sulfur center.

When PS I particles poised at -0.62 V are illuminated by background light, the two bound iron–sulfur centers A and B as well as the electron acceptor $A_{I,2}$ ($=X$) are all reduced, as previously noted. Under these conditions, light-induced absorption changes decay with a halftime of 1.3 msec

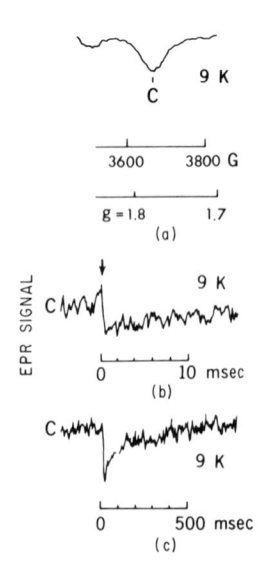

FIG. 3. (Top): Light-minus-dark EPR spectrum at 9 K of TSF-I fragments poised at -0.62 V and in the $g = 1.78$ region, measured with continuous background illumination. (Center and bottom): Kinetics of flash-induced EPR signal change, at point C of the top spectrum, on different time scales.

FIG. 4. (a) Spectrum of light-induced absorbance changes with lifetime of 1.3 msec at 5 K in TSF-I particles poised at -0.62 V. This difference spectrum represents the changes P700‡ − P700 plus $A_{I,1}^{-} - A_{I,1}$. The inset shows the kinetics of the absorbance change at 696 nm induced by 710-nm laser excitation. (b) Solid curve, result of subtraction of the P700‡ − P700 spectrum from the spectrum in (a), assuming equal extinction coefficients (and equal contributions) for P700 and $A_{I,1}$ at 700 nm. Dashed curve, spectrum of Chl^{-} − Chl in solution (taken from Fujita et al., 1978). Wavelength scale for the latter spectrum is shifted to the red region by ~ 25 nm.

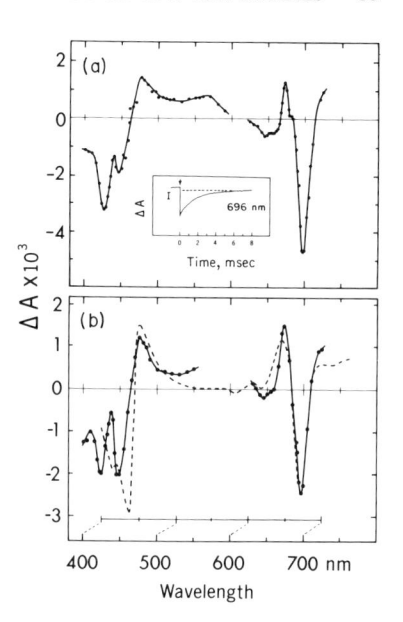

at 5 K (Fig. 4a, inset). The difference spectrum plotted from the 1.3-msec absorption changes was assumed to represent a composite of equal contributions from P700^{+} − P700 and $A_{I,1}^{-} - A_{I,1}$. Thus the spectrum in Fig. 4b is the difference spectrum of $A_{I,1}^{-} - A_{I,1}$ obtained by subtracting the separately obtained difference spectrum of P700^{+} − P700 from the composite spectrum, assuming that the former change contributes approximately half of the total change in the far-red band (Shuvalov et al., 1979a). While the spectrum in Fig. 4b bears some resemblance to that of P700‡ − P700, it is very similar to that for the formation of Chl a anion radical in solution (Fujita et al., 1978), except for a shift toward a longer wavelength by 23 nm. Interestingly, each of the bleaching bands observed at 420, 450, and 700 nm in the derived spectrum coincides with an absorption band previously found for Chl a dimer in solution (Fong et al., 1976). On that basis, the red shift of the $A_{I,1}^{-} - A_{I,1}$ spectrum from the known monomer spectrum was interpreted as due to the Chl a acceptor in PS I being dimeric. However, more recent EPR (Heathcote et al., 1980; Baltimore and Malkin, 1980a) and ENDOR (Fajer et al., 1980) studies on photoaccumulated $A_{I,1}^{-}$ both indicate that it is more likely a monomeric Chl a anion radical.

It should be noted that, besides using background illumination plus a highly reducing environment for isolating $A_{I,1}$, one may also heat the chloroplasts or subchloroplast fragments (at 60–65°C for 5 min) and thus "remove" early acceptors from the active transport chain (Shuvalov et al., 1979c;

Baltimore and Malkin, 1980b). Also, it seems that PS I fragments isolated from chloroplasts after treatment with sodium dodecyl sulfate (SDS) apparently contain only one acceptor, $A_{I,1}$ (Mathis et al., 1978).

Under physiological conditions, i.e., at moderate redox potentials and room temperature, the electron transfer from $A_{I,1}$ to $A_{I,2}$ is expected to be extremely rapid, and therefore the difference spectrum of $A_{I,1}$ can be obtained with flash excitation under these conditions only if the excitation pulse is sufficiently short and the detection method has sufficient time resolution, and also if the overlapping absorption changes due to $P700^{+}$ and $A_{I,1}^{-}$ can be distinguished kinetically. To explore this question, a mode-locked ruby laser pulse (duration ~ 50 psec) at 694.3 nm was used for excitation and for absorption measurement via an optical delay line. The initial absorption change near 700 nm showed a risetime of $\leq 30-50$ psec (Shuvalov et al., 1979b,c; see also Fenton et al., 1979), and the decay kinetics indeed had two components with lifetimes of 200 and 40 msec, respectively, and approximately equal amplitudes (Fig. 5a). The slow component was assigned to the reduction of $P700^{+}$, while the 200-psec component presumably reflected electron transfer from $A_{I,1}^{-}$ to $A_{I,2}$.

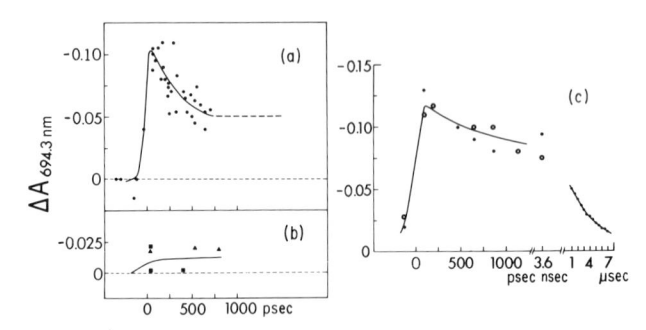

FIG. 5. (a) Kinetics of absorbance changes at 20°C at 694.3 nm in TSF-I fragments (poised at about +0.2 V) induced by single 50- to 60-psec 694.3-nm pulses. (b) Same conditions as in (a), except that the sample was subjected to background illumination in the presence of methyl viologen to cause $P700^{+}$ accumulation. (c) Kinetics of absorbance changes at 694.3 nm in TSF-I fragments poised at −0.62 V and under background illumination to cause $A_{I,2}^{-}$ accumulation; 50- to 60-psec pulses were used for the nano- and subnanosecond time domain, and 300-nsec 700-nm dye laser pulses were used for microsecond measurements.

Both components are dependent on the redox state of P700 and centers A and B. Prior oxidation of P700 decreased the amplitude of absorption change nearly 10-fold, as expected (Fig. 5b), while chemical reduction of centers A and B increased the lifetime of the fast component from 200 psec to 10 nsec and decreased that of the slow component from 40 msec to 3 μsec

(Fig. 5c). This new kinetic pattern might be attributed to a fast reduction of X by $A_{I,1}^-$ (Chl$^-$) in 10 nsec and a slow charge recombination in 3 μsec in the singlet radical pair, ^{1}(P700$^+$·Chl$^-$). The decay time of the latter phase obeys an Arrhenius relationship down to 70 K (activation energy ~ 1 kcal/mole), below which it remains unchanged at 1.3 msec down to the lowest temperature examined (5 K), an invariance which suggests an electron tunneling mechanism. On the other hand, the same kinetic pattern was observed when X was reduced under strong supplementary illumination. Alternatively, the 10-nsec component might rather represent the decay of ^{1}(P700$^+$·Chl$^-$) to either P700·Chl or ^{3}(P700$^+$·Chl$^-$), the latter state then decaying in turn to P700·Chl in 3 μsec, possibly via ^{3}P700. It should be noted, however, that the lifetime of ^{3}Chl is typically ~ 1 msec, even at room temperature.

From the redox potentials of P700$^+$/P700 (+0.5 V) and Chl/Chl$^-$ (−0.8 V), the energy of the state P700$^+$·Chl$^-$ is estimated to be 1.3 eV above the state P700·Chl and, from the minimum energy (1.8 eV) for P700 excitation, 0.5 eV below the state P700*·Chl. The energy of the primary charge transfer state thus appears to be close to energy of the ^{3}P700 state, as estimated by the observed long-wavelength limit (960 nm) in its phosphorescence spectrum.

More recently, a mode-locked neodymium–YAG laser was used to reexamine the kinetics of charge separation in PS I in order to obtain more spectral information (Shuvalov et al., 1979c). A 30-psec pulse at 698 or 708 nm was used for excitation and for generating a picosecond continuum pulse to measure the light-induced absorption change spectrum. As shown in Fig. 6a, the initial absorption decrease at 694 nm induced by a 708-nm pulse occurs in 30 psec and decays in two phases with lifetimes of ~ 45 and ~ 210 psec to the level of P700 photooxidation alone. Figure 6b shows the difference spectra measured at 150 and 800 psec after excitation in the red and blue-green regions. On the assumption that the lifetime of $A_{I,1}^-$ is 200 psec, one can then isolate the difference spectrum for $A_{I,1}^- - A_{I,1}$ from the difference between the 800-psec and 150-psec spectra (Fig. 6b, inset). Note that this difference spectrum also coincides well with the spectrum for the formation of Chl a anion radical in solution (Fujita et al., 1978) (Fig. 4b dashed curve), except for a 30-nm shift toward the red region. The extinction coefficient of the Chl a acceptor ($A_{I,1}$) was estimated to be 46 mM^{-1} cm^{-1} at 695 nm assuming a 1:1 stoichiometry with P700 (Shuvalov et al., 1979c).

The 45-psec component, appearing mostly below 690 nm, is unaffected by prior oxidation of P700 and is probably related to the excitation of antenna chlorophyll molecules rather than to primary photochemistry (Shuvalov et al., 1979c; cf. see also Fenton et al., 1979). Under highly reducing conditions, the 45-psec component is again unaffected, but the reoxidation of $A_{I,1}^-$ is considerably slower (Fig. 6a).

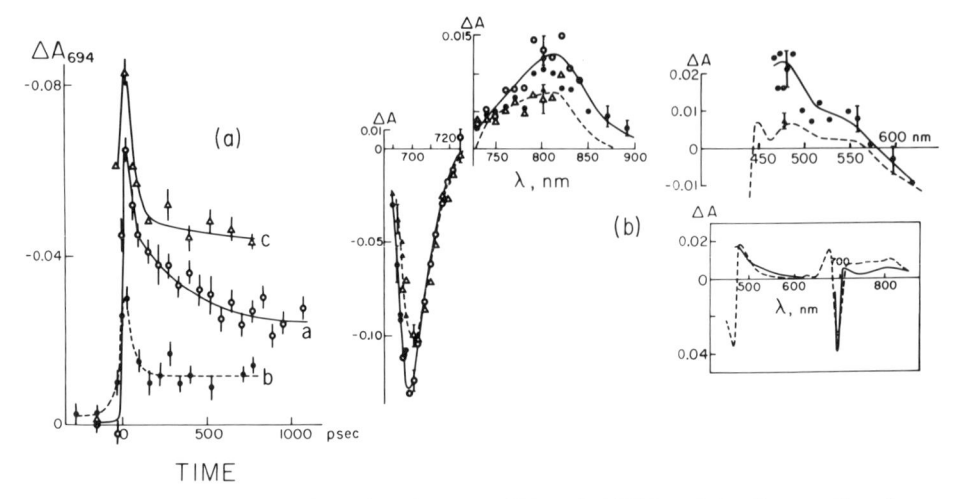

FIG. 6. (a) Kinetics of absorption changes at 694 nm in TSF-I particles. Curve a: Sample contained 1 mM ascorbate, 25 μM DCIP, and 400 μg/ml Chl; curve b: P700 in the sample was preoxidized by illumination in the presence of 0.1 mM methyl viologen; curve c: Heat-treated TSF-1 sample; all other conditions same as for curve a. (b) Spectra of light-induced absorbance changes measured 150 psec (○ and ●) and 800 psec (Δ and ▲) after excitation with 708-nm (● and ▲) or 689-nm (○ and Δ) ps pulses in the red (left) and shorter wavelength (right) regions in TSF-I fragments poised at about +0.2 V. The dashed curves show the difference spectrum of P700 measured in the same sample by continuous illumination. The inset shows the difference spectrum for the 150- and 800-psec spectra; the dashed curve is the difference spectrum for the formation of Chl a anion radical (Fujita et al., 1978), shifted toward the red region to coincide with the measured difference spectrum.

The spectral species and a related EPR species (iron–sulfur centers A and B) were long thought to represent the primary electron acceptor of PS I (Ke, 1978). Of course, as our knowledge progresses and available time resolution becomes more refined, we now recognize that with our present time resolution the acceptor $A_{I,1}$ (Chl a) should be considered the true primary acceptor. On the basis of spectral characteristics, P430 was considered an iron–sulfur protein (Ke, 1973), probably center A, based on various correlations between redox properties (Ke et al., 1973a; Ke, 1975) and decay kinetics (Ke et al., 1974; Visser et al., 1974; Bearden and Malkin, 1974). More recently, a kinetic correlation has revealed that a significant portion of the P430 spectral change might also be due to the acceptor $A_{I,2}$ (or X), which has also been ascribed to an iron–sulfur center (Hiyama and Fork, 1980).

The stable primary electron acceptor of PS II, $A_{II,2}$, has long been considered a membrane-bound plastoquinone (PQ) molecule on the basis of the spectrum of flash-induced absorbance changes in the near ultraviolet, having a rise time and a decay times of <1 μsec and 0.6 msec, respectively. The species responsible for these spectral changes was designated X320 (Stiehl and Witt, 1968). Subsequent studies with subchloroplast particles yielded a difference spectrum which included features of X320 as well as $PQ^{\overline{\cdot}}$–PQ *in vitro* (Van Gorkom *et al.*, 1975). In a Triton-fractionated PS II subchloroplast fragment, the spectrum of flash-induced absorbance changes can be attributed to the formation of $P680^{+\cdot}\cdot PQ^{\overline{\cdot}}$ in ≤ 0.2 μsec (Ke and Dolan, 1980). The decay time, representing charge recombination, displayed an Arrhenius temperature dependence ($E_a = 8.5$ kcal/mole) down to 210 K, below which it remained constant at 1.25 msec down to the lowest measured temperature (10 K), as would be expected for an electron tunneling mechanism.

The location of this PQ species on the reducing, rather than on the oxidizing side of PS II, is supported by the absence of any effect of added 3-(3',4'-dichlorophenyl-1,1-dimethylurea (DCMU) (an inhibitor of the reoxidation of the photoreduced primary electron acceptor) on the risetime of the X320 signal but significant inhibition of its decay. Notably, a concomitant EPR signal assignable to the semiquinone $PQ^{\overline{\cdot}}$ has never been observed under physiological conditions, probably because of a magnetic effect of an interacting iron atom (see below).

Based on the hypothesis that the acceptor molecule Q (Q standing for "quencher") is a fluorescence quencher in its oxidized form (Duysens and Sweers, 1963), the fluorescence yield of PS II increases whenever $A_{II,2}$ is either photochemically or chemically reduced. Redox titration of the PS II fluorescence-quenching property has yielded two quenching transitions at −35 and −270 mV (at pH 7) (Cramer and Butler, 1969; Ke *et al.*, 1976; Horton and Croze, 1979). The exact nature of these multiple quenching centers is not yet understood.

If acceptor Q is chemically reduced beforehand, a flash-induced luminescence is observed at the maximum in the chlorophyll fluorescence spectrum (685 nm). This fluorescence decays in two components, in 150 psec and in 2–4 nsec, the slower phase being absent when Q is in the oxidized state before flash excitation (Klimov *et al.*, 1978). The slower component was thus attributed to excited Chl *a* created via primary charge recombination in the state $(P680^{+\cdot}\cdot A_{II,1}^{-})\cdot A_{II,2}^{-}$. In a sample in which Q was already reduced, maximum-level fluorescence (F_{max}) was observed with only the weak measuring light present. Upon application of intense illumination, however, the fluorescence yield level subsided to the level (F_0) observed when Q was not reduced in the dark (Klimov *et al.*, 1977). These fluorescence results are consistent with the assumption of an intermediary acceptor, $I(A_{II,1})$, which [when Q ($A_{II,2}$) is reduced] participates in a back reaction, $P680^{+\cdot}\cdot I^{-} \rightarrow P680^{*}\cdot I$, following light absorption, giving rise to a delayed fluorescence responsible for $F_{max} - F_0$. Thus the variable fluorescence of PS II appears upon prior reduction of Q but disappears upon photoreduction of I. Measurements of both the activation energy and the lifetime of this luminescence showed that the return of the electron from $I^{\bar{}}$ to $P680^{\stackrel{+}{}}$ to form the excited state of P680 required ~ 0.08 eV (Klimov et al., 1978). The kinetic correlation of changes in the fluorescence yield with absorbance changes over the visible region and also with a free-radical EPR signal (Klimov et al., 1980d) suggests identification of I with development of a chlorophyll-like species (Fig. 7).

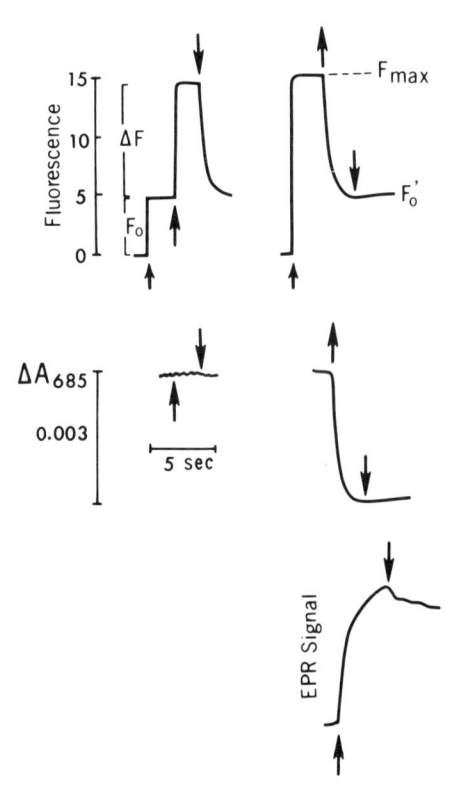

FIG. 7. Light-induced changes in chlorophyll fluorescence yield (top), optical absorption at 685 nm (middle), and EPR signal near $g = 2.00$ (bottom) in a PS II subchloroplast particle (DT-20) at the indicated redox poise.

The identification of I with pheophytin (Pheo), in particular, is suggested by light-minus-dark difference spectra. At -450 mV, spectral features which can be ascribed to the reduction of Pheo in vivo are prominent (Fig. 8): The large decreases at 410, 422, and 685 nm and increases at 450 and 655 nm plus,

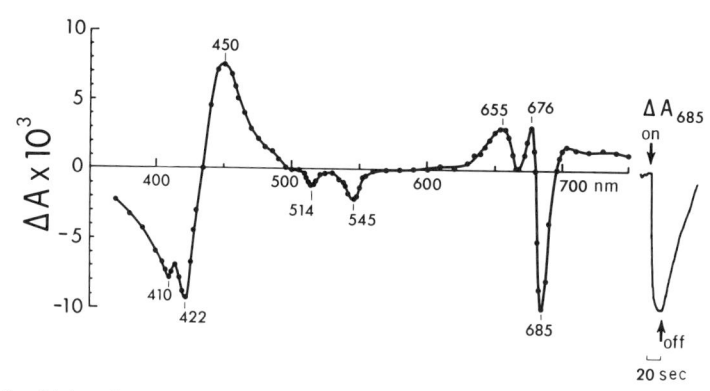

FIG. 8. Light-minus-dark difference spectrum of TSF-IIa fragments at −450 mV and 295 K. The absorption change transient at 685 nm (right) is shown on the same absorbance scale.

notably, the small band bleachings at 515 and 545 nm, are all characteristic of the reduction of Pheo *in vitro* (except for an overall red shift of about 15 nm attributed to environmental influences). Apparently, when Q is reduced, photoaccumulation of the long-lived state P680·Pheo⁻ can occur upon illumination because of a fast reduction (<1 μsec) of P680⁺ by a secondary donor, D, competing with charge recombination in the reaction center. This interpretation is supported, in part, by the agreement of the quantum yield (0.002–0.005 at 295 K) for photoaccumulation of P680·Pheo⁻ with the ratio of the recombination time (2–4 nsec) for P680⁺·Pheo⁻ and the time (∼1 μsec) for reduction of P680⁺ by the secondary donor, D.

A redox titration of the extent of light-induced optical changes attributed to Pheo⁻ accumulation yielded a midpoint potential of −0.61 V (Klimov *et al.*, 1979), consistent with the *in vitro* value of Pheo/Pheo⁻ (−0.64 V) (Fujita *et al.*, 1978).

The interpretation that charge recombination is the source of delayed luminescence is further confirmed by recent nanosecond kinetic experiments (Shuvalov *et al.*, 1980). Flash-induced absorption changes in the PS II fragments (TSF-IIa) poised at −450 mV occur in less than 2 nsec, followed by decay in two phases (Fig. 9a). The slower (33 nsec) component is attributed to a mixture of triplet states of chlorophyll and carotenoids (see below), while the spectrum (Fig. 9b) constructed from the 4-nsec decay component coincides well with the sum of difference spectra for P680 oxidation (Van Gorkom *et al.*, 1975; Ke and Dolan, 1980) and Pheo reduction (Klimov *et al.*, 1980a) (see Fig. 8). Note also that the 4-nsec decay time observed here in absorption changes is in good agreement with the longer fluorescence lifetime induced by a flash in a similarly poised PS II particle (Klimov *et al.*, 1978; Shuvalov *et al.*, 1980).

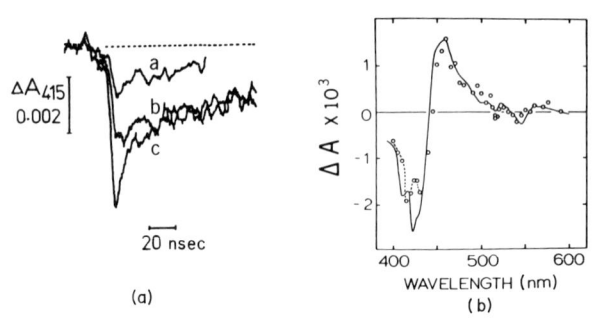

FIG. 9. (a) Absorbance changes in TSF-IIa fragments induced by a 3-nsec laser pulse (694.3 nm) at 295 K. (b) Spectrum of the 4-nsec decay component (open circles); composite spectrum constructed from $P680^+ - P680$ and $Pheo^{\overline{\cdot}} - Pheo$ (continuous curve).

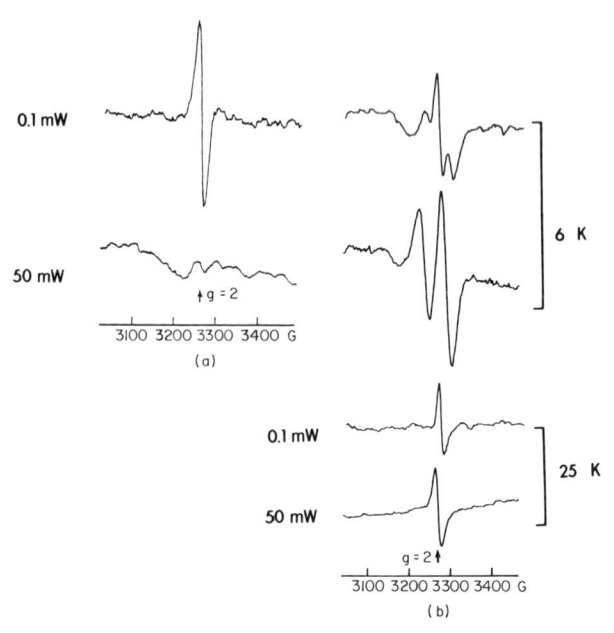

FIG. 10. Influence of illumination temperature on EPR doublet formation in TSF-IIa fragments at -450 mV and dependence of singlet and doublet components on measuring temperature and microwave power. $Pheo^{\overline{\cdot}}$ photoaccumulated at 295 K (a) and 220 K (b).

Additional information concerning the nature of the electron aceptors in PS II reaction centers is provided by EPR measurements. If a sample poised at -450 mV is illuminated at room temperature and then immediately quenched in liquid nitrogen, a narrow, photoinduced signal having $g = 2.0033$ and $\Delta H_{pp} = 13.5$ G is observed (Fig. 10a) (Klimov et al., 1980a), which is consistent with the signal of monomeric Pheo$^{\bar{}}$ in vitro ($g = 2.003$, $H_{pp} = 12.7$ G) (Fujita et al., 1978). However, when Pheo$^{\bar{}}$ is phototrapped at a lower temperature, an additional doublet, split by 52 G and centered at $g = 2.00$, can be detected at 7 K (Klimov et al., 1980a) (Fig. 10b, top). In contrast to the narrow singlet line, the doublet signal reaches a maximum amplitude at a much higher microwave power (20–50 mW) and is detectable only below 15 K. Also, with progressively higher temperatures, the doublet gradually disappears into the noise, while the previously power-saturated radical signal grows in intensity (Fig. 10b, bottom).

By analogy with similar signals found in certain photosynthetic bacteria after illumination, the split signal has been attributed to an exchange interaction between the unpaired electrons of Pheo$^{\bar{}}$ and either PQ$^{\bar{}}$ or the Fe·PQ$^{\bar{}}$ complex, singly reduced by prior chemical reduction. The absence of the "doublet" after room temperature phototrapping is probably due to more rapid electron transfer from Pheo$^{\bar{}}$ to Fe·PQ$^{\bar{}}$, permitting stabilization of the final, light-induced state (P680·Pheo$^{\bar{}}$)Fe·PQ^{2-} and loss of effective exchange interaction.

Similar singlet and doublet signals can also be seen in different types of PS II particles as well as in chloroplasts. On a chlorophyll basis, the signal amplitude is approximately proportional in each case to the enrichment of the sample in reaction center components (Klimov et al., 1980c). Further evidence for monomeric Pheo$^{\bar{}}$ was rendered by hyperfine splitting values obtained by the ENDOR technique from the structureless EPR signal (Fajer et al., 1980).

It has recently been confirmed that development of the EPR doublet requires the presence of both PQ and iron, as demonstrated by extraction–reconstitution experiments (Klimov et al., 1980c). As seen in Fig. 11, the doublet signal was virtually absent in samples subjected to quinone extraction by hexane but could be restored by subsequent readdition of pure PQ. The additional requirement for iron was established also by extraction–reconstitution experiments (Fig. 12) in which the chaotropic agent LiClO$_4$ was used in conjunction with the iron chelator o-phenanthroline for extraction and with excess Fe^{2+} for reconstitution (Klimov et al., 1980c). The doublet was not restored when either Mn^{2+} or Mg^{2+} was added in place of Fe^{2+}. Analysis showed the nonheme iron content in PS-II to be two iron atoms per reaction center in an untreated sample or a sample treated with LiClO$_4$

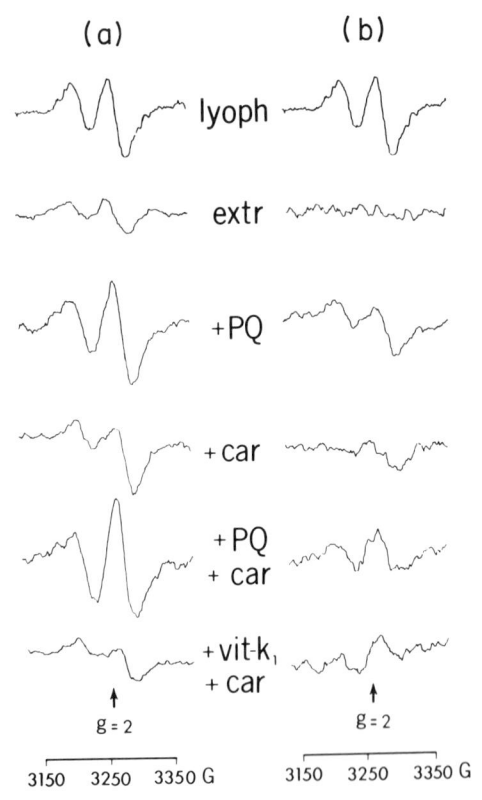

FIG. 11. Loss of EPR doublet of lyophilized TSF-IIa fragments after extraction with hexane containing 0.13% (a) and 0.16% (b) methanol. Restoration by reconstitution with PQ (and partial restoration by β-carotene). EPR singlet not observed at the temperature (7 K) and microwave power (50 mW) used.

only, but less than 0.5 iron atoms per reaction center in samples treated with LiClO$_4$ plus o-phenathroline.

Significantly, in iron-extracted samples it was possible to elicit by illumination at low temperatures a singlet EPR signal, heretofore unobservable, whose g value (2.0045) and linewidth (9.2 G) are very close to those of the anion radical of PQ *in vitro*. Moreover, this signal is not seen if PQ is further reduced to the PQ^{2-} state or if PQ has been extracted (Klimov *et al.*, 1980c).

In reduced PS II reaction centers, when the usual photochemistry is blocked, charge recombination leads to the formation of a triplet state of chlorophyll and, subsequently, of a carotenoid triplet (^{T}Car) by triplet energy transfer (Klimov *et al.*, 1980b). Flash excitation of PS II fragments poised at −450 mV produces transient absorption changes decaying in 7 μsec. The

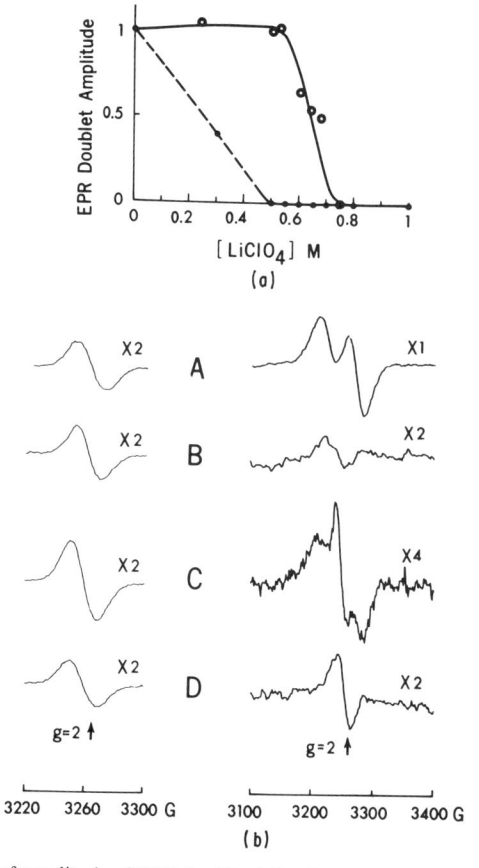

3. GREEN PLANT PHOTOSYNTHESIS: ABSORPTION AND EPR STUDIES 73

FIG. 12. (a) Plot of amplitude of EPR doublet following extraction with different concentrations of LiClO$_4$ with (---) and without (—) 2.5 mM o-phenanthroline. (b) Partial restoration of doublet by Fe^{2+} but not by Mn^{2+}. A, After treatment with 0.4 M LiClO$_4$; B, same as A but with additional 2.5 mM o-phenanthroline during treatment; C and D same as B but after further incubation with 0.2 mM Fe^{2+} (C) or Mn^{2+} (D). Temperature and microwave power (b): 90 K, 50 μW (left); 7 K, 50 mW (right).

spectrum of this absorption change in TSF-11a fragments shows bleaching of three absorption bands in the 400–500 nm region and the development of a broad band with a maximum at 515 nm, all characteristic of the formation of triplet state carotenoid.

According to earlier reports (Wolff and Witt, 1969; Mathis *et al.*, 1979) the formation of triplet states in chloroplasts is not necessarily related to photosynthetic charge transfer reactions but instead results from intersystem crossing in singlet excited molecules of antenna chlorophyll. Our present data show, however, that at least part of the ^{T}Car can be related to the functioning of PS II reaction centers. Figure 13 shows the parallel behavior followed by fluorescence yield, absorption, and the amplitude of flash-induced absorption change, attributed to carotenoid triplet state formation, during the course of Pheo reduction by steady background illumination in TSF-IIa fragments poised at −450 mV. Significantly, phototrapping of Pheo in these fragments at 295 K resulted in a nearly threefold decrease in the flash-induced ^{T}Car signal at 515 nm.

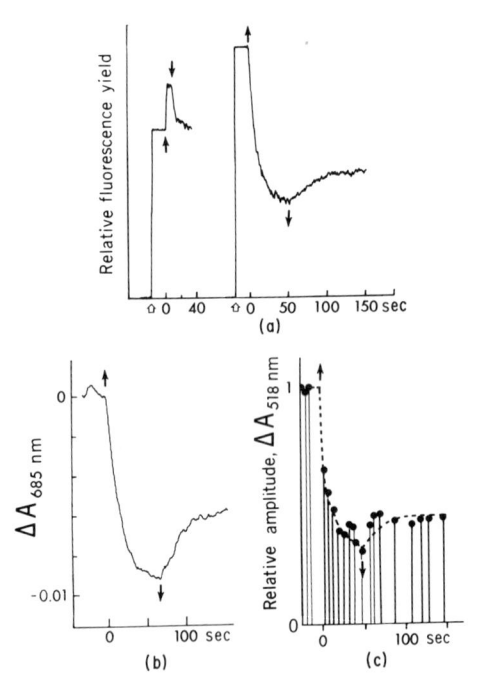

FIG. 13. Effect of continuous illumination of TSF-IIa fragments at −450 mV. (a, left) Chl fluorescence yield in TSF-IIa fragments poised at +400 mV; (a, right) same, except poised at −450 mV. (b) Absorbance change indicating photoaccumulation of Pheo$^-$ at −450 mV. (c) Profile of amplitudes (vertical lines) of flash-induced absorbance changes at 515 nm (−450 mV).

The decay time of the 515-nm change in TSF-IIa at −450 mV was not appreciably changed upon lowering the temperature; at 90 and 7 K it remained at 7–8 μsec. Measurement of the 515-nm change with greater time resolution (∼3 nsec) shows that its risetime in TSF-IIa fragments at −450 mV is 33 nsec at 295 K, in apparent agreement with the decay time of the 415-nm change (attributed to ³Chl). However, it should be pointed out that the occurrence of the 33-nsec decay component apparently is independent of the redox state of the Pheo prior to the activating flash.

From available data, the reaction sequence, kinetics, and energetics of the early reactions in PS II reaction centers can be summarized as in Fig. 14. P680 is the primary electron donor; Pheo and a Fe·PQ complex are the early electron acceptors. Note that the chemical makeup of the PS II acceptor system is very similar to that in photosynthetic bacteria (see Chapter 4). A quantum of absorbed light energy (~1.8 eV) brings P680 into the excited state, P680*, which then transfers an electron in <1 nsec to Pheo, with a redox midpoint potential of −0.61 V. If one assumes an energy difference of 0.08 eV between P680* and P680‡·Pheo$^{\bar{}}$, as established by the activation energy determined for delayed fluorescence, the midpoint potential for the P680‡–P680 couple would be about +1.1 V, quite sufficient for water oxidation (+0.8 V, average midpoint potential) via secondary electron transport components. Of course, the actual midpoint potential value for P680 remains to be determined. When Q is reduced beforehand, the state P680‡·Pheo$^{\bar{}}$ undergoes charge recombination in 4 nsec. Therefore, to obtain a high quantum yield of photosynthesis, the electron transfer from Pheo$^{\bar{}}$ to Fe·PQ must be at least two orders of magnitude faster. If the midpoint potential of Fe·PQ is taken as −0.13 V, the electron transfer from Pheo$^{\bar{}}$ to Q is accompanied by a loss of 0.5 eV, leading to formation of the state (P680‡·Pheo) PQ$^{\bar{}}$·Fe, which has a lifetime of 150 μsec. Under physiological conditions, fast electron transfer from the donor, D, to P680‡ (<1 μsec) prevents charge recombination between P680‡ and PQ$^{\bar{}}$. Thus further electron transfer from Q to more remote acceptors and the simultaneous reduction of D$^+$ by electrons from water leads to eventual stabilization of separated charges.

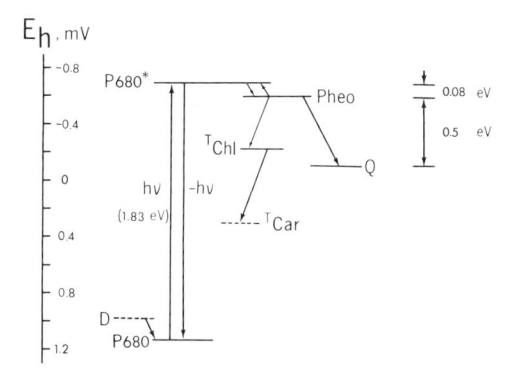

FIG. 14

References

Baltimore, B. G., and Malkin, R. (1980a). *Photochem. Photobiol.* **31**, 485–490.
Baltimore, B. G., and Malkin, R. (1980b). *FEBS Lett.* **110**, 50–52.
Bearden, A. J., and Malkin, R. (1974). *Fed. Proc., Fed. Am. Soc. Exp. Biol.* **33**, 378.
Cramer, W. A., and Butler, W. L. (1969). *Biochim. Biophys. Acta* **172**, 503–510.
Davis, M. S., Formann, A., and Fajer, J. (1979). *Proc. Nat. Acad. Sci. U.S.A.* **76**, 4170–4174.
Döring, G., Stiehl, H. E., and Witt, H. T. (1967). *Z. Naturforsch., Teil. B* **22**, 639–644.
Duysens, L. N. M. (1952). Thesis Univ. Utrecht, Utrecht.
Duysens, L. N. M., and Sweers, H. E. (1963). "Studies on Microalgae and Photosynthetic Bacteria," pp. 353–372. Univ. of Tokyo Press, Tokyo.
Evans, M. C. W., Reeves, S. G., and Cammack, R. (1974). *FEBS Lett.* **49**, 111–114.
Fajer, J., Davis, S. M., Forman, A., Klimov, V. V., Dolan, E., and Ke, B. (1980). *J. Am. Chem. Soc.* **102**, 7143–7145.
Feher, G., Hoff, A. J., Issacson, R. A., and Ackerson, L. C. (1975). *Ann. N.Y. Acad. Sci.* **224**, 239–259.
Fenton, J. M., Pellin, M. J., Govindjee, and Kaufmann, K. J. (1979). *FEBS Lett.* **100**, 1–4.
Fong, F. K., Koester, V. J., and Polles, J. S. (1976). *J. Am. Chem. Soc.* **98**, 6406–6408.
Fujita, I., Davis, M. S., and Fajer, J. (1978). *J. Am. Chem. Soc.* **100**, 6280–6282.
Gläser, M., Wolff, C., Buchwald, H.-E., and Witt, H. T. (1974). *FEBS Lett.* **42**, 81–85.
Haehnel, W., (1976). *Biochim. Biophys. Acta* **423**, 499–509.
Heathcote, P., Williams-Smith, D. L., Sihra, C. K., and Evans, M. C. W. (1978). *Biochim. Biophys. Acta* **503**, 333–342.
Heathcote, P., Timofeev, K. N., and Evans, M. C. W. (1980). *FEBS Lett.* **111**, 381–385.
Hiyama, T., and Fork, D. (1980). *Arch. Biochem. Biophys.* **199**, 488–496.
Hiyama, T., and Ke, B. (1971). *Proc. Natl. Acad. Sci. U.S.A.* **68**, 1010–1013.
Hiyama, T., and Ke, B. (1972). *Biochim. Biophys. Acta* **267**, 160–171.
Holten, D., and Windsor, M. W. (1978). *Annu. Rev. Biophys. Bioeng.* **7**, 189–227.
Horton, P., and Croze, E. (1979). *Biochim. Biophys. Acta* **545**, 188–201.
Katz, J. J., and Norris, J. R. (1973). *Curr. Top. Bioenerg.* **5**, 41–75.
Kaufman, K. J., Dutton, P. L., Netzel, T. L., Leigh, J. S., and Rentzepis, P. M. (1975). *Science* **188**, 1301–1304.
Ke, B. (1972a). *In* "Photosynthesis and Nitrogen Fixation," Part B (A. San Pietro, ed.), Methods in Enzymology, Vol. 24, pp. 25–52. Academic Press, New York.
Ke, B. (1972b). *Arch. Biochem. Biophys.* **152**, 70–77.
Ke, B. (1973). *Biochim. Biophys. Acta* **303**, 1–33.
Ke, B. (1975). *Proc. Int. Congr. Photosynth. 3rd, Rehovot, 1974* **1**, 373–382.
Ke, B. (1978). *Curr. Top. Bioenerg.* **7**, 76–138.
Ke, B., and Dolan, E. (1980). *Biochim. Biophys. Acta* **590**, 401–406.
Ke, B., Hansen, R. E., and Beinert, H. (1973a). *Proc. Natl. Acad. Sci. U.S.A.* **70**, 2940–2945.
Ke, B., Sahu, S., Shaw, E. R., and Beinert, H. (1973b). *Biochim. Biophys. Acta* **347**, 36–48.
Ke, B., Sugahara, K., Shaw, E. R., Hansen, R. E., Hamilton, W. D., and Beinert, H. (1974). *Biochim. Biophys. Acta* **368**, 401–408.
Ke, B., Sugahara, K., and Shaw, E. R. (1975). *Biochim. Biophys. Acta* **408**, 12–25.
Ke, B., Hawkridge, F., and Sahu, S. (1976). *Proc. Natl. Acad. Sci. U.S.A.* **73**, 2211–2215.
Klimov, V. V., Klevanik, A. V., Shuvalov, V. A., and Krasnovsky, A. A. (1977). *FEBS Lett.* **82**, 183–186.
Klimov, V. V., Allakhverdiev, S. I., and Paschenko, V. A. (1978). *Dokl. Akad. Nauk. USSR* **242**, 1204–1207.
Klimov, V. V., Allakhverdiev, S. I., Demeter, S., and Krasnovsky, A. A. (1979). *Dokl. Akad. Nauk USSR* **249**, 227–230.

Klimov, V. V., Dolan, E., and Ke, B. (1980a). *FEBS Lett.* **112**, 97–100.
Klimov, V. V., Ke, B., and Dolan, E. (1980b). *FEBS Lett.* **118**, 123–126.
Klimov, Dolan, E., Shaw, E. R., and Ke, B. (1980c). *Proc. Natl. Acad. Sci. U.S.A.* **77**, 7227–7231.
Klimov, V. V., Allakhverdiev, S. I., and Krasnovsky, A. A. (1980d). *Dokl. Akad. Nauk. USSR* **249**, 485–488.
Kok, B. (1956). *Biochim. Biophys. Acta* **22**, 399–401.
Malkin, R., and Bearden, A. J. (1971). *Proc. Natl. Acad. Sci. U.S.A.* **68**, 16–19.
Malkin, R., and Bearden, A. J. (1973). *Proc. Natl. Acad. Sci. U.S.A.* **70**, 294–297.
Malkin, R., and Bearden, A. J. (1975). *Biochim. Biophys. Acta* **396**, 250–259.
Mathis, P., Sauer, K., and Remy, R. (1978). *FEBS Lett.* **88**, 275–278.
Mathis, P., Butler, W. L., and Satoh, K. (1979). *Photochem. Photobiol.* **30**, 603–614.
Norris, J. R., Scheer, H., Druyan, M. E., and Katz, J. J. (1974). *Proc. Natl. Acad. Sci. U.S.A.* **71**, 4897–4900.
Phillipson, K. P., Satoh, V. L., and Sauer, K. (1972). *Biochemistry* **11**, 4591–4594.
Rockley, M. G., Windsor, M. W., Cogdell, R. J., and Parson, W. W. (1975). *Proc. Natl. Acad. Sci. U.S.A.* **72**, 2251–2255.
Sauer, K., Mathis, P., Acker, S., and Van Best, J. A. (1978). *Biochim. Biophys. Acta* **503**, 120–134.
Seibert, M. (1978). *Curr. Top. Bioenerg.* **7**, 39–75.
Setif, P., and Mathis, P. (1980). *Photochem. Photobiol. Arch. Biochem. Biophys.* **204**, 477–485.
Shuvalov, V. A., Dolan, E., and Ke, B. (1979a). *Proc. Natl. Acad. Sci. U.S.A.* **76**, 770–773.
Shuvalov, V. A., Ke, B., and Dolan, E. (1979b). *FEBS Lett.* **100**, 5–8.
Shuvalov, V. A., Klevanik, A. V., Sharkov, A. V., Kryukov, P. G., and Ke, B. (1979c). *FEBS Lett.* **107**, 313–316.
Shuvalov, V. A., Klimov, V. V., Dolan, E., Parson, W. W., and Ke, S. (1980). *FEBS Lett.* **118**, 279–282.
Stiehl, H. H., and Witt, H. T. (1968). *Z. Naturforsch. Teil B* **23**, 220–224.
Van Best, J. A., and Duysens, L. N. M. (1977). *Biochim. Biophys. Acta* **459**, 187–206.
Van Best, J. A., and Mathis, P. (1978). *Biochim. Biophys. Acta* **503**, 178–188.
Van Gorkom, H. (1974). *Biochim. Biophys. Acta* **347**, 439–442.
Van Gorkom, H., Pulles, M. P. J., and Wessels, J. S. C. (1975). *Biochim. Biophys. Acta* **408**, 331–339.
Visser, J. W. M. (1975). Thesis, Univ. Leiden, Leiden.
Visser, J. W. M., Rijersberg, K. P., and Amesz, J. (1974). *Biochim. Biophys. Acta* **368**, 235–246.
Witt, K., and Wolff, C. (1970). *Z. Naturforsch., Teil B* **25**, 387–388.
Wolff, C., and Witt, H. T. (1969). *Z. Naturforsch., Teil B* **24**, 1031–1037.

CHAPTER 4

Electron Transfer Reactions in Reaction Centers of Photosynthetic Bacteria and in Reaction Center Models

Thomas L. Netzel

Department of Chemistry
Brookhaven National Laboratory
Upton, New York

I.	*In Vivo* ELECTRON TRANSFER REACTIONS	79
	A. Introduction	79
	B. Reaction Center Components	81
	C. Identification of Electron Transfer Intermediate I	82
	D. Summary Data	92
	E. Photosynthetic Overview	97
II.	*In Vitro* ELECTRON TRANSFER REACTIONS	97
	A. Reaction Center Models: Cofacial Diporphyrins	97
	B. Reaction Center Models: Dimers and Trimers of PChl a and PPheo a	105
	REFERENCES	114

I. *In Vivo* Electron Transfer Reactions

A. INTRODUCTION

Both photosynthetic bacteria and green plants harness the sun's energy to drive life-sustaining chemical reactions. Several features distinguish bacterial and plant photosynthetic processes. For example, bacteria use a single photosystem, while plants use two photosystems. Also, because the bacteriochlorophyll (BChl) pigments of bacteria absorb longer wavelengths of light than the chlorophyll (Chl) pigments of plants, bacteria utilize a larger portion of the solar spectrum. However, both bacteria and plants use arrays of pigments as antennae to harvest the sun's energy. This harvested energy is transferred to sites of chemical synthesis called reaction centers (RCs). The elementary chemical reactions in RCs are electron transfers. Because these electron transfers are exceedingly fast, it has been necessary to use the techniques of picosecond spectroscopy (Creutz *et al.*, 1980; Shapiro, 1977; Netzel *et al.*, 1973a,b) to study them. This chapter will concern itself with describing the present understanding of RC electron transfer steps. To provide a deeper appreciation of the intricate mechanism of photosynthetic RCs, some recent results of attempts to construct elementary reaction centers artificially will be described also.

Before addressing the experiments that have probed the detailed operations of the RCs, a few comments on picosecond fluorescence measurements on bacterial antennae are in order (plant antennae are discussed in Chapters 1 and 2). The physical problem is to gather information that will provide insights into the organization of the antennae and RCs and into the energy transfer steps within an antenna complex itself. An antenna is comprised of BChl pigments bound to hydrophobic polypeptides (Clayton and Clayton, 1972; Thornber, 1977). In *Rhodopseudomonas sphaeroides* strain R-26, there are about 60 BChl pigments per RC. Fortunately the antenna–RC complex can be isolated from whole bacterial cells with no apparent disruption of the complex. These cell pieces are called chromatophores and can be thought of as small bubbles. The antenna–RC complex is contained in the wall of the chromatophore.

Campillo *et al.* (1977) studied the variation in both the fluorescence lifetime and quantum yield with changes in the photon density of the excitation pulse for several species of bacteria. The chromatophore fluorescence lifetime was 300 ± 50 psec for strain R-26 and 1.1 ± 0.2 nsec for strain PM-8. These lifetimes were in agreement with studies done by other researchers (Wang and Clayton, 1971; Zankel *et al.*, 1968; Govindjee *et al.*, 1972). The long fluorescence lifetime found for strain PM-8 was due to the fact that this strain of photosynthetic bacteria did not have RCs. That is, the energy harvested by the PM-8 antennae had nowhere to go except to fluoresce or produce heat. For strain R-26, however, the energy harvested by the antennae was captured by the RCs and used to drive chemical reactions.

Because the RC has a finite rate of energy acceptance, increasing the excitation intensity can overload the antenna–RC complex. When this happens, the multiple excitations or excitons present in the antennae can annihilate each other. The exciton annihilation process is viewed as two low-energy excitons combining to form one doubly energetic exciton. However, this doubly energetic exciton rapidly ($\lesssim 1$ psec) degrades to a low-energy exciton and heat, because upper excited states of large organic molecules are usually very short-lived. The result of this annihilation process is to convert two fluorescing excitons into a single exciton. Clearly the rate at which the excitons come together depends both on their number and on their rate of migration from one pigment to another.

Campillo *et al.* (1977) assumed that the pigments of the antenna formed a two-dimensional array. Therefore, the 300-psec fluorescence lifetime of the R-26 chromatophore was viewed as due to the fact that an exciton had to hop 35–40 times at 8 psec per hop to find the RC. Finally, the data on the decrease in fluorescence yield with increasing excitation energy density was fit to a model which viewed an antenna system as capable of supplying energy to several RCs, a "lake" model (Clayton, 1967). Others had suggested that each RC had its own antenna, the "puddle" model (Knox, 1975).

Although picosecond fluorescence studies on isolated RCs have been performed, they have usually found multiple emission lifetimes whose interpretation was ambiguous (Paschenko et al., 1977). Sometimes multiple excitation pulses were used. Other times the RC isolation procedure produced free pigments that were highly fluorescent. In short, the high sensitivity of picosecond fluorescence measurements meant that these experiments frequently detected processes that were not integral to the operation of a RC. In contrast, picosecond absorption measurements were quite myopic. That is, they required both a significant production of a given photoproduct and a large change in molar extinction coefficient to produce an acceptable change-in-absorbance signal. Finally, the various electron transfer intermediates within a RC were not fluorescent themselves. However, as will be described later, they were able to reform the fluorescent singlet state of the primary electron donor and produce delayed fluorescence.

B. REACTION CENTER COMPONENTS

For technical reasons, the details of the operation of RCs from photosynthetic bacteria are understood in much greater detail than those from plants. The chief reason for this is that the associated antenna pigments can be stripped off bacterial RCs while this is not true for plants. So far, particles enriched in RCs of photosystem I of green plants have been prepared with as few as 26 Chl *a* pigments per RC (Vernon et al., 1979). The problem in studying these plant RCs is that the attached antenna pigments have their own optical transients that mask those of the electron transfers occurring at the RC.

The important photosynthetic pigments in a bacterial RC (Feher and Okamura, 1976; Dutton et al., 1976) are four BChl molecules, two bacteriopheophytins (BPheo), and a nonheme ferroquinone complex [$Q(Fe^{2+})$]. (BPheo is a metal-free BChl.) Both the nature of the $Q(Fe^{2+})$ complex and the role of Fe^{2+} are not understood. While the electron paramagnetic resonance (EPR) of the reduced quinone is extensively broadened (300 G) by the presence of the paramagnetic Fe^{2+} (Okamura et al., 1975), no data require Q to be bound directly to Fe^{2+}. The primary electron donor is a special pair (Norris et al., 1971; McElroy et al., 1972) of BChl molecules, $(BChl)_2$. It absorbs at 865 and 605 nm in *R. sphaeroides*. In other species of bacteria the special pair absorbs at slightly different wavelengths: 883 nm in *Chromatium vinosum* and 960 nm in *Rhodopseudomonas viridis*. However, all

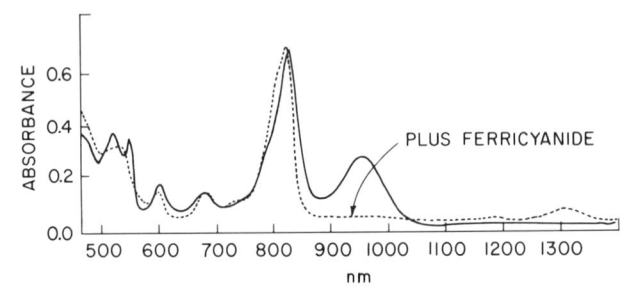

FIG. 1. Ground state absorption spectra of the *R. viridis* reaction center with associated c-type cytochromes. The solid line indicates the spectrum of a normal RC with both cytochromes reduced to their ferro form. When ferricyanide was added, the (BChl b)$_2$ dimer absorbing at 960 nm was oxidized and the spectrum with the dashed line was obtained. Note the increased absorbance beyond 1100 nm characteristic of (BChl b)$_2^{\ddot{+}}$. See Section I,C,3 for an assignment of the other absorption features. [Reprinted with permission from Netzel *et al.*, *Biochim. Biophys. Acta* **460**, 469, Fig. 1 (1977b). Copyright 1977 by Elsevier/North-Holland Biomedical Press.]

these RCs are functionally similar. The other important RC pigments absorb in the following regions in *R. sphaeroides*: The remaining two BChls at 800 and 595 nm and the two BPheos at 760 and 535 nm. A typical ground state absorption spectrum for a RC preparation (Netzel *et al.*, 1977b) is shown in Fig. 1. Later sections will discuss the various spectral features of the RC in greater detail (see especially Section I,C,3 for a discussion of the RC bands for *R. viridis*.).

C. IDENTIFICATION OF ELECTRON TRANSFER INTERMEDIATE I

For a long time the primary electron acceptor was thought to be the Q(Fe^{2+}) complex (McElroy *et al.*, 1970; Leigh and Dutton, 1972). Reduced Q(Fe^{2+}) was stable on a millisecond time scale, and nanosecond laser pulses could not detect any earlier acceptors. The first indication that other RC pigments may be acting as electron transfer relay agents was the detection of a new transient, P^F, with a 10-nsec lifetime at room temperature (Parson *et al.*, 1975). This transient showed absorbance bleaching in the BPheo absorbing regions. However, without picosecond time resolution it could be observed only when normal photosynthetic electron transfers were blocked. In this case the blocking was achieved by prior reduction of the Q(Fe^{2+}) complex to Q$^{\bar{\cdot}}$(Fe^{2+}). To guarantee that P^F was a real and not an artificial intermediate, it had to be detected under conditions in which normal photosynthesis could proceed.

The first picosecond spectroscopic experiment on RCs was reported by Netzel *et al.* (1973a). In it 6-psec-duration pulses at 531 nm excited a BPheo absorption band, and the <10-psec bleaching of the (BChl)$_2$ band at 865 nm was monitored. While this experiment could not distinguish between the

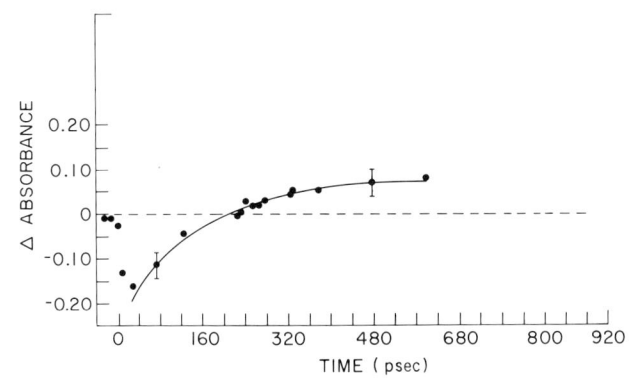

FIG. 2. Change-in-absorbance measurement for RCs from R. sphaeroides. The sample's absorbance at 860 nm in a 1-mm-path-length cell was 2.1. The excitation was produced by a 6-psec pulse at 531 nm, and the decay kinetics are shown at 545 nm, $\tau = 149 \pm 8$ psec.

production of an excited state of $(BChl)_2$ and creation of the oxidized donor, $(BChl)_2^+$, it did demonstrate the extremely rapid energy transfer from BPheo to the electron donor, $(BChl)_2$. This rapid energy transfer showed that 531-nm excitation could be used with little rate limitation for studying RC kinetics. Subsequent work (Kaufmann et al., 1975; Rockley et al., 1975) on RCs from R. sphaeroides R-26 at +200 mV showed absorbance decreases at 600 and 542 nm, as well as a broad absorbance increase in the 610–650 region, 10 psec after flash excitation. Poising the RCs at +200 mV ensured that normal photosynthetic electron transfer processes were operative. The decay kinetics at 545 nm (Fig. 2) and at 640 nm (Fig. 3) (T. L. Netzel, P. M.

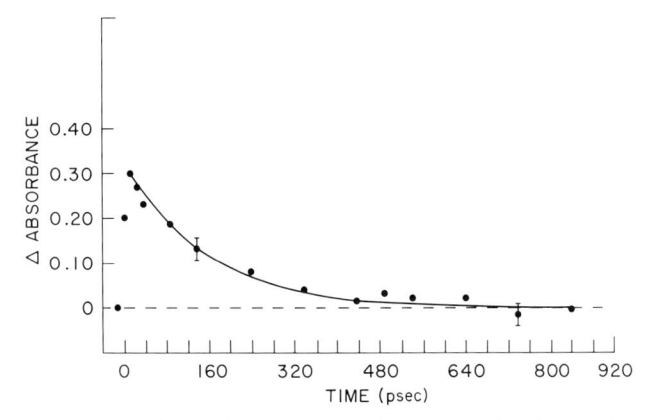

FIG. 3. All experimental conditions were as in Fig. 1, except that the sample's absorbance at 860 nm in a 1-mm-path-length cell was 2.3. The decay kinetics are shown at 640 nm, $\tau = 151 \pm 7$ psec.

Rentzepis, D. M. Tiede, and P. L. Dutton, unpublished data) showed that the lifetime of this state was 150 psec.

This new state, present 10 psec after excitation, was spectrophotometrically similar to the state labeled P^F. Indeed, when $Q^{\bar{}}(Fe^{2+})$ was formed prior to picosecond excitation, the new state showed no decay at all between 10 and 200 psec after excitation (Kaufmann et al., 1975). This was consistent with P^F having a 10-nsec lifetime under these conditions. Since the spectral changes present 250 psec after excitation of a normally operating RC were similar to those expected for the $(BChl)_2^+ Q^{\bar{}}(Fe^{2+})$ photoproduct, the 6.7×10^9 sec^{-1} rate ($1/k = \tau = 150$ psec) was assigned to the reduction of $Q(Fe^{2+})$.

$$(BChl)_2 Q(Fe^{2+}) \xrightarrow[k = 6.7 \times 10^9 \text{ sec}^{-1}]{h\nu} (BChl)_2^+ Q^{\bar{}}(Fe^{2+}). \tag{1}$$

The result left two possibilities: Either the excited singlet of $(BChl)_2$ lasted 150 psec or an intermediate electron acceptor existed prior to the reduction of $Q(Fe^{2+})$. Fortunately, an absorbance band at 1250 nm appeared when $(BChl)_2^+$ was formed either chemically by the addition of an oxidant such as ferricyanide or photochemically by illumination with actinic light at 865 nm. A key picosecond measurement (Dutton et al., 1975a) found that the rate of formation of this infrared absorbance was < 10 psec. Also, the absorbance at 1250 nm did not decay between 10 psec and 1 nsec after excitation. This was true whether $Q(Fe^{2+})$ was oxidized or reduced. Even the removal of the $Q(Fe^{2+})$ complex failed to affect the kinetics of the 1250 nm absorbance band (Netzel et al., 1977a). Clearly, Eq. (1) had to be modified to admit a prior electron acceptor, I:

$$(BChl)_2 IQ(Fe^{2+}) \xrightarrow[k > 10^{11} \text{ p sec}^{-1}]{h\nu} (BChl)_2^+ I^{\bar{}} Q(Fe^{2+}) \xrightarrow[k = 6.7 \times 10^9 \text{ sec}^{-1}]{} (BChl)_2^+ IQ^{\bar{}}(Fe^{2+}). \tag{2}$$

The bleaching at 542 nm in a BPheo absorption region clearly implicated BPheo$^{\bar{}}$ in the formation of I$^{\bar{}}$. Fajer et al. (1975) showed that both BChl$^{\bar{}}$ and BPheo$^{\bar{}}$ had broad absorptions in the 600–650 nm region, as did I$^{\bar{}}$. They measured BChl in vitro to be over 300 mV harder to reduce than BPheo. Therefore, they proposed that, given the existing data, I was best assigned to BPheo rather BChl. What the redox potentials of the various RC pigments in vivo actually are remains unknown. As an example of intermolecular effects on redox potentials, the following can be noted (I. Fujita, J. Fajer, C.-K. Chang, C.-B. Wang, and T. L. Netzel, unpublished data). The sum of the free energy changes (obtained from midpoint potentials) for oxidizing magnesium octaethylporphyin (MgOEP) and reducing free base octaethylporphyin (H$_2$OEP) in methylene chloride (CH$_2$Cl$_2$) with tetrapropylammonium perchlorate is 2.01 eV. The corresponding measurements for a cofacial diporphyrin (comprised of one alkyl-substituted magnesium porphyrin and one similar free base porphyrin joined with two five-atom chains) show that the sum of the free energy changes is only 1.80 eV. Therefore one could be wrong by several hundred millielectronvolts when using monomeric redox values for closely associated RC pigments.

1. *Trapping Studies on I*

Another type of study on I was performed on RCs from *chromatium vinosum*. For this bacterium, isolated RCs still had two c-type cytochromes attached. The one absorbing at 553 nm reduced $(BChl)_2^+$ irreversibly with $t_{1/2} = 10$ μsec at 200 K (DeVault and Chance, 1966). Although this rate was 1000 times slower than that for the reverse electron transfer from $I^{\bar{}}$ to $(BChl)_2^+$, its irreversibility meant that after several minutes of illumination one could build up large amounts of oxidized cytochrome c-553 and $I^{\bar{}}$, if one started with prereduced $Q(Fe^{2+})$ (Tiede et al., 1976a,b; Shuvalov and Klimov, 1976). The net reaction was

$$(C553^{2+})(BChl)_2 IQ^{\bar{}}(Fe^{2+}) \xrightarrow[h\nu]{cw} (C553^{3+})(BChl)_2 I^{\bar{}} Q^{\bar{}}(Fe^{2+}). \quad (3)$$

The spectrum of the trapped $I^{\bar{}}$ produced in this way agreed well with that of P^F when the changes due to $(BChl)_2^+$ were accounted for. All the spectral changes implicating $BPheo^{\bar{}}$ as $I^{\bar{}}$ were present as described above, but additional bleachings at 595 and 800 nm, where BChls absorbed, were also found. Therefore there was no way of eliminating BChl from a role in I.

Figure 4 summarizes the electron transfer events just described. Note that the c-type cytochromes in the interior of the chromatophore reduce the oxidized donor, $(BChl)_2^+$, while an electron is being passed along a chain of acceptors to the aqueous phase surrounding the chromatophore. The lifetime of $BPheo^{\bar{}}$, governed by the rate of reduction of $Q(Fe^{2+})$, is well established (see below). The remainder of Section I of this chapter discusses the events occurring between the time a photon is absorbed and BPheo reduced.

Before reviewing more experimental results that help to define the nature of the intermediate I, it might be useful to show that to a certain extent the measured electron transfer rates are a function of RC preparation. To isolate RC protein complexes, intact cells must be broken and the desired fragments separated from the unwanted ones. Finally detergents such as LDAO and Triton are used to solubilize the RC complex. During these steps assays for photochemical activity, such as supporting cytochrome c oxidation, are performed. However, alteration of other RC processes is always a concern. It is not the case, for instance, that the rate for $Q(Fe^{2+})$ reduction is 6.7×10^9 sec^{-1} ($1/k = t = 150$ psec) or no reduction occurs at all. To illustrate the fact that electron transfer rates can be manipulated, one can add chaotropic

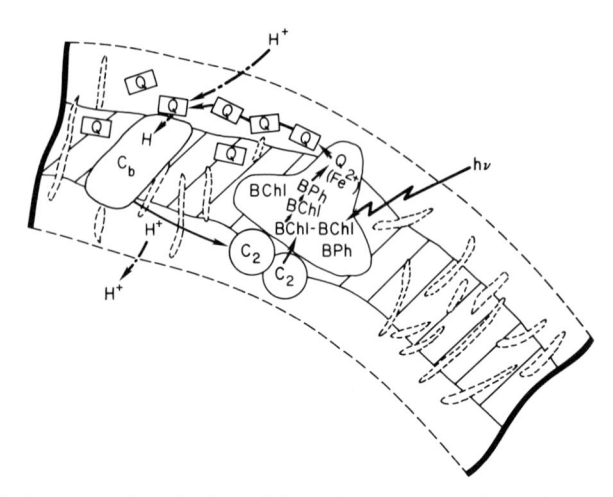

FIG. 4. Artist's conception of a bacterial reaction center spanning a chromatophore membrane. The associated cytochromes are labeled C_b and C_2. Note that the light-driven electron flow is away from C_2 and opposite the proton flow. Also, cytochrome b completes the cyclic electron flow by reducing ferri-C_2 to ferro-C_2. After this the RC is ready to function again. Q denotes ubiquinone, and the elongated dashed shapes denote carotenoids. (Reprinted with permission and some changes from Holten and Windsor, 1978. *Annu. Rev. Biophys. Biochem. Eng.* 7, p. 105, Figure 9. © 1978 *Annual Reviews Incorporated.*) Adaped from an earlier drawing by Dutton et al. (1975b).

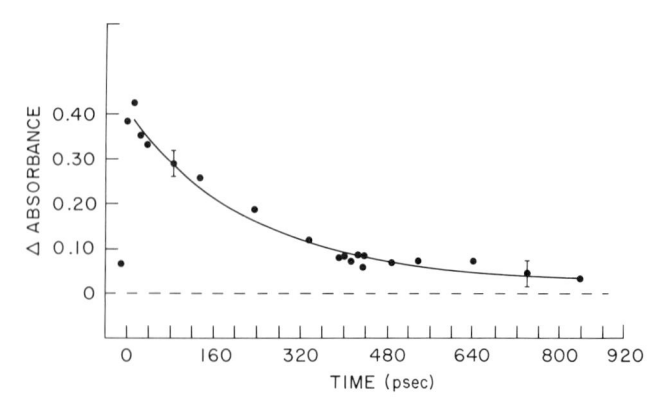

FIG. 5. Change in absorbance measurement for RCs from *R. sphaeroides*. Potassium isocyanate and 4,5-phenanthroline were added to the sample preparation to produce a mild disruption of the RC protein. The sample's absorbance at 860 nm in a 1-mm-path-length cell was 2.0. The decay kinetics are shown at 640 nm, $\tau = 229 \pm 15$ psec.

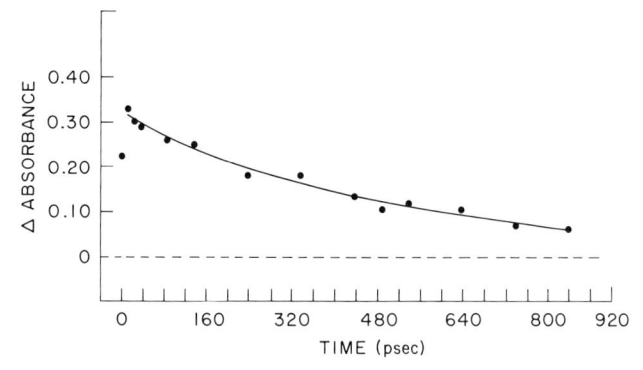

FIG. 6. All experimental conditions were the same as in Fig. 3, except that the sample was not flashed immediately after preparation. Rather, it was stored on crushed ice for 2 days before use. The decay kinetics are shown at 640 nm, $\tau = 506 \pm 22$ psec.

agents such as potassium isocyanate and 4,5-phenanthroline to RC samples. These chemicals disrupt a RC protein complex just as detergents disrupt a chromatophore complex to release RCs. Figures 5 and 6 show the effect that adding potassium isocyanate and 4,5-phenanthroline to RC samples (T. L. Netzel et al., unpublished data) has on the rate of reduction of $Q(Fe^{2+})$. The freshly "treated" samples have a $Q(Fe^{2+})$ reduction rate of 4.3×10^9 sec^{-1} ($\tau = 230$ psec). After storing these samples for 2 days on crushed ice, further denaturing occurs and the reduction rate decreases to 2×10^9 sec^{-1} ($\tau = 500$ psec). After several more days the reduction is no longer a monophasic process. This variation in electron transfer rates shows that a RC has some plasticity and that a variation in measured RC properties can be expected It also suggests that caution should be exercised in interpreting small effects.

2. *Studies on P^F and P^R*

The above discussion showed that the state P^F, found when $Q(Fe^{2+})$ was reduced prior to flashing the RC, was largely the same as the $(BChl)_2^+ I^-$ state found in an operational RC. Therefore studies on the formation and decay of P^F were likely to yield substantially correct information about the nature of I^-. The lifetime of P^F was ~ 10 nsec at 300 K and ~ 30 nsec at 15 K. At 300 K, 10–20% of P^F decayed to a triplet state called P^R (Parson et al., 1975) whose lifetime was 6 μsec. At temperatures <120 K, nearly 100% of P^F decayed to P^R whose lifetime was 120 μsec. Recent delayed fluorescence measurements (Shuvalov and Parson, 1981) from P^R back to the excited singlet of $(BChl)_2$ gave an apparent activation energy of 0.40 ± 0.02 eV. The 0–0 of energy of the RC, which is located midway between the absorption and emission maxima, was at 1.38 eV in R. sphaeroides. Therefore P^R was located at 0.98 eV.

The optical changes characterizing P^R at 77 and 293 K showed important differences in the following regions: 600–700 nm, at 798 nm, and at 1250 nm. These were interpreted as showing that P^R was an equilibrium mixture of two kinds of triplet states (Shuvalov and Parson, 1981). At low temperatures P^R was almost completely in a $(BChl)_2$ excited triplet state. At higher temperatures a biradical triplet $[BChl^{\ddot{+}} I^{\dot{-}}]^3$, only 0.03 eV higher in energy was populated. This accounted for the absorption increases in P^R at higher temperatures at 1250 nm and in the 600–700 nm region. The most striking result, however, was that at 798 nm a pronounced bleach was present at 293 K. This bleach replaced an absorption which was present at 77 K. The location of this bleach was especially noteworthy because it occurred in a region where BChl absorbed and not where BPheo absorbed. In contrast, the spectroscopy of trapped $I^{\dot{-}}$ as well as of P^F always showed a pronounced bleach at 760 nm, where BPheo absorbed. Therefore the higher-energy component of P^R appeared to be $[(BChl)_2^{\dot{+}} BChl^{\dot{-}}]^3$. It should be pointed out that, if $(BChl)_2$ and BChl are close to each other, the label $[(BChl)_2^{\dot{+}} BChl^{\dot{-}}]^3$ may not be a complete description of this state. Since the $(BChl)^T$ lies so close in energy, it (and perhaps other molecular excited states as well) may be mixed with the ionic states. However, the biradical triplet state label will be useful if this caution is kept in mind.

At this point it may be appropriate to describe how an initially excited $(BChl)_2$ singlet state could yield both $[(BChl)_2^{\dot{+}} BChl^{\dot{-}}]^3$ and $(BChl)_2^T$ states with $[(BChl)_2^{\dot{+}} I^{\dot{-}}]^1$ as an intermediate. As already noted, the yield of P^R from P^F varied with temperature. It was also true that application of an external magnetic field at room temperature decreased the yield of P^R (Hoff et al., 1977; Blankenship et al., 1977). Although the magnetic field that was sufficient to cause half of the effect of a saturating field varied with sample preparation, typical values were less than 100 G. The theory of chemically induced magnetic polarization (Lepley and Closs, 1973) accounted for these effects in the following way. When $Q^{\dot{-}}(Fe^{2+})$ was formed prior to photolyzing the RC sample, the lifetime of the radical pair $[(BChl)_2^{\dot{+}} I^{\dot{-}}]^1$ was reasonably long ($t \sim 10$ nsec). This meant in principle that the hyperfine interactions of the nuclear spins on $(BChl)_2^{\dot{+}}$ with its unpaired electron and the hyperfine interactions of the nuclear spins on $I^{\dot{-}}$ with its unpaired electron would be able to switch the biradical's spin state from a singlet to a triplet. The theory showed that the time-dependent solution for the probability of a given initial biradical singlet having a triplet configuration at later time was a periodic function with a characteristic frequency that depended on the spin orientations of all of the radical's nuclei. In practice, many such frequencies exist, and singlet–triplet oscillation of an ensemble of radical pairs is not expected (Haberkorn and Michel-Beyerle, 1979).

The consequences of the above hyperfine interactions on the yield of P^R in the presence and absence of an external magnetic field was understood in the following way. If it is assumed that the exchange interaction between the (BChl)$_2^+$ radical and the I$^-$ radical is zero or nearly zero, the three triplet sublevels, $T_{0, \pm 1}$ will be degenerate in the absence of a magnetic field. Hyperfine interactions within the biradical singlet state would cause each sublevel to be populated to a similar extent and therefore to contribute similarly to P^R formation. Application of a sufficiently strong magnetic field would split the $T_{\pm 1}$ sublevels away from T_0. This would reduce the population of the $T_{\pm 1}$ levels from the biradical singlet state. Thus less P^R would be formed. This same sublevel splitting phenomenon would also account for the observed electron spin polarization of P^R (Dutton et al., 1972, 1973; Thurnauer et al., 1975). The fact that such small magnetic fields produced the reduction in P^R yield put an upper bound of 10^{-6} eV on the spin-exchange interaction within the biradical (Werner et al., 1978). This meant that $[(BChl)_2^+ I^-]^1$ was essentially isoenergetic with $[(BChl)_2^+ I^-]^3$.

Haberkorn et al. (1979) have noted that there is a relationship derived by Anderson (1959), which relates the matrix element governing electron transfer to the exchange integral just discussed. Based on currently available spectroscopic data, the electron transfer matrix element is calculated to be less than 1.2 cm^{-1}. This puts an upper limit on the rate of electron transfer from the (BChl)$_2$ excited singlet to form (BChl)$_2^+ I^-$ of $\sim 8 \times 10^9$ sec^{-1}. The observed rate is equal to $\sim 2 \times 10^{11}$ sec^{-1} (Holten et al., 1980). Therefore the data on the magnetic field dependence of P^R yield are inconsistent with viewing I$^-$ solely as BPheo$^-$.

Jortner (1980) has provided a theoretical treatment for the case in which electron transfer rates are competitive with excited state vibrational relaxation states. The very rapid ($\tau \sim 4$ psec) reduction of BPheo is probably a situation in which this case applies. Certainly a prior reduction of BChl would have to be treated as this type of reaction. An important result of this theoretical calculation, in agreement with others that assumed vibrational relaxation prior to electron transfer, was that to achieve electron transfer rates in excess of 10^{11} sec^{-1} the electronic coupling matrix element must be > 10 cm^{-1}. Thus the inconsistency between the electronic coupling estimated from magnetic field experiments and that required by current electron transfer theories remains. A solution is to have both BChl$^-$ and BPheo$^-$ play roles as electron transfer intermediates between (BChl)$_2^+$ and Q$^-$(Fe^{2+}).

Before reviewing additional evidence for this proposal, the discussion of P^F and P^R energetics can be completed. Delayed fluorescence established P^R as lying 0.4 eV below the (BChl a)$_2$ excited singlet at 1.38 eV. In principle, similar measurements on P^F could yield its location. A convincing measurement was made by van Grondelle et al. (1978). They reported that P^F was 0.13 eV below the (BChl a)$_2$ singlet in R. spaeroides. When they used data obtained by Parson et al. (1975) on P^R yield, they calculated a drop of 0.14 eV. Godik and Borisov (1979) made a combined study of the delayed fluorescence from *Rhodospirillum rubrum*, *Ectothiorhodospira shaposhnikovii*, and *Thiocapsa roseopersicina*. They concluded that P^F was 0.05 ± 0.03 eV below the dimer's singlet. However, their kinetic analysis assumed that only the rate of repopulation of the dimer's singlet varied with temperature. The large variation in P^R yield with temperature suggests that this is not a good assumption. Shuvalov and Parson (1981) also measured the temperature dependence of the same delayed fluorescence in *R. sphaeroides*. They reported as unpublished data that P^F was 0.05 eV below the $(BChl\ a)_2$ singlet. Therefore a range of values, 0.05–0.13 eV, will be used in this chapter. Electrochemical reduction of I in RCs from *R. viridis* placed the midpoint for reduction of $(I/I^{\overline{\cdot}})$ at -0.62 V (Klimov et al., 1977). When combined with data on the oxidation potential and 0–0 excited singlet state energy of $(BChl\ b)_2$, a drop of 0.10 eV was found between P^F and the $(BChl\ b)_2$ excited singlet state.

3. *Evidence for* $BChl^{\overline{\cdot}}$ *as an Electron Transfer Intermediate*

In addition to the magnetic field dependence of P^R yield described above, there is evidence from three other sources to support proposing that BChl mediates electron transfer to I: (1) dichroic studies on pigment–pigment interactions, (2) picosecond spectroscopy, and (3) a comparison with electron transfer reactions observed in RC models.

Circular and linear dichroism (CD and LD) studies afforded the possibility of learning about pigment–pigment interactions within a RC. For example, interaction between the wave functions of two molecules was known to produce the appearance of two CD bands of opposite sign (Kasha et al., 1975). Oxidation or reduction of either of these interacting molecules would eliminate both bands. Several researchers carried out CD and LD measurements as well as normal absorbance measurements on RC preparations from *R. viridis* (Holten et al., 1978b; Shuvalov and Asadov, 1979; Vermeglio and Clayton, 1976). They found that the spectrum of a normal RC at $+250$ mV and 100 K consisted of five bands: 790, 814, 830, 850, and 980 nm. Each of these bands was assigned to an important RC pigment: BPheo at 790 nm, BChl at 814 nm, BChl at 830 nm, and $(BChl)_2$ at 850 and 980 nm. When $I^{\overline{\cdot}}$ was trapped as described earlier, the following RC state was created,

$$(BChl)_2(BChl_{814})(BChl_{830})BPheo^{\overline{\cdot}}Q^{\overline{\cdot}}(Fe^{2+}).$$

In this case CD bands at 790 and 830 nm were attenuated. Also, the normal absorbance bands at 814 and 830 nm were reduced to a single smaller absorbance band at 820 nm. The CD bands at 850 and 980 nm were only slightly perturbed. When the frozen RCs were illuminated at $+250$ mV, the following RC state was trapped,

$$(BChl)_2^{\frac{+}{2}}(BChl_{814})(BChl_{830})BPheoQ(Fe^{2+}).$$

In this case the two CD bands of opposite sign at 850 and 980 nm were very much reduced. Negligible changes were produced at 790, 814, and 830 nm. These results indicated that, while the two BChls at 814 and 830 nm did interact, there was essentially no interaction between BPheo and (BChl)$_2$. This agreed well with the very small spin exchange and electron transfer matrix elements estimated for these two molecules. However, BPheo interacted strongly with either BChl$_{814}$ or BChl$_{830}$, or both. Thus at least one of these BChls was likely to mediate the electron transfer from (BChl)$_2$ to BPheo.

In 1978 Shuvalov *et al.*, using a mode-locked neodymium–YAG oscillator to drive a parametric generator, studied the primary electron transfers in RCs from *R. rubrum*. Their experiment differed from previous picosecond studies in two ways. Their pulse duration was ~ 30 psec (five times broader than earlier picosecond studies), and they excited the RCs at 880 nm. Previous picosecond work had used 531- or 627-nm excitation, which meant that all the RC pigments could have been excited. However, with 880-nm excitation only the electron donor (BChl)$_2$ was excited. Their data showed that during the initial portion of the photolysis pulse the bleachings at 545 and 750 nm characteristic of BPheo$^-$ production were not found. Rather, absorbance increases were found at these wavelengths and bleachings characteristic of BChl$^-$ production were found at 600 and 800 nm. The same experimental observations were made by Akhmanov *et al.* (1980) when the Q(Fe^{2+}) complex was reduced prior to laser excitation. In addition they showed that anomalous kinetics could be obtained in these prereduced RCs by using excessive photolysis pulse energies. Recently, Holten *et al.* (1980) using a synchronously pumped dye laser with a ~ 0.7-psec pulse duration obtained evidence that BPheo was photoreduced in ~ 4 psec.

It should be pointed out that optical transients due to short-lived excited states can be formed during laser excitation. Therefore picosecond experiments have not yet proved the existence of a (BChl)$_2^+$ BChl$^-$ intermediate. The ~ 4-psec reduction of BPheo$^-$ could be preceded solely by the decay of the (BChl)$_2$ excited singlet state. So far no one who has used subpicosecond or 880-nm excitation pulses has monitored the kinetics of the 1250-nm band which is characteristic of (BChl)$_2^+$. If (BChl)$_2^+$ BChl$^-$ is an electron transfer intermediate, the 1250-nm band should be formed in less than 4 psec. If it is not, the early bleaches of BChl bands will have to be interpreted as changes in exciton interactions among the RC pigments due to the formation of the (BChl)$_2$ excited singlet state.

The third type of evidence that BPheo is not the first electron acceptor comes from a comparison of electron spin resonance (ESR) results on RCs and *in vitro* RC modeling studies. The 10^{-6}-eV upper limit for spin exchange interaction within the (BChl)$_2^+$ BPheo$^-$ biradical can be used to estimate that the two radicals must have an edge-to-edge separation of at least 9 Å (Dutton et al., 1979). Supporting results from CD and LD studies place this distance at 10–15 Å (Shuvalov and Asadov, 1979). Therefore, if there are no electron transfer intermediates between (BChl)$_2$ and BPheo, it should be very easy to duplicate the primary charge transfer event in RCs with synthetic models. As we will describe more fully in the next section, this has not been the case. Modeling studies on porphyrin and Chl complexes (Netzel et al., 1979; Bucks et al., 1981) showed that to achieve electron transfer rates $>10^{11}$ sec^{-1} it was necessary to have the electron donor and acceptor less than 7 Å apart. In addition, a nearly cofacial relative orientation was required.

D. Summary Data

1. Kinetic Scheme

Figure 7 summarizes the electron transfer steps that current experiments support. Whether one or two BChls are involved in mediating the electron transfer to BPheo is an open question. Once the electron is transferred to $Q(Fe^{2+})$, the cation and anion radicals are separated sufficiently that the recombination reaction takes 60 msec. As we will see later, the distance of

KINETICS IN R. sphaeroides

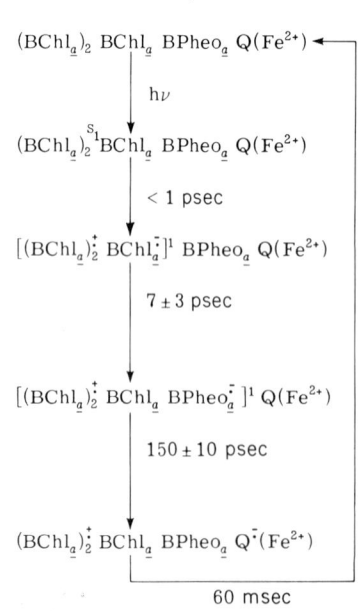

$(BChl_a)_2$ $BChl_a$ $BPheo_a$ $Q(Fe^{2+})$ ←┐
　　　│
　　　$h\nu$
　　　↓
$(BChl_a)_2^{S_1}$ $BChl_a$ $BPheo_a$ $Q(Fe^{2+})$
　　　│
　　　< 1 psec
　　　↓
$[(BChl_a)_2^{\cdot+}$ $BChl_a^{\cdot-}]^1$ $BPheo_a$ $Q(Fe^{2+})$
　　　│
　　　7 ± 3 psec
　　　↓
$[(BChl_a)_2^{\cdot+}$ $BChl_a$ $BPheo_a^{\cdot-}]^1$ $Q(Fe^{2+})$
　　　│
　　　150 ± 10 psec
　　　↓
$(BChl_a)_2^{\cdot+}$ $BChl_a$ $BPheo_a$ $Q^{\cdot-}(Fe^{2+})$
　　　└──── 60 msec ────┘

FIG. 7. Electron transfer reactions in RCs from R. sphaeroides. (BChl $a)_2^{S_1}$ is the lowest-energy, excited singlet state of (BChl $a)_2$.

separation is estimated to be >23 Å (Tiede *et al.*, 1978a,b; Dutton *et al.*, 1979).

2. Biradical Spectroscopy

Figure 8 collects the spectroscopic data discussed above which deals with the location of the various biradical singlets and triplets in RCs from *R. sphaeroides*. Delayed fluorescence data were used to locate [(BChl $a)_2^+$ BPheo a^-]^1. The previous discussion of the need for a BChl a electron transfer intermediate places the [(BChl $a)_2^+$ BChl a^-]^1 between 1.38 and 1.33–1.25 eV. However, P^R was shown to be an equilibrium mixture of [(BChl $a)_2^+$ BChl a^-]^3 and (BChl $a)_2^T$. This means that the singlet–triplet separation for the (BChl $a)_2^+$ BChl a^- biradical is at least 0.32–0.24 eV. This is over five orders of magnitude larger than the separation between the singlet and triplet states of the (BChl $a)_2^+$ BPheo a^- biradical.

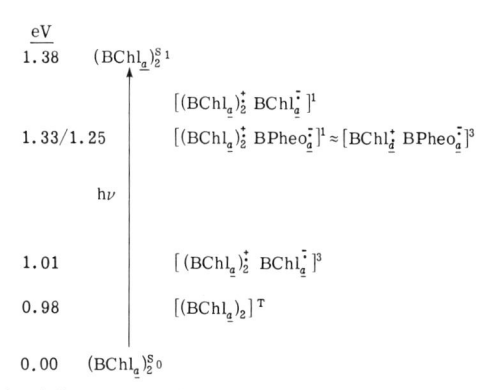

```
eV
1.38      (BChl_a)_2^{S1}

                  [(BChl_a)_2^+ BChl_a^-]^1
1.33/1.25         [(BChl_a)_2^+ BPheo_a^-]^1 ≈ [BChl_a^+ BPheo_a^-]^3

          hν

1.01              [(BChl_a)_2^+ BChl_a^-]^3

0.98              [(BChl_a)_2]^T

0.00      (BChl_a)_2^{S0}
```

FIG. 8. Energy level diagram of various states identified in RCs from *R. sphaeroides*. (BChl $a)^{S_0}$ and (BChl $a)^T$ are, respectively, the ground and lowest-energy triplet electronic states of (BChl $a)_2$.

The large splitting between the singlet and triplet states for the (BChl $a)_2^+$ BChl a^- biradical suggests that molecular states of these two molecules may be mixed with the ionic states. However, an exact description of this state is not available at present. If the entire splitting is attributed to a spin-exchange interaction, various matrix elements, rates, and distances can be calculated. As long as the results of these calculations are viewed with caution, they can provide insights into the likely magnitudes of these quantities. In this spirit, the spin exchange rate is given by

$$k_{ex} = \Delta E^{1,3}/2h \geq 6\text{–}8 \times 10^{13} \text{ sec}^{-1},$$

where h is Plank's constant and $\Delta E^{1,3}$ is the singlet–triplet energy separation. [Shuvalov and Parson (1981), calculated $8.4 \times 10^{13}\ \text{sec}^{-1}$ for k_{ex} using slightly different spectroscopic data.] This range of spin exchange rates can be used to estimate a maximum edge-to-edge distance for the biradical pair (Likhtenstein et al., 1979) of 3.1–3.2 Å. Carrying the analysis one step further with the use of Anderson's relationship between the electron exchange and electron transfer integrals, the minimum $(\text{BChl } a)_2^+ \text{BChl } a^-$ singlet–triplet energy separation of 0.32–0.24 eV yields 0.03–0.07 eV as the minimum range of values for the electron transfer integral. This is more than 200 times larger than one calculated for electron transfer from $(\text{BChl})_2$ to BPheo. This large integral can easily account for a < 1-psec electron transfer rate. It is interesting to note that an electron transfer integral larger than 10^{-2} eV is usually sufficient for such a reaction to be adiabatic (Sutin, 1962). That is, the transmission coefficient for passing over a kinetic barrier from reactants to products is unity. For electron transfer integrals less than 10^{-2} eV there is a chance that the reactants will not be transformed into products every time they reach the top of the kinetic barrier. Rather they can also jump to an upper potential energy surface. It is easy to see that photosynthetic systems can maximize their electron transfer rates by having sufficiently large electronic coupling to achieve adiabatic electron transfers.

3. RC Distance Estimates

At this point we have discussed two RC distance estimates: $(\text{BChl } a)_2$ to BChl a, less than or equal to 3.2 Å, and $(\text{BChl } a)_2$ to BPheo a, about 10–15 Å. Dutton et al. (1976) and Okamura et al. (1979) used the I^- trapping technique to study the ESR interactions between BPheo$^-$ and $\text{Q}^-\text{(Fe}^{2+})$. The edge-to-edge separation in C. vinosum was estimated to be 10–11 Å. In R. sphaeroides, an edge-to-edge distance of 10–12 Å and an electronic coupling integral of 1.3×10^{-4} eV were estimated.

Recently Pachence et al. (1979) incorporated RCs from R. sphaeroides into phosphatidylcholine bilayers and produced one-dimensional vesicles. Fortunately the RCs oriented in a well-defined, rather than random, manner. LD studies showed that the orientation of the RCs in the bilayers was similar to that of RCs in the chromatophore membrane. Also, when cytochrome c was added to performed vesicles containing RCs, as much as 90% of flash-produced $(\text{BChl})_2^+$ was reduced rapidly. Apparently the RCs were oriented so that the cytochrome binding sites were exposed to the vesicle exterior. X-ray diffraction data on oriented membrane multilayers were used to derive an electron density profile of the RC protein. Figure 9 presents a result derived from this work. A key point is that the tip-to-tip RC length is 57 Å. The diffraction data give only radial distribution information, so

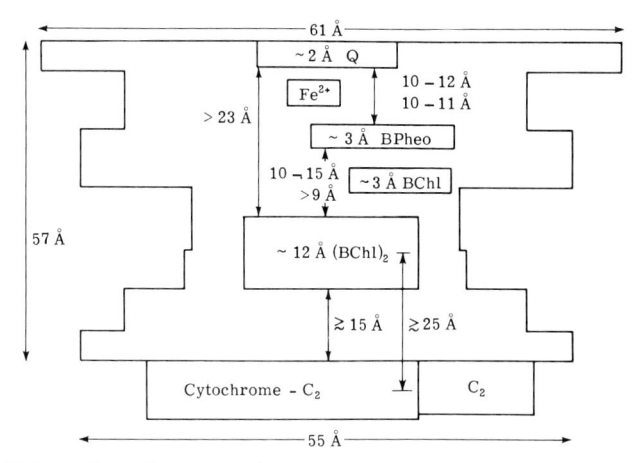

4. IN VIVO AND IN VITRO ELECTRON TRANSFER REACTIONS

FIG. 9. Estimated reaction center distances. A summary of pigment-to-pigment distances within a RC. (See the text for a detailed discussion of each distance calculation.) The indicated thicknesses of the pigments are only estimates. An experimental profile of a RC is shown surrounding the RC pigments. It indicates diameters in a stepwise manner for a RC which is viewed as a solid of revolution. The cytochromes are included for distance comparisons only, no diameter estimates are intended.

Fig. 9 is a cross section through a solid of revolution. The end which binds the cytochrome c molecules is about 10% smaller in diameter than the end in which the quinone is located. No radial information about the cytochromes is implied by Fig. 9. They are included only for electron transfer distance comparisons. Since the photosynthetic membrane in which RCs are located is only 40–45 Å thick, it is clear that a 57-Å RC particle can span it. Therefore the RC could act as an electron pump to drive electrons across the membrane. The resulting electric potential would be positive with respect to the chromatophore membrane interior where the cytochromes are located.

The presence of magnetic interactions between paramagnetic species can give valuable information regarding intermolecular distances. The absence of such interactions is less useful, but it allows minimum distances of separation to be calculated. The lack of interaction between $(BChl)_2^T$ and $Q^{\cdot -}(Fe^{2+})$ suggests that $(BChl)_2$ and $Q(Fe^{2+})$ are at least 23 Å apart (Tiede et al., 1978b; Dutton et al., 1979). These results, along with the previously described distance estimates, are displayed in Fig. 9. Note that no location is specified for Fe^{2+}. Recent X-ray scattering measurements on oriented RC preparations suggest that it is located near the middle of the RC profile (Stamatoff et al., 1978). Yet, it is clearly in close association with the first and second quinone acceptors, which must be near the outer surface of the RC, because proton-binding rates from solution are found to occur on the same time scale as electron transfer from the first to second quinone acceptor. These observations seem to conflict.

The LD work of Vermeglio and Clayton (1976) showed that the (BChl)$_2$ was nearly perpendicular with respect to the plane of the chromatophore membrane. Therefore, it is reasonable to have it span a distance of ~12 Å. Also, 3 Å are a reasonable estimate of the thicknesses of BChl and BPheo. Tiede et al. (1978a,b) used magnetic interactions between the paramagnetic forms of electron transfer pigments in RCs from *C. vinosum* to estimate the distance between cytochrome *c*-553 and (BChl)$_2$. The cyctrochrome-to-(BChl)$_2$ distance estimates shown in Fig. 9 were taken from this work. The small interaction between these RC components suggested that the unpaired spins were at least 25 Å apart. Their estimate of the minimum edge-to-edge separation for the cytochrome heme and (BChl)$_2$ was 15 ± 3 Å.

4. RC Thermodynamics

So far the spectroscopic and electrochemical data on RCs have been examined from kinetic and geometric perspectives. Since a RC is an electron pump or solar-to-chemical energy transducer, it is appropriate to examine the data from a thermodynamic perspective. Figure 10a shows that the

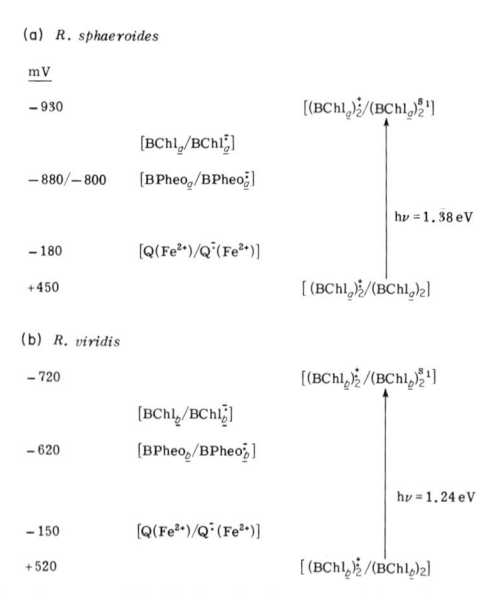

FIG. 10. RC reduction potentials (a) A reduction potential ranking of the electron transfer intermediates in *R. sphaeroides* RCs. For example, (BChl a)$_2^{s_1}$ can reduce BPheo a to BPheo a^- and leave behind (BChl a)$_2^+$. This reaction is exothermic by 50–130 meV. (b) A reduction potential ranking of the electron transfer intermediates in *R. viridis* RCs.

excited singlet state of (BChl a)$_2$ in *R. sphaeroides* has a reduction potential of -930 mV. Studies on delayed fluorescence from P^F imply that only 50–130 meV of this reducing power is lost in forming BPheo a^-. Apparently this transfer is accomplished with such speed and efficiency (energy storage efficiency $>90\%$) because an intermediate electron carrier is used. The next electron transfer to Q(Fe^{2+}), separating the charges >23 Å, takes 150 psec and expends 620–700 mV of energy. The (BChl a)$_2^+$ Q$^-$(Fe^{2+}) redox product stores 630 meV of energy (energy storage efficiency $\sim 46\%$). Of course work was done to separate the cation and anion and is stored in the electric field present within the RC.

Figure 10b shows the same type of data as Fig. 10a for RCs from *R. viridis*. This time the energy drop on going from the (BChl b)$_2$ excited singlet to form BPheo b^- is 100 meV. This was determined by electrochemical titration. The energy storage efficiency is again $>90\%$. In this case also, a large expenditure of energy is made to drive the electron onto the quinone acceptor, ~ 470 meV. Finally, 670 meV of energy (energy storage efficiency $\sim 54\%$) is retained in the (BChl b)$_2^+$ Q$^-$(Fe^{2+}) redox product.

E. PHOTOSYNTHETIC OVERVIEW

The work of the RC protein is complete when the electron has been transferred to the Q(Fe^{2+}) complex. In the absence of other electron acceptors and donors, the reduced quinone is stable for 60 msec. In functioning chromatophores, cytochromes which are bound to the RC in the interior of the chromatophore membrane reduce (BChl)$_2^+$ in ~ 1 μsec. This serves to separate the electron and hole further. On the exterior of the membrane, the electron is transferred to other quinones in the 10–100 μsec time range. Concomitant with this a proton is taken up from the surrounding aqueous medium and bound to a reduced quinone. Thus both pH and electrical gradients are set up across the chromatophore membrane. In a way that is not yet clear, the electrochemical gradient is harnessed to drive the phosphorylation of adenosine diphosphate (ADP) to adenosine triphosphate (ATP). ATP is a stable source of chemical energy for living organisms.

II. *In Vitro* Electron Transfer Reactions

A. REACTION CENTER MODELS: COFACIAL DIPORPHYRINS

It is clear from the previous discussion that bacterial RCs have evolved an elaborate arrangement of BChl and BPheo chromophores to accomplish a rapid and efficient electron transfer to a quinone molecule. The rate of the electron transfer was 6.7×10^9 sec^{-1} and the reduced quinone was stable for 60 msec. Several research groups have investigated the electron transfer properties of excited states of porphyrin-type molecules to quinones and other neutral electron acceptors under bimolecular conditions (Hupert et al., 1976; Holten et al., 1978a; Harriman et al., 1979). The conclusions were that, while both the excited singlet and triplet states were effectively quenched by added quinones, no stable ionic products were found when the reaction proceeded from the singlet state. However, yields of ionic photoproducts as high as 50–60% have been found in high dielectric solvents when the reaction proceeded from the triplet state. Since photosynthetic charge separations proceed from the singlet excited state of the electron donor, a major goal of RC modeling studies was the rapid ($k > 10^{11}$ sec^{-1}) production of a "stable" cation–anion pair from the model's singlet excited state.

Because porphyrin-type electron donors were found in natural photosystems, this kind of electron donor was used in constructing the following RC models. Also, while bacterial RCs used a BChl dimer as the electron donor, PS II of green plants used a chlorophyll monomer (Davis et al., 1979). In view of this, it seemed reasonable to explore the electron transfer properties of a magnesium-containing porphyrin (MgP) and a free base porphyrin (H$_2$P) pair. MgP was the analog of the naturally occurring electron donors, and H$_2$P was the analog of the naturally occurring electron acceptors. Since the electron donor and acceptor pigments were bound in the RC into a fixed geometry, MgP was chemically bound to H$_2$P (Chang, 1977, 1979). Figure 11 shows that by using two covalent chains (R) of five-

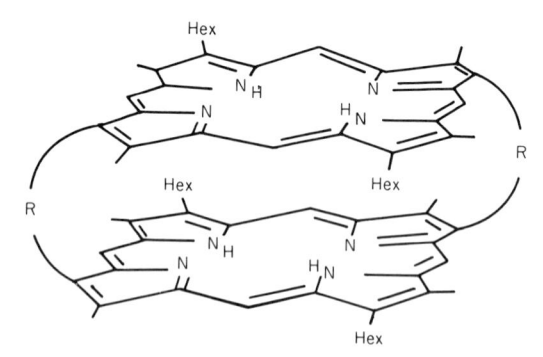

FIG. 11. Structural drawing of a cofacial diporphyrin comprised of two free base units, H$_2$–H$_2$. A dimer with five-atom linking chains was labeled H$_2$–H$_2$(5). In this case the chains were

$$R = -CH_2-\underset{\underset{O}{\|}}{C}-N(n\text{-Bu})-CH_2-CH_2-.$$

[Reprinted with permission from Netzel et al., Chem. Phys. Lett. 67, 224, Structure (I) (1979). Copyright 1979 by North-Holland Publishing Co.]

atoms length each to attach the two porphyrin rings at alternating pyrroles, a cofacial dimer was produced. The interplanar distance was estimated to be ~4 Å on the basis of observed triplet–triplet interactions in a Cu–Cu(5) dimer. Dimers with two copper or magnesium atoms and five-atom bridging chains will be abbreviated as Cu–Cu(5) and Mg–Mg(5), respectively. The monomagnesium and metal-free dimers will be abbreviated as Mg–H$_2$(5) and H$_2$–H$_2$(5), respectively.

The absorption spectrum of Mg–H$_2$(5) is shown in Fig. 12. For comparison, the sum spectrum of the monomers MgOEP and H$_2$OEP is also shown. The cofacial dimer's spectrum is broadened and red-shifted relative to the sum spectrum. This is due to an excitonic interaction between the two porphyrin rings in the dimer. Similar excitonic interactions were described earlier for the case of RC chromophores.

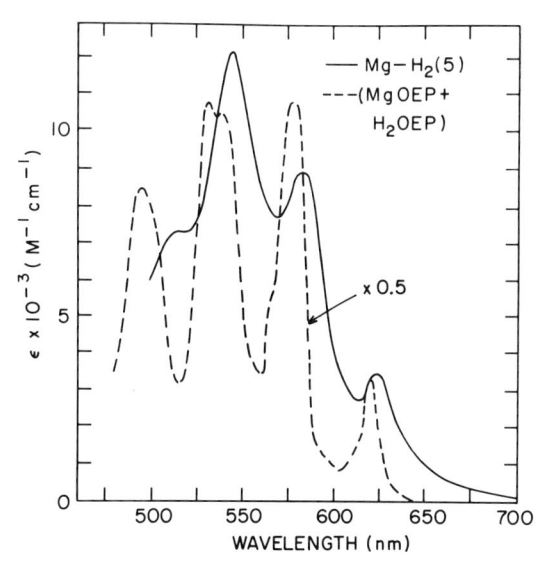

FIG. 12. Absorption spectra for Mg–H$_2$(5), shown by the solid line, and the sum of the absorptions for MgOEP and H$_2$OEP monomers, shown by the dashed line.

The general objective of this modeling work (Netzel et al., 1979; Fujita et al., 1981) was to correlate electrochemical, spectroscopic, and geometric parameters to help predict the likelihood of electron transfer reactions. A zero-order estimate for the free energy change required to produce a Mg^{+}–H$_2^{-}$ electron transfer product is given by the sum of the free energy changes (negative of the midpoint potentials), $\Delta E_{1/2}$, for the one-electron oxidation and reduction of Mg–H$_2$. The free energy change for the excited singlet of the diporphyrin returning to the ground state, E^{S_1}, can be taken as the negative of the 0–0 energy of the excited singlet, assuming the entropy change is negligible. The light-driven electron transfer in reaction (4)

$$(Mg-H_2)^{S_1} \rightarrow Mg^{+\cdot} - H_2^{-\cdot} \tag{4}$$

is favorable if

$$|E^{S_1}| > \Delta E_{1/2}. \tag{5}$$

While $\Delta E_{1/2}$ as an estimate of the free energy change for forming $Mg^{+\cdot}-H_2^{-\cdot}$ neglects competing coulombic and solvation effects, this error is probably constant for a series of diporphyrins with the same linking-chain length. Therefore while the sum of E^{S_1} and $\Delta E_{1/2}$ cannot give a rigorous prediction of whether or not reaction (4) is favorable, the sum can be used to rank a series of diporphyrins in terms of their likelihood of undergoing an electron transfer reaction. Table I lists the excited singlet state (S_1) energies and sums of the free energy changes ($\Delta E_{1/2}$) for oxidation and reduction in the same solvent for the pair of monomers, MgOEP–H$_2$OEP, and three diporphyrins.

Three trends are apparent in the table. First, there is a reduction of 200 meV in $\Delta E_{1/2}$ on going from the monomeric pair, MgOEP–H$_2$OEP, to the cofacial dimer, Mg–H$_2$(5). Second, $\Delta E_{1/2}$ depends both on the solvent and on coordinating ions such as Cl$^-$. Third, these data yield the following

TABLE I

MOLECULAR PROPERTIES

Molecular couple	Solvent[a]	S_1 (eV)[b]	$\Delta E_{1/2}$ (eV)[c]
MgOEP–H$_2$OEP	CH$_2$Cl$_2$	2.00	2.01[d]
	CH$_2$Cl$_2$ + [N(C$_4$H$_9$)$_4$]Cl	2.00	1.83[d]
Mg–H$_2$(5)	THF	1.95	2.03[e]
	CH$_2$Cl$_2$	1.95	1.80[e]
	CH$_2$Cl$_2$ + [N(C$_2$H$_5$)$_4$]Cl	1.95	1.63[e]
Mg–Mg(5)	THF	2.00	2.14[e]
	CH$_2$Cl$_2$ + [N(C$_2$H$_5$)$_4$]Cl	2.05	>2.1[e]
H$_2$–H$_2$(5)	THF	1.95	2.13[e]

[a] In all cases 0.1 M tetraalkylammonium salt served as the electrolyte. If not specified, the perchlorate salt was used. THF, Tetrahydrofuran.

[b] Energy of the long-wavelength edge of the first absorption band.

[c] Sum of the free energy changes (negative of the midpoint potentials) for the one electron oxidation and reduction of the indicated molecule or pair of molecules.

[d] I. Fujita, J. Fajer, C.-K. Chang, C.-B. Wang, and T. L. Netzel, unpublished data.

[e] Netzel et al. (1979).

ranking of the likelihood of particular dimers undergoing electron transfer reactions, in decreasing order:

Mg–H$_2$(5)(CH$_2$Cl$_2$,Cl$^-$) > Mg–H$_2$(5)(CH$_2$Cl$_2$) >

Mg–H$_2$(5)(THF) > Mg–Mg(5)(THF) ≈ H$_2$–H$_2$(5)(THF).

Figure 13 shows the change-in-absorbance (ΔA) spectra for three five-chain dimers in THF. The dimers were excited at 527 nm with a 6-psec pulse and the resulting change-in-absorbance spectra immediately after excitation ($t = 0$ psec) and 5 nsec later are shown. The ΔA spectra for the symmetric dimers, Mg–Mg(5) and H$_2$–H$_2$(5), showed very little decay during the first 5 nsec. The small decay of the absorbance increase in the 600–700 nm region and the small growth in absorbance increase beyond 750 nm indicated a slow singlet-to-triplet (S$_1$ → T$_1$) intersystem crossing with a lifetime

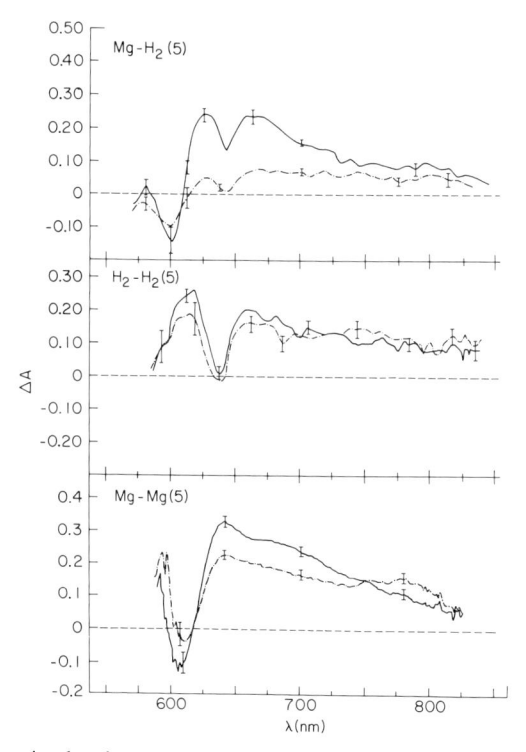

FIG. 13. Change-in-absorbance spectra for three cofacial diporphyrins in tetrahydrofuran (THF). They were excited at 527 nm with a 6-psec laser pulse. The ΔA are presented for two times: immediately after excitation ($t = 0$) and 5 nsec after excitation. ——, $t = 0 \pm 4$ psec; –·–, $t = 5$ nsec. [Reprinted with permission from Netzel et al., Chem. Phys. Lett. **67**, 225, Fig. 1 (1979). Copyright 1979 by North-Holland Publishing Co.]

>5 nsec. While the spectra of Mg–H$_2$(5) changed more rapidly in the first 5 nsec, they had the same overall intensity and shape as those of the symmetric dimers. The biggest differences were the locations of band dips, which were due to differences in the positions of the ground state absorption bands of the dimers. As a result there was no basis for assigning these states to anything other than singlets and triplets.

Figure 14 shows the absorption spectra of Mg–H$_2$(5), MgOEP$^+$, and H$_2$OEP$^-$. While it was unlikely that a Mg$^+$–H$_2^-$(5) electron transfer product would look simply like the sum of these two ionic spectra, it was reasonable to expect it to absorb strongly in the 660-nm region. The spectral response of the SIT-vidicon prevented measurements above 800 nm. The ΔA spectra found after exciting Mg–H$_2$(5) in CH$_2$Cl$_2$ are shown in Fig. 15. Indeed, an intense absorption was found in the 660-nm region immediately after excitation. It decayed with a lifetime of 200 psec largely to the ground state. However, a residual absorption present in the 1–5 nsec time interval showed that some of the electron transfer product, Mg$^+$–H$_2^-$(5), also decayed to a longer-lived state. When chloride ion was added to the CH$_2$Cl$_2$ solvent, the same electron transfer state was populated in <6 psec. However, it decayed with a lifetime of 600 psec completely to the ground state. Table I shows that the electron transfer product in this case was expected to be 170 meV lower in energy than it was in neat CH$_2$Cl$_2$.

To explore solvent effects on the electron transfer product's formation and decay, N,N-dimethylformamide (DMF) was also used. In this case, the electron transfer product was still formed in <6 psec, but its lifetime was 1.3 nsec. Thus, although the lifetime of the electron transfer product

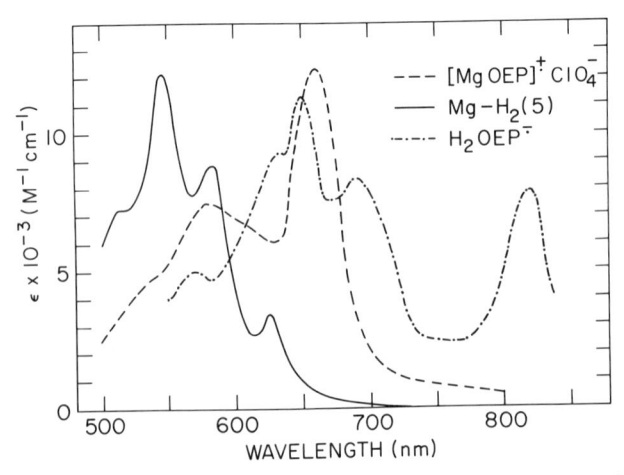

FIG. 14. Absorbance spectra for (1) Mg–H$_2$(5), shown by the solid line, (2) MgOEP$^+$, shown by the dashed line, and (3) H$_2$OEP$^-$, shown by the dash–dot line.

4. IN VIVO AND IN VITRO ELECTRON TRANSFER REACTIONS 103

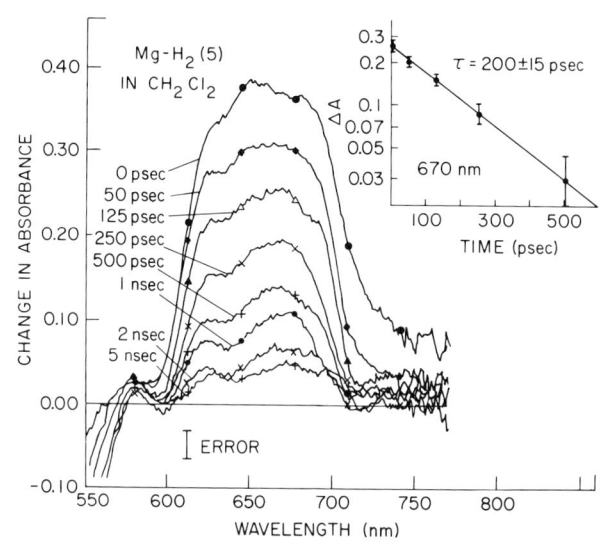

FIG. 15. Change-in-absorbance spectra for Mg–H$_2$(5) in CH$_2$Cl$_2$ at the indicated times after excitation at 527 nm with a 6-psec laser pulse. The decay lifetime of the optical transient was 200 ± 15 psec.

varied with the solvent, its rate of formation remained faster than the glass laser's pulses could resolve.

To the extent that the electronic coupling matrix element in Mg–H$_2$(5) for excited state electron transfer is similar to that of a pair of naphthalene or anthracene molecules, calculations by Jornter and Rice (1975) suggest that it should be >100 cm^{-1}. At present there is no reason to suppose that the electronic coupling is significantly less for the reverse electron transfer to re-form the ground state of Mg–H$_2$(5). Therefore, the slowness of the reverse electron transfer in contrast to the <6 psec forward electron transfer indicates a large kinetic barrier. The source of this barrier is probably weak electron–vibration coupling resulting from a small displacement between the minima of the potential energy surfaces for Mg^{+}–H$_2^{-}$(5) and Mg–H$_2$(5). This near nesting of the potential energy surfaces of the reactant and product means that the reverse electron transfer will be nonadiabatic and that its rate will be governed by Frank–Condon factors which control the frequency nuclear tunneling from one surface to the other.

Of course a large kinetic barrier can exist for the case of very strong electron–vibration coupling. In this case the minima of the Mg^{+}–H$_2^{-}$(5) and Mg–H$_2$(5) surfaces would be shifted apart such that the point of intersection of the two surfaces was between the two minima. However, strong electron–vibration coupling implies large changes in bond lengths, force constants, or solvent reorganization energy. Existing data on monomeric porphyrins

(Felton and Yu, 1978; Gouterman, 1978; Spaulding et al., 1974) indicate that bond lengths and force constants do not change very much on going from a neutral to an ionic form. Also, the large size of a diporphyrin reduces the magnitude of the solvent reorganization energy (Netzel et al., 1981). While additional work is necessary to establish an accurate quantum mechanical description of the electron transfer processes in these diporphyrins, a good working hypothesis is that the electronic couplings are large enough to permit rapid electron transfer reactions in both the forward and reverse directions and that the observed difference in forward and reverse rates is a consequence of the sizes of the energy gaps and the strengths of the electron–vibration couplings in the two directions. Discussions of the interplay of electronic and electron–vibration couplings can be found in the following papers: Jortner (1976, 1980); Hopfield (1974); Warshel (1980).

The geometry of the Mg–H$_2$(5) diporphyrin was altered by attaching the linking chains to adjacent pyrroles on one porphyrin and to alternating pyrroles on the other. This shifted one porphyrin over the top of the other and allowed the porphyrin subunits freedom to assume noncofacial orientations. Three distinct decay regimes for the photoproduct were observed in CH$_2$Cl$_2$: almost no decay between 10 and 300 psec, and fast and slow decays, respectively, between 0.3 and 1 nsec and 1 and 5 nsec. The magnitude of the ΔA increase at 660 nm for this photoproduct suggested that electron transfer product states were present at early times ($t \leq 6$ psec). Whether or not any conformations had longer-lived ($t > 6$ psec) S$_1$ states was not clear. However, the triphasic relaxation did demonstrate that relative orientation measurably affected the diporphyrin's ET reactions. Geometrical control of electron transfer rates may also be operative in natural photosystems where the quantum efficiency of primary electron transfer reactions is $\sim 100\%$ (Fujita et al., 1981).

The work on the cofacial diporphyrins showed that the Mg–H$_2$(5) RC model could reproduce the $>10^{11}$ sec^{-1} electron transfer reaction from a singlet excited state found in natural photosystems. The difference in free energy between the electron transfer product and the reactive S$_1$ state, 150 meV, was comparable to the 50–130-meV drop found in R. sphaeroides RCs for the formation of (BChl)$_2^+$ BPheo$^-$. Its energy storage efficiency, $\sim 92\%$, was similar to that found in bacterial RCs and its short distance of electron transfer was similar to the ≤ 3.2 Å distance estimated for the (BChl)$_2$ to BChl electron transfer in RCs.

Recent work (Netzel et al., 1982) has shown that Mg$^+_\cdot$–H$^-_{2\cdot}$(5) can reduce added p-benzoquinone (BQ) molecules. This reaction in both CH$_2$Cl$_2$ and DMF produced separated Mg$^+_\cdot$–H$_2$ and BQ$^-_\cdot$ products which recombined under diffusional control. Similar results were not obtained when BQ was reacted with photoexcited meso-tetraphenylporphin (TPP) or Mg–H$_2$(5) in THF. These studies established the similarity of the (π, π^*) excited states of TPP and Mg–H$_2$(5) in THF and confined the prior assignment of Mg^{+}–H$_2^{-}$(5) formation upon photoexcitation of Mg–H$_2$(5) in CH$_2$Cl$_2$ and DMF.

B. Reaction Center Models: Dimers and Trimers of PChl *a* and PPheo *a*

1. Single-Chain Dimers

Although there is good evidence that BPheo is not the first electron acceptor in RCs, one might ask whether there must be another acceptor between (BChl)$_2$ and BPheo to make rapid electron transfer possible? The best estimates placed (BChl)$_2$ and BPheo >9 and 10–15 Å apart. Therefore a dimeric model of pyrochlorophyll *a* (PChl *a*) and pyropheophorbide *a* (PPheo *a*) bound together with a single 10-atom chain (Bucks *et al.*, 1981; Netzel *et al.*, 1980) was studied. In most experiments pyridine was added to prevent aggregation. However, in some experiments alcohol was added because it caused intramolecular bonding of the model's subunits through ROH bridges (Boxer and Closs, 1976). The oxygen of the alcohol coordinated to the magnesium in PChl *a*, while its proton hydrogen-bonded to the keto group that was present on both PChl *a* and PPheo *a*. PPheo *a* did not contain a magnesium atom. In this way, the relative orientation of the donor and acceptor subunits was altered. As in the work on cofacial dimers, the surrounding medium was changed by using the following solvents: toluene, CH$_2$Cl$_2$, and CH$_3$CN. Since pheophorbide *a* (Pheo *a*) was 90 meV easier to reduce than PPheo *a*, both were used as electron acceptors. Also, electrochemical studies showed that the addition of Cl$^-$ decreased the $\Delta E_{1/2}$ for these models as it had for the cofacial diporphyrins.

Figure 16a shows the absorption spectra of PChl *a*⤳PPheo *a* "half-folded" with ethanol. A computed difference spectrum for the transformation of PChl *a*⤳PPheo *a* into Chl *a*$^{+}$ and Pheo *a*$^{-}$ is shown in panel 10b. Note that the absorptions of the cation and anion were quite broad and featureless. This contrasted strikingly with the case for porphyrin cations and anions. Therefore spectral identification of electron transfer products in these models was difficult. Figure 16c shows the ΔA spectrum found for this sample +3 psec after excitation. There was reasonable agreement between this experimental result and the calculated spectrum. However, this was fortuitous, because the singlets and triplets of chlorophylls and pheophytins also had broad, featureless, visible absorptions.

Figure 17 shows detailed kinetic measurements on this same dimer in an open configuration (addition of pyridine) in two solvents. The ground state was repopulated with lifetimes of 1.8 ± 0.2 nsec and 0.85 ± 0.10 nsec in

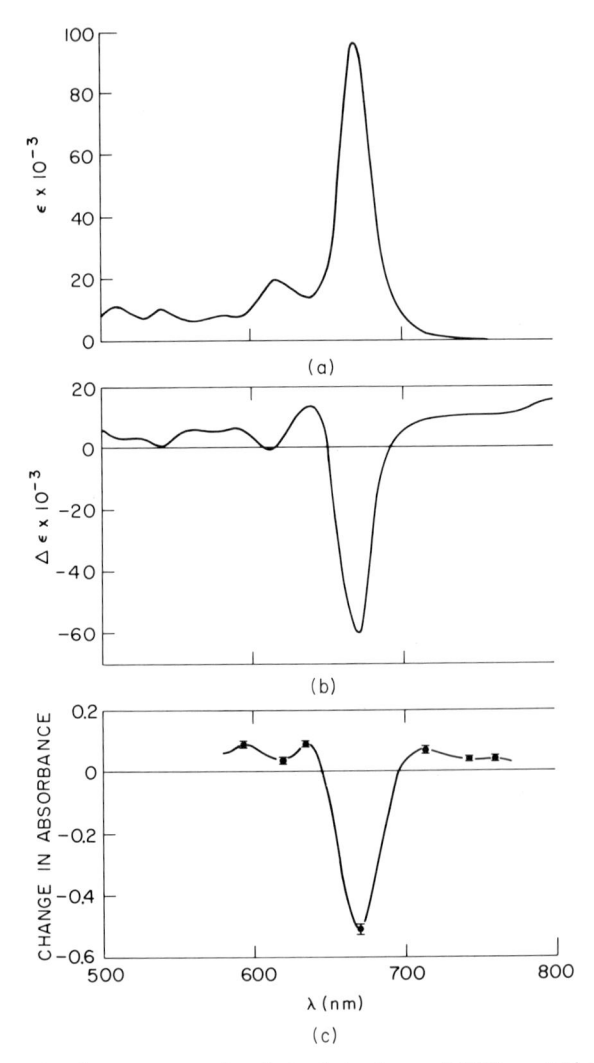

FIG. 16. (a) Absorbance spectra of a single-chain dimer of PChl a∿PPheo a in toluene with 0.2 M ethanol. The ethanol formed an intramolecular R—OH bridge to produce a "half-folded" conformation. (See text for details.) (b) Calculated change-in-absorbance spectra for 100% production of Chl $a^{\ddot{+}}$ and Pheo $a^{\bar{\cdot}}$ from PChl a∿PPheo a. (c) Experimentally observed change-in-absorbance spectrum 3 psec after excitation at 527 nm with a 6-psec laser pulse.

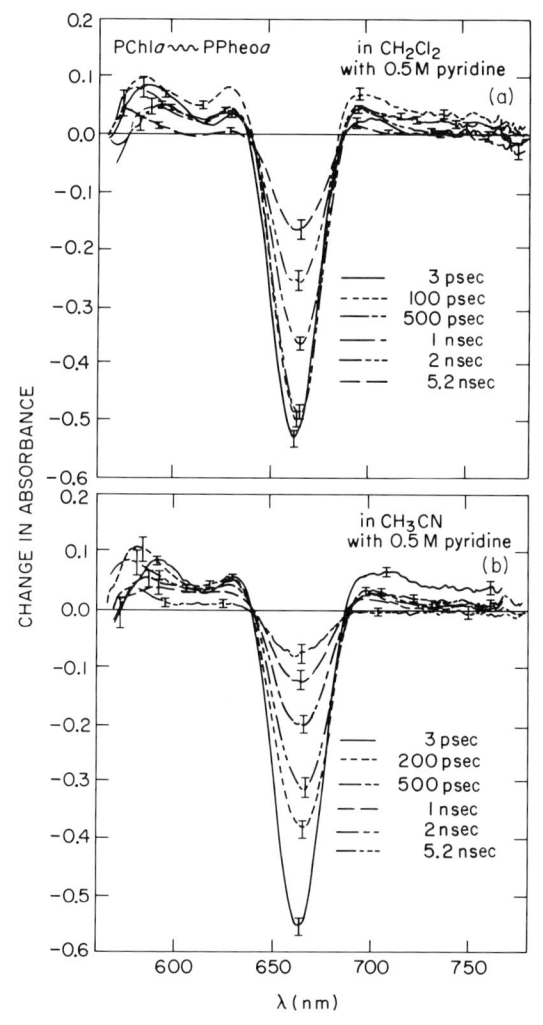

FIG. 17. (a) Change in absorbance spectra for a single-chain dimer of PChl a ~~ PPheo a in CH$_2$Cl$_2$ with 0.5 M pyridine. The pyridine prevented aggregation and produced an open conformation. The ground state absorbance was restored with a lifetime of 1.8 ± 0.2 nsec. (b) Change in absorbance spectra for the same molecule as in (A), except that the solvent was acetonitrile (CH$_3$CN). The ground state absorbance was restored with a lifetime of 0.85 ± 0.10 nsec.

CH$_2$Cl$_2$ and CH$_3$CN, respectively. All the single chain dimer results were assigned to S$_1$ → T$_1$ processes, because the lowest fluorescence quantum yields observed implied only a factor of 5 reduction in singlet lifetime relative to those of monomeric PChl *a* and PPheo *a*. If the optical transients present <6 psec after excitation were due to electron transfer products, the dimers' fluorescence quantum yields would have to have been ~1000 times smaller than those of the PChl *a* and PPheo *a* monomers. Table II presents data on relative fluorescence quantum yields and observed absorption transient lifetimes for the single-chain dimer models. For comparison it is helpful to note that both PChl *a* and PPheo *a* gave essentially the same fluorescence yield in these three solvents, ~100 arbitrary units. Also, they exhibited no trend in yield of fluorescence with respect to change in solvent or exchange of ethanol for pyridine.

TABLE II

RELATIVE FLUORESCENCE QUANTUM YIELDS[a] AND LIFETIMES[b]

Model compound[c]	Toluene[d]	CH$_2$Cl$_2$[d]	CH$_3$CN[d]
PChl *a*⤳PChl *a* + Py	98 (6.2 ± 0.8)	78	27
PChl *a*⤳PPheo *a* + Py	100 (7.5 ± 1)	33 (1.8 ± 0.2)	18 (0.85 ± 0.10)
PChl *a*⤳PPheo *a* + EtOH	47 (3.8 ± 0.8)	36	18
PChl *a*⤳Pheo *a* + Py	80	38	23
PChl *a*⤳Pheo *a* + EtOH	56	40	27
Cl$^-$: PChl *a*⤳PPheo *a*		38	20
Cl$^-$: PChl *a*⤳Pheo *a*		40	24

[a] Arbitrary units.
[b] Bucks *et al.* (1981), Netzel *et al.* (1980).
[c] Py, 0.5 *M* pyridine, EtOH, 0.5 *M* ethanol; Cl$^-$, 0.1 *M* [N(C$_2$H$_5$)$_4$]Cl.
[d] Lifetimes (in nsec) obtained from the decay of an absorption transient are shown in parentheses.

While other interesting trends were present in the data of Table II, the key points were (1) the fluorescence quenching relative to the fluorescence yields of PChl *a* and PPheo *a* monomers was small and (2) the lifetimes for ground state repopulation from picosecond absorption measurements correlated very well with the relative fluorescence yields. It was clear that only excited singlet states and not electron transfer product states were responsible for the absorption transients. In short, these models were spectacular failures. This conclusion is not surprising for the first entry in the table, PChl *a*⤳PChl *a* because in this case the formation of an electron transfer product from the dimer's S$_1$ state is endothermic. However, electron transfer reactions become increasingly exothermic as one goes down the table. The exothermicity is

~100 meV for entries 2 and 3, ~190 meV for entries 4 and 5, ~320 meV for entry 6, and ~430 meV for entry 7. Apparently, a kinetic bottleneck prevents an $S_1 \to$ ET (electron transfer) reaction during the S_1 state's lifetime.

There was a trend of decreased S_1 lifetime with increased solvent dielectric constant. While this may have been due to increased interaction of an electron transfer product state with the dimer's S_1 state, this seemed unlikely because no similar decrease in S_1 lifetime was found as the free energy of formation $\Delta E_{1/2}$ of the electron transfer product state was decreased. Note that half-folding with ethanol in toluene caused a comparable reduction in the dimers' lifetimes and quantum yields. Thus, although the mechanism causing the increased nonradiative decay of the dimers' S_1 states is unclear, it is reasonable to propose that it involves increased association of the dimers' subunits in solvents of higher static dielectric constant.

The results for cofacial diporphyrins and singly linked PChl *a*-containing dimers bound the electron transfer problem for RC models. The former work and the latter do not. This implies that either the rate of electron transfer in RCs from (BChl)$_2$ to BPheo cannot be $>10^{11}$ sec^{-1} without an intermediate electron acceptor for separations >9 Å, or that a particular orientation between the electron donor and acceptor is crucial. This latter possibility seems unlikely at large donor-to-acceptor separations.

2. *A Trimeric Reaction Center Model*

To define further the distance and orientational constrains on electron transfers between porphyrin-type donors and acceptors, the photophysics of an elegant RC model constructed by Boxer and Bucks (1979) was studied. The model is shown in Fig. 18. Note that two significant changes were present in the trimer relative to the single-chain dimers: (1) a dimeric rather than a monomeric electron donor was used, and (2) the electron acceptor was joined to the donor with a five-atom chain. In addition, the $\Delta E_{1/2}$ of the hoped for electron transfer products was changed by using both PPheo *a* and Pheo *a* as electron acceptors.

Unfortunately, the fluorescence of this folded trimer (addition of ethanol) in toluene was only quenched a factor of 2 relative to the fluorescence of a PChl *a* or PPheo *a* monomer. Thus an electron transfer did not occur in <6 psec. In addition, Fig. 19 shows that 3 psec after excitation, the apparent bleaching ($-\Delta A$) at 710 nm was larger than the sample's initial absorbance. At 200 psec the discrepancy was even greater. This was evidence of the trimer amplifying the probe light through stimulated emission. Figure 20 presents a detailed kinetic history of this phenomenon. The repopulation of the PPheo *a* ground state at 640 nm proceeded with a 60-psec lifetime. This was due to an energy transfer process in which the lower-energy S_1 state of (PChl *a*)$_2$ was populated. The maximum of the gain occurred, therefore, in

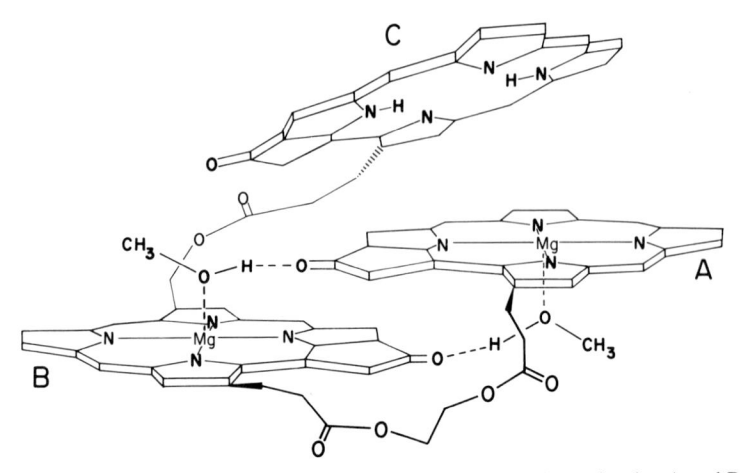

FIG. 18. Structural drawing of a trimeric RC model. Two PChl *a* molecules, A and B, were joined by a 10-atom chain and by double R—OH bridges to form a folded configuration. In addition, a five-atom chain covalently bound PPheo *a*, C, to the folded dimer. (From Boxer and Bucks, 1979. Reprinted with permission from *Journal of the American Chemical Society* **101** (No. 7), 1885. Copyright 1979 American Chemical Society.)

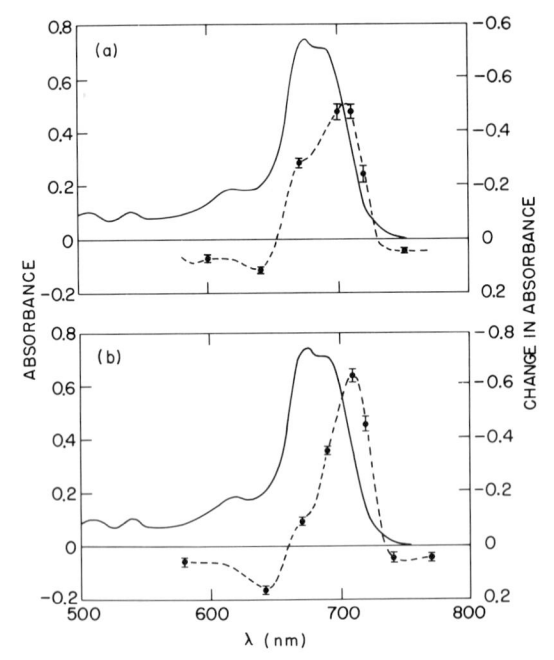

FIG. 19. (a) A comparison of the absorbance spectrum and change-in-absorbance spectrum 3 psec after excitation at 527 nm for the folded trimeric RC model shown in Fig. 18. The solvent was toluene with 0.2 M ethanol. (b) A similar comparison as at the top, except that the change-in-absorbance spectrum was measured 200 psec after excitation with a 6-psec pulse.

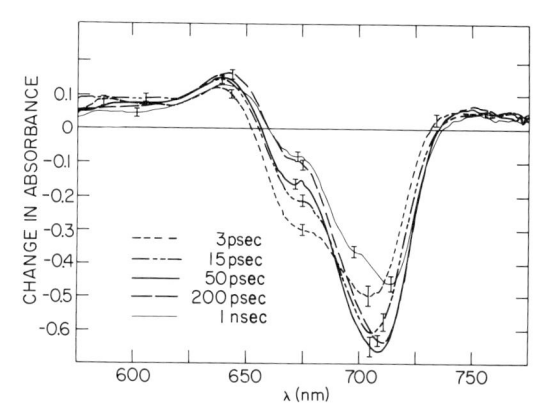

FIG. 20. Change-in-absorbance spectra for the same sample as in Fig. 19 at the indicated times after excitation.

the 50–100 psec time interval. The recovery of the trimer's ground state after 1 nsec proceeded with a 3.5 nsec lifetime. This was about one-half the lifetime of the S_1 state of PChl *a*. This evidence for stimulated emission also demonstrated that the trimeric model in toluene did not undergo an $S_1 \to$ ET reaction.

Prior work by Pellin *et al.* (1980) on (PChl *a*)$_2$ alone in CH_2Cl_2 (with no acceptor present) showed that the S_1 lifetime was 110 psec. Freed (1980) provided a quantum mechanical description of the unusual properties of the excited singlet state of this folded dimer in terms of symmetric and antisymmetric "exciton-like" states. The ~60-fold decrease in S_1 lifetime in going from toluene to CH_2Cl_2 was striking enough to warrant asking whether or not the folded trimer would undergo a rapid $S_1 \to$ ET reaction in this solvent.

Absorbance work on (PChl *a*)$_2$ in CH_2Cl_2 (Bucks *et al.*, 1981; Netzel *et al.*, 1980) reproduced the 110-psec transient ground state repopulation that streak camera work (Pellin *et al.*, 1980) had found earlier. This state was survived by a long-lived state that did not decay in the 1- to 5-nsec time interval. Periasamy *et al.* (1978) showed that at times > 30 nsec after photoexcitation folded (PChl *a*)$_2$ yielded an unfolded configuration with one subunit a triplet and the other a ground state singlet. The picosecond absorbance data were consistent with a 110-psec S_1 state decaying to a long-lived T_1 state.

When the PPheo *a* electron acceptor was attached to the (PChl *a*)$_2$ donor and dissolved in CH_2Cl_2, the data shown in Fig. 21 were obtained. The 65-psec transient at 660 nm was assigned to S_1 state decay because the fluorescence quantum yield of this trimer was ~60% of that found for (PChl *a*)$_2$ alone, respectively, 8 and 13 arbitrary units. This transient was

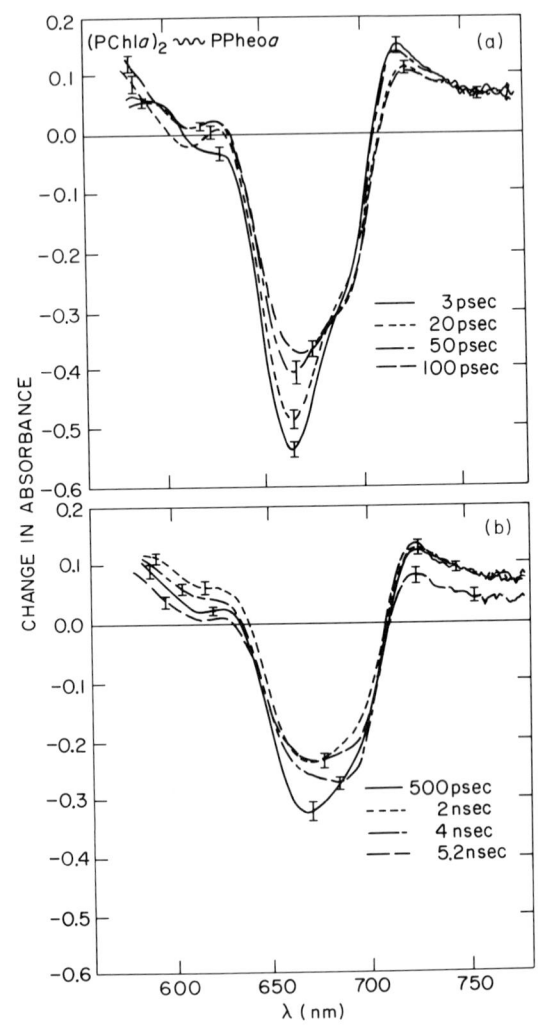

FIG. 21. Change-in-absorbance spectra for a folded trimeric RC model with PPheo *a* as the potential electron acceptor (structure shown in Fig. 18). The solvent was CH_2Cl_2 with 0.5 M ethanol. The time after excitation for each spectrum is indicated.

followed by a long-lived state whose ΔA varied but did not decay, in the 1–5 nsec time interval. Since this was essentially the same result as was obtained for (PChl *a*)$_2$ without an acceptor, there was no basis for assigning any process other than $S_1 \rightarrow T_1$ decay to this trimer.

Substitution of Pheo *a* for PPheo *a* as the electron acceptor increased the exothermicity of an electron transfer reaction from the S_1 state of (PChl *a*)$_2$

4. IN VIVO AND IN VITRO ELECTRON TRANSFER REACTIONS 113

from ~40 to ~130 meV. If one took into account that a doubly linked (PChl a)$_2$ molecule was easier to oxidize (Wasielewski et al., 1978) than a PChl a monomer by 70 meV, the exothermicity of the reduction of Pheo a may have been as large as 200 meV. The resulting kinetics following 6 psec excitation at 527 nm are shown in Fig. 22. The 110-psec transient at 660 nm was assigned to S$_1$ state decay because the fluorescence quantum yield was

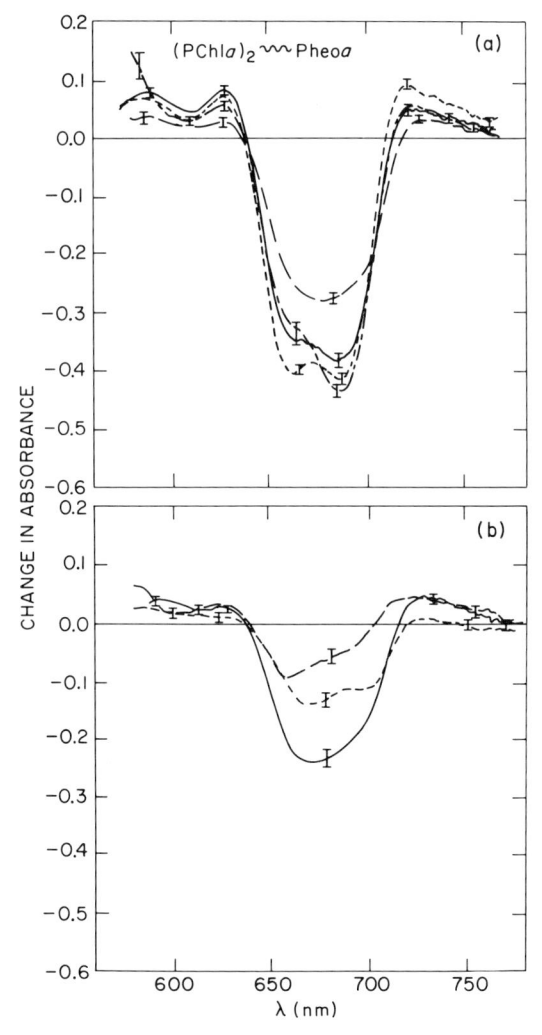

FIG. 22. Change-in-absorbance spectra for a folded trimeric RC model with Pheo a as the potential electron acceptor. The solvent was CH$_2$Cl$_2$ with 0.5 M ethanol. The time after excitation for each spectrum is indicated. (a) ———, 3 psec; ----, 20 psec; —·—, 50 psec; ——, 200 psec. (b) ———, 1 nsec; ---- 3 nsec; —·— 5.2 nsec.

the same as that of $(PChl\ a)_2$ alone. However, the state following this transient decayed with a lifetime of 3.0 ± 0.7 nsec. This may have been a triplet state of the trimeric RC model, but this seemed unlikely. Another possibility was that the S_1 state decayed into an electron transfer product state. The back electron transfer rates in cofacial diporphyrins were found to be 0.2–1.3 nsec. However, nearly 50% of the initial ground state bleaching had decayed by 1 nsec. This suggested that the yield of the electron transfer product was not high. In short, if this trimeric RC model worked at all, it did not work very well. Instead of reacting at a rate $>10^{11}$ sec^{-1}, the $S_1 \to$ ET rate was less than 9×10^9 sec^{-1} and the yield was much less than 100%.

To summarize the modeling work, it is fair to say that the Mg–H$_2$(5) cofacial diporphyrin model compares favorably with natural photosystems. The electron transfer distances are similar, both have rates of electron transfer that are $>10^{11}$ sec^{-1}, and both store >90% of the incident photon's energy in electron transfer products. Attempts to produce rapid ($>10^{11}$ sec^{-1}) electron transfer in singly linked dimers and trimers of PChl a and Pheo a were unsuccessful. This was true in spite of (1) using both 5-atom and 10-atom linking chains, (2) using a dimeric as well as a monomeric electron donor, (3) decreasing the average donor-to-acceptor distance by adding ethanol, (4) varying the solvent, and (5) increasing the exothermicity of the potential $S_1 \to$ ET reaction to over 400 meV.

These results lend strong support to RC electron transfer schemes that employ an electron transfer intermediate between $(BChl)_2$ and BPheo, because they show that ET processes between porphyrin-type molecules are unlikely to proceed at rates $>10^{11}$ sec^{-1} at distances >7 Å. Finally, the modeling studies suggest that a close to cofacial arrangement may be necessary for rapid electron transfer even at donor to acceptor separations of <7 Å.

Acknowledgments

This article was prepared at Brookhaven National Laboratory under contract with the Office of Basic Energy Sciences of the U.S. Department of Energy, Washington, D.C. (Contract No. DE-AC02-76CH00016).

I would like to thank Drs. P. L. Dutton, R. Cogdell, and W. W. Parson for the helpful discussions I had with them, Drs. J. Fajer and I. Fujita for supplying the spectra of MgOEP$^{+\cdot}$ and H$_2$OEP$^{-\cdot}$, and Dr. M. A. Bergkamp for measuring the kinetics of Mg$^{+\cdot}$–H$_2^{-\cdot}$(5) decay in CH$_2$Cl$_2$.

References

Akhmanov, S. A., Borisov, A. Y., Danielius, R. V., Gadonas, R. A., Kozlowski, V. S., Piskarskas, A. S., Razjivin, A. P., and Shuvalov, V. A. (1980). *FEBS Lett.* **114**, 149.

Anderson, P. W. (1959). *Phys. Rev.* **115**, 2.

Blankenship, R. E., Schaafsma, T. J., and Parson, W. W. (1977). *Biochim. Biophys. Acta* **461**, 297.
Boxer, S. G., and Bucks, R. R. (1979). *J. Am. Chem. Soc.* **101**, 1883.
Boxer, S. G., and Closs, G. L. (1976). *J. Am. Chem. Soc.* **98**, 5406.
Bucks, R. R., Netzel, T. L., Fujita, I., and Boxer, S. G. (1981). In preparation.
Campillo, A. J., Hyer, R. C., Monger, T. G., Parson, W. W., and Shapiro, S. L. (1977). *Proc. Natl. Acad. Sci. U.S.A.* **74**, 1997.
Chang, C. K. (1977). *J. Hetercycl. Chem.* **14**, 1285.
Chang, C. K. (1979). *Adv. Chem. Ser.* No. 173, p. 162.
Clayton, R. K. (1967). *J. Theor. Biol.* **14**, 173.
Clayton, R. K., and Clayton, B. J. (1972). *Biochim. Biophys. Acta* **283**, 492.
Creutz. C., Chou, M., Netzel, T. L., Okumura, M., and Sutin, N. (1980). *J. Am. Chem. Soc.* **102**, 1309.
Davis, M. S., Forman, A., and Fajer, J. (1979). *Proc. Natl. Acad. Sci. U.S.A.* **76**, 4170.
DeVault, D. C., and Chance, B. (1966). *Biophys. J.* **6**, 825.
Dutton, P. L., Leigh, J. S., and Seibert, M. (1972). *Biochim. Biophys. Res. Commun.* **46**, 406.
Dutton, P. L., Leigh, J. S., and Wraight, C. A. (1973). *FEBS Lett.* **36**, 169.
Dutton, P. L., Kaufmann, K. J., Chance, B., and Rentzepis, P. M. (1975a). *FEBS Lett.* **60**, 275.
Dutton, P. L., Petty, K. M., Bonner, H. S., and Morse. S. D. (1975b). *Biochim. Biophys. Acta* **387**, 536.
Dutton, P. L., Prince, R. C., Tiede, D. M., Petty, K. M., Kaufmann, K. J., Netzel, T. L., and Rentzepis, P. M. (1976). *Brookhaven Symp. Biol.* **28**, 213.
Dutton, P. L., Leigh, J. S., Prince, R. C., and Tiede, D. M. (1979). In "Tunneling in Biological Systems" (B. Chance, D. C. DeVault, H. Frauenfelder, R. Marcus, J. R. Schrieffer, and N. Sutin, eds.), p. 319. Academic Press, New York.
Fajer, J., Brune, D. C., Davis, M. S., Forman, A., and Spaulding, L. D. (1975). *Proc. Natl. Acad. Sci. U.S.A.* **72**, 4956.
Feher, G., and Okumura, M. Y. (1976). *Brookhaven Symp. Biol.* **28**, 183.
Felton, R. H., and Yu, N.-T. (1978). In "The Porphyrins" (D. Dolphin, ed.), Vol. 3 p. 347. Academic Press, New York.
Freed, K. F. (1980). *J. Am. Chem. Soc.* **102**, 3130.
Fujita, I., Bergkamp, M. A., Chang, C.-K., Wang, C.-B., and Fajer, J., Netzel, T. L. (1981). In preparation.
Godik, V. I., and Borisov, A. Y. (1979). *Biochim. Biophys. Acta* **548**, 296.
Gouterman, M. (1978). In "The Porphyrins" (D. Dolphin, ed.), Vol. 3, p. 1. Academic Press, New York.
Govindjee, Hammond, J. A., and Merkelo, H. (1972). *Biophys. J.* **12**, 809.
Haberkorn, R., and Michel-Beyerle, M. E. (1979). *Biophys. J.* **26**, 489.
Haberkorn, R., Michel-Beyerle, M. E., and Marcus, R. A. (1979). *Proc. Natl. Acad. Sci. U.S.A.* **76**, 4185.
Harriman, A., Porter, G., and Searle, N. (1979). *J.C.S. Faraday II* **75**, 1515.
Hoff, A. J., Rademaker, H., van Grondelle, R., and Duysens, L. N. M. (1977). *Biochim. Biophys. Acta* **460**, 547.
Holten, D., Windsor, M. W., Parson, W. W., and Gouterman, M. (1978a). *Photochem. Photobiol.* **28**, 951.
Holten, D., Windsor, M. W., Parson, W. W., and Thornber, J. P. (1978b). *Biochim. Biophys. Acta* **501**, 112.
Holten, D., Hoganson, C., Windsor, M. W., Schneck, C. C., Parson, W. W., Migus, A., Fork, R. L., and Shank, C. V. (1980). *Biochim. Biophys. Acta* **592**, 461.
Hopfield, J. J. (1974). *Proc. Natl. Acad. Sci. U.S.A.* **71**, 3640.

Hupert, D., Rentzepis, P. M., and Tollin, G. (1976). *Biochim. Biophys. Acta* **440**, 356.
Jortner, J. (1976). *J. Chem. Phys.* **64**, 4860.
Jortner, J. (1980). *J. Am. Chem. Soc.* **102**, 6676.
Jortner, J., and Rice, S. (1975). *In* "Physics of Solids at High Pressures" (C. T. Tomizuka and R. M. Emrich, eds.), p. 63. Academic Press, New York.
Kasha, M., Rawls, H. R., and El-Bayomi, M. A. (1975). *Pure Appl. Chem.* **11**, 371.
Kaufmann, K. J., Dutton, P. L., Netzel, T. L., Leigh, J. S., and Rentzepis, P. M. (1975). *Science* **188**, 1301.
Klimov, V. V., Shuvalov, V. A., Krakhmaleva, I. N., Klevanik, A. V., and Krasnovsky, A. A. (1977). *Biokhimiya* **42**, 519.
Knox, R. S. (1975). *In* "Bioenergetics of Photosynthesis" (Govindjee, ed.), p. 183. Academic Press, New York.
Leigh, J. S., and Dutton, P. L. (1972). *Biochem. Biophys. Res. Commun.* **46**, 414
Lepley, A. R., and Closs, G. L. (1973). "Chemically Induced Magnetic Polarization," Wiley, New York.
Likhtenstein, G. I., Kotelinkov, A. I., Kulikov, A. W., Syrtsova, L. A., Bogatyrenko, A. I., Melnikov, A. I., Frolov, E. N., and Berg, A. I. (1979). *Int. J. Quantum Chem.* **16**, 419.
McElroy, J. D., Feher, G., and Mauzerall, D. C. (1970). *Biophys. J.* **10**, 204a.
McElroy, J. D., Feher, G., and Mauzerall, D. C. (1972). *Biochim. Biophys. Acta* **267**, 363.
Netzel, T. L., Rentzepis, P. M., and Leigh, J. S. (1973a). *Science* **182**, 238.
Netzel, T. L., Struve, W. S., and Rentzepis, P. M. (1973b). *Annu. Rev. Phys. Chem.* **24**, 473.
Netzel, T. L., Dutton, P. L., Petty, K. M., Degenkolb, E. O., and Rentzepis, P. M. (1977a). *Adv. Mol. Relaxation Interact. Processes* **11**, 217.
Netzel, T. L., Rentzepis, P. M., Tiede, D. M., Prince, R. C., and Dutton, P. L. (1977b). *Biochim. Biophys. Acta* **460**, 467.
Netzel, T. L., Kroger, P., Chang, C. K., Fujita, I., and Fajer, J. (1979). *Chem. Phys. Lett.* **67**, 223.
Netzel, T. L., Bucks, R. R., Boxer, S. G., and Fujita, I. (1980). *In* "Picosecond Phenomena II" (R. Hockstrasser, W. Kaiser, and C. V. Shank, eds.), p. 322. Springer-Verlag, Berlin and New York.
Netzel, T. L., Bergkamp, M. A., and Chang, C.-K. (1982). *J. Am. Chem. Soc.* **104**, in press.
Norris, J. R., Uphaus, R. A., Crespi, H. L., and Katz, J. J. (1971). *Proc. Natl. Acad. Sci. U.S.A.* **68**, 625.
Okamura, M. Y., Isaacson, R. A., and Feher, G. (1975). *Proc. Natl. Acad. Sci. U.S.A.* **72**, 3491.
Okamura, M. Y., Isaacson, R. A., and Feher, G. (1979). *Biochim. Biophys. Acta* **546**, 394.
Pachence, J. M., Dutton, P. L., and Blasie, J. K. (1979). *Biochim. Biophys. Acta* **548**, 348.
Parson, W. W., Clayton, R. K., and Cogdell, R. J. (1975). *Biochim. Biophys. Acta* **387**, 265.
Paschenko, V. Z., Kononenko, A. A., Protasov, S. P., Rubin, A. B., Rubin, L. B., and Uspenskaya, N. Y. (1977). *Biochim. Biophys. Acta* **461**, 403.
Pellin, M. J., Wasielewski, M. R., and Kaufmann, K. J. (1980). *J. Am. Chem. Soc.* **102**, 1868.
Periasamy, N., Linschitz, H., Closs, G. L., and Boxer, S. G. (1978). *Proc. Natl. Acad. Sci. U.S.A.* **75**, 2563.
Rockley, M. G., Windsor, M. W., Cogdell, R. J., and Parson, W. W. (1975). *Proc. Natl. Acad. Sci. U.S.A.* **72**, 2251.
Shapiro, S. L., ed. (1977). "Ultrashort Light Pulses," Springer-Verlag, Berlin and New York.
Shuvalov, V. A., and Asadov, A. A. (1979). *Biochim. Biophys. Acta* **545**, 296.
Shuvalov, V. A., and Klimov, V. V. (1976). *Biochim. Biophys. Acta* **440**, 587.
Shuvalov, V. A., and Parson, W. W. (1981). *Proc. Natl. Acad. Sci. U.S.A.* **78**, 957.
Shuvalov, V. A., Klevanik, A. V., Sharkov, A. V., Matveetz, J. A., and Krukov, P. G. (1978). *FEBS Lett.* **91**, 135.

Shuvalov, V. A., Dolan, E., and Ke, B. (1979a). *Proc. Natl. Acad. Sci. U.S.A.* **76**, 770.
Shuvalov, V. A., Klevanik, A. V., Sharkov, A. V., Kryukov, P. G., and Ke, B. (1979b). *FEBS Lett.* **107**, 313.
Spaulding, L. D., Eller, P. G., Bertrand, J. A., and Felton, R. H. (1974). *J. Am. Chem. Soc.* **96**, 982.
Stamatoff, J., Eisenberger, P., Brown, G., Pachence, J., Dutton, P. L., Leigh, J., and Blasie, K. (1978). *In* "Frontiers of Biological Energetics" (P. L. Dutton, J. S. Leigh, and A. Scarpar, eds.), Vol. 1, p. 760. Academic Press, New York.
Sutin, N. (1962). *Annu. Rev. Nucl. Sci.* **12**, 285.
Thornber, J. P. (1977). *In* "The Photosynthetic Bacteria" (R. K. Clayton and W. R. Sistrom, eds.) Plenum, New York.
Thurnauer, M. C., Katz, J. J., and Norris, J. R. (1975). *Proc. Natl. Acad. Sci. U.S.A.* **68**, 625.
Tiede, D. M., Prince, R. C., and Dutton, P. L. (1976a). *Biochim. Biophys. Acta* **449**, 447.
Tiede, D. M., Prince, R. C., Reed, G. H., and Dutton, P. L. (1976b). *FEBS Lett.* **65**, 301.
Tiede, D. M., Leigh, J. S., and Dutton, P. L. (1978a). *Biochim. Biophys. Acta* **503**, 524.
Tiede, D. M., Leigh, J. S., and Dutton, P. L. (1978b). *Biophys. J.* **21**, 196a.
Van Grondelle, R., Holmes, N. G., Rademaker, H., and Duysens, L. N. M. (1978). *Biochim. Biophys. Acta* **253**, 187.
Vermeglio, A., and Clayton, R. K. (1976). *Biochim. Biophys. Acta* **449**, 500.
Vernon, L. P., and Shaw, E. R. (1971). "Photosynthesis and Nitrogen Fixation" (A. San Pietro, ed.), Methods in Enzymology, Vol. 23, p. 277. Academic Press, New York.
Wang, R. T., and Clayton, R. K. (1971). *Photochem. Photobiol.* **13**, 215.
Warshel, A. (1980). *Proc. Natl. Acad. Sci. U.S.A.* **77**, 3105.
Wasielewski, M. R., Svec, W. A., and Cope, B. T. (1978). *J. Am. Chem. Soc.* **100**, 1961.
Werner, H. J., Schultan, K., and Weller, A. (1978). *Biochim. Biophys. Acta* **502**, 255.
Zankel, K. L., Reed, D. W., and Clayton, R. K. (1968). *Proc. Natl. Acad. Sci. U.S.A.* **61**, 1243.

CHAPTER 5

Photoprocesses in Chlorophyll Model Systems*

Joseph J. Katz and J. C. Hindman

Chemistry Division
Argonne National Laboratory
Argonne, Illinois

I.	Introduction	119
II.	Chlorophyll and the Photosynthetic Unit	121
III.	Chlorophyll Properties Relevant to Models	123
	A. Coordination Interactions and the Self-Assembly of Chlorophyll Systems	123
	B. Correlation of the Optical Properties of in Vitro and in Vivo Chlorophyll Systems	126
	C. Fluorescence, Concentration Quenching, and Energy Trapping	129
	D. The Chlorophyll Laser	133
	E. Photophysics of the Chlorophylls	135
IV.	Covalently Linked Pairs as Photoreaction Center Models	138
V.	Energy Transfer in Self-Assembled Systems	144
VI.	Energy Transfer in Chl_{sp} Antenna Model	150
VII.	A Model for Electron Transfer from the Primary Donor	152
	References	152

I. Introduction

The processes initiated by the absorption of light are of paramount importance in phenomena that range from nuclear fusion to vision in animals and photosynthesis in green plants. The reemission of light in the form of fluorescence and phosphorescence provides important clues to the early consequences of light absorption. In pursuit of the understanding that will one day make it possible to replicate the primary events of photosynthesis outside the living cell, a prime concern must be the study of the photopro-

---

* Work performed under the auspices of the Office of Basic Energy Sciences, Division of Chemical Sciences, U.S. Department of Energy.

cesses attending the absorption of light by the primary photoreceptors in photosynthesis.

Even a cursory survey of the literature shows that most of what is known or presumed to be known about green plant and bacterial photosynthesis is based on optical spectroscopy. Because photosynthetic pigments absorb light in the visible region of the spectrum, they are intensely colored and lend themselves to examination by visible absorption and emission spectroscopy. Instrumentation for recording visible absorption and fluorescence spectra is highly developed and widely available. Consequently, much of the huge photosynthetic literature consists of absorption spectra, chromatic transients under different light regimes, fluorescence spectra at temperatures from liquid helium to well above room temperature, and linear and circular dichroism. Optical studies have been carried out on many different systems: more-or-less intact microalgae, photosynthetic bacteria, and leaves of higher plants; chloroplasts, bacterial chromatophores, and thylakoids; bacterial photoreaction centers; and pigment–protein complexes isolated from green plants and algae or cyanobacteria. Modern developments in laser photochemistry and new techniques for highly time-resolved absorption and emission spectroscopy have added new dimensions to optical studies of photosynthesis, and the time domain from a few picoseconds to several minutes after the initial light absorption event can now be explored. Other spectroscopic techniques, perhaps the most important of which is electron paramagnetic resonance (EPR) spectroscopy, have complemented and supplemented optical studies but have not replaced them.

The kind of information readily derivable from optical studies does not by itself provide structural information about photosynthesis on the molecular level. Even the simplest and best characterized photosynthetically active systems isolated from living organisms are still exceedingly complex. Often even a knowledge of the components of the experimental system is lacking. It is not surprising therefore that the interpretation of optical experiments in simplified systems, let alone intact photosynthetic organisms, often has an ambiguous and tentative quality.

One way of dealing with such a difficult experimental situation is to use model systems. Properly selected model systems can greatly facilitate the interpretation of experiments carried out on complicated biological systems. Properly chosen model systems can provide the feedback necessary to guide and refine the choice, execution, and interpretation of *in vivo* experiments. It is the object of this chapter to describe some basic aspects of model systems appropriate to the study of photosynthesis and ways in which such systems can contribute to the understanding of the primary photoprocesses of photosynthesis.

II. Chlorophyll and the Photosynthetic Unit

Chlorophyll has long been recognized as the primary photoacceptor and the essential mediator of the primary light conversion act in photosynthesis. The view that each chlorophyll molecule was capable of undergoing excitation by light absorption, and then reducing carbon dioxide and oxidizing water, was shown about 50 years ago to be invalid. Instead, chlorophyll function in photosynthesis is now considered to be a cooperative phenomenon in which a large number of chlorophyll molecules, say on the order of 200–300, act in concert to convert the energy of a single photon to an electron (reducing capacity) and a positive hole (oxidizing capacity). The bulk of the chlorophyll acts only as a photoacceptor and is therefore known as light-harvesting or antenna chlorophyll. Electronic excitation resulting from the capture of a photon by a chlorophyll molecule in the antenna is transferred to a few highly specialized chlorophyll molecules where charge separation is effected; these specialized chlorophylls are therefore referred to as photoreaction center chlorophyll. Antenna chlorophyll in green plants absorbs maximally at ~ 680 nm, and in photosynthetic bacteria ~ 855 nm. Antenna chlorophyll, in addition to its role in photon capture, provides just the right amount of energy dissipation (from an energy level of 680 to 700 nm) to ensure highly efficient energy trapping by photoreaction center chlorophylls.

It is important to note that very little if any of the chlorophyll in the photosynthetic unit (PSU) has the optical, redox, and magnetic properties of monomeric chlorophyll molecules present in a chlorophyll solution in a polar (nucleophilic) solvent. While the chlorophyll units, of which the antenna and photoreaction center are constituted, have identical molecular structures, the organized entities that perform the specialized tasks of the PSU have different structures and different optical properties. An understanding of the photochemical and photophysical properties of monomeric chlorophyll is a necessary preliminary, but it is the photoprocesses occurring in the different *in vivo* chlorophyll species that are the real objectives of photosynthesis research.

Photoreaction center chlorophyll in green plants and cyanobacteria is usually referred to as P700 [the pigment (P) that absorbs light at 700 nm, the characteristic absorption maximum of the photoreaction center]; the reaction center chlorophyll in purple photosynthetic bacteria is designated P865. It is generally considered that there are two kinds of reaction centers in green, oxygen-evolving plants: photosystem I (PS I) or P700, and photosystem II(PS II) or P680. P680 undergoes photobleaching when the oxygen-evolving photosystem is in operation. So little is known about the structure and composition of PS II that it is scarcely possible to discuss it in terms of models at this time.

Excitation energy trapped in the reaction center produces charge separation, which may be viewed as electron ejection and positive hole formation:

$$P700 + h\nu \longrightarrow P700^{+} + e^{-}, \tag{1}$$

$$P865 + h\nu \longrightarrow P865^{+} + e^{-}. \tag{2}$$

There is no reason of course to suppose that free electrons (in the sense of a metallic conductor) are involved in charge separation; the usage is metaphorical. Concomitant with the photooxidation of P700 and P865 is a (reversible) photobleaching at these wavelengths. The ejected electron is captured by a primary acceptor and then transferred along a chain of electron transport agents until the ultimate reductant NADPH is formed in the aqueous regions of the chloroplast, where carbon dioxide is reduced to form the organic components of the cell by a complicated set of enzymatic reactions. Another electron transport chain conducts electrons that are somehow abstracted from water (in green plants) or organic compounds (in photosynthetic bacteria) back into the reaction center to restore the oxidized $P700^{+}$ or $P865^{+}$ to its resting state. The species $P700^{+}$ and $P865^{+}$ have an unpaired electron in contrast to the precursors P700 and P865 and are therefore paramagnetic. Whether $P700^{+}$ and $P865^{+}$ are actually cations is not known. The photooxidized species, however, are free radicals and they are therefore readily detected in an EPR experiment.

The photo-EPR signal of $P700^{+}$ and $P865^{+}$ has a Gaussian line shape and a g value of 2.0025, indicative of a "free" electron delocalized over a framework of carbon and hydrogen atoms. The signal from $P700^{+}$ has a peak-to-peak linewidth of ~ 7 G, whereas that of $P865^{+}$ is ~ 9.5 G. The EPR signals of Chl $a \cdot L_1^{+}$ and Bchl $a \cdot L_1^{+}$ prepared by chemical oxidation of the respective monomeric chlorophyll and bacteriochlorophyll a also have a free electron G value and a Gaussian line shape, but Chl $a \cdot L_1^{+}$ has a linewidth of 9.3 G and Bchl $a \cdot L_1^{+}$ a ~ 13-G signal. These are significantly broader than the corresponding *in vivo* signals.

The narrowing of the $P700^{+}$ and $P865^{+}$ signals relative to those of the corresponding monomeric cations *in vitro* has long been known, but the discrepancy was attributed to the biological environment of the photoreaction center. However, another explanation can be used to reconcile the deviations between the *in vivo* and *in vitro* EPR cation signals. Spin sharing by two or more equivalent molecules narrows an EPR signal to an extent determined by the number of chlorophyll molecules over which the unpaired spin is shared or delocalized (Norris *et al.*, 1971; Bard *et al.*, 1976). Spin sharing by two chlorophyll molecules satisfactorily accounts for the narrowing of the P700⁺ and P865⁺ signals. This is the basis of the chlorophyll special pair (Chl$_{sp}$) concept, which holds that green plant PS I and the bacterial P865 reaction centers are composed of two chlorophyll molecules that experience a one-electron loss when photoexcited. Further details and supporting evidence for the Chl$_{sp}$ concept can be found in the papers of Katz and Norris (1973), Norris et al. (1974, 1975), Thurnauer et al. (1975), and Norris and Bowman (1980).

III. Chlorophyll Properties Relevant to Models

Modern investigations by visible, infrared (IR), NMR (Katz and Brown, 1981), EPR, and mass spectroscopy (Hunt et al., 1980) have revealed previously unsuspected aspects of chlorophyll behavior that we believe are highly relevant to models of chlorophyll function in photosynthesis. In this section we now describe a number of these.

A. Coordination Interactions and the Self-Assembly of Chlorophyll Systems

The central magnesium atom of chlorophyll with coordination number 4 (as shown in Fig. 1) is coordinatively unsaturated (i.e., it is electron-deficient), and therefore a strong driving force exists for the acquisition and insertion of electron donor functions (nucleophiles) at one or both of the axial positions of the magnesium atom. Dissolved in a nucleophilic (Lewis base, polar) solvent such as acetone, diethyl ether, pyridine, or the like, Chl a occurs predominantly as a monomer, with one or two molecules of nucleophile (depending on the base strength) in the magnesium axial positions. Monomeric Chl $a \cdot L_1$ or Chl $a \cdot L_2$ absorbs maximally in the red region (the lowest S_1 transition) at 660–670 nm, depending on the polarizability of the bulk solvent. Dilute solute solutions (< 0.001 M) of these species are intensely fluorescent.

The chlorophyll molecule in addition to its electrophilic center also has nucleophilic groups that can function as electron donors in coordination interactions. Chl a contains two ester carbonyl functions at positions 7c and 10a and a keto C=O function at position 9 (Fig. 1). From ir, ^{1}H-NMR, and ^{13}C-NMR, the keto C=O group is found to be the strongest nucleophile in the molecule (Katz et al., 1978a). BChl a contains two strong donor groups: the keto C=O and an acetyl C=O at position 2 (Fig. 1). In Chl b and BChl a, b, and c, where two strong nucleophilic groups are present, the relative donor strengths are still undetermined.

In the absence of extraneous nucleophiles, one chlorophyll molecule can act as a donor, by way of its keto C=O group, to the magnesium atom of

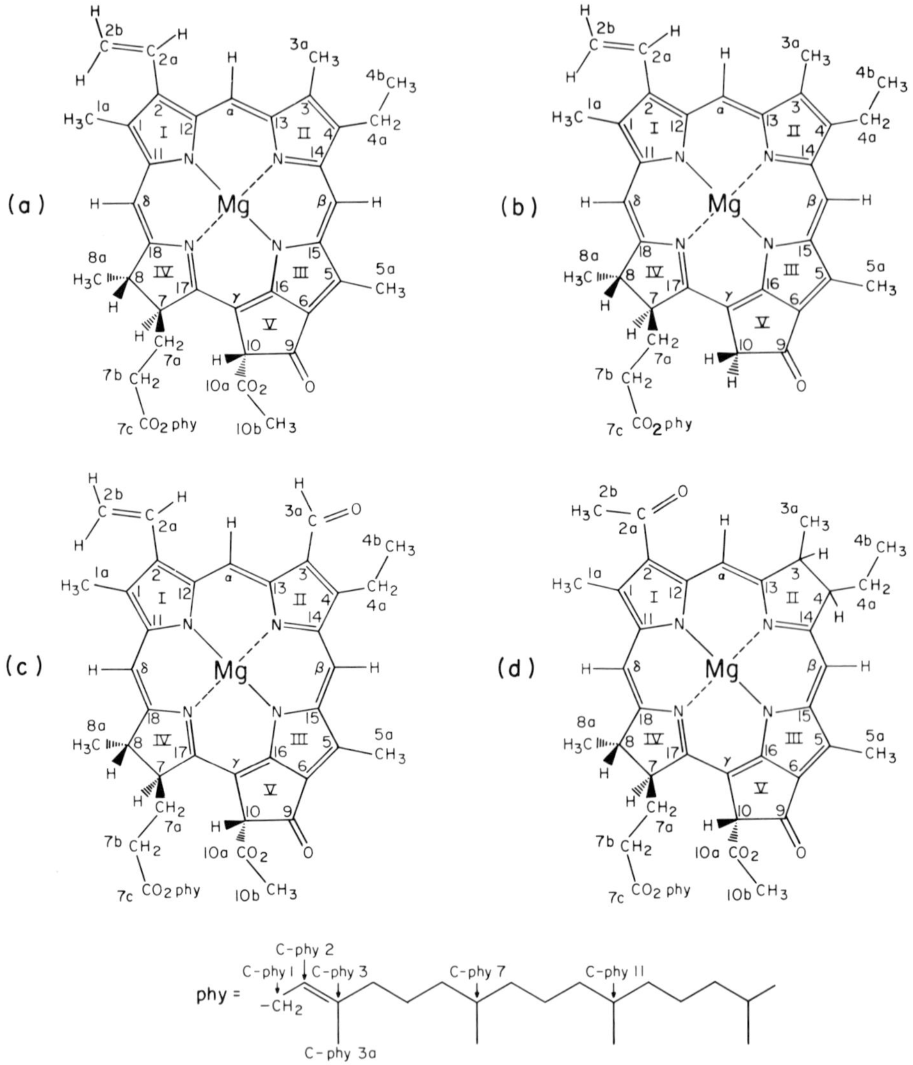

FIG. 1. Structures of the chlorophylls. (a) Chlorophyll a (Chl a); (b) pyrochlorophyll a (Pyrochl a); (c) chlorophyll b (Chl b); (d) bacteriochlorophyll a (Bchl a). Replacement of the magnesium atom by two hydrogens forms the corresponding pheophytin. Replacement of the phytyl group by a methyl (CH_3) group forms the magnesium-containing methyl chlorophyllides or the magnesium-free methyl pheophorbides. Chl a and b, and Bchl a occur in green plants and photosynthetic bacteria, respectively; Pyrochl a is a synthetic derivative of Chl a in which the carbomethoxy group (CO_2CH_3) normally present at position 10 is replaced by a hydrogen.

another. In carbon tetrachloride, such an interaction forms a dimer (Chl $a)_2$; in difficulty polarizable (hard) organic solvents such as aliphatic hydrocarbons, oligomers of the general formula (Chl $a)_n$, with $n > 20$, can be formed by keto C=O $\cdots$ Mg coordination interactions between Chl a molecules. The (Chl $a)_n$ species absorbs maximally at ~ 678 nm, and such solutions are only feebly if at all fluorescent.

Bifunctional nucleophiles with two donor centers may interact with chlorophyll to form monomers (in the presence of a large stoichiometric excess of the nucleophile) or linear polymers of high molecular weight (when the nucleophile is present in a 1:1 stoichiometric ratio). Bifunctional ligands can thus acts as cross-linking agents to form species of the structure (—L—Chl—L—Chl—L—Chl—)$_n$. Dioxane, or its nitrogen analog pyrazine, acts in this fashion to form adducts of colloidal dimensions. (Chl $a)_2$ can also be cross-linked to form adducts of the composition (Chl-L-Chl)$_n$, which contain both 5- and 6-coordinated magnesium. Chlorophyll adducts with organic bifunctional ligands have larger optical red shifts than those observed in (Chl $a)_2$ or (Chl $a)_n$. For a considerable number of organic bifunctional ligands that lack hydrogen-bonding properties the optical spectrum is red-shifted to 680–700 nm, depending on the distance between the nucleophilic centers in the ligand (Katz, 1973).

Bifunctional ligands of the generic formula RXH, where R = H or alkyl, and X = 0, S, or NH, are particularly important. The oxygen, sulfur, and nitrogen atoms in these ligands have lone electron pairs not otherwise involved in chemical bonding interactions, which are available for nucleophilic interaction with the central magnesium atom of chlorophyll. The hydrogen atom attached to the nucleophilic center is available for hydrogen-bonding interactions with the keto C=O function of another chlorophyll molecule. The preferred partner in hydrogen bonding is the keto C=O group, as judged from IR studies (Cotton et al., 1978). Thus such ligands as ethanol (CH$_3$CH$_2$OH), ethanethiol (CH$_3$CH$_2$SH), and n-butylamine (n-C$_4$H$_9$NH$_2$), and, most important, water (HOH) are able to cross-link chlorophyll molecules to form self-assembled systems with remarkable properties. The (Chl $a \cdot$ H$_2$O)$_n$ adduct dispersed in n-octane absorbs maximally in the red region at 740 nm and is the only Chl a species so far prepared that is photoactive in red light by the EPR criterion. BChl a–water adducts have been prepared with optical properties that closely resemble those of intact, living purple photosynthetic bacteria (Katz et al., 1976a). The most faithful model systems for P700 that have so far been prepared have been based on the properties of chlorophyll–water adducts prepared by self-assembly in the laboratory. Other model systems, also formed as a result of the operation of coordination interactions implicit in the chlorophyll structure, are described in Section IV.

B. Correlation of the Optical Properties of *In Vitro* and *In Vivo* Chlorophyll Systems

The optical properties of chlorophyll *in vivo* have long been known to be anomalous in the sense that essentially none of it has the optical properties characteristic of monomeric chlorophyll in a polar solvent. Nevertheless, chlorophyll in the monomeric form is still by far the most common species invoked to explain the optical properties of *in vivo* chlorophyll. The usual justification when this incongruity is pointed out is that a biological context is responsible for the red shifts in the optical spectra of the monomer. More specifically, it is suggested that there is something inherent in the interaction of chlorophyll with a protein that provides the red shift. Or, long-range order in the chlorophyll, or intense local electric fields produced in a lipid bilayer or by charged groups in a protein, are held responsible. Now, any or all of these hypotheses may be true, but to our knowledge they have not been demonstrated in the laboratory. What has been found in laboratory studies is that chlorophyll molecules in an appropriate geometric arrangement can experience π–π or coordination interactions that produce optical red shifts. At this writing, we consider that a red shift in the optical spectrum is indicative of interactions between chlorophyll molecules. Further, we think it reasonable to equate a laboratory chlorophyll system with an *in vivo* system if the two have similar optical properties. It is, of course, entirely possible that new ways will be discovered that provide red shifts in chlorophyll, but until such are demonstrated to exist in the laboratory we will not invoke them. We think it useful, even essential, in the present discussion to attempt a correlation between the optical properties of chlorophyll systems and molecular structures, even though there is the very real possibility that future work may require radical revisions in interpretation.

The visible absorption spectra in the spectral region 500–700 nm of monomeric, dimeric, and oligomeric Chl *a* in solution have been analyzed by Shipman *et al.* (1976a), and the energy of the transitions, extinction coefficients, and oscillator and dipole strengths for these Chl *a* species obtained. By a refined exciton theory on the effect of molecular aggregation on the visible spectra (Shipman *et al.*, 1976b), relationships have been developed among the spectra of Chl $a \cdot L_1$, Chl $a \cdot L_2$, (Chl a)$_2$, and (Chl a)$_n$. The principal tool in the spectral assignments have been computer-assisted deconvolution of the spectra. Although the spectra are at first glance featureless, there is considerable information in the line shapes that is not visible to the naked eye but which can be extracted by computer manipulation of the spectral data. The reader is referred to the papers of Shipman *et al.* (1976a) and Katz *et al.* (1976b) for a detailed discussion of the rationale of the deconvolution procedure.

The ground electronic state and the first and second excited singlet states are designated S_0, S_1, and S_2, respectively. The transitions $S_0 \to S_1$ (Q_y) and $S_0 \to S_2$ (Q_x) occur in the red–orange and yellow–green regions of the visible spectrum. The well-known Soret absorption band of chlorophylls occurs in the 400–435 nm region. It should be noted that the S_1 and S_2 transitions of Chl *a* are both at the low-energy end of the spectrum; the transitions in the Soret region are highly excited states that are at much higher energies than the S_2 (Q_x) transition. A somewhat unusal feature of the Q_y and Q_x transitions is the small energy gap, in the vicinity of 1100 cm^{-1}, between the S_1 and S_2 singlet states.

In the monomeric species Chl $a \cdot L_1$ and Chl $a \cdot L_2$, the chlorophyll functions as an *acceptor*: The magnesium atom has the coordination number 5 or 6 and acts as an acceptor for one or two molecules of nucleophile. The energy of the Q_y transition is not sensitive to the coordination number, but Q_x is red-shifted in 6-coordinated species; the effect is not obvious in Chl *a* systems, where Q_x and Q_y are close, but in Bchl *a* systems, where Q_x and Q_y are well-separated, the coordination number of the magnesium can be deduced from the position and shape of the Q_x band (Evans and Katz, 1975). The keto C=O donor function in Chl $a \cdot L_1$ and Chl $a \cdot L_2$ is free; that is, it does not participate in a coordination interaction. In a diethyl ether solution of Chl $a \cdot L_1$, acceptor Chl *a* has its lowest energy Q_y transition at ~ 660 nm.

When monomeric Chl *a* functions as a donor, that is, when its keto C=O participates in a coordination interaction, the aromatic π electron system of the macrocycle is perturbed (the keto C=O group is conjugated to the π system) and the electronic transition energy is altered. Acceptor and donor Chl *a* do not have the same Q_y transition energy.

The most common acceptor for Chl *a* acting as a donor is the magnesium atom of another Chl *a* molecule. Monomeric donor Chl *a* can be formed and its properties observed by adding an electrophile to a solution of Chl $a \cdot L_1$ or Chl $a \cdot L_2$. Suitable electrophiles are magnesium hexafluoroacetylacetonate and tris(1,1,1,2,2,3,3-heptafluoro-7,7-dimethyl-4,6-octane dionato) europium [Eu(fod)$_3$]. In these metal chelate compounds, the metal ion is coordinatively unsaturated, much like the magnesium atom with coordination number 4 in chlorophyll. A donor Chl *a* molecule may also act as an acceptor at its magnesium atom; it is the condition of the keto C=O group that determines whether a chlorophyll molecule acts as donor. Interaction with these electrophiles shifts the Q_y transition in Chl *a* to ~ 676 nm. That acceptor and donor Chl *a* have different electron transition energies and that donor Chl *a* is red-shifted relative to acceptor Chl *a* are important considerations in the interpretation of *in vivo* chlorophyll spectra.

In a solution of dimer (Chl *a*)$_2$, half of the Chl *a* is acceptor and half is donor, existing in a mobile equilibrium. The time scale of a NMR experiment is long relative to the kinetics of the equilibrium, and only the weighted average of the two interconverting forms is observed. In an electronic transition experiment, the transitions occur on a time scale short relative to the interchange, and both populations are observed. It is the population of red-shifted donor Chl *a* species that is responsible for the observed red shifts in (Chl *a*)$_2$ and (Chl *a*)$_n$, not electronic interactions between the macrocycles.

In oligomer systems, the amount of donor Chl *a* increases with the size of the oligomer; it is only the first Chl *a* molecule in a linear oligomer that is acceptor-only Chl *a*. As the donor concentration increases, the red shift approaches 680 nm as a limit. The red shift in these oligomers is not size-dependent. The absorption maximum depends on the relative amounts of donor and acceptor chlorophyll and, once a small oligomer is formed, a further increase in size has only a small effect on the absorption maximum.

These considerations are very important for the analysis of green plant visible absorption spectra. Most of the Chl *a* in the PSU absorbs maximally at ~ 678 nm. If in fact the antenna chlorophyll occurs as (Chl *a*)$_n$, then it is not possible to deduce the size of the aggregate from the wavelength maxima alone. The seductive notion that a chlorophyll aggregate will necessarily be red-shifted to an extent determined by size is not true of Chl *a* oligomers. Exciton considerations appear not to be important in (Chl *a*)$_n$ oligomers, presumably because the orthogonal arrangement of the macrocycles allows exciton interactions to make only a small contribution to the red shift. Similar considerations probably apply to the circular dichroism spectra, which may well be better accounted for in terms of donor and acceptor Chl *a* contributions of opposite sign to the spectra rather than by the usual exciton analysis.

The most strongly red-shifted chlorophyll species so far observed in the laboratory are chlorophyll–water adducts prepared by hydration of chlorophyll oligomers in solution or in a film. The macrocycles in these adducts are presumably arranged as in a crystal of ethyl chlorophyllide *a*·2 H$_2$O in which a water molecule is coordinated to the magnesium atom of a chlorophyll molecule and simultaneously is hydrogen-bonded to the keto C=O function of another to form a stack (Kratky and Dunitz, 1977; Chow et al., 1975). The Chl *a*–water adduct absorbs maximally at 740 nm, and the Bchl *a*–water adduct may be red-shifted from 780 nm to as much as 1 μm. The red shifts in these aggregates are best accounted for by exciton interactions between the chlorophyll macrocycles, which are parallel and in contact at their van der Waals radii (Shipman and Katz, 1977).

Any absorption maximum *in vivo* below 670 nm we assign to monomeric Chl *a*. Chl *a* species absorbing in the region 675–680 nm are assigned to (Chl *a*)$_2$ and (Chl *a*)$_n$ formed by keto C=O $\cdots$ Mg interactions. In the laboratory, Chl a-bifunctional ligand adducts absorbing between 680 and 705 nm can be prepared (Katz, 1973); it is possible that the ~680-nm-absorbing species *in vivo* could be organized by bifunctional ligands, or consists of a mixture of oligomer and bifunctional ligated species. Any Chl a species absorbing *in vivo* at wavelengths >680 nm is most likely the result of aggregate formation by a bifunctional ligand such as water or RXH (R = alkyl; X = O, N, or S); the bifunctional ligating agent may in principle be nucleophilic functions in protein side chains. Stacks of Chl a cross-linked by water can range in absorption maxima from 695 nm for a stack of two to 720 nm for an infinite stack (Shipman and Katz, 1977). Aggregates prepared from Chl a and pheophytin a (Pheo a) cross-linked by water also have absorption maxima between 705 and 720 nm (Norris et al., 1970).

To add to the complexity, all higher green plants contain Chl b. Hydration of Chl b forms an aggregate absorbing at 680 nm, the same region where Chl a oligomers absorb maximally and where the P680 involved in oxygen evolution also absorbs. Because the nature of the chlorophyll species giving rise to a particular absorption maximum can only be guessed at, a model can be used for falsification of a hypothesis, but not as proof. A model that has optical properties similar to those of an *in vivo* species may be correct, but not necessarily so. A model that has optical properties different from those of its presumed prototype in nature cannot be correct in the absence of a convincing reason for ignoring the optical requirements.

C. FLUORESCENCE, CONCENTRATION QUENCHING, AND ENERGY TRAPPING

Many of the conclusions about the chlorophyll species present *in vivo* and their involvement in the energy transfer process in photosynthesis have been based on fluorescence observations. At room temperature *in vivo* fluorescence is characterized by a low quantum yield and a short lifetime. *In vitro*, chlorophylls in dilute solution are highly fluorescent; fluorescence quantum yields and lifetimes are listed in Table I. Both the quantum yield and lifetime of fluorescence are determined by the relaxation channels available to the excited state and by the rate of energy transfer to other molecules. The original observations by Watson and Livingston (1950) that the quantum yield of chlorophyll fluorescence *in vitro* was concentration-dependent led to the suggestion that the low quantum yield of fluorescence could be related to the high ($\geq 0.1\ M$) chlorophyll concentration in the thylakoid membrane (Rabinowitch, 1945).

Recently, a systematic investigation of the quenching of pigment fluorescence in pyridine solution was carried out for Chl a, Pheo a, pyropheophytin

TABLE I

FLUORESCENCE QUANTUM YIELDS AND LIFETIMES OF CHLOROPHYLLS

Species	Solvent	Fluorescence Quantum Yield[a,b]	Lifetime (sec × 10^9)[b,c]
Chlorophyll a	Diethyl ether	0.32	5.1
Chlorophyll b	Ethanol	0.23	5.0
	Pyridine	0.36	5.3
Bacteriochlorophyll a	Acetone	0.25	4.7

[a] Weber and Teale (1957).
[b] Livingston (1960).
[c] Dmetrievsky et al. (1957).
[d] Parson (1974).

a (Pyropheo a), and their covalently linked pairs (Yuen et al., 1980b). The fluorescence quenching of Chl a and Pheo a was found to have a quadratic dependence on the bulk pigment concentration, suggesting that the quenching mechanism may involve energy transfer from excited monomers to weakly or nonfluorescent traps. The kinetic results are in accordance with the observations of others who also noted a quadratic dependence of the quantum yield on concentration (Watson and Livingston, 1950) and the fluorescence lifetime in different media (Beddard et al., 1976; Beddard and Porter, 1976).

Intermolecular quenching constants were computed by fitting the measured decay rate constant versus concentration to a power series in the concentration of the various pigments. A detailed analysis of the results indicated that the quenching mechanisms differed for the various species. For Pheo a and Chl a the observed kinetics were consistent with a quenching mechanism involving transfer of singlet excitation to relatively nonfluorescent traps. The pyropheophytin data were interpretable in terms of a mechanism involving energy transfer between an excited monomer and a ground state monomer. For all three of the covalently linked pairs (Section IV), the decay kinetics could be interpreted in terms of transfer of singlet excitation between fluorescent "open" forms and relatively nonfluorescent "closed" forms.

Since the empirical rate equations indicated that energy transfer to dimers was important in Chl a–pyridine solutions, the fluorescence data were treated in terms of energy transfer to monomer [M], and dimer [M$_2$], i.e.,

$$1/\tau_f = k_0 + k'_1[\text{M}] + k'_2[\text{M}_2], \tag{3}$$

where k_0 includes all processes affecting the fluorescence decay other than energy transfer from excited singlet to monomer k'_1, and transfer from excited singlet to dimer, k'_2. For Chl a this analysis indicates that the rate of energy transfer to the trap is fast, $k'_2 \cong 7.8 \times 10^{10} \, M^{-2} \, \text{sec}^{-1}$, and that a significant concentration of trap should be present in the solution at high chlorophyll concentrations. The equilibrium constant for the process that forms the trap, assuming these are dimeric species, is 3.2.

In an effort to answer the question whether these are true chemical species or "statistical pair" traps as suggested by Beddard and Porter (1976), difference absorption spectroscopy was used to obtain direct evidence for the formation of ground state aggregates at higher chlorophyll concentrations.

Calculations on model systems were carried out (Yuen et al., 1980b) by techniques described by Shipman (1980). Two kinds of species were assumed: fluorescent molecules and relatively nonfluorescent traps. It was concluded that the experimental data could be interpreted by a mechanism where the traps are quenching dimeric species in thermodynamic equilibrium with fluorescent species in the ground state, and where $k_q \ll k_{\text{detrap}}$ and $k_f \ll k_q$; where k_q is the quenching rate constant, k_f is the fluorescence rate, and k_{detrap}, $\sim 3 \times 10^{13} \, [M]s^{-1}$, is the sum of the Förster rate constants for transfer of excitation from the trap to all fluorescent species. It is estimated that a value of $k_q \cong 3 \times 10^9 \, \text{sec}^{-1}$ is consistent with the shape of the experimental k_f-versus-concentration curves.

No unambiguous assignment of structure for the quenching species has as yet been made, but whatever its structure the trap must have a structure that provides energy transfer by a nonradiative channel. It is suggested that nonequivalence of the two molecules in the trap is necessary for trap function. Nonequivalence should facilitate internal charge transfer in the excited state, and this in principle can provide a very efficient return to the ground state without light emission.

Some insight into the role of symmetry in energy transfer in dimeric systems can be obtained from the experiments of Netzel et al. (1979) on cofacially linked diporphyrins. From picosecond optical absorption measurements these workers conclude that excitation of symmetrical diporphyrins containing either magnesium or free base on both faces decays by a singlet-to-triplet mechanism but that excitation of a magnesium porphyrin macrocycle coupled to a demetallated one (Mg–2 H) yields charge transfer species that decay back to the ground state with a lifetime of 620 ± 20 psec. These results are in agreement both with the hypothesis that charge transfer provides the channel for nonradiative internal conversion to the ground state (Seeley, 1978; Gutschick, 1978) and that nonequivalence of the two macrocycles may be important in facilitating charge transfer-enhanced internal conversion (Yuen et al., 1980b).

TABLE II

SPECTROSCOPIC AND LASING PROPERTIES OF CHLOROPHYLLS AND DERIVATIVES[a]

Species	Solvent[b]	λ_a (nm)	λ_f (nm)	λ_l (nm)[c]	$\Delta\lambda_l$ (nm)[c,d]	Lasing Threshold (MW/cm²)[e]
Chlorophyll a	Pyridine	670	681, 737	680.6	4.6	1.9
	Acetone (2.2 × 10⁻³ M)	663	677, 728	676	3.5	2.5
Chlorophyll b	Pyridine (2.2 × 10⁻³ M)	655	666, 720	664.9	2.8	20.2
	Acetone (2.0 × 10⁻³ M)	645	658, 713	656	1.8	7.0
Bacteriochlorophyll a	Pyridine	782	799	799	5.7	3.0
Bacteriochlorophyll c	Pyridine	670	680, 738	678	3.5	2.5
Pyrochlorophyll a	Acetone	661	675, 727	673.5	2.1	2.3
Pheophytin a[f]	Pyridine (2.0 × 10⁻³ M)	670	682.5, 724	682.1, 727.0[h]	2.6, 2.1	18.1
Methylpheophorbide a[g]	Pyridine	669	680, 724	681.3	5.3	9.7

[a] λ_a, λ_f, and λ_l are the wavelengths of the absorption, fluorescence, and lasing bands.
[b] In a 1 × 10⁻³ M solution unless otherwise specified.
[c] Varies with input power.
[d] $\Delta\lambda_l$ is the half-width (FWHM) of the lasing band.
[e] Excitation at 337 nm.
[f] Chlorophyll a in which the magnesium has been replaced by 2 H.
[g] Pheophytin a in which the phytyl chain has been replaced by a methyl group.
[h] The larger of two simultaneously observed lasing wavelengths.

D. The Chlorophyll Laser

Optical pumping by a nitrogen laser results in population inversion and laser output from solutions of chlorophylls and related compounds in a variety of solvents (Leupold et al., 1977; Hindman et al., 1977).

The properties of these lasers are of considerable interest because they directly reflect the optical characteristics of the ground and excited states of the molecular species. Furthermore, they provide a unique tool for the study of rate processes associated with light absorption and energy transfer in photosynthetic pigments. Some experimental observations on the spectroscopic and lasing properties of chlorophylls are summarized in Table II. There are a number of properties of chlorophylls and chlorophyll-related species with respect to their behavior as laser dyes that require comment. One is that the lasing frequency is near the maximum of the fluorescence band. This implies that the population of the excited S_1 states is large (Leupold et al., 1977; Hindman et al., 1977). A second point to be noted is that the chlorophyll derivatives Pheo a and methylpheophorbide a (Fig. 2) lase both in the short- and long-wavelength bands, whereas Chl a lases only in the short-wavelength band (Fig. 3).

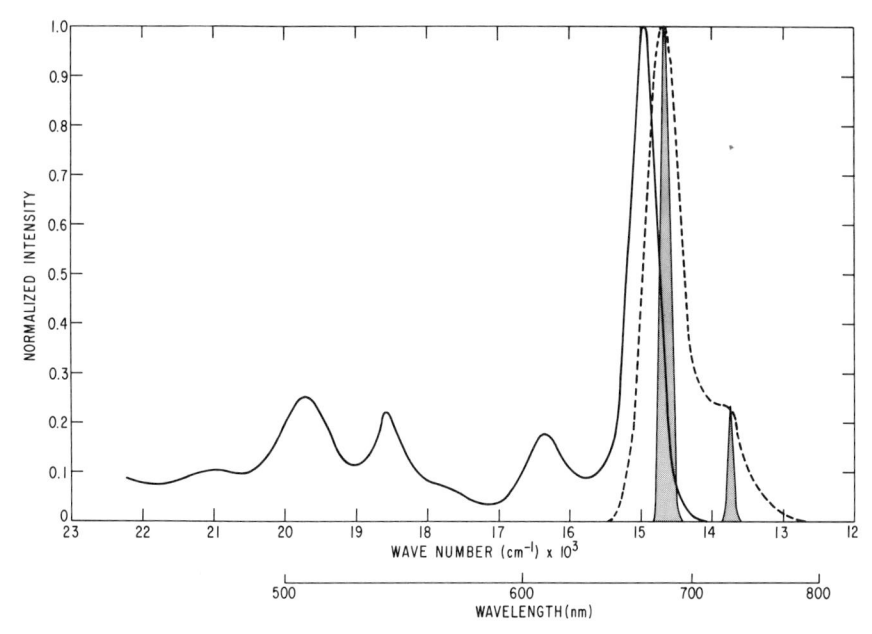

FIG. 2. Normalized absorption, fluorescence, and lasing spectra for methyl pheophorbide a in pyridine solution. ———, Absorption spectrum; -----, fluorescence spectrum; shaded area, lasing region. Note that lasing can occur in either of the two fluorescence bands.

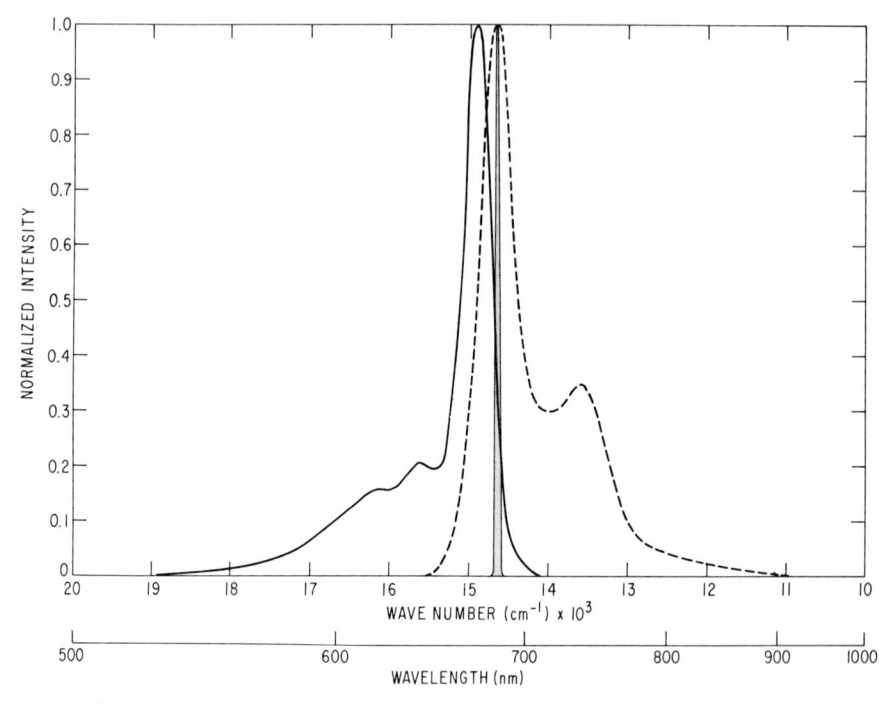

FIG. 3. Normalized absorption, fluorescence, and lasing spectra for Chl *a* in pyridine solution. ———, Absorption spectrum; -----, fluorescence spectrum; shaded area, lasing region. Lasing occurs at the maximum of the fluorescence band. This implies a large S_1/S_0 ratio.

1. *The Threshold Condition*

One of the factors that strongly influences the laser behavior of these compounds is the significant thermal population of one of the vibrational sublevels of the lower laser level resulting from the small Stokes shift between the absorption and emission bands. Before amplification can occur, it is necessary that

$$N_2/N_1 \geq e^{-h(v_0 - v_f)/kT} \geq e^{-u/kT}, \qquad (4)$$

where N_1 and N_2 are the populations of the lower and upper levels respectively, v_0 the frequency of the 0–0 transition, v_f the fluorescence frequency, and $u = h(v_0 - v_f)$. The minimum N_2/N_1 ratios are 0.55 and 0.0028, respectively for laser action at the maxima of the short wavelength (band 1) and long wavelength (band 2) fluorescence bands. The Stokes shifts are not significantly different for the different compounds, hence the minimum population inversion is approximately the same in all cases. The power requirements for lasing action differ considerably for the different chlorophylls,

Chl *a* and Bchl *a* being easily induced to emit laser light, whereas Chl *b* requires very high pump power.

A more detailed consideration of the conditions for laser action indicates that the inversion at threshold is much higher than suggested by Eq. (4). The conditions for oscillation of a dye laser were discussed by Schafer (1972). The lasing threshold as a function of frequency is

$$N_2/N \geq [\delta_{12}(v) + (S/N)]/[\delta_{21}(v) + \delta_{12}(v)] \tag{5}$$

for the condition that no triplets are formed and there is no absorption of light by the excited singlet. In Eq. (5), $\delta_{12}(v)$ and $\delta_{21}(v)$ are the wavelength-dependent cross sections (in cm^2) for absorption and stimulated emission, respectively, S the cavity loss, and N the total concentration (molecules/cm^3). It has been pointed out that lasing does not occur at the wavelength where a plot of the right side of Eq. (5) versus frequency shows a minimum (Leupold et al., 1977; Hindman et al., 1977). The implication is that the population in the first excited singlet (S_1) state of Chl *a* is very large, much above the level found by Lessing et al. (1970) and Lessing (1976) to decrease significantly the fluorescence lifetime of a dye molecule as a result of the stimulated emission process. This may also be taken to indicate that photon absorption by the excited singlet state may be an important factor in determining the lasing properties (Leupold et al., 1977). As the rate k_{ST} of intersystem crossing to form the triplet is relatively fast for Chl *a*, triplet absorption of the excitation may also be important, particularly where long pulse lasers are used. In Chl *a* in pyridine $N_T/N_2 \cong 0.4$, and for Chl *b* in the same solvent, $N_T/N_2 \cong 0.9$. The high power required to produce population inversion in a solution of Chl *b* is most likely the result of a large triplet population and a large triplet cross section.

Computer-modeling of the laser behavior of chlorophylls and their derivatives provides an entry into aspects of the excited states not otherwise easily accessible to investigation. Such investigations are in progress and can be expected to reveal information about the excited states of chlorophyll relevant to its role in photosynthesis.

E. Photophysics of the Chlorophylls

To understand how chlorophyll functions in photosynthesis requires a knowledge of the properties of the excited electronic states. A schematic energy level diagram for Chl *a* is shown in Fig. 4. The first and second excited singlet states differ by only $3.5kT$ in energy (Shipman et al., 1976a). Excitation by lasers emitting in the visible and near-ultraviolet region will result in the production of either excited S_1 or higher S_N states. The use of laser excitation in the study of light-induced photophysical processes associated with photosynthesis may be complicated by the significant populations of excited

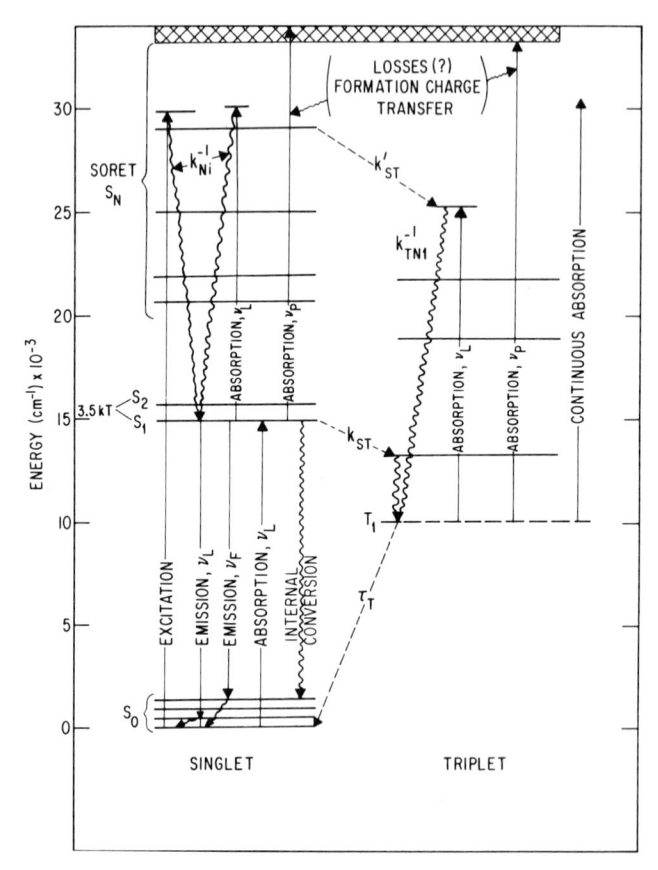

FIG. 4. Schematic energy level diagram for Chl *a*. Note small energy difference between S_1 and S_2. Excitation in the Soret band results in the population of higher S_N levels and cascading to low vibrational levels of S_1 before fluorescence v_f or laser emission v_L occurs.

states produced by the intense photon fluxes. The higher S_N states are expected to decay by rapid internal conversion to the lowest excited singlet state, S_1. Experimental observations with nanosecond pulse lasers will usually be unaffected by the formation of these higher singlet states except to the extent that the pumping process will be affected by the photon loss associated with cycling between S_1 and S_N states. On the other hand, the decay times of the higher states for many substances are on the order of 2–10 psec, and the interpretation of picosecond excitation data can be complicated by the processes associated with the decay of these higher excited states. No rates for these internal conversion processes are available for chlorophyll species.

5. PHOTOPROCESSES IN CHLOROPHYLL MODEL SYSTEMS

The role of internal conversion in energy loss in chlorophylls is generally ignored. Energy loss from the S_1 states is usually treated entirely in terms of fluorescence and intersystem crossing to the triplet state. No real data exist on mechanisms or rates of internal conversion, although it is undoubtedly an important pathway for energy loss in nonfluorescent dimers and oligomers and may play a significant role in the energy conversion of covalently linked dimers in the open configuration.

Triplet absorption spectra have been measured for Chl a and Chl b (Linschitz and Sarkanen, 1958; Pekkarinen and Linschitz, 1960; Baugher et al., 1979) and for Pheo a (Zanker et al., 1970; Prell and Zanker, 1970). There is even less information on S_1 absorption. Baugher et al. (1979) have obtained the absorption spectra of the triplet and first excited singlet states of Chl a in pyridine. S_1 absorption is characterized by an absorption at 440 nm and only a small absorptivity at the S_0 absorption maximum (Fig. 5). Picosecond absorption measurements on chlorophyll in ethanol showed a broad, structureless band in the 460–550 nm region and bleaching in the 665-nm band (Huppert et al., 1976). Flourescence emission in the wavelength region of the bleached band complicates direct measurement of the cross sections in this region. Recently, however, Shepanski and Anderson (1981) have used picosecond excitation to obtain the transient (S_1) spectrum on a very short timescale. They report that the excited singlet state is characterized by broad bands near 670 and 450 nm.

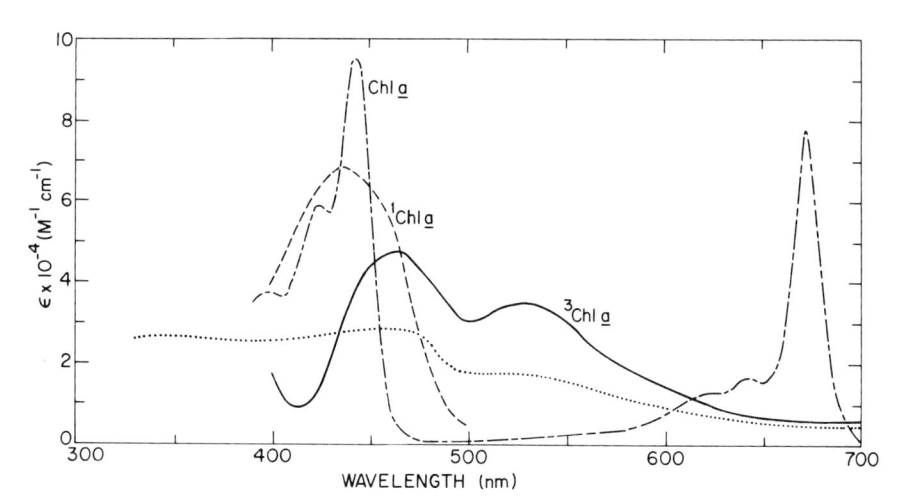

FIG. 5. Absorption spectra of the ground state (Chl a), the first excited singlet state (^{1}Chl a), and the triplet state (^{3}Chl a) of Chl a in pyridine solution. ———, Chl a; -----, ^{1}Chl a spectrum (Baugher et al., 1979); ———, ^{3}Chl a; ......, ^{3}Chl a spectrum (Linschitz and Sarkanen, 1958).

IV. Covalently Linked Pairs as Photoreaction Center Models

From the optical and EPR data we seek in a model two chlorophyll molecules with an absorption maximum near 700 nm and a suitably narrowed EPR signal. We also wish to mimic the redox properties of P700, but as these vary considerably according to the method of measurement (Setif and Mathis, 1980) we do not make this as stringent a requirement as that for the optical and EPR properties. A variety of models have been suggested (Katz and Norris, 1973; Katz et al., 1976a,b; Fong, 1974; Shipman et al., 1976c). For a discussion of the merits and disadvantages of the various Chl_{sp} models that have been proposed the reader is referred to the papers of Katz (1979a,b) and Katz et al. (1976b, 1978b, 1979).

The model of Shipman et al. (1976c) and of Boxer and Closs (1976) has overall been the most successful. In this model (Fig. 6) the chlorophyll macrocycles are arranged in parallel orientation with C_2 symmetry (a feature first suggested explicitly in Fong, 1974). Unlike the earlier Fong version, the Shipman et al. (1976c) model uses the keto C=O functions for cross-linking by two water molecules coordinated to the two magnesium atoms. By virtue of this arrangement, the macrocycles are at their van der Waals radii, 3.6 Å. Optimum π–π overlap is ensured, and the optical and EPR requirements are

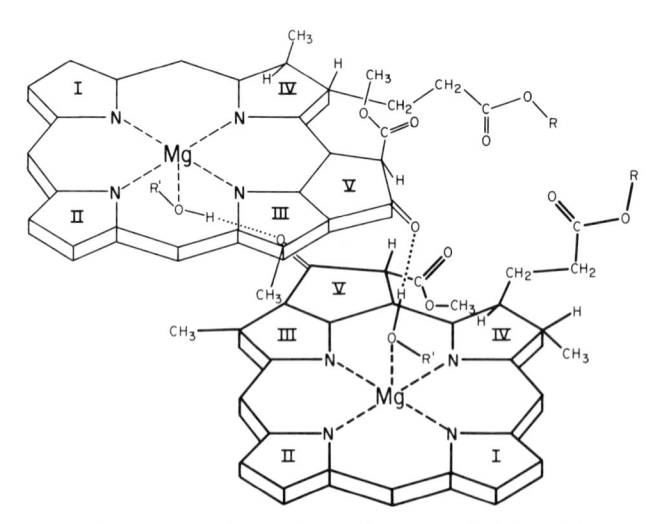

FIG. 6. Chlorophyll special pair model of Shipman et al. (1976c). The two chlorophyll macrocycles are cross-linked in this figure by two molecules of a primary alcohol. Other nucleophiles such as water, primary amines, and thiols can be used for the same purpose, as can presumably nucleophilic side chain groups characteristically present in proteins.

TABLE III

COMPARISON OF IN VIVO AND SYNTHETIC PHOTOREACTION CENTERS

System	Absorption (nm)	Fluorescence (nm)	Laser Light Emission (nm)[a]	EPR Signal (G)[b]
In vivo				
P700	~700			7.0
P865	~865			9.6
Covalently linked pair				
Bis(chlorophyllide *a*)ethylene glycol diester				
Open	662	683, 734	NL	
Folded	697	730	733	7.5
Bis(pyrochlorophyllide *a*)ethylene glycol diester				
Open	666	680	NL	
Folded	696	730	731	
Magnesium tris(pyrochlorophyllide *a*)-1,1,1-tris(hydroxymethyl)ethane triester				
Open	670	684	NL	
Folded	670, 690	726	735	
Bis(bacteriochlorophyllide *a*)ethylene glycol diester				
Open	760			
Folded	803			10.6
Self-assembled pairs				
Chl *a*, ethanol at 181 K	702	730	732	7.1

[a] NL, No lasing.
[b] EPR linewidth (ΔH_{pp}) of the cation.

satisfied on theoretical grounds. They are satisfied also on experimental grounds, for model systems based on this concept have been actualized and found to have the optical and EPR spin-sharing properties required for a P700 model (Table III).

Two Chl *a* molecules, each with a molecule of water coordinated to its magnesium atom, will not spontaneously assemble themselves into the structure of Fig. 6 because of the unfavorable entropy loss associated with the ordering of the macrocycles. The entropy loss can be successfully overcome by linking the two macrocycles by a covalent chemical bond. Replacing the monohydride alcohol phytol at the propionic acid side chain by a dihydric alcohol such as ethylene glycol ($HOCH_2CH_2OH$) makes a chemical link that prevents the two macrocycles from diffusing away from each other (Fig. 7). The chemical manipulations required to produce the link cause a loss of the central magnesium atom, and its reinsertion is required. The simplest magnesium reinsertion procedure requires elevated temperatures, and this results in a loss of the carbomethoxy group from position 10. Consequently, pyrochlorophyll *a* (Pyrochl *a*), a synthetic derivative of Chl *a* in which the carbomethoxy group at position 10 is replaced by a hydrogen, is frequently used to make linked pairs (Boxer and Closs, 1976). Chl *a* and Bchl *a* macrocycles have been covalently linked by Wasielewski *et al.* (1976, 1977), and details of the syntheses are given by Wasielewski and Svec (1980).

Solutions of the linked pairs in a nonpolar, nonnucleophilic solvent have optical properties very similar to those of dimeric (Chl *a*)$_2$ or (Pyrochl *a*)$_2$;

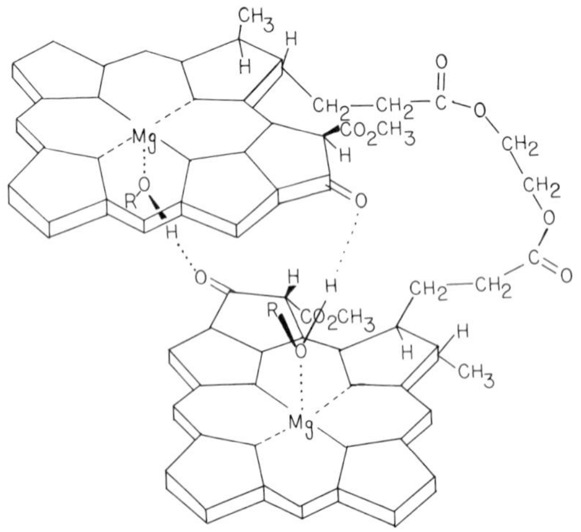

FIG. 7. A covalently linked pair in its folded configuration. The ethylene glycol bridge is to the right.

these species can in principle be formed by either inter- or intramolecular keto C=O ··· Mg interactions between linked pairs. The extent of aggregation in these systems has not been measured; it appears likely that linked species containing an internal keto C=O ··· Mg bond make a substantial contribution. The addition of water or ethanol causes a red shift to ~700 nm, which can reasonably be attributed to assumption of the folded configuration of Fig. 7. For linked Pyrochl a pairs the red shift is complete, but for linked Chl a pairs the red shift is only partial. The failure of linked Chl a pairs to fold completely is very likely a consequence of the presence of ~15–20% diastereoisomeric Chl a', formed when Chl a or its linked derivatives are dissolved in a polar solvent of even weak basicity (Katz et al., 1968; Hynninen et al., 1979). The scrambling of the stereochemistry at position 10 creates a steric barrier to folding, as the carbomethoxy group in the unnatural configuration in one of the macrocycles prevents assumption of a parallel configuration. From the optical spectra it is judged that all the linked Chl a pairs that can fold do so. In Pyrochl a-linked pairs, the carbomethoxy groups are missing and no steric barriers to folding exist. The conclusion that the optical red shift is due to folding is supported by ^{1}H-NMR data. The ^{1}H-NMR spectra of the Pyrochl a-linked pair in dry benzene are extremely broad, but in water-saturated benzene the lines become narrow and the spectrum is well-defined (Boxer and Closs, 1976). Only one set of resonances is observed, indicating that the two macrocycles are equivalent on the NMR time scale. The resonance of the 5-CH$_3$ groups shows a large diamagnetic (upfield) shift that can be attributed to a ring current effect on the 5-CH$_3$ group, which is positioned above the plane of the other macrocycle in the structure of Fig. 7. Similar conclusions have been reached by Wasielewski et al. (1977) from NMR studies on the linked Bchl a pair.

The linked pairs in their folded configuration exhibit the property of spin sharing. A 1-mM solution of the linked Pyrochl a pair in water-saturated carbon tetrachloride containing an equimolar amount of iodine as oxidant develops a signal when irradiated with red light in the microwave cavity of an EPR spectrometer. The intense signal is Gaussian and has a 7.5-G linewidth, suitably narrowed from the ~9-G linewidth characteristic of Chl $a \cdot L_1^{\ddot{+}}$ and in satisfactory agreement with the signal from P700$^{\ddot{+}}$. The geometry and electronic properties of the folded pair evidently are conducive to the delocalization of an unpaired spin over both macrocycles.

The redox properties of this model system are also encouraging, for the folded pair is easier to oxidize than monomeric Chl $a \cdot L_1$ itself. The redox properties of P700 and P865, as determined by experiment, are not thermodynamic quantities; they are more like energies of activation, for the experimentally determined redox potentials involve rate processes. Setif and Mathis (1980) have found that the measured redox potential of P700 in chloroplast lamellae and in subchloroplast particles depends on the detergent treatment used in the preparation of the test system and on the temperature at which the measurement is made. In addition, it is hardly likely that the environment and the electron acceptor in the model systems are in any way similar to those found *in vivo*. We consider the easier removal of an electron from the folded linked pair relative to monomer Chl $a \cdot L_1$ to indicate that the model tends in the proper direction, even if it does not fully duplicate the *in vivo* redox behavior of P700.

The linked Bchl *a* pair (Wasielewski *et al*., 1977) in its folded configuration gives a narrowed EPR signal on oxidation, but it does not have an 865-nm absorption. In its open configuration, the linked Bchl *a* pair has an absorption maximum at 760 nm; when folded, it is red-shifted only to 803 nm. The bacterial reaction center contains a total of four Bchl *a* and two bacteriopheophytin *a* (Bpheo *a*) molecules, and it is therefore a reasonable assumption that additional interactions with these molecules in the reaction center are responsible for the optical shift to P865 nm. It is not immediately obvious why photobleaching is observed at 865 nm rather than 803 nm in the photooxidized P865. It is evident that there are some serious problems with our P865 model that will need to be addressed to resolve the optical discrepancy.

Chl *a*- and Pyrochl *a*-linked pairs in the open configuration have light absorption and emission properties essentially identical with those of the monomers from which they are made. We have calculated for a 0.1 *M* solution of linked Chl *a* pairs a fluorescence lifetime significantly shorter (0.36 nsec versus 0.75 nsec) than that for Chl $a \cdot L_2$ in the same solvent and at the same concentration (Yuen *et al*., 1980b). The fluorescence decay kinetics are consistent with a mechanism in which singlet excitation is tranferred from fluorescent open forms to relatively nonfluorescent closed forms, which is the mechanism that also seems operative in Chl *a* solutions.

The covalently linked Chl *a* pair exhibits remarkably different optical properties in the open and folded configurations. In the folded configuration the absorption maximum is at 695 nm and the fluorescence maximum is at 730 nm. Coherent (laser) light output is obtained at 733 and 735 nm in wet benzene and in 0.1 *M* ethanol–toluene solutions, respectively. The expected fluorescence lifetime shortening associated with stimulated emission from appreciable concentrations of molecules in the S_1 excited state is observed. However, the open linked pair shows no laser emission or fluorescence lifetime shortening even at pump powers much higher than those needed to induce lasing in the folded pair. It does not appear that the behavior of the open pair can be attributed to excimer or triplet formation. A possible explanation is that absorption of the excitation radiation (337 nm) by open pairs in the S_1 state forms an exciplex or a charge transfer state that prevents attainment of the population inversion necessary for laser emission (Hindman *et al.*, 1978). Flourescence emission from the open linked pair and laser emission from the linked pair in its folded configuration indicate that the simple proximity of two chlorophyll macrocycles does not necessarily lead to fluorescence quenching and lifetime shortening, as suggested by Beddard and Porter (1976). The lasing behavior of the folded linked pair is particularly interesting, for the wavelength of the laser emission suggests that the long-wavelength fluorescence near 735 nm observed in higher-plant chloroplasts (Brody and Brody, 1961; Govindjee and Yang, 1966; Konishi *et al.*, 1973) and attributed to PS I (Kochubei, 1980; Kochubei and Guliev, 1980) originates in P700 (or in a Chl *a* species formed at 77 K that has a structure similar to that of P700). Under conditions of normal photosynthesis, no fluorescence can be expected from P700 (Lavorel and Etienne, 1977), but where the normal channel for dissipation of excitation energy from P700* is not available, for whatever reason, fluorescence should occur. We consider that one of the criteria for a successful P700 model is that it be capable of fluorescence when electron transfer cannot occur.

Periasamy *et al.* (1978) have studied the photochemical behavior of the folded pair and suggest that photoexcitation results in the formation of a transient containing a triplet and a ground state moiety in the unfolded configuration. However, the finding that folded covalently linked pairs can be induced to lase in the 730–740 nm region shows that a significant population of the excited S_1 state of the folded pair is produced by optical excitation (Hindman *et al.*, 1978), and this appears incompatible with the proposal of Periasamy *et al.* (1978).

In an effort to reconcile the two views on the consequences of excitation of the folded pair we have repeated the Periasamy *et al.* experiment with two lasers of 4- to 5-nsec pulse width to excite and probe the excited state. The photobleaching of the folded pair at zero time (coincident excitation and probe pulses) is accompanied by an increased transient absorption in the 630–676 nm region. That this transient absorption may be due to two species is suggested by the observation that transient absorption is highest at zero time delay and decays very rapidly, as is expected for an excited singlet state. This decay is accompanied by a decrease in the bleaching at the absorption maximum of the folded pair. The residual absorption at approximately 670 nm corresponds to that observed by Periasamy *et al.* (1978) and may be due to a triplet species.

The differences in the two sets of observations may be reconciled if we take into account the difference in excitation conditions. For the folded pair, the rate of intersystem crossing k_{ST} to form the triplet is 0.16 × 10^9 sec^{-1} (Hindman *et al.*, 1978). For the 30-nsec excitation pulse used by Periasamy *et al.* (1978), the triplet-to-excited singlet ratio at the maximum of the pulse would be approximately 3.2, and essentially complete conversion to the triplet state would have occurred by the end of the excitation pulse. On the other hand, for a 4-nsec excitation pulse the triplet-to-excited state ratio would only be 0.5 at the maximum of the absorption pulse. The large population of triplets produced in a 30-nsec pulse may well unfold as claimed by Periasamy et al. The high ratio of singlet states produced in a short pulse makes it possible to achieve a population inversion and subsequent laser action. This conclusion is supported by picosecond studies on transients produced in the folded pyrochlorophyllide pair in ethanol–carbon tetrachloride solutions (Pellin et al., 1980). It was concluded that on the picosecond time scale the decay of an optical transient at 660 nm and the decrease in bleaching at 700 nm could be interpreted in terms of repopulation of the ground state of the folded pair.

Pellin et al. (1980) reported that the behavior of the folded dimer in methylene chloride was anomalous in that the fluorescence lifetime was greatly shortened and the fluorescence yield greatly reduced in this solvent. Temperature-dependent quantum yields and fluorescence lifetimes were observed and interpreted in terms of a dual excited state model with one fluorescent and one nonfluorescent singlet state, and Freed (1980) has developed a theory to explain the results of Pellin et al. (1980).

We have examined the behavior of the linked Pyrochl a pair in methylene chloride with results significantly different from those of Pellin et al. (1980). A nitrogen laser with a 1.1-nsec pulse was used for excitation. The fluorescence lifetime was somewhat shortened relative to other solvent systems, but not remarkably so, and in other respects we found nothing particularly unusual. Pellin et al. (1980) used a high-powered picosecond pulse for excitation; there is thus the possibility that their observations relate to excited states higher than S_1 resulting from multiphoton processes. Further work is clearly needed to resolve the situation.

V. Energy Transfer in Self-Assembled Systems

It has long been recognized that simple chlorophyll systems can be readily prepared in the laboratory that have optical properties closely resembling those of in vivo antenna and photoreaction center chlorophyll. Krasnovsky and Bystrova (1980) have reviewed work dating back to the early 1950s on chlorophyll films that simulate the salient features of antenna chlorophyll present in photosynthetic cells. Other workers have noted that solutions of chlorophyll in various solvents develop long-wavelength absorption and fluorescence maxima at low temperatures (Freed and Sancier, 1954; Brody, 1958; Brody and Brody, 1963; Shipman et al., 1976c; Fong and Koester, 1976; Cotton et al., 1978). Although early workers were unable to give a molecular interpretation of the genesis of the red-shifted species in their systems, it is now evident that the observed optical shifts result from coordination and/or hydrogen-bonding interactions that juxtapose the chlorophyll macrocycles in ways that perturb their π systems. These systems can be described as self-assembled: they have a short or intermediate range order to varying degrees that result from the operation of coordination properties innate to chlorophyll, and the ordering does not involve the formation of true chemical bonds.

The loss of entropy on forming an ordered system can be compensated for by reducing the temperature. Many self-assembled systems are therefore prepared at cryogenic temperatures. At 0.005 M solution of Chl a in methylcyclohexane containing 0.05 M ethanol is largely converted to Chl a species absorbing near 700 nm at cryogenic temperatures (Fig. 8). From computer deconvolution of the visible absorption spectra recorded at different temperatures it is evident that this system at room temperature largely contains monomer (Chl $a \cdot L_1$) and dimer [(Chl $a)_2$]; on cooling, red-shifted (Chl $a \cdot C_2H_5OH)_n$ species are formed. The oscillator strength of the self-assembled species is very significantly larger than that of Chl $a \cdot L_1$, and the optical data cannot be used to deduce the stoichiometry of the system. The data are consistent with operation of the following equilibria:

$$(\text{Chl } a)_2 + 2\,\text{ROH} \rightleftharpoons 2\,\text{Chl } a \cdot \text{ROH}, \qquad (6)$$

$$n(\text{Chl } a \cdot \text{ROH}) \rightleftharpoons (\text{Chl } a \cdot \text{ROH})_n. \qquad (7)$$

Lowering the temperature displaces the equilibria to the right. Solutions at intermediate temperatures consequently consistent of a mixture of Chl $a \cdot C_2H_5OH$, (Chl $a)_2$, and (Chl $a \cdot C_2H_5OH)_n$, where n is a small number. Whether the (Chl $a \cdot C_2H_5OH)_n$ species has a structure identical (except for

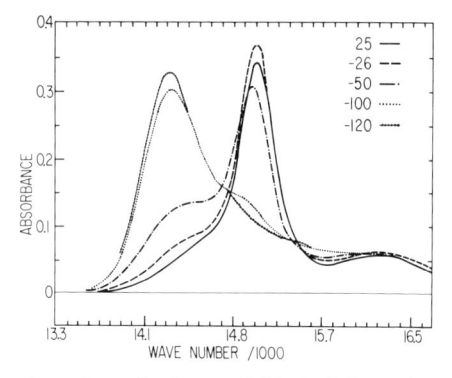

FIG. 8. Temperature dependence (in degrees Celsius) of absorption spectrum for 0.005 M Chl a in methylcyclohexane containing 0.05 M ethanol.

the chemical link) with that formed by folding the linked pairs is unknown. It is likely that the extreme red-shifted species in this system are a mixture of short ($n = 2, 3$) stacks with the Kratky and Dunitz (1975) ethyl chlorophyllide $a \cdot 2\,H_2O$ structure. Exciton calculations indicate that short stacks could likely account for the experimental observations (Shipman and Katz, 1977). A stack of two has no symmetry element, and a stack of three has two equivalent Chl a molecules and translational symmetry (Katz et al., 1976b). We will refer to these species as self-assembled special pairs.

The predominant fluorescence emission is in the 719–730 nm region, presumably arising from self-assembled special pairs, and emission in the 671–680 nm region that can be assigned to monomeric Chl $a \cdot$ L. The (Chl a)$_2$ species is essentially nonfluorescent. Lasing action is observed in the 724–735 nm spectral region, but not near 675 nm, even though monomeric Chl a is known to lase. Population inversion, then, is achieved not only by direct pumping of (Chl $a \cdot C_2H_5OH$)$_2$ but also by energy transfer from the monomer. The lasing behavior of the self-assembled system is very similar to that of the folded linked pairs (Hindman et al., 1978), which argues for a basic similarity between the configuration of the folded linked pairs and that of the self-assembled pairs.

Because the self-assembled systems contain a mixture of species, they are especially useful for energy transfer studies. We have found that fluorescence emission from these systems is dependent on the excitation wavelength. Irradiation of a self-assembled Chl a system in ethanol–methylcyclohexane at 450 or 700 nm should excite principally the self-assembled special pairs (cf. Fig. 9). Irradiation at 661 and 672 nm should excite for the most part

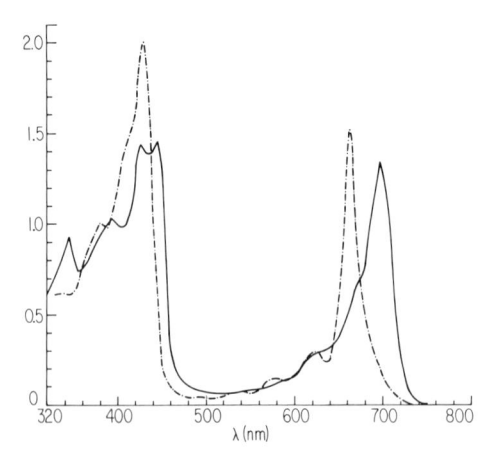

FIG. 9. Absorption spectra of a 0.005 M Chl a solution in methylcyclohexane containing 0.05 M ethanol. $-\cdot-\cdot-$, Monomer Chl a (26°C); ———, self-assembled special pairs ($-123°$C).

monomeric Chl a species, and irradiation at 681 nm should excite Chl $a \cdot L_1$, (Chl $a)_2$, and the self-assembled species. Irradiation at 700 nm should excite only the longest-wavelength self-assembled species. With 450-nm excitation at room temperature the primary emission is at 713–714 nm; cooling results in a red shift of this emission to 721 nm. Similar fluorescence is obtained with 700-nm excitation, which supports the idea that optical pumping at these wavelengths primarily excites the self-assembled species. With 450-nm excitation at room temperature, emission is also observed at 673 nm, near the emission maximum for monomeric Chl $a \cdot L_1$. On cooling, this emission band shifts to the red and decreases in intensity, and below $-100°C$ the emission maximum is near 685 nm. This suggests that emission occurring from a species other than the monomer as excitation at 661 nm (in the monomer band) results mainly in fluorescence near 667 nm at all temperatures. At $-123°C$, there is an indication of emission at 683–685 nm.

Excitation with 700 nm at room temperatures leads to fluorescence centered near 714 nm. At $-62°C$ and lower, this emission band is shifted to 720 nm. In the temperature range -101 to $-123°C$ a shoulder at 694 nm is observed. When excitation with 672- and 681-nm light at room temperature is used, emission with a maximum near 705 nm is observed. With 672-nm excitation, in addition to the 705-nm emission, fluorescence in the 670-nm region assignable to the monomer is present in the spectrum, and there is a definite indication of a third species that emits in the 685-nm region. With 681-nm excitation, much less monomer fluorescence is evident. A discrete peak in the fluorescence spectrum at 685 nm is not observed, but it is evident from the shape of the emission curve that more than one component is required for deconvolution. On cooling, the red emission shifts to 720–725 nm, and well-defined bands in the monomer region near 685 nm are observed.

At least three emitting species are present in the Chl a self-assembled system in ethanol–methylcyclohexane: monomeric Chl $a \cdot L_1$, self-assembled special pairs, and a species emitting in the region intermediate between monomer fluorescence and the emission from the self-assembled pairs. There is also some indication of a fourth component that fluoresces in the 694-nm region.

We have carried out a parallel set of experiments to ascertain the effects of different excitation frequencies on the fluorescence of *Tribonema aequale*. This alga belongs to the Xanthopycaeae and contains only Chl a; no other chlorophyll or phycobilin is present in these organisms, which nevertheless evolve oxygen during photosynthesis. The absence of any other chlorophyll-like or phycobilin pigments makes these organisms of particular interest in photosynthesis studies as the absence of Chl b and phycobilin simplifies interpretation of the fluorescence spectra.

Irradiation of intact *Tribonema* at room temperature with 337-nm light from a nitrogen laser produces a broad fluorescence emission with maxima at 690 and 740 nm. At 77 K, the fluorescence emission is narrowed and the principal emission band shifts to 720 nm, with a distinct shoulder in the 690-nm region. Returning the algae to room temperature restores the original fluorescence, suggesting that self-assembled species form reversibly in the organism at low temperatures. This behavior is consistent with assignment (Butler, 1978) of the *in vivo* long-wavelength fluorescence to PS I.

When 450-nm radiation is used for excitation, the room temperature fluorescence has a shoulder at 673 nm, the wavelength expected for emission from monomeric Chl *a*, and the principal emission band is at 686 nm, tailing to the red region. The 77 K fluorescence spectrum is quite similar to that obtained with 337-nm excitation and has a maximum at 715 nm. With excitation at 672 and 681 nm, primary emission at room temperature is centered at 695 nm, which is the fluorescence usually considered to originate from the PS II reaction center. In green algae, the 695-nm emission is only seen at low temperatures, but in *Tribonema* it is observed at room temperature. Excitation with 672-nm light reveals a shoulder in the spectrum at 685 nm, and the principal emission at 695 nm is relatively narrow, suggesting that the species responsible for this emission is well-defined in a structural sense. At 77 K, the primary emission is near 720 nm, with a shoulder in the 695-nm region.

The primary emission band resulting from 650- and 662-nm irradiation (excitation into the S_1 monomer absorption band) is at ~ 687 nm. From the asymmetry in the red tail it appears that at least two additional emission bands contribute to the spectrum, one at 670–700 nm and the other in the 720-nm region. Cooling to $-120°C$ shifts the emission maximum to 715 nm, with a shoulder at 690 nm.

The *in vivo* results with *Tribonema* and the *in vitro* experiments with self-assembled systems are compared in Table IV. Fluorescence at 685 and 695 nm is detectable in this organism at room temperature by selective optical pumping. Fluorescences at these wavelengths are widely considered to arise from PS II as a result of changes in excitation energy transfer at low temperatures (Lavorel and Etienne, 1977; Butler, 1978). The considerable similarity between the fluorescence emitted by the self-assembled systems and that of intact *Tribonema* suggests that the origin of the different fluorescence should be sought in different chlorophyll species rather than in changes in the pattern of energy flow caused by the opening and closing of PS II traps. We believe further that the data can be best interpreted by the assumption that self-assembly processes occur when intact organisms or chloroplasts are strongly cooled; this implies that the room temperature chlorophyll species are not

TABLE IV

FLUORESCENCE OF SELF-ASSEMBLED IN VITRO AND IN VIVO CHLOROPHYLL SYSTEMS

Species	Absorption Temp. (K)	λ_{max} (nm)	Fluorescence Temp. (K)	Excitation (nm)	Fluorescence (nm)[e]
In vitro chlorophyll species					
Chl $a \cdot L_1$	298	663	298	337	676 (730)
Chl $a \cdot L_2$	149	672	149	337	685 (740)
(Chl $a)_2$	298	675	298	337	None
			77		(735–737)
(Chl $a)_n$	298	678	298	337	(698, 745)[f]
Covalently linked pairs[b]	298	695	298	337	730
Self-assembled pairs, (Chl $a \cdot$ ROH)$_2$[c]	150	698	150	337	720, 694 (sh)
			150	450	685, 721
			150	672	683, 720–725
Self-assembled system[d]			150	681	682–685, 720–725
Tribonema aequale			298	337	690
			77		740
			298	450	686, 673 (sh)
			77		715
			298	672, 681	695, 685 (sh)
			77		715
			298	650, 662	687, 690–700 (sh)
			140		715, 690 (sh)

[a] The $S_1 \rightarrow S_0$ transition in the red region of the spectrum.
[b] In the folded configuration (cf. Fig. 7).
[c] See Fig. 8.
[d] This system contains monomeric, dimeric, and self-assembled pairs of Chl a. See Section V.
[e] Numbers in parentheses indicate very weak fluorescence.
[f] The very weak fluorescence may not be intrinsic and may arise from impurities in the system.

necessarily identical either in quantity or kind with those present at cryogenic temperatures. That fluorescence at 685 and 685 nm can be obtained from a system containing only Chl *a*, ethanol, and methylcyclohexane, which is remarkably similar to the fluorescences emitted from intact *T. aequale* needs, in our opinion, to be taken into account in the interpretation of the fluorescence from living organisms.

Only partial assignment of the fluorescences can be made at this time. From the similarities between the self-assembled and *Tribonema* fluorescences (Table IV) it is evident that chlorophyll–protein interactions need not necessarily be invoked. As *Tribonema* does not contain Chl *b*, an assignment of the 695-nm emission in this organism to a light-harvesting Chl *a* or Chl *b* protein, as has been done by Butler (1978), is not required. While the nature of the chlorophyll species emitting in the regions 670–680 nm and 720–730 nm is reasonably clear, the nature of the 685- and 695-nm species can only be the subject of speculation. Chlorophyll in PS II is an unlikely candidate for self-assembled systems, although the possibility that chlorophyll species present in *in vivo* PS II may have been formed in self-assembled systems cannot be excluded. The ~700-nm-absorbing species in the self-assembled system is likely to have a configuration very similar to that shown in Fig. 6; the 685- and 695-nm species may be similar, but cross-linked with only one molecule of ethanol, so that the transition dipoles of the two macrocycles are not exactly parallel as they are in the structure of Fig. 6. The low-temperature 685- and 695-nm-emitting species in *Chlorella* may be formed by new coordination interactions with nucleophiles present in the thylakoid but which are not accessible for cross-linking chlorophyll molecules at room temperature for either steric or thermodynamic reasons. Butler and Katajima (1975a,b) have in fact suggested that the 695- and 720-nm emissions are indeed artifacts produced by cooling the chloroplasts. The 720-nm fluorescence can now be assigned with some confidence to a 700-nm-absorbing entity with the Shipman *et al.* (1976c) structure. Whatever the merits of the assignments we suggest here, they can at least be tested by experiment. With additional data from *in vitro* systems, we may look forward to assignment of the *in vivo* fluorescences in terms of molecular structure and to a better understanding of what fluorescence says about photosynthesis.

VI. Energy Transfer in the Chl_{sp} Antenna Model

In the quest for a more complete model of a photoreaction center, it is necessary to attach electron donors, acceptors, and an antenna. A structurally well-defined model has been developed (Yuen *et al.*, 1980a) that can be used to study energy transfer from a light-harvesting chlorophyll to a Chl_{sp}.

5. PHOTOPROCESSES IN CHLOROPHYLL MODEL SYSTEMS 151

The zinc and magnesium tris(pyrochlorophyllide *a*)-1,1,1-tris(hydroxymethyl ethane) triesters can be prepared by esterifying three Pyrochl *a* macrocycles to a trihydric alcohol. These compounds are thus similar to the Chl *a* and Pyrochl *a* pairs covalently linked by ester bonds to the dihydric alcohol ethylene glycol (Section IV). Solutions of the triester in polar (nucleophilic) solvents are in an open configuration, as in the case of the covalently linked pairs, and all three macrocycles have optical properties essentially identical with those of monomeric Pyrochl $a \cdot L_1$. Dissolved in a nonpolar solvent to which a small amount of hydrogen-bonding nucleophile has been added, folding occurs between two of the macrocycles and the third remains free (Fig. 10). This is a dynamic situation, and there is a rapid interchange between the folded pair and the free macrocycle; on the time scale of an electronic transition, there is a population of pairs and a population of monomeric macrocycles. The two macrocycles in the folded configuration have an absorption maximum at ∼700 nm, and the free macrocycle absorbs light of higher energy near ∼665 nm. The free macrocycle thus stands in the position of an antenna relative to the cross-linked pair.

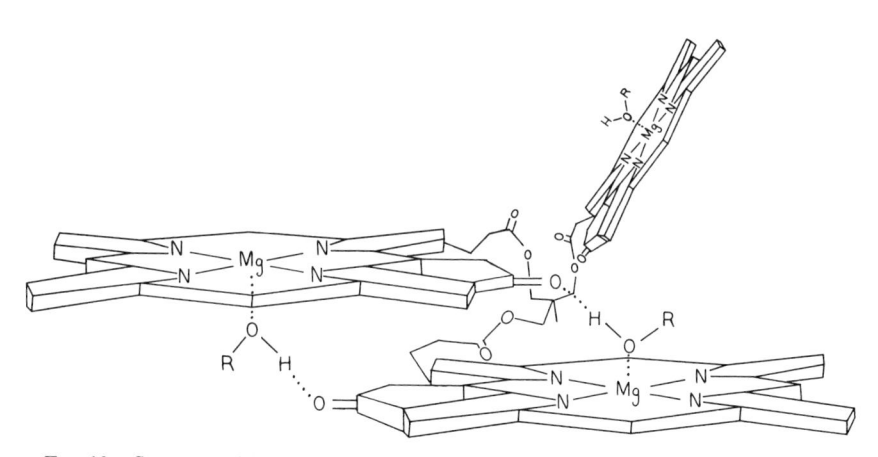

FIG. 10. Structure of folded form of covalently linked magnesium tris(pyrochlorophyllide *a*)-1,1,1-tris(hydroxymethyl) ethane triesters. In this configuration the solitary macrocycle acts as an antenna for the two macrocycles cross-linked by two ethanol molecules.

In the open configuration, the triesters have absorption and fluorescence properties similar to those of monomeric Pyrochl *a* and, like the open covalently linked pairs, have a mechanism for dissipation of excitation energy that prevents the buildup of a concentration of molecules in the S_1 state high enough to attain a population inversion and laser action. In the folded configuration (Fig. 10), the triesters emit coherent (laser) light when optically pumped. Regardless of excitation frequency, lasing occurs in the folded system at 735 nm. Even when folding has occurred to only a small extent, as judged from absorption spectra, lasing is observed only at long wavelengths. In solutions containing both monomeric Chl a and self-assembled pairs not linked to each other as they are in the triesters, fluorescence in both the monomer and the folded pair emission regions occurs. In the triester system, excitation energy is transferred efficiently from the monomer to the folded pair, and such models are expected to provide useful information relevant to antenna function *in vivo*.

VII. A Model for Electron Transfer from the Primary Donor

Pellin *et al.* (1979) have described a very interesting model for electron transfer from Chl_{sp}. Usually, folding the linked pairs is accomplished by a simple monohydric alcohol such as ethanol. Pellin *et al.* folded the linked Pyrochl *a* pair with a pyropheophytin derivative in which the phytyl chain was replaced by transesterification with ethylene glycol; one hydroxyl group of the glycol is esterified at the propionic acid side chain, and the other hydroxyl group is then available for folding the linked pair. In its folded configuration the linked pair has two Pyropheo *a* macrocycles fixed at a known distance from the folded pair. In such an assembly, the characteristic ~730-nm fluorescence from the folded pair is almost completely quenched. If Pyropheo *a* is added to a linked pair folded by ethanol, the fluorescence yield at 730 nm is enchanced rather than quenched. In the Pellin *et al.* (1979) system, spectral changes in the pair are observed to occur in less than 6 psec after excitation, which is interpreted to indicate that either a radical pair or a charge transfer state has been produced on a very short time scale, very much as is postulated to happen *in vivo*. The spectral changes last for 10–20 nsec before a return to the ground state of the pair. The reverse reaction in this system thus is slower than the forward reaction by a factor of at least 10^3. As the reasons for the high directionality of electron transfer from the primary donor to the acceptor is one of the most important unsolved problems in photosynthesis, the ability to mimic this aspect of electron transfer in a simple system may indicate that the prevention of electron back transfer to the primary donor can be achieved by simpler means than those that have been contemplated.

REFERENCES

Bard, A. J., Ledwith, A., and Shine, H. J. (1976). *Adv. Phys. Org. Chem.* **13**, 155–278.
Baugher, J., Hindman, J. C., and Katz, J. J. (1979). *Chem. Phys. Lett.* **63**, 159–162.
Beddard, G. S., and Porter, G. (1976). *Nature (London)* **260**, 366–367.

Beddard, G. S., Carlin, S. E., and Porter, G. (1976). *Chem. Phys. Lett.* **43**, 27–32.
Boxer, S. G., and Closs, G. L. (1976). *J. Am. Chem. Soc.* **98**, 5406–5408.
Brody, S. S. (1958). *Science* **128**, 838–839.
Brody, S. S., and Brody, M. (1961). *Arch. Biochem. Biophys.* **95**, 521–525.
Brody, S. S., and Brody, M. (1963). *In* "Photosynthetic Mechanisms of Green Plants" (B. Kok and A. T. Jagendorf, eds.), Publ. No. 1145, pp. 455–478. Natl. Acad. Sci.–Natl. Res. Counc., Washington, D.C.
Butler, W. L. (1978). *Annu. Rev. Plant Physiol.* **29**, 345–378.
Butler, W. L., and Katajima, M. (1975a). *Biochim. Biophys. Acta* **376**, 116–125.
Butler, W. L., and Katajima, M. (1975b). *Biochim. Biophys. Acta* **396**, 72–85.
Chow, H.-C., Serlin, R., and Strouse, C. E. (1975). *J. Am. Chem. Soc.* **97**, 7230–7237.
Cotton, T. M., Loach, P. A., Katz, J. J., and Ballschmiter, K. (1978). *Photochem. Photobiol.* **27**, 735–749.
Dmetrievsky, O. D., Ermolsev, U. L., and Terenin, A. N. (1957). *Dokl. Akad. Nauk SSSR* **114**, 468–470.
Evans, T. A., and Katz, J. J. (1975). *Biochim. Biophys. Acta* **396**, 414–426.
Fong, F. K. (1974). *Proc. Natl. Acad. Sci. U.S.A.* **71**, 3692–3695.
Fong, F. K., and Koester, V. J. (1976). *Biochim. Biophys. Acta* **123**, 52–64.
Freed, K. (1980). *J. Am. Chem. Soc.* **102**, 3130–3135.
Freed, S., and Sancier, K. (1954). *J. Am. Chem. Soc.* **76**, 198–205.
Govindjee, and Yang, L. (1966). *J. Gen. Physiol.* **49**, 763–780.
Gutschick, V. P. (1978). *J. Bioenerg. Biomembr.* **10**, 153–170.
Hindman, J. C., Kugel, R., Svirmickas, A., and Katz, J. J. (1977). *Proc. Natl. Acad. Sci. U.S.A.* **74**, 5–9.
Hindman, J. C., Kugel, R., Wasielewski, M. R., and Katz, J. J. (1978). *Proc. Natl. Acad. Sci. U.S.A.* **75**, 2076–2079.
Hunt, J. E., Macfarlane, R. D., Katz, J. J., and Dougherty, R. C. (1980). *Proc. Natl. Acad. Sci. U.S.A.* **77**, 1745–1748.
Huppert, D., Rentzepis, P. M., and Tollin, G. (1976). *Biochim. Biophys. Acta* **440**, 1356–1364.
Hynninen, P. H., Wasielewski, M. R., and Katz, J. J. (1979). *Acta Chem. Scand., Sect. B* **33**, 637–648.
Katz, J. J. (1973). *In* "Inorganic Biochemistry" (G. L. Eichhorn, ed.), Vol. 2, pp. 1022–1066. Elsevier, Amsterdam.
Katz, J. J. (1979a). *In* "Advances in Biochemistry and Physiology of Plant Lipids" (L. Appelqvist and C. Liljenberg, eds.), pp. 37–56. Elsevier/North-Holland, Amsterdam.
Katz, J. J. (1979b). *In* "Light Induced Charge Separation in Biology and Chemistry" (H. Gerischer and J. J. Katz, eds.), pp. 331–359. Verlag Chemie, Weinheim.
Katz, J. J., and Brown, C. E. (1981). *Bull. Magn. Reson.* (in press).
Katz, J. J., and Norris, J. R. (1973). *Curr. Top. Bioenerg.* **5**, 41–75.
Katz, J. J., Norman, G. D., Svec, W. A., and Strain, H. H. (1968). *J. Am. Chem. Soc.* **90**, 6841–6845.
Katz, J. J., Oettmeier, W., and Norris, J. R. (1976a). *Philos. Trans. R. Soc. London, Ser. B* **273**, 227–253.
Katz, J. J., Norris, J. R., and Shipman, L. L. (1976b). *Brookhaven Symp. Biol.* **28**, 16–55.
Katz, J. J., Shipman, L. L., Cotton, T. M., and Janson, T. R. (1978a). *In* "The Prophyrins" (D. Dolphin, ed.), Vol. 5, pp. 401–458. Academic Press, New York.
Katz, J. J., Norris, J. R., Shipman, L. L., Thurnauer, M. C., and Wasielewski, M. R. (1978b). *Annu. Rev. Biophys. Bioeng.* **7**, 393–434.
Katz, J. J., Shipman, L. L., and Norris, J. R. (1979). *Chlorophyll Organ. Energy Transfer Photosynth., Ciba Found. Symp.* No. 61, pp. 1–34.

Kochubei, S. M. (1980). *Photosynthetica* **14**, 8–11.
Kochubei, S. M., and Guliev, F. A. (1980). *Photosynthetica* **14**, 182–188.
Konishi, K., Ogawa, T., Inonue, Y., and Shibata, K. (1973). *Plant Cell Physiol.* **14**, 227–236.
Krasnovsky, A. A., and Bystrova, M. I. (1980). *BioSystems* **12**, 181–194.
Kratky, C., and Dunitz, J. D. (1975). *Acta Crystallogr.* Sect. B, **B-31**, 1586–1589.
Kratky, C., and Dunitz, J. D. (1977). *J. Mol. Biol.* **113**, 431–442.
Lavorel, J., and Etienne, A.-L. (1977). *In* "Primary Processes in Photosynthesis" (J. Barber, ed.), pp. 206–268. Elsevier, Amsterdam.
Lessing, H. E. (1976). *Opt. Quantum Electron.* **8**, 309–315.
Lessing, H. E., Lippert, E., and Rapp, W. (1970). *Chem. Phys. Lett.* **7**, 247–253.
Leupold, D., Mory, S., Konig, R., Hoffman, P., and Hieke, B. (1977). *Chem. Phys. Lett.* **45**, 567–571.
Linschitz, H., and Sarkanen, K. (1958). *J. Am. Chem. Soc.* **80**, 4826–4832.
Livingston, R. (1960). *Q. Rev., Chem. Soc.* **14**, 174–199.
Netzel, T. L., Kroger, P., Chang, C. K., Fujita, I., and Fajer, J. (1979). *Chem. Phys. Lett.* **67**, 232–228.
Norris, J. R., and Bowman, M. K. (1980). *Int. Congr. Photosynth., 5th, Halkidiki, Greece* Abstr., p. 419.
Norris, J. R., Uphaus, R. A., Cotton, T. M., and Katz, J. J. (1970). *Biochim. Biophys. Acta* **223**, 446–449.
Norris, J. R., Uphaus, R. A., Crespi, H. L., and Katz, J. J. (1971). *Proc. Natl. Acad. Sci. U.S.A.* **68**, 625–628.
Norris, J. R, Scheer, H., Druyan, M. E., and Katz, J. J. (1974). *Proc. Natl. Acad. Sci. U.S.A.* **71**, 4897–4900.
Norris, J. R., Scheer, H., and Katz, J. J. (1975). *Ann. N.Y. Acad. Sci.* **244**, 260–280.
Parson, W. G. (1974). *In* "Chemical and Biochemical Applications of Lasers" (C. B. Moore, ed.), pp. 339–372. Academic Press, New York.
Pekkarinen, L., and Linschitz, H. (1960). *J. Am. Chem. Soc.* **82**, 2407–2411.
Pellin, M. J., Kaufmann, K. J., and Wasielewski, M. R. (1979). *Nature (London)* **278**, 54–55.
Pellin, M. J., Wasielewski, M. R., and Kaufmann, K. J. (1980). *J. Am. Chem. Soc.* **102**, 1868–1873.
Periasamy, N., Linschitz, H., Closs, G. L., and Boxer, S. G. (1978). *Proc. Natl. Acad. Sci. U.S.A.* **75**, 2563–2566.
Prell, G., and Zanker, V. (1970). *Ber. Bunsenges. Phys. Chem.* **74**, 985–988.
Rabinowitch, E. (1945). "Photosynthesis," Vol. 1, p. 421. Wiley (Interscience), New York.
Schafer, F. P. (1972). *In* "Laser Handbook" (F. T. Arecchi and E. O. Schulz-DuBois, eds.), pp. 369–423, Vol. 1, Chap. B3. North-Holland Publ., Amsterdam.
Seely, G. R. (1978). *Curr. Top. Bioenerg.* **7A**, 3–37.
Setif, P., and Mathis, P. (1980). *Arch. Biochem. Biophys.* **204**, 477–485.
Shepanski, J. F. and Anderson, R. W. Jr. (1981) *Chem. Phys. Lett.* 78, 165–173.
Shipman, L. L. (1980). *Photochem. Photobiol.* **31**, 156–167.
Shipman, L. L., and Katz, J. J. (1977). *J. Phys. Chem.* **81**, 577–581.
Shipman, L. L., Cotton, T. M., Norris, J. R., and Katz, J. J. (1976a). *J. Am. Chem. Soc.* **98**, 8222–8230.
Shipman, L. L., Norris, J. R., and Katz, J. J. (1976b). *J. Phys. Chem.* **80**, 877–882.
Shipman, L. L., Cotton, T. M., Norris, J. R., and Katz, J. J. (1976c). *Proc. Natl. Acad. Sci. U.S.A.* **73**, 1791–1794.
Thurnauer, M. C., Katz, J. J., and Norris, J. R. (1975). *Proc. Natl. Acad. Sci. U.S.A.* **72**, 3270–3274.
Wasielewski, M. R., and Svec, W. A. (1980). *J. Org. Chem.* **45**, 1969–1974.

Wasielewski, M. R., Studier, M. H., and Katz, J. J. (1976). *Proc. Natl. Acad. Sci. U.S.A.* **73**, 4282–4286.
Wasielewski, M. R., Smith, U. H., Cope, B. T., and Katz, J. J. (1977). *J. Am. Chem. Soc.* **99**, 4172–4173.
Watson, W. F., and Livingston, R. (1950). *J. Chem. Phys.* **18**, 802–809.
Weber, G., and Teale, F. W. J. (1957). *Trans. Faraday Soc.* **53**, 646–655.
Yuen, M. J., Closs, G. L., Katz, J. J., Roper, J. A., Wasielewski, M. R., and Hindman, J. C. (1980a). *Proc. Natl. Acad. Sci. U.S.A.* **77**, 5598–5601.
Yuen, M. J., Shipman, L. L., Katz, J. J., and Hindman, J. C. (1980b). *Photochem. Photobiol.* **32**, 281–296.
Zanker, V., Rudolph, E., and Prell, G. (1970). *Z. Naturforsch., Teil.* **B25**, 1137–1143.

CHAPTER 6

Exciton Annihilation and Other Nonlinear High-Intensity Excitation Effects

Nicholas E. Geacintov

Chemistry Department
New York University
New York, New York

and

Jacques Breton

Departement de Biologie, Service de Biophysique
Centre d'Etudes Nucleaires de Saclay
Gif sur Yvette, France

I.	INTRODUCTION	158
II.	PHOTOPHYSICAL PHENOMENA	158
	A. Two-Photon Absorption	159
	B. Ground State Depletion	159
	C. Excited State Absorption	160
	D. Stimulated Emission and Lasing	161
	E. Photochemical Effects	161
	F. Exciton–Exciton Annihilation	162
III.	LASER CHARACTERISTICS IN RELATION TO THE DIFFERENT PHOTOPHYSICAL PROCESSES	162
IV.	SINGLE PICOSECOND PULSE EXCITATION	163
	A. Fluorescence of Chlorophyll in Vivo	163
	B. Distribution of Excitation Energy and Topology of the Photosynthetic Apparatus	168
	C. Distribution of Excitation Energy in the Red Alga *Porphyridium cruentum* and the Blue-Green Alga *Nostoc* sp.	173
V.	NANOSECOND PULSE EXCITATION	174
	A. General Kinetic Equations	174
	B. The Steady State Approximation	175
	C. The Three-Level System in the Steady State Approximation	176
	D. Inclusion of Exciton Annihilation Effects—General Time-Dependent Case	177
	E. Criteria for Distinguishing between Exciton Annihilation and Ground State Population and Excited State Absorption Effects	178

F. Application to Chlorophyll-Containing Systems—
Dilute Solutions of Chlorophyll a and Chloroplasts 179
G. Singlet–Singlet versus Singlet–Triplet Annihilation
Rates on the Nanosecond Time Scale 181
VI. MICROSECOND PULSE EXCITATION 183
A. Effect of Integrated Pulse on Triplet Yield 184
B. Buildup of Triplets as a Function of Time within a
Microsecond Pulse 184
C. Unequal Fluorescence Quenching Observed for the Two
Fluorescence Bands F685 and F735 at Low Temperatures
in Chloroplasts 185
D. Picosecond Pulse Train Excitation 186
VII. PICOSECOND PULSE EXCITATION STUDIES ON ABSORPTION
CHANGES IN BACTERIAL REACTION CENTERS 187
REFERENCES 189

I. Introduction

High-intensity pulsed laser light sources have come into increased use in photobiology research within the last ten years. When intense laser pulses are incident on a system of chromophores embedded in a biological membrane, several photophysical phenomena, which can be usually neglected when more conventional weak-intensity light sources are utilized for excitation, may manifest themselves. In properly designing an experiment in which laser pulses are utilized, the characteristics of the laser excitation in relation to the critical parameters of the system under investigation must be examined. In this chapter some of the most important nonlinear effects are discussed, and their occurrence in photosynthetic systems exposed to intense laser pulses is described. Recent fluorescence studies on energy transfer characteristics, topology, and exciton–exciton annihilation in green plant photosynthetic membranes are reviewed. The possible occurrence of nonlinear processes in absorption kinetic studies on primary events in bacterial reaction centers RCs is also discussed.

II. Photophysical Phenomena

Rather than attempt an exhaustive listing of the nonlinear photophysical processes which can occur, we will limit ourselves here to phenomena which we have found to be of importance in the study of systems containing chlorophyll either *in vivo* or *in vitro* or of related model systems exposed to intense pulsed laser light.

A. Two-Photon Absorption

Arsenault and Denariez-Roberge (1976) observed a nonlinear absorption process for chlorophyll *a* in solution excited by a Q-switched ruby laser pulse, which they attributed to a two-photon excitation phenomenon. From their data they calculated a two-photon absorption cross section σ_{02} of $\approx 10^{-43}$ cm^4 sec for 694-nm radiation, which is several orders of magnitude higher than the value of σ_{02} measured normally for organic molecules (Birks, 1970); they attributed this unusually high value of σ_{02} for chlorophyll *a* to "resonance phenomena." There is, in general, a lack of reliable information concerning the values of two-photon absorption coefficients for chlorophyll *a* and related molecules. Some general principles regarding the occurrence of this effect can nevertheless be advanced here.

The relative rate of two-photon versus one-photon absorption is

$$\sigma_{02} I^2(t)/\sigma_{01} I(t), \tag{1}$$

where $I(t)$ is the instantaneous laser photon flux. If we take the range of values of $\sigma_{02} = 10^{-49}$–10^{-50} cm^4 sec observed for several organic chromophores (Birks, 1970), while σ_{01}, the one-photon absorption cross section is in the range of 10^{-16}–10^{-18} cm^2, then the ratio in Eq. (1) is 10^{-7}–10^{-3}, taking typical (maximum) values of $I(t) \approx 10^{27}$ photons cm^{-2} sec^{-1} in picosecond pulse excitation experiments (Yu *et al.*, 1977; Campillo and Shapiro, 1977; Porter *et al.*, 1977; Geacintov *et al.*, 1977a,b, 1979a). Equation (1) thus shows that two-photon absorption processes are usually not important, particularly when the excitation is carried out within an absorption band of the chromophore (large σ_{01}). Outside an absorption band, however, when σ_{01} is very small, two-photon absorption processes may become important.

B. Ground State Depletion

This effect, frequently termed bleaching or induced transparency, can become important when the number of photons absorbed by the system approaches several percent or more of the number of molecules present in the irradiated volume. Under these conditions the absorption of photons becomes nonlinear with increasing pulse intensity (Schafer, 1976). The specific conditions under which this effect becomes important depend on the laser pulse duration, instantaneous photon flux, absorption cross section, and excited state lifetime, as will be illustrated in more detail below. Recently Doukas *et al.* (1981) have utilized the recovery of the ground state in phycobiliproteins excited with a single picosecond laser pulse to monitor the lifetime of the singlet excited states in these systems.

C. Excited State Absorption

The importance of this effect depends on the relative absorption cross sections σ_{01}, σ^S_{12}, and σ^T_{12} (in square centimeters) at the excitation wavelength of the laser (Fig. 1). These absorption cross sections are related to the following processes:

$$S_0 + h\nu_l \xrightarrow{\sigma_{01}} S_1, \quad (2)$$

$$S_1 + h\nu_l \xrightarrow{\sigma^S_{12}} S_2, \quad (3)$$

$$T_1 + h\nu_l \xrightarrow{\sigma^T_{12}} T_2, \quad (4)$$

where the subscripts 0, 1, and 2 denote the ground state and first and second excited states of the molecules, respectively, and the superscripts S and T denote singlet and triplet excited states, respectively. Excited state absorption effects become important when

$$\sigma^S_{12}[S_1] \gtrsim \sigma_{01}[S_0], \quad (5)$$

$$\sigma^T_{12}[T_1] \gtrsim \sigma_{01}[S_0]. \quad (6)$$

Therefore excited state absorption effects may manifest themselves when ground state depopulation occurs. For chlorophyll *a* in solution, the wavelength dependence of the exicted state absorption cross sections σ^S_{12} and σ^T_{12} have been recently published (Baugher *et al.*, 1979). Recently Shepanski and Anderson (1981) measured the transient absorbance spectra from 400 to 700 nm of chlorophyll *a* in pyridine at 10 and 300 psec after excitation by a 7-psec 355-nm pulse. The excited singlet absorption cross sections in the regions of 530 and 694 nm are 2.5 and 7 times that of the ground state absorption cross section. The occurrence of excited state absorption effects *in vivo* can be estimated if the pulse intensity, pulse duration, intersystem crossover constant K_{IC}, and excited state lifetimes are known.

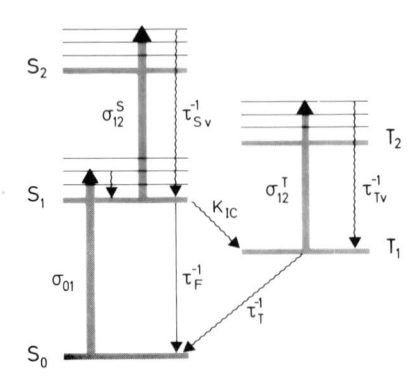

FIG. 1. Energy-level diagram. S, Singlet states; T, triplet states, σ, absorption cross sections, τ, excited state lifetimes. Heavy vertical arrows indicate optical transitions induced by a laser, the thin vertical arrow indicates fluorescence, and wavy arrows indicate nonradiative transitions.

D. Stimulated Emission and Lasing

Stimulated emission and lasing for chlorophyll and bacteriochlorophyll in solution have been observed by Hindman et al. (1978a). The conditions for lasing of chlorophyll have been discussed (Hindman et al., 1977) and depend on the configuration of the reflecting surfaces, path length, etc. For chlorophyll a in solution, in a 1 × 1 cm cuvet, lasing occurs at relatively high excitation intensities (nitrogen laser photon flux $I(t) \gtrsim 4 \times 10^{25}$ photons cm^{-2} sec^{-1} at 337 nm). The stimulated emission exhibits highly directional properties. Viewing the fluorescence along the direction of the excitation beam gives rise to an enhancement of the signal, while viewing the signal in a perpendicular direction gives rise to a decrease in the fluorescence yield F with increasing pulse intensity, where F is defined as

$$F = \text{(fluorescence intensity)}/\text{(incident light intensity)}.$$

Such directional effects have been demonstrated for dilute solutions of chlorophyll a (Hindman et al., 1978a) and of rhodamine 6G (R6G) (Husiak, 1980). Amplification of the fluorescence by stimulated emission is accompanied by a decrease in the fluorescence lifetime (Penzkofer and Falkenstein, 1978).

Lasing action and stimulated emission for chlorophyll a *in vivo* have not been observed (Geacintov et al., 1979b). The fluorescence lifetime of chlorophyll a *in vivo* is considerably shorter than that of chlorophyll in solution. It is therefore more difficult, in principle, to observe lasing effects in photosynthetic systems. Furthermore, at the high intensities needed to build up a sufficient population of excited states, bimolecular exciton annihilation processes occur, which decrease the singlet excited state lifetime still further (see below) and thus render the observation of lasing *in vivo* even less likely.

E. Photochemical Effects

The chromophores may not be stable under repeated exposure to high-intensity pulsed laser flashes because of photochemical reaction pathways. Thus, during the course of a given series of experiments, the occurrence of such irreversible photochemical processes must be monitored. Chlorophyll a *in vivo* exposed to either intense nanosecond (Mauzerall, 1976a) or single picosecond pulses has been found to be remarkably stable. We have nevertheless noted that the fluorescence of spinach chloroplasts decreases when either intense mode-locked picosecond pulse trains or microsecond-duration laser pulses are utilized when the total photon flux exceeds 10^{18} photons/cm^2 per pulse. Such effects are attributed to irreversible chemical transformations of chlorophyll a whose nature is not well understood. Experimentally, we have found it convenient to monitor the fluorescence yield at low intensities after exposure of chloroplast samples to high pulse intensities. An irreversible decrease in this yield indicates that photochemical bleaching of chlorophyll has occurred.

F. Exciton–Exciton Annihilation

When the average distance between molecules is on the order of 50–100 Å or less, resonance energy transfer can occur between neighboring molecules. Such effects are thus possible when the concentration of molecules exceeds 10^{-3}–10^{-2} M. *In vivo*, the chlorophyll concentration is about 0.1 M, thus facilitating the efficient transfer of excitation energy from the antenna molecules to the RCs. If a sufficient density of excitons is present, bimolecular interactions with the annihilation of one or both excitons is possible. Such annihilations lead to nonlinear fluorescence-quenching effects at high laser pulse intensities. Both singlet–singlet and singlet–triplet exciton annihilations are readily observed in organic crystals and solutions (for review, see Swenberg and Geacintov, 1973) and in photosynthetic systems (Campillo *et al.*, 1976b; Campillo and Shapiro, 1978; Geacintov *et al.*, 1977a,b, 1979a). In photosynthetic membranes, decreased fluorescence yield and exciton lifetimes constitute the dominant nonlinear high-intensity effects.

III. Laser Characteristics in Relation to the Different Photophysical Processes

In assessing which of the nonlinear photophysical phenomena are important, it is necessary to consider the characteristics of the laser pulse, such as its wavelength, intensity, and duration. The important properties of the system under investigation are the absorption cross sections of the ground and excited states, the intersystem crossover rate constant K_{IC}, and the lifetimes of the singlet (τ_S) and triplet (τ_T) excited states.

The output of a mode-locked laser typically consists of a train of ~300-psec pulses spaced about 5–10 nsec apart. Using an electrooptic shutter it is possible to isolate a single pulse for excitation, whose duration is typically 10–30 psec. The laser flash repetition rate is low in this case. More recently, subpicosecond pulse excitation at high repetition rates has become available by passively mode-locking a (continuous-wave) cw dye laser pumped by a continuous argon ion laser (Ippen and Shank, 1977). There are, as of this writing, few applications of subpicosecond excitation techniques to photosynthetic systems, except those described in the recent paper by Beddard *et al.* (1979). Because of the low intensity of these picosecond pulses (10^{11} photons/cm^2 per pulse; Beddard *et al.*, 1979), the high repetition rate, and the time resolution, there is no doubt that this type of excitation will become of increasing importance in the study of photosynthetic and other biological systems. In this chapter, we will confine our discussion mainly to the classic picosecond pulse excitation mode which has been used extensively in photosynthesis research thus far. In single picosecond pulse excitation the role of triplets can be usually neglected. The intersystem crossover constant K_{IC} being on the order of 10^8 sec^{-1} for chlorophyll a, very few triplets are produced in the immediate time interval following the excitation pulse. Therefore, the evolution of the fluorescence within the first few hundreds of picoseconds is thus not affected by triplet excited states. However, a finite number of triplets are nevertheless produced for each picosecond pulse. Thus when a pulse train is utilized, there is a progressive buildup of triplet excited states with increasing time. The triplet lifetime (carotenoids) in chloroplasts is on the order of microseconds, and the triplets thus persist from pulse to pulse within the mode-locked train; their concentration increases progressively with time (Breton and Geacintov, 1976; Geacintov et al., 1978; Breton et al., 1979). Similar effects occur when a non-mode-locked microsecond-duration pulse is used (Geacintov and Breton, 1977).

In the case of single picosecond pulse excitation singlet–singlet annihilations dominate over singlet–triplet annihilations in reducing the fluorescence yield at high excitation intensities, while for microsecond pulsed excitation singlet–triplet annihilations dominate. Thus different nonlinear high-intensity phenomena occur depending on the duration of the laser pulse. It is convenient to classify the different experiments according to the relative values of the laser pulse duration t_p, the singlet exciton lifetime τ_S, and K_{IC}. In studies on the kinetic response of photosynthetic systems to pulsed laser excitation we distinguish three cases:

(a) $t_p \ll \tau_S, K_{IC}^{-1}$, single picosecond pulse excitation,
(b) $t_p \approx \tau_S, K_{IC}^{-1}$, nanosecond pulse excitation, ruby laser, nitrogen laser,
(c) $t_p \gg \tau_S, K_{IC}^{-1}$, microsecond-duration dye laser, or mode-locked laser pulse train.

Each of these cases will now be discussed separately.

IV. Single Picosecond Pulse Excitation

A. FLUORESCENCE OF CHLOROPHYLL *IN VIVO*

1. *Intensity-Dependent Quenching*

In single picosecond pulse excitation an important consideration is, besides the instantaneous photon flux, the total number of photons per square centimeter per pulse (I) in relation to the absorption coefficient σ_{01} at the excitation wavelength. When $\sigma_{01}I \ll 1.0$, ground state depopulation, and thus excited state absorption, can usually be neglected if excitation is taking place within an absorption band. At the end of the pulse, if deactivation of the excited states during the pulse by either decay or absorption is neglected, the fraction of molecules in the excited (singlet) state is given by

$$[S_1]/([S_1] + [S_0]) = 1 - \exp(-\sigma_{01}I). \tag{7}$$

An important assumption in the derivation of this formula is that the sample is optically thin, i.e., that $\sigma_{01}[S_0]l \ll 0.1$, where $[S_0]$ is the total number of molecules present (when $[S_1] = 0$) and l is the optical path length. This ensures that the intensity of the laser beam is approximately constant within the sample.

We have utilized in our own studies on the fluorescence of spinach chloroplasts an excitation of 610 nm, for which $\sigma_{01} = 2.5 \times 10^{-17}$ cm². A plot calculated according to Eq. (7) for this wavelength of excitation of chloroplasts as a function of the total number of photons per square centimeter per pulse is shown in Fig. 2. At an intensity of 10^{16} photons/cm² per pulse, about 20% of the molecules are in the excited state, increasing to about 40% at the maximum pulse intensity of 2×10^{16} utilized in our single-pulse, intensity-dependent fluorescence-quenching experiments (Geacintov et al., 1977a,b, 1979a). A typical fluorescence-quenching curve is shown in Fig. 3. The fluorescence yield F begins to decline from its limiting low-intensity value when there are as few as 0.1 hits per photosynthetic unit (PSU)—simply defined here as a collection of 300 chlorophyll molecules

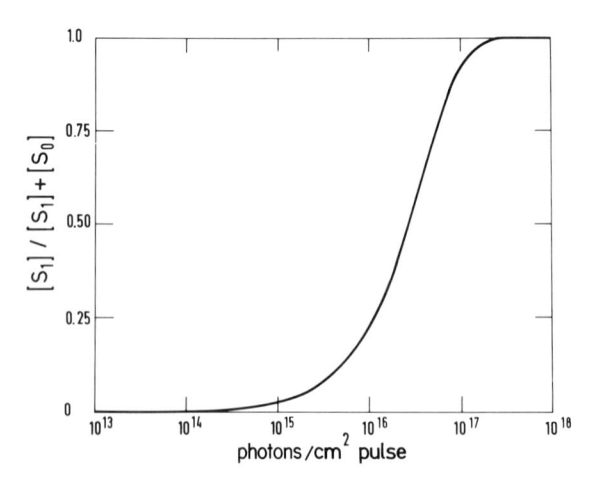

FIG. 2. Fraction of chlorophyll molecules in chloroplasts promoted to the singlet excited state by a single picosecond laser pulse at 610 nm.

6. EXCITON ANNIHILATION—NONLINEAR EXCITATION EFFECTS 165

FIG. 3. Fluorescence-quenching curves (yields versus picosecond laser pulse intensity). Spinach chloroplasts, room temperature, 610-nm excitation wavelength. The inset shows the shape of the fluorescence spectrum.

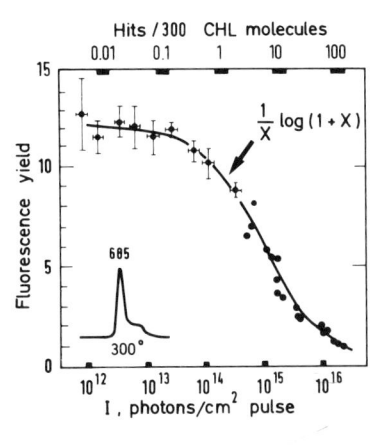

per reaction center. The yield F is reduced by a factor of ~ 2 when only 1% of the molecules are in the excited state, and by a factor of ~ 10 when 10% of the molecules have absorbed a photon. Thus, the phenomenon observed in Fig. 3 cannot be due to ground state depletion or excited state absorption effects. It is now generally accepted that this reduction in the fluorescence yield is due to singlet–singlet exciton annihilation according to the scheme

$$S_1 + S_1 \xrightarrow{\gamma_{ss}} S_0 + S_n, \qquad (8)$$

where γ_{ss} is the bimolecular annihilation constant. Thus, because of their mobility, few excitons need be present for these bimolecular exciton annihilation effects to occur. Swenberg et al. (1976) have proposed a simple equation, based on Stern–Volmer kinetics (Birks, 1970) to account for the results shown in Fig. 3:

$$F = X^{-1} \log(1 + X), \qquad (9)$$

where $X = \gamma_{ss}\tau_s I/2$. A fit of this equation to the experimental data is also shown in Fig. 3. The fit is excellent, and from the value of the parameter X which is necessary to fit the experimental data as shown, a value of $\gamma_{ss} \approx 10^{-8}$ cm^3/sec^1 is deduced. This value of γ_{ss} is characteristic of singlet–singlet exciton annihilation in organic crystals (Swenberg and Geacintov, 1973; Campillo and Shapiro, 1978).

Equation (9) predicts that the decrease in F will occur at lower pulse intensities for longer singlet exciton lifetimes τ_S; qualitatively, the longer the lifetime, the greater the probability that two excitons will encounter one another, i.e., find themselves within the interaction radius of ~ 50 Å (Rahman and Knox, 1973) and give rise to an annihilation. Conversely, the

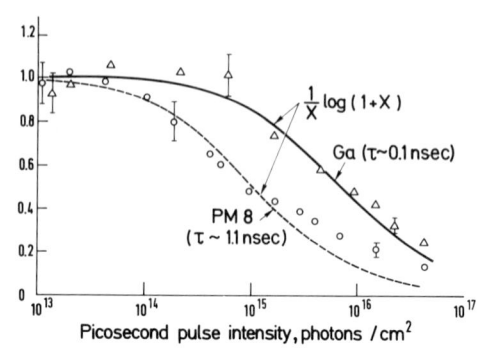

FIG. 4. Fluorescence-quenching curves for two strains of bacterial chromatophores (*R. sphaeroides*) exhibiting two different exciton lifetimes (see text) at room temperature and 530-nm excitation. (Adapted from the results of Campillo *et al.*, 1977a; reproduced with permission of the publisher National Academy of Sciences, U.S.A.)

shorter the lifetime, the less probable an annihilation event at a given intensity. Such an effect can be seen in the data of Campillo *et al.* (1977a) for their single picosecond pulse study of F as a function of pulse intensity for two strains of the photosynthetic bacterium *Rhodopseudomonas sphaeroides* (Fig. 4). Strain Ga exhibits a short exciton lifetime of 0.1 nsec which is probably due to a fast energy transfer to the reaction centers which quench the exciton. Strain PM-8 *dpl* lacks reaction centers, and its exciton lifetime is considerably longer (1.1 nsec). As expected from the exciton-quenching model, the decline in the F curve sets in at much lower pulse intensities for strain PM-8 *dpl* than for strain Ga (Fig. 4). Equation (9) provides a good fit to the data, at least for strain Ga.

2. *Connectivity between Photosynthetic Units*

The simple kinetic model based on $S_1 + S_1$ annihilation [Eq. (9)] has been refined and considerably extended by Paillotin *et al.* (1979) using a master equation approach. In addition to providing information on the dynamics of singlet excitons in photosynthetic membranes (Geacintov *et al.*, 1977b, 1979a), fluorescence-quenching curves such as those shown in Fig. 3 can provide information on the size of the molecular domain over which the singlet excitons can migrate.

The principle of the method can be illustrated by the following example. Consider a system of isolated, unconnected domains consisting of n molecules per domain. Fluorescence quenching by bimolecular exciton annihilation can occur only if there are two or more photon hits per domain. Now we consider that there is one hit per domain and that all domains consisting of groups of n molecules each are connected to one another, so that excitons can diffuse freely from domain to domain. In such a case, fewer than one hit per n molecules may suffice to produce a decrease in F by annihilation, an effect which, of course, depends on the mobility of the excitons, their lifetimes, and thus their ability to encounter and to annihilate one another. According to Paillotin's theory, the relative position of the F curve on the horizontal (pulse intensity) axis in Fig. 3 gives an indication of the number of molecules (or domain size) over which an exciton can migrate in chloroplast membranes. Detailed analysis of experimental data for chloroplasts (Geacintov et al., 1977b, 1979a) indicates that a typical domain size consists of 600–1200 molecules or more and that the connectivity between adjacent photosynthetic units is not markedly dependent on the temperature in the range 20–300 K.

Under certain conditions, Poisson statistics (Mauzerall, 1976a,b) may be more appropriate to describe the fluorescence-quenching data of the type shown in Fig. 3. In this case, the F curves decrease more abruptly with increasing pulse intensity. A detailed analysis shows that Poisson statistics describe the picosecond pulse experimental fluorescence-quenching data in chloroplasts less well than the exciton annihilation model (Geacintov et al., 1977b, 1979a; Paillotin et al., 1979). Paillotin et al. (1979) have shown that the Swenberg equation [Eq. (9)] and the Poisson statistics cases (Mauzerall, 1976a,b) are the two limiting cases of the master equation approach. Poisson statistics can provide an adequate explanation of high-intensity pulse fluorescence-quenching data when the parameter $r = 2K/\gamma \ll 1.0$, i.e., when the total unimolecular decay rate K of the exciton is much lower than the bimolecular decay rate γ. In the opposite limit, $r \gg 1.0$, the intensity dependence of the fluorescence yield is described by the Swenberg equation [Eq. (9)].

3. *Chlorophyll a Singlet Exciton Lifetime in Vivo*

The strong fluorescence yield decrease first observed with picosecond pulses by Campillo et al. (1976a) and by Mauzerall (1976a) with nanosecond-duration pulses, are accompanied by a strong reduction in the fluorescence decay time *in vivo*. This effect, shown in Fig. 5, was first demonstrated by Campillo et al. (1976b) with single picosecond pulse excitation of the fluorescence of *Chlorella pyrenoidosa*.

Because this exciton annihilation effect was relatively unappreciated before 1976, fluorescence lifetime values measured by single picosecond pulse excitation methods before this time were considerably shorter than the ones measured by conventional techniques. With single picosecond pulse excitation, singlet–singlet annihilation is responsible for this lifetime decrease, while for excitation with a mode-locked pulse train, singlet–triplet annihilation also plays a role in reducing the lifetime. With mode-locked pulse

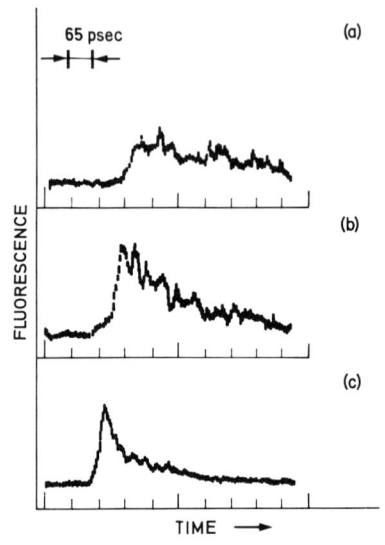

FIG. 5. Fluorescence decay profiles of *C. pyrenoidosa* at different intensities of single picosecond excitation pulses (530 nm) at room temperature. The lifetimes (τ) shown are approximate $1/e$ values. (a) 10^{14} photon/cm^2 pulse ($\tau \sim 375$ psec), (b) 3×10^{14} photon/cm^2 pulse ($\tau \sim 175$ psec), (c) 3×10^{15} photon/cm^2 pulse ($\tau < 50$ psec). (Adapted from Campillo *et al.*, 1976b; reproduced with permission of the publisher. Copyright 1976 by the American Association for the Advancement of Science.)

trains, a particular pulse, usually in the middle of the train, was selected for exciting the fluorescence and for monitoring its decay. In such experiments a considerable concentration of triplets may have accumulated from the previous pulses (Breton and Geacintov, 1976; Geacintov *et al.*, 1978), giving rise to singlet–triplet annihilation.

Since 1976, it has been increasingly recognized that the pulse intensity in single-pulse excitation experiments must be kept below $\sim 10^{14}$ photons/cm^2 per pulse in order to avoid distortions in the decay curves arising from exciton annihilation effects. However, for mode-locked pulse trains, even when the intensity of the individual pulses is this low, a significant buildup of triplet excited states may occur, an effect which should be carefully assessed in such experiments.

B. Distribution of Excitation Energy and Topology of the Photosynthetic Apparatus

While exciton–exciton annihilation may be viewed as an undesirable artifact in measurements of fluorescence decay times, it may be utilized to probe energy transfer pathways on a time scale which is otherwise inaccessible. By increasing the picosecond pulse intensity, the lifetime of singlet excitons may be reduced to almost any value below the normal ~ 500-psec decay time. Thus time domains on the tens or hundreds of picosecond time scales may be probed before the excitation energy has been delivered either to the reaction centers or to the groups of molecules from which fluorescence emission normally takes place, i.e., before equilibration of the exciton distribution has occurred. Before reviewing what has been learned by such techniques, we briefly discuss the fluorescence properties of the photosynthetic apparatus of green plants.

1. *Fluorescence Properties of Chloroplasts*

The fluorescence emission spectra of chloroplasts at low temperatures is shown in Fig. 6. According to the model of the photosynthetic apparatus proposed by Butler (1978), the emission band at 685 nm at room temperature, and at low temperatures, is due to emission from the antenna light-harvesting chlorophyll *a* or *b* proteins. The shoulder at 695 nm which appears upon cooling is due to the chlorophyll antenna molecules closely associated with the reaction centers of photosystem II (PS II), while the 735-nm band is due to a specific form of chlorophyll *a* termed C705, which is present in photosystem I (PS I) only.

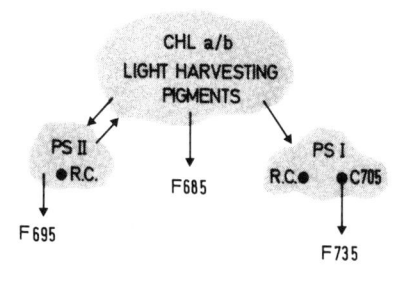

FIG. 6. Model of the photosynthetic apparatus of chloroplasts based on Butler's (1978) model, and a typical fluorescence spectrum of spinach chloroplasts at low temperatures.

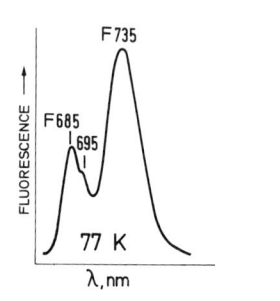

Moya (1979) has made the most detailed study of the dependence of the fluorescence decay time on the fluorescence emission wavelength. In the 670–750 nm region at 77 K, five different fluorescence decay times can be discerned, the lifetimes increasing as the emission wavelength increases. The fluorescence lifetime measured at 685 nm displays little temperature dependence and lies in the range 0.5–0.8 nsec (for review, see Breton and Geacintov,

1980). The 735-nm (PS I) fluorescence decay time, on the other hand, strongly depends on the temperature and varies from about 0.20 nsec at 210 K to 2.0 nsec at 77 K (Butler et al., 1979); however, lifetimes as short as 1.5 ± 0.4 nsec (Geacintov et al., 1977a, 1978) and as long as 2.9–3.5 nsec at 77 K (735 nm) have been reported (Moya, 1979; Butler et al., 1979). Despite these discrepancies in the exact lifetimes measured at 735 nm, it is clear that the fluorescence decay time for the PS I emission at 735 nm varies with temperature and is significantly longer than the lifetime measured at 685–695 nm due to the light-harvesting PS II antenna chlorophyll *a* molecules.

2. *Picosecond Pulse Intensity Dependence of the 685- and 735-nm Fluorescence Yields*

Because of the differences in the fluorescence decay times for the light-harvesting PS II emission (F685 fluorescence band) and the PS I emission (F735 band) it was initially expected that a difference in the fluorescence-quenching curves (plots of fluorescence yield versus pulse intensity) should be observed. Such an effect, due to differences in exciton lifetimes, has been observed with chromatophores of photosynthetic bacteria (Fig. 4). Fluorescence-quenching curves for the F685 and F735 emission bands in chloroplasts, obtained by excitation with a single picosecond pulse (610 nm) and a microsecond-duration laser pulse, are shown in Fig. 7. With microsecond pulse excitation there is indeed a difference between the F685 and F735 fluorescence-quenching curves; this effect will be discussed in detail below. Surprisingly the two fluorescence-quenching curves coincide in the single picosecond pulse excitation case.

According to the considerations outlined in Section IV, a difference in these fluorescence-quenching curves obtained with picosecond pulse

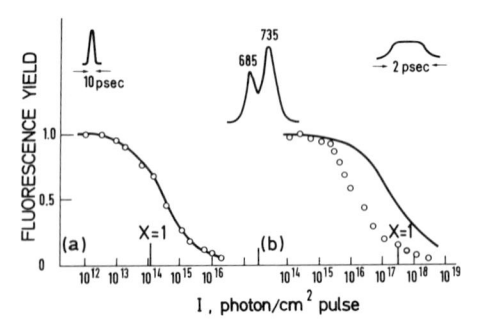

FIG. 7. Fluorescence-quenching curves for the two major emission bands of spinach chloroplasts at 100 K (○,685; ———, 735). (a) Excitation with a single (610-nm) picosecond pulse. (b) Excitation with a single 2 μsec (600-nm) pulse. (From Geacintov et al., 1977a; reproduced with permission of Elsevier/North-Holland Biomedical Press.)

excitation was expected for at least two different reasons: (1) Because of their longer lifetime in PS I, excitons have a higher probability of encountering and annihilating one another; and (2) the pool of pigment molecules (domain size) over which the excitons can migrate are likely to be different for PS I than for the pool of chlorophyll molecules constituting the light-harvesting and PS II pigment system, which emit in the 685–695 nm region.

The fact that the F685 and F735 fluorescence-quenching curves coincide, in spite of these considerations, provides information about the nature of the pigment pools and energy transfer to the groups of molecules in PS I responsible for the 735-nm emission band. The experimental result in Fig. 7a can be explained if it is assumed that singlet–singlet exciton annihilation occurs within the light-harvesting pool of pigments, that energy is transferred to the C705 fluorescence-emitting molecules, and that the pool of pigments (or domain size) surrounding these C705 molecules is sufficiently small so that exciton–exciton annihilations occurring there are much less probable than in the light-harvesting antenna pigment bed. Thus, the 735-nm fluorescence emission, in this model, is a probe of the exciton density in the light-harvesting antenna pool. This model predicts two other effects which can be verified experimentally:

(1) While the fluorescence decay time should strongly depend on the picosecond pulse intensity for the F685 band, the lifetime for the F735 band should be independent of pulse intensity. This prediction is based on the assumption that there are no singlet–singlet annihilations in the immediate environment of the fluorescence-emitting C705 molecules.

(2) The transfer of energy from the light-harvesting pigment bed to the C705 chlorophyll molecules might be finite and measurable on the picosecond time scale.

These predictions of the model were investigated experimentally.

a. *Picosecond Excitation Pulse Dependence of the F735 Fluorescence Decay Time at 77 K.* The pulse intensity dependence of the decay time of the fluorescence of spinach chloroplasts at 77 K was investigated by Geacintov *et al.* (1977a). For a pulse energy (530 nm) of 2.5×10^{14} photons/cm^2 per pulse the decay time at 685 nm was 0.35 nsec and decreased to less than 130 psec for a pulse energy of 7.4×10^{14} photons/cm^2 per pulse. For the F735 fluorescence, on the other hand, the decay time (1.1 nsec, somewhat shorter than that reported by other workers) was independent of the pulse intensity in the intensity range 3.5×10^{14} to 2.6×10^{15} photons/cm^2 per pulse. Thus, in the singlet–singlet annihilation case, the pulse intensity has no effect on the decay time of the 735-nm fluorescence, indicating that the domain size of the pigment pool surrounding the C705 molecules in PS I is small relative to the light-harvesting pigment pool. These pigments thus appear to derive most of their energy by the transfer of excitons from the light-harvesting pigment system.

b. *Risetime of the 735-nm PS I Fluorescence at 77 K.* Campillo et al. (1977b) have shown that, at low picosecond pulse intensities (2–5 × 10^{14} photons/cm^2 per pulse at 530 nm), the risetime of the fluorescence measured at the 735-nm emission wavelength is 150 ± 20 psec (see also Breton and Geacintov, 1980). The risetime of the 685-nm fluorescence, on the other hand, is less than 50 psec, which is the resolution time of the apparatus. For excitation energies in excess of 5 × 10^{14} photons/cm^2 per pulse, the risetime of the 735-nm fluorescence diminishes to the value of the resolution time of the apparatus. Typical signal-averaged traces of the risetimes for the F685 and F735 fluorescence bands are shown in Fig. 8. At the higher pulse energies, the risetime of the 735-nm fluorescence appears to be limited by the exciton lifetime (shortened by annihilation effects) in the light-harvesting pool of pigments.

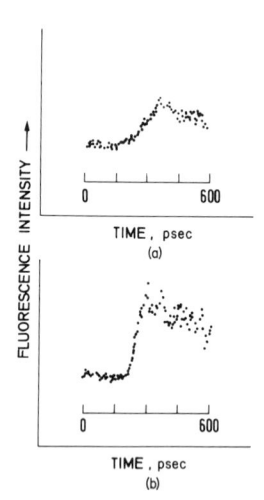

FIG. 8. Risetime of the F735 fluorescence in spinach chloroplasts at 77 K at two different pulse intensities. Excitation with a single picosecond laser pulse (530 nm). (Unpublished results of A. J. Campillo, S. L. Shapiro, K. R. Winn, J. Breton, and N. E. Geacintov.)

The presence of such a risetime of the F735 fluorescence has been questioned by Butler et al. (1979). However, additional support for this PS I fluorescence delayed risetime may be deduced from the time-resolved emission spectra described by Pellegrino et al. (1978); in this work the fluorescence spectrum was determined at $t \approx 0$, i.e., during the picosecond excitation pulse. In this case the intensity of F735 was found to be negligible relative to that of the F685 band, in agreement with the result that this fluorescence builds up after 140 psec. In the time-integrated spectrum, the

F735 band is present. However, the risetime of the 730-nm component was measured to be ~ 13 ps.

In summary of this section, it is evident that picosecond pulse-induced exciton annihilation studies have provided new information on how energy is distributed among the pigment pools in green plant photosynthetic membranes. The results described indicate further that the C705 molecules obtain their energy from a specialized set of light-harvesting molecules (Campillo et al., 1977b) which well may be the so-called PS I antenna pigment molecules (Butler, 1979). However, additional verification of this model is needed. When technically feasible, a more detailed investigation of the pulse intensity dependence of the risetimes and decay profiles of the fluorescence as a function of wavelength should be undertaken. Similar studies on photosynthetic bacteria, in which there is also considerable heterogeneity of the fluorescence, may prove interesting.

C. DISTRIBUTION OF EXCITATION ENERGY IN THE RED ALGA
  Porphyridium cruentum AND BLUE-GREEN ALGA *Nostoc sp.*

An interesting study on the fluorescence risetimes in a photosynthetic system on a picosecond time scale was performed by Porter et al. (1978). The red alga *P. cruentum* possesses water-soluble accessory light-harvesting pigments contained within structures called phycobilisomes attached to the thylakoid membranes. Phycobilisomes contain three pigments: B-phycoerythrin (B-Ph), R-phycocyanin (R-PC), and allophycocyanin (APC). The thylakoid membranes contain chlorophyll *a*. Each of these pigments emits fluorescence, and the emission is thus highly heterogeneous. By selecting the proper wavelength, the emission of each of these pigments was monitored as a function of time following excitation of the B-Ph pigments with a single picosecond pulse of 530 nm. The results are shown in Fig. 9. The risetime of B-Ph is instantaneous (limited by the resolution of the apparatus), as expected. The risetimes of the other pigments, however, is longer, with the slowest risetime being exhibited by chlorophyll *a*. From these results the

FIG. 9. Risetimes of the wavelength-resolved pigment fluorescence in *P. cruentum*. Single (530-nm) picosecond pulse excitation; see text. (From Porter et al., 1978; reproduced with permission of Elsevier/North-Holland Biomedical Press.)

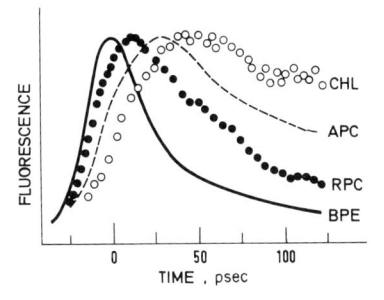

energy transfer pathways B-Ph → R-PC → APC → chlorophyll *a* was established.

Exciton–exciton annihilation effects were not observed even at relatively high picosecond pulse energies (10^{14}–10^{15} photons/cm^2 per pulse); it was concluded that the absence of annihilation was due to rapid energy transfer along the chain of the different accessory pigments. In isolated phycobilisomes in which energy transfer to chlorophyll *a* was no longer possible, an intensity-dependent quenching of the fluorescence was observed (Searle *et al.* 1978). More recently Wong *et al.* (1981) and Pellegrino *et al.* (1981) have carried out more extensive studies on the fluorescence relaxation kinetics and quantum yield in phycobilisomes and isolated phycobiliproteins as a function of the intensity of single (530 nm) picosecond pulses. Doukas *et al.* (1981) have utilized picosecond absorption spectroscopy (partial depletion of the ground state and its recovery as a function of time) to monitor the decay of excitons back to the ground state in phycobiliproteins isolated from a blue-green alga. In all cases (Wong *et al.* 1981; Pellegrino *et al.*, 1981; Doukas *et al.*, 1981) the pulse intensity-dependent decay kinetics were attributed to exciton–exciton annihilation.

V. Nanosecond Pulse Excitation

Nitrogen lasers and Q-switched ruby lasers are commonly utilized in photosynthesis research and have pulse durations in the 7–20 nsec range. Flash lamp-pumped dye lasers are also utilized and have pulse durations typically in the range of one to several microseconds. With electrooptic shutters it is possible to isolate portions of different lengths of such pulses (1–2 nsec). The pulse durations can thus be comparable to the lifetimes of the singlet and triplet excited states; the relationship between these lifetimes, the absorption cross sections of the ground and excited states, the instantaneous intensity of the laser pulses, and the duration of the pulse are now considered.

A. General Kinetic Equations

The important limiting cases may be understood by considering the kinetics of the creation and evolution of the different electronic states of the excited molecules. In reference to Fig. 1, a set of kinetic equations may be written for the different electronic states of the molecule. In writing down the following sets of equations, it was assumed that $dI(t)/dx = 0$, i.e., that the intensity of the excitation laser beam is uniform within the sample (x is the absorption depth), which corresponds to the limit of an optically thin sample (optical density <0.05):

$$d[S_0]/dt = -\sigma_{01}I(t)[S_0] + [S_1]\tau_S^{-1} + \tau_T^{-1}[T_1] \tag{10}$$

$$d[S_1]/dt = \sigma_{01}I(t)[S_0] - \tau_S^{-1}[S_1] - \sigma_{12}^S I(t)[S_1] + (\tau_{21}^S)^{-1}[S_2], \tag{11}$$

$$d[T_1]/dt = K_{IC}[S_1] - \tau_T^{-1}[T_1] - \sigma_{12}^T I(t)[T_2] + (\tau_{21}^T)^{-1}[T_2], \tag{12}$$

$$d[S_2]/dt = \sigma_{12}^S I(t)[S_1] - (\tau_{21}^S)^{-1}[S_2], \tag{13}$$

$$d[T_2]/dt = \sigma_{12}^T I(t)[T_1] - (\tau_{21}^T)^{-1}[T_2], \tag{14}$$

$$[S_0] + [S_1] + [S_2] + [T_1] + [T_2] = \text{const.} \tag{15}$$

In the above equations, other, less likely, photophysical processes have been neglected for simplicity. For the time being, exciton annihilation terms have not been included, to emphasize only the phenomena associated with ground state depletion.

B. THE STEADY STATE APPROXIMATION

It is useful to consider this approximation, which is valid when $t_P > \tau_S, \tau_T$. Calculations by Husiak (1980) indicate that this is a reasonably good approximation as long as the pulse duration is 5–10 times longer than the excited state lifetimes. When each of the Eqs. (10)–(14) is set equal to zero, it can be shown that the fluorescence yield is equal to

$$F = \frac{F[I(t)]}{F[I(t)\to 0]} = \left[1 + \frac{I(t)}{I_S} + \frac{I(t)}{I_T} + \frac{\sigma_{12}^S \tau_{21}^S}{\sigma_{01}\tau_S}\left(\frac{I(t)}{I_S}\right)^2 + \frac{\sigma_{12}^T \tau_{21}^T}{\sigma_{01}\tau_T \phi_{IC}}\left(\frac{I(t)}{I_T}\right)^2\right]^{-1}, \tag{16}$$

where the parameters I_S and I_T are defined by

$$I_S = (\sigma_{01}\tau_S)^{-1}, \tag{17}$$

$$I_T = (\sigma_{01}\tau_T \phi_{IC})^{-1}, \tag{18}$$

and ϕ_{IC} is the quantum efficiency of triplet formation given by

$$\phi_{IC} = \tau_S K_{IC}. \tag{19}$$

For a square wave excitation, Eq. (16) predicts that F will decrease when $I(t) \sim I_S$ or $I(t) \sim I_T$, an effect which can be attributed to depopulation of the ground state. Therefore, the important parameters to consider are the instantaneous incident intensity in relation to the quantities $\sigma_{01}\tau_S$ and $\sigma_{01}\tau_T \phi_{IC}$. In order to avoid effects due to ground state depopulation, the laser beam intensity must be such that

$$I(t) \ll I_S \quad \text{and} \quad I(t) \ll I_T. \tag{20}$$

A consideration of the fourth and fifth terms in Eq. (16) indicates that excited state absorption effects do not significantly influence the fluorescence yield F except at very high laser intensities, i.e., when $I(t) \gg I_S$ or $I(t) \gg I_T$. This is because τ_{21}^S and τ_{21}^T are internal conversion lifetimes ($\sim 10^{-12}$–10^{-13} sec), and thus

$$\tau_{21}^S/\tau_S \ll 1 \quad \text{and} \quad \tau_{21}^T/\tau_T \ll 1. \tag{21}$$

These terms can thus usually be neglected in Eq. (16), particularly if excitation is carried out within an absorption band of the molecule such that σ_{01} is not significantly smaller than the excited state cross sections σ_{21}^S and σ_{12}^T.

C. THE THREE-LEVEL SYSTEM IN THE STEADY STATE APPROXIMATION

It is instructive to consider a simplified model in which triplet formation can be neglected, i.e., $\phi_{IC} = 0$. This has been specifically done by Husiak (1980), who considered all these photophysical processes for the dye R6G in which the triplet yield is known to be very small. For dilute solutions of R6G, Eq. (16) reduces to

$$F = [1 + (I(t)/I_S)]^{-1}, \tag{22}$$

since excited state absorption effects can be neglected in Eq. (16). However, if the transmittance T of the sample is considered, where

$$T = \frac{\text{intensity of transmitted light}}{\text{intensity of incident light}},$$

excited state absorption effects become quite important. The transmittance may now depend strongly on the intensity of the laser pulse and, in addition, on the ratio of the absorption coefficients $\sigma_{01}/\sigma_{12}^S$. The appropriate equation describing the intensity-dependent transmittance is (Husiak, 1980):

$$T = \ln\left[\frac{T(I(t))}{T_0}\right] = \left(\frac{\sigma_{01}}{\sigma_{12}^S} - 1\right)\ln\left[\frac{\sigma_{01}/\sigma_{12}^S + T_0(I(t)/I_S)}{\sigma_{01}/\sigma_{12}^S + I/I_S}\right], \tag{23}$$

where $T_0 = T(I(t) \to 0)$, the low-intensity transmittance of the dye solution.

A typical result for dilute solutions of R6G is shown in Fig. 10. The fluorescence yield decreases as predicted by Eq. (22); from the fit of Eq. (22) to the data, a value of $I_S = 2.2 \times 10^{25}$ cm^2 sec is obtained. From the independently known values of σ_{01} and τ_S for R6G, a value of $I_S = (\sigma_{01}\tau_S)^{-1} = 1 \times 10^{25}$ cm^2 sec is obtained, which can be considered to be in excellent agreement in view of the difficulties associated with accurate measurements of absolute light intensities per unit area.

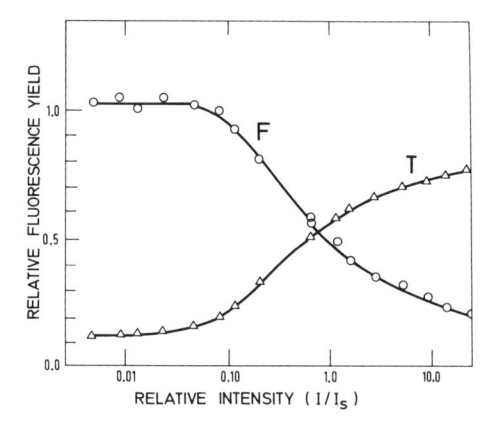

FIG. 10. Fluorescence yield (F) and transmittance of R6G in ethanol (9.0×10^{-5} M, 1.0-mm cuvet) as a function of laser pulse intensity (nitrogen-pumped dye laser operating at 527 nm, 10 nsec FWHM). (Adapted from the data of Husiak, 1980.)

Husiak has also shown that, while F is independent of the wavelength of excitation (utilizing a nitrogen pulsed laser with a tunable dye laser module), T depends strongly on the wavelength of excitation. Thus, at 337 nm, T decreases with increasing pulse intensity ($\sigma_{12}^S > \sigma_{01}$), at 480 nm it is independent of pulse intensity ($\sigma_{12}^S = \sigma_{01}$), while at 525 nm it increases with increasing intensity ($\sigma_{12}^S < \sigma_{01}$), i.e., a bleaching effect is observed. Such wavelength-dependent changes in T as a function of pulse intensity are in accord with Eq. (23).

D. Inclusion of Exciton Annihilation Effects—General Time-Dependent Case

For a concentrated solution ($>10^{-2}$ M) singlet–singlet and singlet–triplet annihilation effects can occur. The kinetic terms $-\frac{1}{2}\gamma_{SS}[S_1]^2$ and $-\gamma_{ST}[S_1][T_1]$ must now be added to Eq. (11), and their appropriate positive counterparts should be added to Eq. (10). It should be noted that, according to our present knowledge of these processes (Breton and Geacintov, 1980), the triplet concentration remains unaffected by annihilation, triplets are not generated by singlet–singlet annihilation (Breton et al., 1979), and triplet–triplet annihilation terms can be neglected. The kinetic Eqs. (10)–(14) can no longer be solved analytically. Husiak (1980) has developed a computer program capable of solving these equations numerically under non-steady state conditions and with inclusion of the exact laser pulse profile ($t_p \approx$ 10 nsec in Husiak's case). Both the quantities F and T were calculated as a function of laser pulse intensities for different concentrations of R6G in solution and fitted to the experimental data by an appropriate choice of parameters.

The basic conclusions of this work are as follows: For dilute solutions of R6G (in ethanol), the F and T data are qualitatively reasonably well described by the steady state model [Eqs. (22) and (23)], even though the duration of the nitrogen laser pulse [$t_p = 10$ nsec at FWHM (full width at half maximum)] is only ~ 2.5 times greater than τ_S (4 nsec). For concentrated solutions of R6G, the fluorescence lifetime decreases with increasing concentration, a common phenomenon known as a concentration-quenching effect (Bojarski et al., 1978) which is a low-intensity excitation effect. For such concentrated solutions the calculated values of I_S are therefore higher than for dilute solutions. Because of this effect, ground state depopulation occurs less readily. Nevertheless it is observed experimentally that the decrease in F occurs at lower laser intensities than for dilute solutions, an effect which is attributed to singlet–singlet exciton annihilation. However, the T curves begin to deviate from their low-intensity values at higher pulse intensities than in the dilute case, as predicted by the lower singlet excited state lifetime.

E. CRITERIA FOR DISTINGUISHING BETWEEN EXCITON ANNIHILATION AND GROUND STATE DEPOPULATION AND EXCITED STATE ABSORPTION EFFECTS

The above effects observed by Husiak (1980) with dilute and concentrated solutions of R6G suggest a useful experimental method for attributing intensity-dependent fluorescence-quenching effects to exciton–exciton annihilation rather than to the other nonlinear effects described here. This method consists of determining both transmittance curves and fluorescence-quenching curves as a function of laser pulse intensity. Since exciton–exciton annihilation effects generally occur at pulse intensities for which $I(t) \ll I_S, I_T$, fluorescence yield decreases should be observed at much lower laser intensities than transmittance changes. However, care must be exercised to choose a wavelength of excitation such that $\sigma_{01} \neq \sigma_{12}^S$ or σ_{12}^T. Utilizing microsecond pulse excitation of chloropasts at 600 nm, Geacintov and Breton (1981) have indeed shown that the transmittance is intensity-independent, while the fluorescence decreases by a factor of nearly 10.

In addition, a simple experimental test for the occurrence of ground state depopulation, excited state absorption, and stimulated emission effects is to study the behavior of the fluorescence yield and transmittance changes as a function of laser intensity utilizing dilute solutions (10^{-4}–10^{-5} M). Under these conditions exciton annihilation effects cannot occur, but the other nonlinear effects, which do not depend on the transfer of energy from molecule to molecule, will manifest themselves (Swenberg et al., 1978; Geacintov et al., 1979b).

F. APPLICATION TO CHLOROPHYLL-CONTAINING SYSTEMS—
DILUTE SOLUTIONS OF CHLOROPHYLL a AND CHLOROPLASTS

1. *Intensity Thresholds for Ground State Depopulation by the Formation of Excited Singlet States*

The molecular absorption cross section σ_{01} of chlorophyll can be calculated from the relationship

$$\sigma \, (\text{cm}^2/\text{molecule}) = 3.81 \times 10^{-21} \varepsilon \, (\text{liter mole}^{-1} \, \text{cm}^{-1}),$$

where ε is the usual decadic molar extinction coefficient and σ is defined according to the Beer–Lambert absorption law $I(x) = I(x = 0)\exp(-\sigma[S_0]x)$. Absolute values of ε for chloroplasts are given by Schwartz (1972) at some wavelengths; with the use of published absorption spectra of chloroplasts (Breton et al., 1973), values of ε and of σ can thus be obtained at any wavelength of excitation. The wavelength dependence of ε for chlorophyll a in solution may be found in the article by Goedheer (1966), for example.

Thus, for excitation of chloroplasts at 337 nm (nitrogen laser) we find $\sigma_{01}^{CP}(337) = 1 \times 10^{-16} \, \text{cm}^2$. With lifetimes of $\tau_S^{SOL} \approx 6.0$ nsec (Hindman et al., 1978b) for chlorophyll a in solution, and $\tau_S^{CP} = 0.5$ nsec in chloroplasts, we obtain

$$I_S^{CP}(337) = 2 \times 10^{25} \, \text{cm}^{-2} \, \text{sec}^{-1}, \tag{24}$$

$$I_S^{SOL}(337) = 2.3 \times 10^{24} \, \text{cm}^{-2} \, \text{sec}^{-1}. \tag{25}$$

2. *Intensity Thresholds for Ground State Depopulation by Formation of Excited Triplet States*

Since triplet formation also occurs in chlorophyll, I_T must be estimated. In solution, from the data of Hindman et al., (1978b) and using $\phi_{IC} = 1 - \phi_f$ (ϕ_f is the fluorescence quantum yield), $\phi_{IC}^{SOL} = 0.70$; however, the triplet lifetime of chlorophyll a in air-saturated solution is not known. In air-saturated organic solutions the triplet lifetime is determined by the diffusional encounters and quenching by oxygen molecules. For molecules of relatively low triplet energy, about $\frac{1}{10}$ of such bimolecular collisions with oxygen lead to quenching. Under these conditions the triplet lifetime in solution is estimated to be ~ 200 nsec (Birks, 1975). In chloroplasts, τ_T^{CP} (chlorophyll triplets) ≈ 10 nsec (Kramer and Mathis, 1980); evidence has also been provided that the carotenoid triplet yield in chloroplasts is 5–15%, depending on the redox state of the PS II reaction center. Since carotenoid triplets are derived by the transfer of energy from chlorophyll triplets, we assume that $\phi_{IC}^{CP} \approx 0.05-0.15$. Utilizing these parameters for chloroplasts and for chlorophyll a in air-saturated solution, we find

$$I_T^{CP}(337) = (\sigma_{01}^{CP}\tau_T^{CP}\phi_{IC}^{CP})^{-1} \approx 7\text{-}20 \times 10^{24} \text{ cm}^{-2} \text{ sec}^{-1}, \quad (26)$$

$$I_T^{SOL}(337) = (\sigma_{01}^{SOL}\tau_T^{SOL}\phi_{IC}^{SOL})^{-1} \approx 1 \times 10^{23} \text{ cm}^{-2} \text{ sec}^{-1}. \quad (27)$$

Thus, according to the values estimated in Eqs. (24)–(27), chlorophyll (but not carotenoid) ground state depopulation effects in chloroplasts will occur at photon fluxes of 10^{25} cm^{-2} sec^{-1}, as long as the pulse duration $t_p > 50\text{-}100$ nsec, so that a steady state is reached. The shorter the duration of the laser pulse, the lower the triplet concentration, and thus the photon flux can be higher than given in Eq. (30) before ground state depopulation effects become significant. The above calculations apply to absorption wavelengths ($\lambda > 600$ nm) for which the absorption by carotenoid molecules is negligible.

In chlorophyll a solutions, it is predicted that depopulation of the ground state due to promotion to the triplet state becomes important when the laser intensity exceeds 10^{23} photons/cm^2 per pulse and the pulse duration is in the microsecond range. Again, for shorter pulses, such depopulation effects due to storage of energy in the triplet level will be less pronounced.

A comparison of fluorescence-quenching curves for chloroplasts and chlorophyll a in ethanol solution is shown in Fig. 11. In the solution case, F begins to decline for pulse intensities of $\sim 10^{15}$ photons/cm^2; by extrapolation, it is estimated that F decreases by 50% at $\sim 10^{17}$ photons/cm^2, in good agreement with the above estimate of I_S, indicating that depopulation by triplet formation is not significant for this pulse duration (10 nsec). Hindman et al. (1978a) have found that, for chlorophyll a in solution, lasing action in a 1×1 cm cuvet, which requires a significant fraction of the molecules to be in the excited singlet state, sets in near 4×10^{16} photons/cm^2 per pulse. This value is reasonable in view of the result given by Eq. (24),

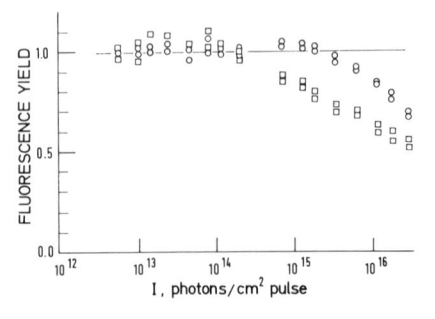

FIG. 11. Fluorescence yield curves as a function of laser pulse intensity at room temperature. □, Spinach chloroplasts, 337-nm nitrogen laser excitation, 10 nsec FWHM; ○, 3×10^{-5} M chlorophyll a in ethyl alcohol solution, 337-nm nitrogen laser excitation; ——, 10^{-4} M chlorophyll a in ethanol; single picosecond laser pulse excitation (530 nm). (Adapted from Geacintov et al., 1979b; reproduced with permission of North-Holland Publishing Company.)

which shows that ~50% of the molecules are excited at pulse intensities of $I = I_S \times t_p \approx 2 \times 10^{24} \times 7 \times 10^{-9} = 1.4 \times 10^{16}$ photons/cm² per pulse.

For chloroplasts, fluorescence quenching sets in at significantly lower intensities than for chlorophyll *a* in solution. A similar difference between fluorescence-quenching curves in chloroplasts and for chlorophyll *a* in solution has been found in the case of single picosecond pulse excitation at 610 nm (Swenberg et al., 1978) and at 530 nm (Geacintov et al., 1979b). A typical fluorescence yield curve for chlorophyll *a* in solution, indicating no change as a function of energy, is also shown in Fig. 11. These differences in the onset of quenching in the solution and chloroplast cases once again are in accord with a model in which exciton–exciton annihilation take place in chloroplasts but not in dilute chlorophyll *a* solutions.

G. Singlet–Singlet versus Singlet–Triplet Annihilation Rates on the Nanosecond Time Scale

Monger et al. (1976) and Monger and Parson (1977) utilized 15–20 nsec FWHM ruby laser excitation to study triplet formation and fluorescence yields in chromatophores of photosynthetic bacteria. They found that the fluorescence yield decreased with increasing pulse intensity and attributed this effect to singlet–triplet exciton annihilation. At relatively low pulse intensities the fluorescence first rises, reaches a maximum, and then declines due to annihilation effects. The initial rise is observed under conditions when the reaction centers are active. RC are quenchers of singlet excitons (and fluorescence) when they are open (not having previously trapped an exciton). When in the oxidized state (closed), the fluorescence yield and the lifetimes of the excitons increase, since the quenching efficiency of closed RCs is less than that of open ones. The fluorescence yield curves thus exhibit a characteristic peak occurring at pulse intensities at which about 80% of the RCs have been oxidized. At higher intensities the decrease normally associated with exciton annihilation effects becomes dominant.

Mauzerall (1972, 1976a,b) has also studied fluorescence yield curves in *C. pyrenoidosa* as a function of pulse intensities using 7-nsec FWHM nitrogen laser pulse excitation (337 nm). He carried out detailed studies on the fluorescence yield as a function of time after the initial (actinic) laser pulse by triggering a weaker flash at some time Δt after the actinic flash to probe the fluorescence. He noted that the fluorescence yield measured during the actinic pulse was strongly intensity-dependent, an effect which he attributed to quenching effects due to triplets.

In the nanosecond pulse excitation case $t_p \approx K_{IC}^{-1}$ for chlorophyll *in vivo*, and triplets are formed on this time scale. The chlorophyll triplets initially formed decay rapidly (~10 nsec; Kramer and Mathis, 1980) to carotenoid triplets (lifetimes in the microsecond range) which have been shown to be the long-lived quenchers in chloroplasts (Mathis et al., 1979; Breton et al., 1979). While chlorophyll triplets can also act as relatively short-lived fluorescence quenchers, this has not yet been established with certainty (Breton and Geacintov, 1981).

It is not obvious a priori whether the fluorescence yield decrease upon excitation with strong nanosecond pulses is due to singlet–singlet or singlet–triplet annihilation or to both types of interactions. Using known and estimated parameters in chloroplasts, Breton et al. (1979) have sought to estimate the relative rates of singlet–singlet to singlet–triplet annihilation in chloroplasts without specifying the exact nature of the triplets. This relative rate depends on the excitation intensity. A typical time dependence of this rate in chloroplasts, calculated for a square wave excitation at a photon flux of 10^{24} photons cm^{-2} sec^{-1}, is shown in Fig. 12. For a 10-nsec FWHM pulse, this photon flux would correspond to approximately 10^{16} photons/cm^2 per pulse. At short times, particularly on a picosecond time scale, singlet–singlet annihilations occur. On longer time scales, for $t_p > 6$ nsec, singlet–triplet annihilations become increasingly dominant.

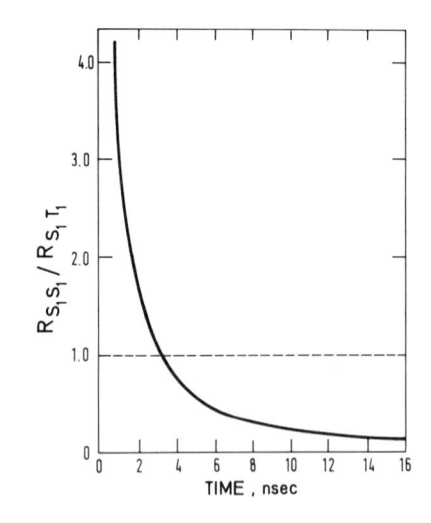

FIG. 12. Calculated relative rates of singlet–singlet and singlet–triplet annihilation in chloroplasts as a function of time within a square wave excitation pulse. Pulse intensity $I(t) = 10^{24}$ photons cm^{-2} sec^{-1}, 600-nm excitation. Calculated according to the methods described by Breton et al. (1979) with the parameters $\tau_S^{-1} \approx 10^9$ sec^{-1}, $\tau_T^{-1} \approx 10^5$ sec^{-1}, $K_{IC} = 10^8$ sec^{-1}, γ_{SS} and $\gamma_{SST} \approx 6 \times 10^{-9}$ cm^8/sec^1.

While this calculation (Fig. 12) is approximate, it indicates that, in the case of nanosecond pulse excitation, fluorescence quenching by both types of annihilations must be taken into account. For pulses of longer duration, particularly in the microsecond case, the rate of absorption of photons is sufficiently low so that singlet–singlet annihilations are minimized and and singlet–triplet fusion effects dominate.

6. EXCITON ANNIHILATION—NONLINEAR EXCITATION EFFECTS

However, there exists a threshold intensity $I(t, S_1-S_1)$ for singlet–singlet exciton annihilation. The photon flux and the exciton density must be sufficiently high so that two singlet excitons may encounter and annihilate one another during their lifetimes of ~ 0.5 nsec. This threshold intensity $I(t, S-S)$ may be estimated from the data in Fig. 3. From this figure, which shows the fluorescence yield as a function of picosecond pulse intensity (quenching due to singlet–singlet annihilation only), it is evident that 25% quenching occurs when there is about one photon hit per 250 chlorophyll molecules. For an exciton lifetime of about 0.5 nsec, $I(t, S_1-S_1)$ is given by

$$I(t, S-S) = (250\sigma_{01}\tau_S)^{-1} \approx 8 \times 10^6 \sigma_{01}^{-1} \quad \text{cm}^{-2}\text{sec}^{-1}. \tag{28}$$

In chloroplasts, $I(t, S_1-S_1)$ turns out to be 8×10^{22} photons cm^{-2} sec^{-1} for 337-nm excitation and 4×10^{23} photons cm^{-2} sec^{-1} for 600-nm excitation. With 10-nsec-duration pulses, these values correspond to 8×10^{14}–4×10^{15} photons/cm^2 per pulse.

The threshold for singlet–triplet annihilation $I(T_1-S_1)$ can be estimated also, but in a different manner. Since the lifetime of the carotenoid triplet quenchers (Mathis *et al.*, 1979; Breton *et al.*, 1979) is in the range of several microseconds, triplet excited states do not decay to any significant extent for pulse duration $t_p < 1$ μsec. Since the singlet exciton can migrate over at least 300–1000 chlorophyll molecules (Paillotin *et al.*, 1979), quenching can set in when there is one triplet present per ~ 1000 molecules. Thus, singlet–triplet quenching does not depend so much on the instantaneous photon flux but on the integrated number of photons absorbed by the sample previously. This integrated absorbed number of photons is given by

$$I(S_1-T_1) = I(t)\Delta t = (\phi_{\text{IC}}\sigma_{01})^{-1} 1/1000, \tag{29}$$

where Δt is the appropriate excitation interval and ϕ_{IC} is the quantum efficiency for carotenoid triplet formation. The latter has been estimated to be about 10% in chloroplasts (Kramer and Mathis, 1980). For 337-nm excitation, $I(S_1-T_1)$ is then estimated to be equal to 10^{14} photons/cm^2 (incident) and about 4×10^{14} photons/cm^2 for 600-nm excitation.

VI. Microsecond Pulse Excitation

Under these conditions of excitation of chloroplasts τ_S, $K_{\text{IC}}^{-1} \ll t_p < \tau_T$. Thus, as long as $I(t) < I(t, S_1-S_1)$, singlet–singlet annihilations are unimportant. From Eq. (28) it is evident that the singlet–singlet annihilation intensity threshold increases with decreasing singlet exciton lifetime. As more and more triplets accumulate, the singlet exciton lifetime decreases as a result of singlet–triplet annihilation, and the probability of singlet–singlet annihilation decreases even further. Thus for microsecond-duration pulses singlet–singlet annihilation can, in most cases, be totally neglected, and the effects of singlet–triplet annihilation account for the decrease in the fluorescence yield.

A. Effect of Integrated Pulse Energy on Triplet Yield

The yield of triplets as a function of integrated pulse energy is itself nonlinear. Utilizing a simple kinetic model in which triplets are produced by intersystem crossing from singlets and in which singlets are quenched by triplets, Breton et al. (1979) have calculated the triplet yield as a function of pulse energy (Fig. 13). The quantum yield tends to decrease with increasing pulse energy. This is a consequence of the decreasing singlet exciton lifetime at higher pulse energies, which results in a lowered probability of triplet formation since $\phi_{IC} = \tau_S K_{IC}$. Experimentally, the yield of carotenoid triplets does indeed show the behavior indicated in Fig. 13 in chloroplasts (Mathis, 1969; Breton et al., 1979) and in bacterial chromatophores (Monger and Parson, 1977).

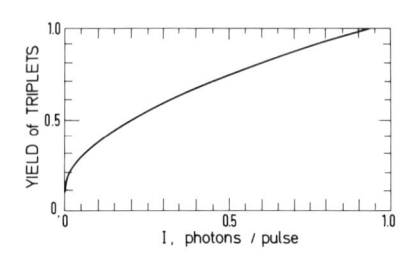

Fig. 13. Calculated yield of triplet excited states in chloroplasts as a function of laser pulse intensity (in relative units). Calculated according to the methods described by Breton et al. (1979).

B. Buildup of Triplets as a Function of Time within a Microsecond Pulse

With a gated optical multichannel analyzer (OMA) for detecting the fluorescence and transmittance (at 515 nm, where carotenoid triplets exhibit an absorption maximum), Breton et al. (1979) have followed the fluorescence yield and triplet buildup as a function of time within a 500-nsec laser pulse at ambient temperature. The results are shown in Fig. 14. The yield diminishes drastically within the initial portions of the pulse, while the carotenoid triplets increase gradually in concentration. A simple kinetic scheme with the appropriate parameters can well account for the drop in the fluorescence yield in Fig. 14b; however, the leveling off of the yield at 300 nsec should correspond to a similar leveling off of the carotenoid triplet concentration.

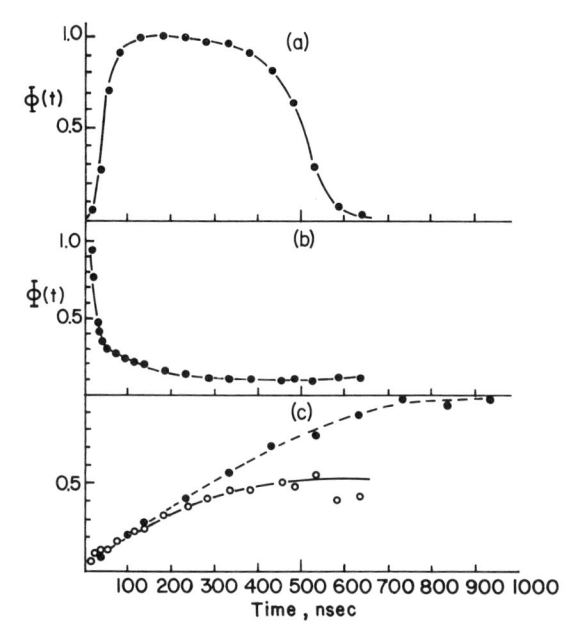

6. EXCITON ANNIHILATION—NONLINEAR EXCITATION EFFECTS   185

FIG. 14. Time dependence of (a) laser excitation pulse profile, (b) fluorescence yield, (c) build-up of carotenoid tripets (●), of quencher concentration Q (○) calculated from (b) according to Stern–Volumer kinetics, and calculated triplet concentration (——). Spinach chloroplasts excitation at 600 nm, room temperature. (From Breton *et al.*, 1979; reproduced with permission of Elsevier/North-Holland Biomedical Press.)

The experimentally determined triplet yield continues to increase right up to the end of the pulse; this is puzzling, because it indicates that not all carotenoid triplets formed have the same efficiency for quenching the fluorescence. A possible explanation may be that the carotenoid triplets are formed at different membrane locations, where their quenching efficiencies may be different.

C. UNEQUAL FLUORESCENCE QUENCHING OBSERVED FOR THE TWO FLUORESCENCE BANDS F685 AND F735 AT LOW TEMPERATURES IN CHLOROPLASTS

Fluorescence quenching experiments at low temperatures in which the PS I emission (F735) can be resolved from the light-harvesting PS II emission (F685–F695) indicate that triplets accumulate preferentially in PS I. Thus, with increasing time within a 500-nsec pulse, the quenching of the F735 fluorescence becomes stronger than that of the F685 fluorescence, and the character of the emission spectrum changes with time (Fig. 15); however, as

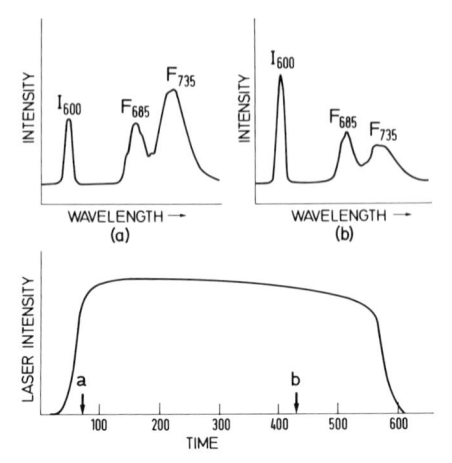

FIG. 15. Time dependence of fluorescence emission spectra of spinach chloroplasts at 100 K measured during different portions of a 500-nsec laser excitation (600 nm) pulse. Fluorescence viewed with a spectrograph-gated OMA system (2.8-nsec-wide viewing gate) opened at (a) 70 nsec and (b) 425 nsec. The pulse profile is also shown (c), and I_{600} refers to the signal due to scattered excitation laser light detected simultaneously with the fluorescence during the 2.8-nsec opening time of the gate.

the relative height of the fluorescence bands compared to the intensity of the laser beam (scattered light) indicates, the quenching of the fluorescence due to triplets is operative for both emission bands. The preferential fluorescence quenching due to triplets manifests itself in Fig. 7B, in which the yield for F685 and F735 is plotted as a function of the energy of the microsecond-duration pulse. This effect is not observed in the picosecond pulse excitation case (Fig. 7A), since triplets play no role in decreasing the yield.

That the preferential quenching for the F735 band is due to greater accumulation of triplets in PS I, rather than to differences in triplet lifetimes, can be concluded on the basis of the work by Mathis *et al.* (1979). These workers have shown that carotenoid triplets having the same lifetimes are formed in all the different topological parts of the chloroplast membrane.

D. PICOSECOND PULSE TRAIN EXCITATION

Analogous effects occur when chloroplasts at low temperatures are excited with trains of picosecond pulses rather than with a non-mode-locked microsecond-duration laser pulse. Thus, Breton and Geacintov (1976) have shown that the F735 fluorescence yield relative to that of the F685 band decreases progressively as the number of picosecond pulses in a pulse train is increased. Geacintov *et al.* (1978) have provided evidence that triplet excited states flow into PS I on a time scale of ~50 nsec and that the F735 fluorescence is quenched dynamically rather than statically; the lifetime of the F735 fluorescence decreases with increasing pulse number within a given pulse train. These experiments indicate that triplet states are indeed formed by picosecond pulses. The triplet quenchers, because of their longer lifetimes, persist from pulse to pulse, while the singlets, whose lifetimes are considerably shorter than the 5- to 10-nsec spacing between adjacent picosecond pulses, do not persist from pulse to pulse. Such techniques, i.e., monitoring fluorescence yields and decay times determined for different pulse numbers in a given pulse sequence, can thus serve to distinguish short-lived singlets from longer-lived triplet quenchers.

VII. Picosecond Pulse Excitation Studies on Absorption Changes in Bacterial Reaction Centers

We close this chapter with a brief discussion of possible high-intensity effects in the study of fast absorption kinetics induced by strong picosecond laser pulses in bacterial RCs.

Bacterial RCs, essentially free of antenna pigments, are extracted from bacterial chromatophores by suitable treatment with detergents. Primary electron transfer events in RC preparations have been studied by fast picosecond absorption spectroscopy techniques (Holten and Windsor, 1978). A bacterial RC contains two bacteriopheophytin (BPh) molecules, one quinone molecule, and four bacteriochlorophyll (Bchl) molecules, two of which constitute the special dimer pair P. Excitation with strong 530-nm picosecond pulses indicates that an electron is ejected rapidly from P (<10 psec) and resides for a short time (100–200 psec) on an intermediate electron acceptor (I) before it is transferred to the quinone (PI + hv → P + T). It has been proposed that I is one of the BPh molecules. This was concluded on the basis of the spectra of the absorbance changes following the actinic picosecond laser flash.

However, most of the absorption studies so far have been conducted under conditions of rather high energies of excitation. At these high-excitation levels, the possibility of absorption of more than one photon within several picoseconds by a given RC is quite probable. Thus, it was shown (Akhmanov et al., 1978; 1979a,b) that the quantum yield of P^+ was significantly reduced as the number of photons incident per reaction center (PRC) was varied from 0.1 to 60. Moskowitz and Malley (1978) had previously demonstrated that the yield of P^+ under intense (PRC = 13) picosecond pulse excitation was less than under conditions of low-level, steady state excitation. They interpreted this effect in terms of stimulated emission, while Mauzerall (1979) proposed that photoreversal or multitrap effects were involved in this phenomenon.

It appears that ground state depletion, excited state absorption, and interaction and annihilation between the different excited states in each RC must be considered. Akhmanov et al. (1979a,b) utilized excitation within the absorption band of P at 870 nm rather than within the absorption band of BPh at 530 nm (see also Shuvalov et al., 1978); they showed that only absorption changes (bleaching) occurred at low levels of excitation (PRC $\lesssim$ 3) within the BPh 750-nm absorption band, while bleaching occurred within the 800-nm BChl and 870-nm P absorption bands. They concluded on the basis of these experiments that BChl rather than BPhe was the primary electron acceptor from P. At high levels of excitation, strong bleaching within the 750-nm BPhe bands is also observed; this high-intensity phenomenon may account for the previously observed (Holten and Windsor, 1978) absorption changes within the BPh absorption bands, on which the conclusion that the primary electron acceptor is BPh was based. Of course, in the earlier experiments (Holten and Windor, 1978), excitation was carried out within the BPh absorption band, which also accounts for at least some of the bleaching of the BPh bands. These effects should be verified further before definite conclusions regarding the nature of the primary electron acceptor I can be made.

Akhmanov et al. (1979b, 1980) have shown that there are not just differences in the absorption changes as a function of absorption wavelengths, but that the kinetics of the bleaching changes may also be distorted by high-intensity effects. For example, the kinetics of the absorbance changes observed at 802 nm (BChl) at very low intensities (PRC $\lesssim$ 1) and at high intensities (PRC $\approx$ 50–60) are shown in Fig. 16. The decay of this absorbance change is seen to be much faster at the high intensity than at the lower intensity. This effect may account for the puzzling fast kinetic phase observed at 790 nm ($\sim$35 psec) in RCs of R. sphaeroides (R-26) (Holten and Windsor, 1978) and at 800 nm by Shuvalov et al. (1978); the latter authors have interpreted this rapid phase in terms of the decay kinetics of BChl$^-$, conjectured

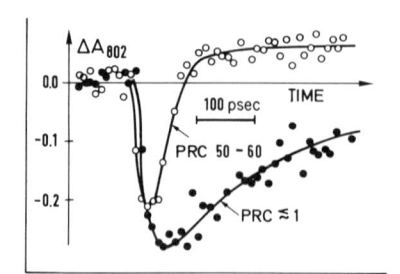

FIG. 16. Kinetics of absorption changes at 802 nm induced by a single picosecond pulse (870 nm) of two different intensities incident on bacterial RCS. PRC, Photons incident per reaction center particle. (Adapted from Akhmanov et al., 1979b; reproduced with permission of Springer-Verlag.)

to be the primary electron acceptor. In view of Fig. 16, this rapid phase is clearly due to multiple excitation of the RC, and the kinetic data must be reinterpreted after careful consideration of these high-intensity effects.

Finally, it seems clear that the primary electron transfer from P occurs on a fast time scales (<10 psec) and is probably not influenced by multiple excitation effects as long as P receives at least one excitation. The subsequent kinetic changes, as well as other wavelength changes associated with the different electron acceptors should be investigated at low levels of excitation. On the other hand, multiple excitation processes of RCs, when properly understood and interpreted, may also provide new information on the structure and functioning of bacterial RCs.

Acknowledgments

The writing of this chapter was supported by National Science Foundation grant PCM-8006109, and in part by the Department of Energy.

References

Akhmanov, S. A., Borisov, A. Y., Danielius, R. V., Kozlovsky, V. S., Piskarkas, A. S., and Razjivin, A. P. (1978). *In* "Picosecond Phenomena" (C. V. Shank, E. P. Ippen, and S. L. Shapiro, eds.), Springer Series in Chemical Physics, Vol. 4, pp. 134–139. Springer-Verlag, Berlin and New York.
Akhmanov, S. A., Borisov, A. Y., Danielius, R. V., Gadonas, R. A., Kozlowski, V. S., Piskarkas, A. S., and Razjivin, A. P. (1979a). *Stud. Biophys.* **77**, 1–3.
Akhmanov, S. A., Borisov, A. Y., Danielius, R. V., Gadonas, R. A., Kozlovskskij, V. S., Piskarkas, A. S., and Razjivin, A. P. (1979b). *Proc. Int. Conf. Laser Spectrosc. 4th, Munich* pp. 387–398.
Akhmanov, S. A., Borisov, A. Y., Danielius, R. V., Gadonas, R. A., Kozlowski, V. S., Piskarkas, A. S., Razjivin, A. P., and Shuvalov, V. A. (1980). *FEBS Lett.* **114**, 149–152.
Arsenault, R., and Denariez-Roberge, M. M. (1976). *Chem. Phys. Lett.* **40**, 84–87.
Baugher, J., Hindman, J. C., and Katz, J. J. (1979). *Chem. Phys. Lett.* **63**, 159–162.
Beddard, G. S., Fleming, G. R., Porter, G., Searle, G. F. W., and Synowiec, J. A. (1979). *Biochim. Biophys. Acta* **545**, 165–174.
Birks, J. B. (1970). "Photophysics of Aromatic Molecules." Wiley (Interscience), New York.
Birks, J. B. (1975). *In* "Organic Molecular Photophysics" (J. B. Birks, ed.), Vol. 2, pp. 545–556. Wiley, New York.
Bojarski, C., Bujko, R., and Twardowski, R. (1978). *Acta Phys. Pol. A* **54**, 713–720.
Breton, J., and Geacintov, N. E. (1976). *FEBS Lett.* **69**, 86–89.
Breton, J., and Geacintov, N. E. (1980). *Biochim. Biophys. Acta* **594**, 1–32.
Breton, J., Michel-Villaz, M., and Paillotin, G. (1973). *Biochim. Biophys. Acta* **314**, 42–56.
Breton, J., Geacintov, N. E., and Swenberg, C. E. (1979). *Biochim. Biophys. Acta* **548**, 616–635.
Butler, W. L. (1978). *Annu. Rev. Plant Physiol.* **29**, 345–378.
Butler, W. L. (1979). *In* "Chlorophyll Organization and Energy Transfer in Photosynthesis," Ciba Foundation Symposium, No. 61, pp. 237–256. Excerpta Med. Found., Elsevier/North-Holland, Amsterdam.

Butler, W. L., Tredwell, C. J., Malkin, R., and Barber, J. (1979). *Biochem. Biophys. Acta* **545**, 309–315.
Campillo, A. J., and Shapiro, S. L. (1977). *In* "Ultrashort Light Pulses" (S. L. Shapiro, ed.), Topics in Applied Physics, Vol. 18, pp. 317–376. Springer-Verlag, Berlin and New York.
Campillo, A. J., and Shapiro, S. L. (1978). *Photochem. Photobiol.* **28**, 975–989.
Campillo, A. J., Shapiro, S. L., Kollmann, V. H., Winn, K. R., and Hyer, R. C. (1976a). *Biophys. J.* **16**, 93–97.
Campillo, A. J., Kollman, V. H., and Shapiro, S. L. (1976b). *Science* **193**, 227–229.
Campillo, A. J., Hyer, R. C., Monger, T. G., Parson, W. W., and Shapiro, S. L. (1977a). *Proc. Natl. Acad. Sci. U.S.A.* **74**, 1997–2001.
Campillo, A. J., Shapiro, S. L., Geacintov, N. E., and Swenberg, C. E. (1977b). *FEBS Lett.* **83**, 316–320.
Doukas, A. G., Stefancic, V., Buchert, J., Alfano, R. R., and Zelinskas, B. A. (1981). *Photochem. Photobiol.* **34**, 505–510.
Geacintov, N. E., and Breton, J. (1977). *Biophys. J.* **17**, 1–15.
Geacintov, N. E., and Breton, J. (1981). *In* "Trends in Photobiology" (C. Helene and M. Charlier, eds.), Plenum, New York. In press.
Geacintov, N. E., Breton, J., Swenberg, C., Campillo, A. J., Myer, R. C., and Shapiro, S. L. (1977a). *Biochim. Biophys. Acta* **461**, 306–312.
Geacintov, N. E., Breton, J., Swenberg, C. E., and Paillotin, G. (1977b). *Photochem. Photobiol.* **26**, 619–638.
Geacintov, N. E., Swenberg, C. E., Campillo, A. J., Hyer, R. C., Shapiro, S. L., and Winn, K. R. (1978). *Biophys. J.* **24**, 347–359.
Geacintov, N. E., Breton, J., Swenberg, C. E., and Paillotin, G. (1979a). *Photochem. Photobiol.* **29**, 651–652.
Geacintov, N. E., Husiak, D., Kolubayev, T., Breton, J., Campillo, A. J., Shapiro, S. L., Winn, K. R., and Woodbridge, P. K. (1979b). *Chem. Phys. Lett.* **66**, 154–158.
Goedheer, J. C. (1966). *In* "The Chlorophylls" (L. P. Vernon and G. R. Seely, eds.), Chap. 6. Academic Press, New York.
Hindman, J. C., Kugel, R., Svirmickas, A., and Katz, J. J. (1977). *Proc. Natl. Acad. Sci. U.S.A.* **74**, 5–9.
Hindman, J. C., Kugel, R., Svirmickas, A., and Katz, J. J. (1978a). *Chem. Phys. Lett.* **53**, 197–200.
Hindman, J. C., Kugel, R., Wasielewski, M. R., and Katz, J. J. (1978b). *Proc. Natl. Acad. Sci. U.S.A.* **75**, 2076–2079.
Holten, D., and Windsor, M. W. (1978). *Annu. Rev. Biophys. Bioeng.* **7**, 189–227.
Husiak, D. (1980). Ph.D. Thesis, New York Univ., New York.
Ippen, I. P., and Shank, C. V. (1977). *In* "Ultrashort Light Pulses" (S. L. Shapiro, ed.), Topics in Applied Physics, Vol. 18, pp. 80–119. Springer-Verlag, Berlin and New York.
Kramer, M., and Mathis, P. (1980). *Biochim. Biophys. Acta* **593**, 319–329.
Mathis, P. (1969). *In* "Photosynthesis Research" (H. Metzner, ed.), Vol. 2, pp. 818–822. Univ. of Tübingen, Tübingen.
Mathis, P., Butler, W. L., and Satoh, K. (1979). *Photochem. Photobiol.* **30**, 603–614.
Mauzerall, D. (1972). *Proc. Natl. Acad. Sci. U.S.A.* **69**, 1358–1362.
Mauzerall, D. (1976a). *Biophys. J.* **16**, 87–91.
Mauzerall, D. (1976b). *J. Phys. Chem.* **80**, 2306–2309.
Mauzerall, D. (1979). *Photochem. Photobiol.* **29**, 169–170.
Monger, T. G., and Parson, W. W. (1977). *Biochim. Biophys. Acta* **460**, 393–407.
Monger, T. G., Cogdell, R. J., and Parson, W. W. (1976). *Biochim. Biophys. Acta* **449**, 136–153.
Moskowitz, E., and Malley, M. M. (1978). *Photochem. Photobiol.* **27**, 55–59.

Moya, I. (1979). Thèse Doctorat d'Etat, Univ. of Paris-Sud, Paris.

Paillotin, G., Swenberg, C. E., Breton, J., and Geacintov, N. E. (1979). *Biophys. J.* **25**, 513–534.

Pellegrino, F., Yu, W., and Alfano, R. R. (1978). *Photochem. Photobiol.* **28**, 1007–10012.

Pellegrino, F., Wong, D., Alfano, R. R., and Zilinskas, B. A. (1981). *Photochem. Photobiol.* In press.

Penzkofer, A., and Falkenstein, W. (1978). In "Picosecond Phenomena" (C. V. Shank, E. P. Ippen, and S. L. Shapiro, eds.), Springer Series in Chemical Physics, Vol. 4, pp. 71–74. Verlag, Berlin and New York.

Porter, G., Synowiec, J. A., and Tredwell, C. J. (1977). *Biochim. Biophys. Acta* **459**, 329–336.

Porter, G., Tredwell, C. J., Searle, G. F. W., and Barber, J. (1978). *Biochim. Biophys. Acta* **501**, 232–245.

Rahman, T. S., and Knox, R. S. (1973). *Phys. Status Solidi B* **58**, 715–720.

Schafer, F. P. (1976). In "Physical and Chemical Applications of Dyestuffs" (F. L. Borchke, ed.), Topics in Current Chemistry, 61, pp. 1–30. Springer-Verlag, Berlin and New York.

Schwartz, M. (1972). In "Photosynthesis and Nitrogen Fixation," Part B (A. San Pietro, ed.), Methods in Enzymology, Vol. 24, pp. 139–146. Academic Press, New York.

Searle, G. F. W., Barber, J., Porter, G., and Tredwell, C. J. (1978). *Biochim. Biophys. Acta* **501**, 246–256.

Shepanski, J. F., and Anderson, R. W. (1981). *Chem. Phys. Lett.* **78**, 165–173.

Shuvalov, V. A., Klevanik, A. V., Sharkov, A. V., Matveatz, J. A., and Krukov, P. G. (1978). *FEBS Lett.* **91**, 135–139.

Swenberg, C. E., and Geacintov, N. E. (1973). In "Organic Molecular Photophysics" (J. B. Birks, ed.), Vol. 1, pp. 489–564. Wiley, New York.

Swenberg, C. E., Geacintov, N. E., and Pope, M. (1976). *Biophys. J.* **16**, 1447–1452.

Swenberg, C. E., Geacintov, N. E., and Breton, J. (1978). *Photochem. Photobiol.* **28**, 999–1006.

Wong, D., Pellegrino, F., Alfano, R. R., and Zilinskas, B. A. (1980). *Photochem. Photobiol.* **33**, 651–662.

Yu, W., Pellegrino, F., and Alfano, R. R. (1977). *Biochim. Biophys. Acta* **460**, 171–181.

CHAPTER 7

Fluorescence Decay Kinetics and Bimolecular Processes in Photosynthetic Membranes

Charles E. Swenberg

Laboratory of Preclinical Studies
National Institute of Alcohol Abuse and Alcoholism
Rockville, Maryland

I.	INTRODUCTION	193
II.	FLUORESCENCE DECAY PROFILES: LOW-INTENSITY REGIME	194
III.	EXCITON–EXCITON ANNIHILATION: HIGH EXCITATION INTENSITIES	201
IV.	CONCLUSION	212
	REFERENCES	213

I. Introduction

Photosynthetic membranes have, in addition to chlorophyll *a* molecules which use only a small portion of the solar spectrum, accessory pigment molecules (e.g., carotenoids, phycobilins, other forms of chlorophyll, bacteriochlorophylls) whose absorption spectra are symbatic with respect to the spectral distribution of sunlight. These arrays of accessory pigments, sometimes loosely termed the *antennae*, enhance the capturing of incident light and, by a series of nonradiative transfers, this collected energy is funneled to a special pair of chlorophyll molecules, the reaction center, where a series of electron transfer processes is initiated. This migration of electronic states from the antenna to the reaction center has been the subject of intensive study both theoretical (Franck and Teller, 1938; Robinson, 1967; Pearlstein, 1967, 1982; Knox, 1977; Borisov, 1978) and experimental (Govindjee, 1975; Barber, 1978; Geacintov *et al.*, 1977b). The transfer of electronic energy from its site of absorption to the reaction center is very efficient as compared to radiative and other nonradiative decay channels (Knox, 1975). Fortunately, the antenna molecules exhibit weak fluorescence.

An analysis of its decay characteristics, yield, and polarization as a function of the wavelength of excitation and emission and excitation source characteristics (pulse duration and intensity) constitutes one of the major techniques, in addition to transient optical absorbance changes, for elucidating the microscopic routes of connectiveness among the pigment moieties of photosynthetic membranes.

This chapter will review briefly (1) energy migration and trapping in the photosynthetic antenna, focusing attention on the decay characteristics of the donors, and (2) bimolecular processes which result when intense laser sources are utilized; in particular, a detailed discussion of Paillotin's theory of singlet fusion is given, which hopefully will clarify its distinction from two previous formulations of singlet–singlet annihilation.

II. Fluorescence Decay Profiles: Low-Intensity Regime

Electronic excited states may lose their excitation energy either by emitting radiation or by nonradiative deexcitation processes. All rates depend on the molecule and its local environment. A particular nonradiative channel for excited state decay is that of energy transfer to another molecule. A theory for this transfer process in the very weak dipole–dipole approximation was developed by Forster (1965) and has been reviewed extensively in the photosynthetic literature (see, e.g., Knox, 1977). In this limit, the transfer rate $W(R_n)$ between donor and acceptor separated by a distance R_n can be written as

$$W(R_n) = (1/\tau^0)(R_0/R_n)^6 = (1/\tau)(\bar{R}_0/R_n)^6, \tag{1}$$

where τ^0 is the natural lifetime of the donor in the absence of quenchers, τ is the observed donor lifetime, and R_0 is the distance at which the energy transfer rate from donor to acceptor is equal to the radiative decay. The related quantity $\bar{R}_0$ is the donor–acceptor separation at which transfer competes equally with the *total* rate of removal of energy from the donor by all *other* means. If ϕ_F is the fluorescence yield, then $\bar{R}_0 = \phi_F^{1/6} R_0$. The critical distance $\bar{R}_0$ is related to the spectroscopic properties of the molecules involved:

$$\bar{R}_0 = \left[\frac{3\phi_F f}{4\pi} \int \frac{c^4}{\omega^4 n_0} F(\omega)\sigma(\omega)\, d\omega \right]^{1/6}, \tag{2}$$

where $F(\omega)$ is the normalized fluorescence emission spectrum, $\sigma(\omega)$ is the normalized acceptor absorption cross section in units of square centimeters, n_0 is the index of refraction of the solvent, c is the speed of light, and the angular frequency (ω) integration is over all frequencies. The orientation factor f allows for the anisotropy of the dipolar interaction (Knox, 1975). Equation (2) explicitly demonstrates that no transfer is possible unless the donor fluorescence and acceptor absorption spectra overlap.

A major premise in deriving Eq. (1) is the assumption of incoherence, namely, that energy received by the acceptor is quickly dispersed among a dense set of degenerate vibrational states, each of which can relax to vibronic states of lower energy in a time fast enough so that back transfer becomes very inefficient. Also implicit is the assumption that the two molecules are coupled to different vibrational fields (see Soules and Duke, 1971, for a discussion of this point). For extensions of the theory of resonance energy transfer to more general situations, the articles by Dexter (1953) and Kenkre and Knox (1974) should be consulted.

Photosynthetic membranes contain arrays of Chl-protein (CP) complexes in addition to other accessory pigment moieties; together they operate as a collective group in delivering the absorbed energy to the photochemical reaction center. A reaction center and its associated group of antenna pigments is called a photosynthetic unit (PSU). In green plants, there are two types of reaction centers and thus two distinct types of photosystems, PS I and PS II. The average number of chlorophylls found in one PS I plus one PS II is approximately 500, of which ~ 400 are more strongly associated with the reaction center of PS II, at least in green plants. In algae, the number of chlorophylls associated with PS II is considerably smaller. Theoretical treatments of energy transfer have considered various limiting degrees of connectiveness between the PSUs. The term "lake model" is often used if energy absorbed in one PS II is allowed to transfer (efficiently) to another PS II. At the opposite extreme, where energy migration is prohibited between different PS II's, one speaks of a "puddle model." Theoretical interpretation of the fluorescence yield as a function of excitation intensity provides evidence in support of the lake model for photosynthetic membranes, as discussed later in this chapter. Transfer of energy from PS II to PS I is called spillover and is well known to be modulated by the cation Mg^{2+}. For a more refined discussion of the kinetics of the partition of energy between PS I and PS II, and how the degree of energy connectiveness between photosystems depends on the state of the reaction centers, the work of Butler (1980) should be consulted. The scope of our discussion of energy transfer will be restricted to processes within a single photosystem, although in discussing exciton annihilation migration between different PSUs is implicitly allowed.

In Fig. 1 a PSU is illustrated schematically as a two-dimensional array of molecules where the trap (i.e., the reaction center) is denoted by T. For simplicity, we have neglected the structure associated with CP complexes and have furthermore considered only two types of molecules, represented by open circles and boxes. Energy absorbed by molecules A and B can transfer to neighboring molecules (as shown in Fig. 1) or to non-nearest neighbors. Such transfer rates are given by the Forster expression, Eq. (1). With different environments, the total nonradiative rates for molecules A and B are different, and thus even though they fluoresce at the same wavelength

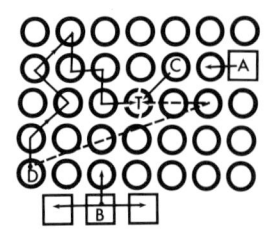

FIG. 1. Schematic two-dimensional array of pigment molecules with trap T that illustrates pathways to trapping and site heterogeneity of resonance transfer rates. Circles and square boxes denote two types of molecules. Pathway represented by the solid line from molecule D to trap T denotes incoherent motion. Dashed pathway denotes coherent motion before trapping (see text). Note that, although molecules A and B are the same, they have different environments and therefore different fluorescence lifetimes (see text for details).

(neglecting small local site energy shifts, i.e., diagonal disorder), their lifetimes can be quite different. Site heterogeneity therefore can result in a fluorescence decay function which is nonexponential, even though the decay for individual donors is exponential. In the case where the donor is surrounded by a large number of acceptors (N), its fluorescence decay, in the absence of back transfer, is given by

$$F(t) = \exp\left(-\frac{t}{\tau^0}\right) \prod_{n=1}^{N} \exp[-tW(R_n)], \qquad (3)$$

where $W(R_n)$ is the transfer rate to an acceptor located a distance R_n from the donor. For a *fixed* given configuration of acceptors, the behavior of $F(t)$ versus t as given by Eq. (3) is always an *exact* exponential function of time t. However, the quantity of interest is generally the ensemble average of $F(t)$ over all possible acceptor configurations. In the low-concentration regime and for randomly distributed acceptors, it can be shown that the decay of the donor's excited state following a brief excitation for times not too short (i.e. $t \gtrsim 0.2\tau^0$) is

$$F_1(t) \propto \exp(-k_1 t^{1/6} - t/\tau^0), \qquad (4)$$

$$F_2(t) \propto \exp(-k_2 t^{1/3} - t/\tau^0), \qquad (5)$$

$$F_3(t) \propto \exp(-k_3 t^{1/2} - t/\tau^0), \qquad (6)$$

for one-, two-, and three- dimensional systems, respectively, provided the transfer rate is given by Eq. (1). The constants k_1, k_2, and k_3 depend on τ^0, the lifetime of the donor in the absence of acceptors, the concentration of acceptors, and the critical distance R_0 (Wolber and Hudson, 1979; Dewey and Hammer, 1980; Blumen and Manz, 1979). At very short times, estimated to be less than $0.2\tau^0$, although the exact value depends on the concentration of acceptors, the approximation involved in performing the ensemble averages leading to Eqs. (4)–(6) is invalid. For this time regime the donor's fluorescence is exponential in time. Furthermore, numerical calculations by Blumen and Manz (1979) show that, in addition to these smooth decays, the decay function, $\exp(t/\tau^0)F(t)$ has a superimposed wavy structure. This additional damped oscillatory structure of the decay function bears information on the local lattice structure. From the above remarks, it should be evident that the fluorescence decay from a donor can be quite complex and can be highly nonexponential, and that its forms depend on the dimensionality and spatial arrangement of the acceptor array. Thus, caution should be exercised in interpretating emission decays in the fast time regime. For example, an analysis of the decay kinetics and yield from the phycobiliproteins in red and blue-green algae necessitates careful consideration of the special aggregated structure of the phycobilisomes (i.e., whether it is two- or three-dimensional with regard to each of the phycobiliproteins) (see Gantt, 1975).

Unfortunately, the above results are not applicable to the photosynthetic chlorophyll antenna (denoted by open circles in Fig. 1), since the concentration of donors (chlorophylls) is larger than the concentration of acceptors (traps). This limit, where the number of donors is significantly larger than the number of acceptors, is a problem of great complexity, and most investigations have resorted to numerical solutions of the master equation governing the probabilities as functions of time that each chlorophyll molecule is in its first excited state, S_1 (Pearlstein, 1967, 1982; Robinson, 1967; Shipman, 1980).

The motion of electronic excitation energy (loosely called an exciton) in the chlorophyll antenna can be viewed in two distinct limits: (1) coherent transfer where the exciton motion is treated as a traveling wave (the dashed curve in Fig. 1 from D to T illustrates coherent motion with one scattering event en route), and (2) complete incoherent transfer where motion is a series of uncorrelated hops (the solid line from D to T in Fig. 1). The first theoretical treatment of energy migration within the PSU and its subsequent trapping at the reaction center was given more than 40 years ago by Franck and Teller (1938). These authors adopted a highly simplified one-dimensional model consisting of approximately 1000 molecules per trap. Because of the size and dimensionality they assumed for the PSU, their calculated trapping efficiency at the active site was far too small. This is directly traceable to the well-known result that a random walker in one dimension visits only $\sim\sqrt{n}$ distinct sites in n hops.

From a theoretical standpoint the observed fluorescence quantum yield and lifetime must be consistent with reasonable rates of energy transfer among the light-harvesting pigments and with the geometrical size of the PSU. This consistency was first established in an important theoretical calculation by Pearlstein (1967, 1982). Consider the low-intensity regime so that nonlinear processes can be neglected and assume the trap is a perfect energy sink. Under these conditions, the probability $P(R_n, t)$ that the nth donor will be excited at time t after the excitation source has terminated obeys the equation

$$\frac{dP(R_n,t)}{dt} = -\left[\beta + X(R_n) + \sum_{n'} W(|R_n - R_{n'}|)\right]P(R_n,t) + \sum_{n'} W(|R_n - R_{n'}|)P(R_{n'},t), \quad (7)$$

where β is the monomolecular decay rate in the absence of trapping and energy transfer, taken to be the same for all pigment molecules, $X(R_n)$ is the transfer rate from the donor at position R_n to the reaction center, and $W(|R_n - R_{n'}|)$ is the transfer rate between donor sites R_n and $R_{n'}$ and is given by Eq. (1). The time decay of the entire antenna $\Omega(t)$, after termination of the excitation is thus a simple sum over the array

$$\Omega(t) \propto \sum_n P(R_n,t). \quad (8)$$

For uniform initial excitation conditions (i.e., all molecules in the array have equal probability of being excited) Pearlstein (1967) showed that the higher-order terms in the exponential series

$$\Omega(t) = \exp(-\gamma_0 t) + \text{higher order terms} \quad (9)$$

could be neglected. In this case, the network decays exponentially when t is large. For a discussion of the range of validity of this "zero-mode" approximation the recent review by Pearlstein (1982) should be consulted. γ_0 is called the zero-mode relaxation rate, and its value depends not only on the spatial arrangement of the pigment molecules but also on the Forster transfer rate and its range. Therefore γ_0^{-1} is the mean lifetime for an exciton, and to a good approximation it is equal to the reciprocal of the trapping rate K_t, since under most conditions the intrinsic exciton lifetime contribution is small; i.e., $\beta \ll K_t$. The time K_t^{-1} can be written as a sum of the mean time required to visit a trap (K_0^{-1}) and the time needed to photoconvert at the trap (K_x^{-1}). A system is called *diffusion-limited* if $K_0^{-1} \gg K_x^{-1}$; otherwise the reaction is termed *quencher* (or trap)-*limited* $(K_x^{-1} \gg K_0^{-1})$. For photosynthetic membranes theoretical calculations by Pearlstein (1967, 1982) and Paillotin (1976a,b) have clearly established that exciton migration is quite rapid (trap-limited) and that an exciton visits several reaction centers during its intrinsic lifetime β^{-1}.

To illustrate the compatibility of the fluorescence yield and its lifetime with geometrical and energy transfer rate constraints consider the case of a regular two-dimensional array with N antenna pigment molecules. In this case (Pearlstein, 1967),

$$\gamma_0^{-1} = K_0^{-1} + K_x^{-1} \approx K_x^{-1} + 0.08 N \ln N / W(R), \quad (10)$$

where only nearest-neighbor transfers, at a rate $W(R)$, are allowed. A *lower bound* for $W(R)$ can be obtained by neglecting the trapping factor K_x^{-1}. With a "typical" lifetime of 1 nsec and an antenna of 300 molecules, Eq. (10), when the revisiting term is neglected, implies that $W(R) \geq 10^{11}$ sec^{-1}. Although this calculation involves several approximations (e.g., dimensionality, nearest neighbors only, square regular lattice), it nevertheless indicates the compatibility of geometrical constraints and reasonable transfer rates. A more detailed analysis for chromatophores of *Rhodopseudomonas sphaeroides* R26 has been given by Pearlstein (1982). His analysis reveals that the exciton visits the reaction center, either the same one or a different one, about six times; i.e., $K_x^{-1} = 6K_0^{-1}$.

In the short time regime the zero-mode approximation to Eq. (7) is invalid, and thus the inclusion of higher order terms in Eq. (9) is necessary to describe properly the emission from the donor array. Current knowledge of donor fluorescence when there are many donor–donor transfer steps is meager, although some recent progress has been made in solving Eq. (7) (Huber, 1979, 1980). The following discussion will be restricted to the incoherent transfer limit. For a discussion of modifications when the exciton motion is coherent, the papers by Huber (1979, 1980) should be consulted. The fluorescence emission can be expressed as an appropriate average (denoted by the angle brackets) over all donor and acceptor configurations,

$$F(t) = \sum_n \langle P_0(R_n, t) \rangle = e^{-\beta t} f(t),$$

where $P_0(R_n, t)$ is the solution of Eq. (7) with initial conditions $P(R_n, 0) = \delta_{R_n, R_{n'}}$. An exact expression for $f(t)$ is possible in only two limits: (1) the *static limit* [donor–donor transfer is not allowed, and only transfer from a donor to an acceptor is permitted (see Fig. 1); the transfer is from molecule C to the trap] where

$$f(t) = \prod_n \{1 - C_A + C_A \exp[-X(R_n)t]\} \tag{11}$$

and C_A is the concentration of acceptors (or the probability that a site is occupied by an acceptor); and (2) the *fast transfer limit* which applies when transfer among the donors is very rapid in comparison with the rate of transfer to the trap(s). In this limit, the fluorescence decays exponentially, and its dependence is

$$F(t) = \exp(-\beta t) f(t) = \exp(-\beta t) \exp\left(-C_A \sum_n X(R_n) t\right), \tag{12}$$

The behavior of $f(t)$ in these two limits as given by Huber (1979) is shown in Fig. 2. In the static limit, the function $F(t)$ is nonexponential for all times in contrast to the exponential behavior in the fast transfer limit. The physical situation falls somewhere between these two extremes, and curve (c) in Fig. 2

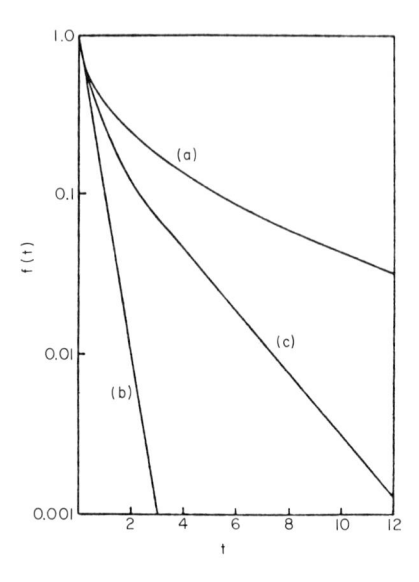

FIG. 2. Schematic semilog plot of $f(t)$-versus-t. (a) $f(t)$ in the absence of donor–donor transfer (static limit); $f(t)$ is nonexponential at all times. (b) $f(t)$ in the rapid transfer limit where it is exponential for all t (see text). (c) $f(t)$ in the intermediate regime where donor–donor and donor–trap rates are comparable; in this limit $f(t)$ becomes exponential as $t \to \infty$. (Redrawn from Huber, 1979.)

illustrates schematically the expected behavior. At short times, $f(t)$ is nonexponential after a brief excitation pulse, whereas for longer times the decay becomes exponential. Support for this type of fluorescence decay behavior is evident from computer simulation studies on energy migration on a two-dimensional random lattice by Altmann et al. (1978). With Monte Carlo techniques and considering lattices of 150–300 chlorophylls per trap, where the trap was taken as a perfect energy sink, the calculated chlorophyll emission was nonexponential for short times and could be fitted to an analytical expression of the form

$$F(t) \propto \exp(-At - Bt^n),$$

where $n \sim \frac{1}{2}$ and A and B are constants which depend on the intrinsic lifetime of the exciton, the donor/trap ratio, and the nearest-neighbor transfer rates. This result offers some theoretical support for the previously used empirical $\exp(-Ct^{1/2})$ decay law in the absence of exciton annihilation (Porter et al., 1978; Searle et al., 1979), although its applicability in determining rates of transfer using the relationship

$$\text{Rate} \propto \exp(Ct^{1/2}) \frac{d}{dt} \exp(-Ct^{1/2})$$

within the pigments of the phycobilisomes and from the phycobilisomes to the chlorophylls is tenuous. There is the additional complication that a $\sqrt{t}$ behavior can also be approximated by a sum of two exponentials (Searle *et al.*, 1979) and thereby could signal the existence of two different emitting species. Furthermore, at higher excitation intensities where bimolecular exciton annihilation is effective, Paillotin *et al.* (1979) have shown that, for some excitation intensity regimes, the fluorescence displays the functional form $\ln F(t) \propto \sqrt{t}$. It should thus be apparent that without additional experimental information the appropriate form of $F(t)$ versus t is uncertain and could be different for different physical situations.

III. Exciton–Exciton Annihilation: High Excitation Intensities

With increasing excitation intensities, various bimolecular processes become increasingly important mechanisms for the removal of singlet excitations. In most cases, these processes manifest themselves as a decrease in the quantum efficiency and lifetime of the fluorescence. Chapter 6 summarizes the effects of these nonlinear processes. The effect of singlet–singlet and singlet–triplet processes on the fluorescence yield and lifetime can easily be separated experimentally. This follows since in single picosecond pulse excitation, singlet–triplet annihilation is ineffective because of the long time required for triplet formation. However, if pulse train or microsecond laser excitation sources are employed, the time scale can be sufficiently long so that triplet states accumulate in the antenna and result in singlet–triplet fusion becoming the major quenching mechanism for fluorescent singlet states (Breton *et al.*, 1979; Monger and Parson, 1977). We restrict ourselves to a discussion of singlet–singlet annihilation.

The most general theoretical model for exciton annihilation in confined domains consisting of a homogeneous array of molecules is that of Paillotin *et al.* (1979; Paillotin and Swenberg, 1979). This formulation of singlet exciton fusion contains the limiting cases: (1) the continuum model of Swenberg *et al.* (1976) and (2) the statistical model of Mauzerall (1976a,b, 1978). The following discussion will attempt to clarify when each of the two limiting cases is realized. As currently formulated, Paillotin's theory applies only when the incident excitation source can be approximated by a delta function, realized in single picosecond pulse experiments. The four basic assumptions of the theory are: (1) The exciton distribution randomization time is significantly shorter than any deactivation process, including the characteristic exciton annihilation time; (2) the depletion of the ground state population can be neglected; (3) the exciton coherence time is extremely short, so that memory effects can be neglected; and (4) upper excited state absorption is assumed to be negligible.

Let us briefly comment on these assumptions. An indication that the first assumption is reasonable comes from random walk calculations for regular lattices where fast trapping times are predicted, at least for regular lattices

(Knox, 1977; Pearlstein, 1967); i.e., the diffusion of singlet excitons in the PSU is not rate-limiting ($K_x^{-1} \gg K_0^{-1}$). The validity of the first assumption allows the spatial variables of the distribution functions to be neglected (Teramoto and Shigesada, 1967), thereby characterizing the state of the system solely by the number of excitons it contains. The validity of the second assumption depends not only on the incident intensity but also on the absorption coefficient at the excitation wavelength, as discussed in Chapter 6. The condition that only a small fraction of molecules are excited is satisfied over most intensity ranges. However, for the single picosecond pulse experiment by Geacintov et al. (1977b) the second assumption fails at the highest intensity, $I \approx 2 \times 10^{16}$ photons/cm^2 per flash. At this intensity, Geacintov and Breton (see p. 164) estimate that approximately 40% of the molecules are in the first excited electronic state. Discrepancies between experimental data and theoretical predictions are therefore expected for intensities 2×10^{16} photons/cm^2 per flash and higher. Evidence for the coherence time being short, at least an order of magnitude less than the annihilation time, derives mainly from the theoretical formulation of excitation transfer by Kenkre and Knox (1974). In this formalism the inverse of the spectral linewidth provides a measure of the exciton coherence time τ_c, estimated to be less than 10^{-13} sec at room temperature (Kenkre and Knox, 1976). This is in accord with estimates of coherence times in organic solids where τ_c at room temperature is less than 10^{-13} sec because of the large exciton–phonon interaction energies characteristic of these materials (Pope and Swenberg, 1982). Thus the assumption of incoherence appears amply justified, since nearest-neighbor transfer rates are of the order of 10^{12} sec^{-1}. This approximation allows the master equation to be written in the Pauli form. Furthermore, this lack of coherence is implicit in the randomization approximation. Note also that with a two-dimensional annihilation rate constant, $\bar{\gamma}^{(2)} \approx 10^{-2}$ cm^2 sec^{-1} (Geacintov et al., 1977b) and an average high-intensity excited state density $\bar{n} \approx 10^{13}$ cm^{-2}, the annihilation time is approximately $(\bar{\gamma}^{(2)}\bar{n})^{-1} \sim 10^{-11}$ sec, a value considerably larger than that for the coherence time.

In Paillotin's theory, excitons are created within a domain of unspecified size and are allowed to diffuse unimpeded within a given domain, although motion between domains is prohibited. A domain is defined as the effective number of pigment molecules over which excitons can diffuse and may consist of many PSUs. A schematic illustration of a domain consisting of four PSUs, each represented by a square box, is shown in Fig. 3. On collision, the following excitonic reactions may occur

$$S_1 + S_1 \xrightarrow{\gamma^{(1)}} S_0 + S_1, \tag{13}$$

$$S_1 + S_1 \xrightarrow{\gamma^{(2)}} \begin{cases} S_0 + T_1 \\ S_0 + S_0, \\ T_1 + T_1 \end{cases} \tag{14}$$

FIG. 3. Lake model representation of singlet exciton annihilation in a domain consisting of four PSUs. In the case illustrated here singlet excitons are created in two different PSUs with annihilation and emission occurring in different units.

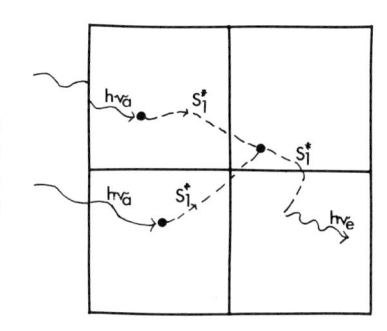

where S_1 and T_1 denote the first excited singlet states, respectively, and S_0 is a molecule in its ground state. The calculated observables of the theory are the fluorescence decay $F(t)$ and the quantum yield ϕ as a function of excitation intensity. The parameters inferred from the theory are the domain size and the total bimolecular rate constant, $\gamma = \gamma^{(1)} + \gamma^{(2)}$; the latter quantity can be related to the diffusion coefficient and the rate of exciton transfer. For the bimolecular processes given by Eqs. (13) and (14), the general (Pauli) master equation can be written as

$$\frac{dp_i(n,t)}{dt} = \left[K(i+1) + \frac{\gamma^{(1)}i(i+1)}{2}\right]p_{i+1}(n,t) + \left[\frac{\gamma^{(2)}(i+1)(i+2)}{2}\right]p_{i+2}(n,t)$$
$$- \left[Ki + \frac{\gamma^{(1)}i(i-1)}{2} + \frac{\gamma^{(2)}i(i-1)}{2}\right]p_i(n,t), \quad (15)$$

where K is the rate of all unimolecular processes and $p_i(n,t)$ is the probability that at time t there will be i excitons in a given domain which initially contained n excitons [see Paillotin et al., 1979, for a derivation of Eq. (15) and its exact solution]. Under normal experimental conditions there are many noninteracting domains, each containing a different number of excitons at $t = 0$. The distribution function for the ensemble of domains is governed by Poisson statistics. Thus, the measurable parameters, the fluorescence decay and quantum yield, are appropriate weighted averages

$$F(t) = \sum_{n=1}^{\infty} \frac{y^{n-1}\exp(-y)}{(n-1)!}\left[\sum_{i=1}^{n} \frac{ip_i(n,t)}{n}\right] \quad (16)$$

$$\phi(y) \propto K_F \int_0^{\infty} F(t)\,dt, \quad (17)$$

where K_F denotes the radiative decay rate and y is the mean number of excitons per domain created at $t = 0$ (y is proportional to the incident laser intensity). The sum in parentheses in Eq. (16) is the fluorescence decay for a domain with a known number, n excitons at $t = 0$. The second summation denotes the ensemble average.

It is convenient to define the parameters

$$\gamma = \gamma^{(1)} + \gamma^{(2)}, \qquad r = 2K/\gamma, \qquad Z = y(1 + \gamma^{(2)}/\gamma) \qquad (18)$$

in discussing the behavior of the fluorescence intensity $F(t)$ and quantum yield ϕ on excitation intensity. The parameter r equals the ratio of the total monomolecular decay rate to the total bimolecular decay rate;

$$r = Ki(i-1)/[\gamma^i(i-1)/2] = 2K/\gamma.$$

For three-dimensional systems the usual bimolecular rate constant $\bar{\gamma}^{(3)}$ (in cubic centimeters per second) can be written as V (in cubic centimeters) × γ (per second), where V is the size of the domain. Thus

$$\bar{\gamma}^{(3)} = 2VK/r.$$

This equation demonstrates that for a fixed value of $\bar{\gamma}^{(3)}$ a small r value corresponds to small domains; i.e., V is small. The major predictions of Paillotin's theory can be summarized as follows:

(1) The smaller the value of r, the more rapid the fluorescence yield decreases with increasing intensity I. The shape of ϕ versus I is quite insensitive to r for r greater than 5.

(2) In the limit where $r = 2K/\gamma < 1$, the quantum yield is described by the expression initially derived by Mauzerall (1976a,b);

$$\phi(Z) = \{[1 - \exp(-Z)/Z]\}, \qquad (19)$$

where Z is proportional to the number of hits per domain xd; i.e., $Z = (1 + \gamma^{(2)}/\gamma)xd$, where d is the number of PSUs per domain and $x = y/d$. In this limit, $F(t)$ drops rapidly at $t = 0$, whereas for $t > 0$, $F(t)$ is exponential for all I, although ϕ decreases with increasing intensity. This behavior for $Z = 10$ is evident in Fig. 4; note the rapid drop in $F(t)$ versus t for the curve $r = 0.1$. For $t > 0.1/K$, $F(t) \propto \exp(-Kt)$. Thus, if $F(t)$ is not experimentally resolvable for $t \leq 0.1/K$, if $K^{-1} \sim 10^{-9}$ nsec, then for $t > 100$ psec, the observed exponential behavior would be taken to indicate the lack of bimolecular exciton annihilation, whereas in fact exciton annihilation is very effective since complete annihilation of excitons occurs as long as there is more than one exciton in the domain. Only the remaining exciton may decay radiatively in the limit $r \ll 1$, and it does so with the normal low-intensity rate K. This Mauzerall limit applies primarily to *very* small domains.

(3) When $r = 2K/\gamma > 1$ (a *very* weak inequality), the predictions of the theory are in accord with the continuum model of Swenberg et al. (1976). This limit applies to *large* domains, that is, where the dimensions are larger than the exciton diffusion length. In this limit, the fluorescence decay is non-

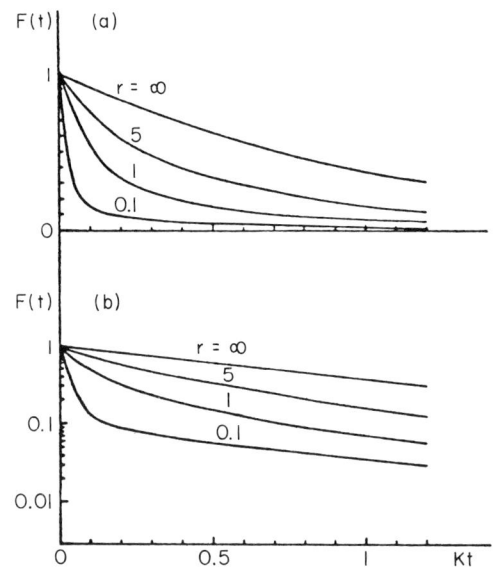

7. DECAY KINETICS AND BIMOLECULAR PROCESSES 205

FIG. 4. Fluorescence decay after a delta function excitation pulse at $t = 0$ for the general case $0 < r < \infty$ (proportional to the number of hits per domain, $Z = 10$) and different values of $r = 0.1, 1, 5,$ and ∞. (a) Linear fluorescence intensity $F(t)$ scale. (b) Logarithmic $F(t)$ scale. (From Paillotin et al., 1979.)

exponential at high excitation intensities,

$$F(t) \propto \exp(-Kt)/[1 + Z/r - (Z/r)\exp(-Kt)], \qquad (20)$$

and the quantum yield ϕ decreases with increasing intensity,

$$\phi \propto (r/Z)\log(1 + Z/r). \qquad (21)$$

Figure 4 with $r = 1$ and 5 illustrates the nonexponential behavior of $F(t)$ at $Z = 10$, i.e., moderately high intensities. For spinach chloroplasts excited at $\lambda = 610$ nm and assuming approximately five PSUs per domain, $Z \approx 10$ corresponds to an intensity of 2×10^{14} photons/cm^2 per flash. The solid line in Fig. 5 illustrates the excellence of the fit of Eq. (21) to spinach chloroplast fluorescence yield (at $T = 300$ K) when a single picosecond laser pulse at 610 nm is utilized (Geacintov et al., 1977b). The correspondence between Z/r and the usual bimolecular rate constant $\bar{\gamma}$ (in cubic centimeters per second) is $Z/r = I_a\bar{\gamma}/2K = \alpha I_0\bar{\gamma}/2K$, where I_a is the number of photons absorbed (per cubic centimeter per pulse) and I_0 the peak height of the single picosecond pulse, α is the absorption coefficient (per centimeter). With α (610 nm) = 1400 cm^{-1}, the theoretical curve gives $\bar{\gamma}$ (at $T = 300$ K) = $(5.2 \pm 1.6) \times 10^{-9}$ cm^3/sec if the lifetime (K^{-1}) is taken as 0.8 nsec (Hervo et al., 1975). This value

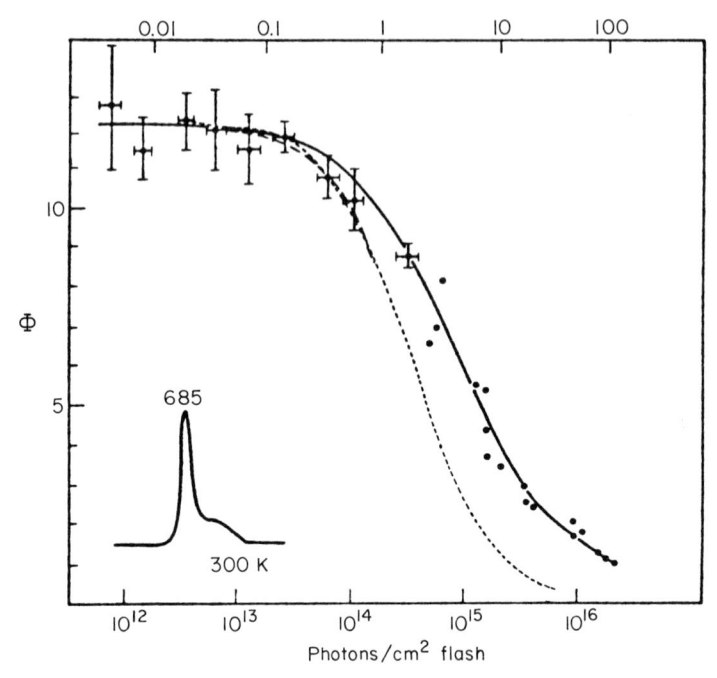

FIG. 5. Relative fluorescence yield ϕ from spinach chloroplast as a function of a single picosecond laser pulse intensity at 300 K. Here the upper abscissa variable represents the number of photon hits per PSU consisting of 200–300 molecules, and ●, experimental data. Inset illustrates a typical fluorescence spectrum obtained with the OMA spectrograph. Solid line denotes calculated quenching curve using lake model of exciton–exciton annihilation; i.e., $r \gg 1$. Dashed curve is calculated quenching curve using the Poisson distribution of photon hits per PSU in the limit where $r \to 0$. (Reprinted with permission from *Photochemistry and Photobiology* **26**, 629–638. Geacintov et al. "A Single Pulse Picosecond Laser Study of Exciton Dynamics in Chloroplasts." Copyright 1977, Pergamon Press, Ltd.)

for the singlet–singlet annihilation rate constant is typical of those observed in organic crystals (Swenberg and Geacintov, 1973). The calculated fluorescence-quenching curve utilizing the Mauzerall limit for ϕ versus I, Eq. (19), is denoted by the dotted line in Fig. 5. The large reduction in ϕ with increasing intensity predicted in this limit is a direct consequence of the dominance of the bimolecular rate. No shift in the dotted curve along the intensity axis can provide an improved fit, since the form of ϕ-versus-I curves for values of $r < 1$ is quite distinct from the behavior of ϕ versus I at higher r values (Paillotin et al., 1979; Paillotin and Swenberg, 1979, Fig. 5). A further indication of the validity of fitting ϕ versus I to the continuum limit (large r) comes from measurements of the fluorescence decay times at 690 mm where strong variations are observed as a function of the intensity of the picosecond pulse (Geacintov et al., 1977a). In particular, the emission at 690 nm from spinach chloroplast at 77 K having a lifetime of ≈ 0.38 nsec at $I_0 \approx 2.5 \times 10^{14}$ photons/cm² decreases to less than 0.13 nsec at $I_0 = 7.4 \times 10^{14}$ photons/cm². Such a dependence of the fluorescence decay on the intensity cannot be accounted for within the context of the Mauzerall limit ($r = 0$). Unfortunately there are presently no existing detailed fits to fluorescence decay curves (emission from chlorophyll *a* pigments) for *different* values of incident intensity using either the continuum limit, Eq. (20) or the more general functional dependence, Eq. (16).

An important distinction exists between the derivation of Eq. (19) from Paillotin's theory and that given by Mauzerall (1976a,b). In Mauzerall's formulation, the quenching is due to the annihilation of excitons by reaction centers that have already captured an exciton, whereas the effects of reaction center processes on the quantum yield are absent in Paillotin's theory. In this formalism, the reaction centers are assumed closed and are taken to be ineffective. Because of the large number of internal states associated with reaction centers and annihilation processes which can occur there (Paillotin, 1976a,b), the effects of the reaction centers on the antennae yield are formidable. Support for the dominance of annihilation within the antennae (which is called singlet homofusion) comes from fluorescence-quenching experiments on strain PM-8 of *R. sphaeroides* which lacks reaction centers. For this system, Campillo and co-workers (1977) showed that the lake model (continuum limit) provided a reasonable description of yield versus intensity. However, a more detailed analysis appears to be needed, since for intensities greater than 10^{15} photons/cm² per flash, the theoretical curve falls somewhat more rapidly with increasing intensity than the observed results, an observation which is currently unexplained.

The equivalence of the Poisson saturation model for ϕ versus I with one trap per unit (Mauzerall, 1976a, 1978) and the $r = 0$ limit of Paillotin's theory is only a consequence of the dominance of exciton annihilation, either at the trap (heterofusion) or within the light-harvesting antenna (homofusion). If a particular system displays a yield dependence on I as given by Eq. (19), the microscopic mechanism responsible for it cannot be inferred from these data only. A physical system where Eq. (19) appears to apply is the protein complex CP II which contains three chlorophyll *a* molecules, three chlorophyll *b* molecules, and a single polypeptide chain (Thornber et al., 1979) and therefore constitutes a *small* domain. Measurements by Nordlund and Knox (1981) give an exponential fluorescence decay for spinach chlorophyll–protein complex II (CP II) with a lifetime of 3.1 ± 0.3 nsec independent of laser intensity over the range 10^{14} to 3×10^{18} photons/cm² as expected for the $r = 0$ limit. Unfortunately, single picosecond pulse quantum yield curves for CP II have not been reported, although evidence that they obey

Poisson saturation statistics comes from the microsecond quenching experiments of Breton and Geacintov (1980). For microsecond pulse excitation, annihilation of singlet excitons by triplet excited states is the dominant quenching mechanism. Under these conditions, Breton and Geacintov (1980) found that the quenching curves were very steep and obeyed Mauzerall's equation as expected for small domains. Obviously, careful experiments involving ϕ versus I for CP II, and a fast time resolution of the decay characteristics (on the order of a picosecond or less), are necessary for determination of the r value and possibly observation of the rapid dependence of $F(t)$ at short time (see Fig. 4). Such experiments and their analysis could necessitate extensions of the existing theory to include the effects of upper excited state absorption. In the light-harvesting complex (LHC), the complex prepared by Arntzen's method (Burke et al., 1978), the domain size over which exciton migration can occur has increased and the effects of the bimolecular process have decreased relative to monomolecular rates. In this case, the fluorescence decay is expected to be intensity-dependent, and such an intensity effect has been reported by Nordlund and Knox (1981). This is shown in Fig. 6. A similar shortening of the lifetime (emission at 700 nm) with increasing laser intensity for *Chlorella pyrenoidosa* has been reported by Campillo et al. (1976a). Thus, experimental results for CP II and the LHC appear to substantiate the effects of domain size on fluorescence decay and quantum yield as predicted by Paillotin's theory (Paillotin and Swenberg, 1979; Paillotin et al., 1979).

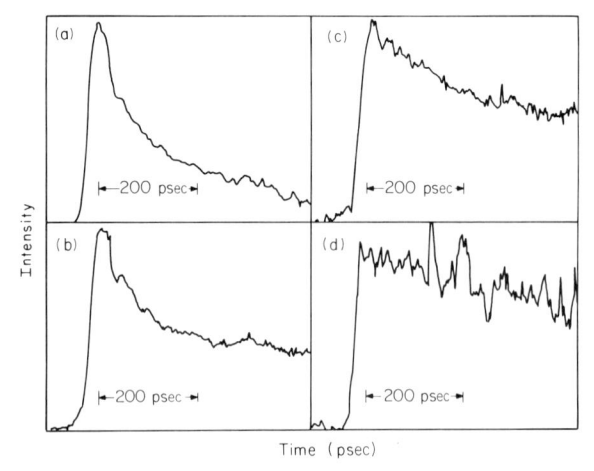

FIG. 6. Fluorescence decay of LHC as a function of a single picosecond pulse intensity I. (a) $I = 1.4 \times 10^{17}$ photons/cm^2; (b) 5×10^{16} photons/cm^2; (c) 1.4×10^{16} photon/cm^2; (d) 7×10^{14} photons/cm^2. (From Nordlund and Knox, 1981.)

There are cases, however, where at a particular emission wavelength the fluorescence yield decreases with increasing intensity and where the fluorescence lifetime is intensity-independent but does not correspond to the $r = 0$ limit. For example, the 735-nm emission from spinach chloroplasts has a lifetime of 1.1 nsec at 77 K for intensities ranging from 3.5×10^{14} to 2.6×10^{15} photons/cm^2 per flash, even though the fluorescence quantum yield decreases sharply in this excitation energy range and obeys the large r limit (Geacintov et al., 1977a,b). In this case the fluorescence moiety, a special form of chlorophyll a termed C705 (Butler et al., 1979), acts like a trap operating with little back transfer into the light-harvesting pigment antenna and with very ineffective exciton heterofusion. The homofusion of singlet excitons within the light-harvesting antenna merely depletes the number of excitons that can arrive at C705—thus the decrease in fluorescence yield. In such cases a finite risetime representing the average trapping time is expected (Breton and Geacintov, 1980; Butler et al., 1979).

In principle the bimolecular rate constant can be utilized to calculate the singlet exciton diffusion coefficient. However, the inferred value can be in error by at least a factor of 5 because of incomplete knowledge of the range of the annihilating interaction R and the effectiveness of each collision. Since photosynthetic pigment molecules are most likely arranged in a two-dimensional network, the values of $\bar{\gamma}$ (in cubic centimeters per second) calculated from the fluorescence-quenching curves must be divided by the membrane thickness, estimated to be ~ 50 Å, to give appropriate two-dimensional values ($\bar{\gamma}^{(2)}$, in square centimeters per second). The expression relating $\bar{\gamma}^{(2)}$ to the diffusion constant D has been given by Suna (1970) as

$$\bar{\gamma}^{(2)} = -4\pi Dp/\ln(\tfrac{1}{2}\sqrt{R^2/D\tau^0}) + 0.577 \qquad (22)$$

provided $\sqrt{D\tau^0} > R$, where p allows for collisions which do not lead to annihilation. The value of R, the range over which annihilation is effective, is not known and could be as small as the nearest-neighbor distance, ~ 20 Å, or as large as $\bar{R}_0$, the Förster critical distance, estimated to be ~ 50–80 Å. For $p \sim 1$ Geacintov and co-workers (1977b; see also Swenberg et al., 1978) estimated a lower bound of D (for $T = 300$ K) $\approx 10^{-3}$ cm^2/sec. This value of D corresponds to a diffusion length $l \sim \sqrt{4D\tau^0} \sim 200$ Å (if $\tau^0 \sim 0.8$ nsec). A value of $p \sim 0.5$ would yield a diffusion length of ~ 800 Å. Thus the singlet exciton could easily migrate over several PS II units and visit several reaction centers. Quenching experiments by Geacintov and co-workers (1977b) also have shown that $\bar{\gamma}^{(2)}$ is relatively insensitive to temperature over the range 21–300 K. This observation and the lack of a strong dependence on temperature of the fluorescence lifetime (from the light-harvesting antenna), 0.8 nsec at 300 K to 1.1 nsec at 21 K, implies that D depends very weakly on temperature. As indicated by these investigators, these results favor the completely incoherent model for exciton migration among the chlorophyll antenna, since for coherent transfer $D \propto T^{-1/2}$ (Pope and Swenberg, 1982). In the incoherent limit with $D \geq 10^{-3}$ cm^2/sec, $R \approx 20$ Å, a value for the nearest-neighbor transfer rate $W(R) > 3 \times 10^{10}$ sec^{-1} was inferred. This value could be in error by at least a factor of 10 because of uncertainties in R and p; in fact Paillotin (1976a,b) estimated a value for W approximately 10 times larger.

The lack of a strong temperature dependence of the diffusion coefficient and therefore of the Forster transfer rate is supported by the theoretical calculations of Sarai and Yomosa (1980). By considering the excitation transfer as a weak perturbation and employing the multiphonon formulation previously utilized in electron transfer reactions (Sarai, 1979) the transfer rate could be written as

$$W(R_n) = [2\pi/\hbar(\hbar\omega_s)]|J(R_n)|^2 \exp(-G(T))F(T, \Delta E). \qquad (23)$$

Here ω_s is the average frequency of a group of low-frequency soft modes, $J(R_n)$ is the electronic dipole–dipole interaction between donor and acceptor separated by a distance R_n, ΔE is the energy gap between the relaxed initial and final electronic states, and both $F(T, \Delta E)$ (a complex function of Bessel functions) and $G(T)$ (the conventional Huang–Rhee factor) are functions of coupling strengths (S) and frequencies of the normal modes. Using a single average (high-frequency) mode, $\hbar\langle\omega\rangle \sim 500$ cm^{-1} and setting $S = E_m/\hbar\langle\omega\rangle$, where E_m is the nuclear rearrangement energy, a very weak temperature dependence for $W(R_n)$ among the different chlorophyll forms was predicted for $100 \leq T \leq 300$ K. This lack of a strong temperature dependence of $W(R)$ is consistent with the picosecond fluorescence decays measured by Yu et al. (1977).

The additional ingredient in the master equation formalism for exciton–exciton annihilation is that the theory allows for determination of the domain size. This can be seen as follows. The theoretical quenching curves $\phi(y)$ are given in terms of the number of hits per domain, y (a calculated parameter) or more generally Z, whereas the experimental fluorescence curves are measured as a function of x, the number of hits per PSU (an experimentally accessible parameter provided the absorption coefficient is known) or more generally as a function of the number of hits per molecule. The shape of ϕ versus x determines a *lower* bound to r, lower because for large r the theoretical quenching curves are very similar. From r an *upper* bound to the total annihilation rate constant γ can be calculated provided the monomolecular decay rate is known. Demanding that ϕ(theoretical) fit ϕ(experimental) allows the multiplicative factor d relating y and x to be inferred ($y = xd$). This is accomplished by sliding along the abscissa the ϕ(experimental) curves relative to the family of theoretical curves. In this manner the conservative value for the number of PSUs over which a given exciton can migrate during its lifetime was inferred for spinach chloroplasts; $d > 2$ (at 300 K) and at slightly lower temperatures, $d > 4$ (below 200 K) (for details, see Paillotin et al., 1979; Paillotin and Swenberg, 1979). Thus the nonlinear effects on fluorescence yield provide support for the lake model of photosynthetic membranes. Unfortunately this more general theory of singlet exciton fusion has not been applied to other photosynthetic systems.

Photosynthetic systems where the small r limit might be applicable (r on the order of unity) are in the phycobilisomes of the blue-green and red algae. Phycobilisomes are multiprotein aggregated structures containing phycobiliproteins, e.g., phycoerythrins (PE), phycocyanins (PC), and allophycocyanins (APC), in a highly organized manner (Gantt et al., 1976; Grabowski and Gantt, 1978a,b), which are attached to the outer surface of the photosynthetic lamellae (Gantt, 1975). It is well-known from time-resolved picosecond experiments (Porter et al., 1978) and structural studies (Gantt et al., 1976) that the route of energy transfer,

$$\text{PE} \to \text{PC} \to \text{APC} \to \text{chlorophyll antenna (PS II primarily)}, \qquad (24)$$

within the phycobilisomes is quite efficient (estimated to be $\sim 99\%$ between different pigments), presumably because of the hemispherical layerlike organization of the pigments. The fast transfer of energy to the chlorophyll antenna has been suggested for the lack of bimolecular annihilation within phycobilisomes for intact systems (Porter et al., 1978), at least when $I \leq 10^{15}$ photons/cm^2 per flash. However, if energy transfer to the chlorophylls is prevented by dissociation of the phycobilisomes from the photosynthetic lamellae, then exciton–exciton annihilation is observed in isolated phycobilisomes as was shown by Searle et al. (1978) for the red algae, *Porphyridium cruentum*. Single picosecond pulse studies (excitation at 530 nm) of fluorescence kinetics and quantum yield from both suspensions of phycobilisomes and individual phycobiliproteins isolated from the blue-green alga *Nostoc* sp. by Alfano and co-workers (Pellegrino et al., 1982; Wong et al., 1981) revealed a host of nonlinear processes. For suspensions of phycobilisomes the 575-nm PE emission decay was exponential (a lifetime of $\sim 31 \pm 4$ psec) and depended weakly (if at all) on source intensity for $I < 10^{15}$ photons/cm^2 per flash, whereas its fluorescence yield decreased by approximately 60% over the same intensity range. Although interpreted in terms of the continuum (large r) model of exciton annihilation, these observations are those expected for small r values (approximately unity). This observation should be contrasted with the decay of APC fluorescence from phycobilisomes which was strongly intensity-dependent—a characteristic of the large r limit. Nevertheless, caution should be exercised in this interpretation, since formally Paillotin's theory does not apply to exciton annihilation in PC and APC complexes in phycobilisomes since the source conditions no longer are approximated by a delta function and transfer between domains occurs. In isolated phycobiliproteins, where Paillotin's theory should apply, the separated pigment aggregates exhibit fluorescence yield decreases and decay kinetics (Wong et al., 1981) indicative of small r values.

IV. Conclusion

This chapter has restricted itself primarily to the theoretical aspects of fluorescence decay and singlet bimolecular annihilation in photosynthetic membranes. The major point emphasized is the complexity of the fluorescence decay function $F(t)$ in the short time regime. In particular it was noted that $\ln F(t)$ proportional to $\sqrt{t}$ can occur for a variety of microscopic conditions; for example, (1) donor emission in the case where a random three-dimensional distribution of acceptors (low concentration limit) is available to which the donor can transfer its excitation energy, (2) at particular source intensities where bimolecular processes are important, the emission from an array of donors can give decays which behave as $\ln F(t) \propto \sqrt{t}$ (Paillotin et al., 1979), (3) emission from a donor array where donor–donor transfer is competitive with donor trap rates. Furthermore since a sum of two exponential functions can approximate quite well the behavior of $F \propto \exp(-\alpha\sqrt{t})$ (Searle et al., 1979), it should be evident that without additional information, for example, the expected dependence of $F(t)$ on extrinsic cations, poisons, or temperature, there is no formal method of distinguishing the appropriate decay function form.

This chapter has also discussed the major predictions and assumptions in the master equation formulation of exciton–exciton annihilation for single picosecond sources. Although the theory describes quite well the behavior of the fluorescence yield from spinach chloroplasts as a function of intensity, its applicability to other photosynthetic membranes has not been tested. There are indications that extensions of the theory are necessary. In particular the single-pulse quenching data from *C. pyrenoidosa* (Campillo et al., 1976b) and the larger (greater than expected on the bases of theory) fluorescence yield from strain PM-8 of *R. sphaeroides* at high excitation intensities are unaccounted for by current theories. In addition the master equation theory needs to be extended to include the effects of *open* traps.

ACKNOWLEDGMENTS

The author is grateful to Prof. N. E. Geacintov (New York University) and Dr. Parviz Yavari (NIAAA) for a critical reading of the manuscript and is very indebted to Dr. T. Nordlund and

Mr. W. Knox (University of Rochester) for experimental information on CP II and the LHC prior to publication and use of their results given in Fig. 6. The author is also indebted to Dr. R. Pearlstein for a copy of his recent review on chlorophyll singlet excitons prior to publication.

REFERENCES

Altmann, J. A., Beddard, G. S., and Porter, G. (1978). *Chem. Phys. Lett.* **58**, 54–57.
Barber, J. (1978). "Primary Processes in Photosynthesis." Elsevier/North-Holland, New York.
Blumen, A., and Manz, J. (1979). *J. Chem. Phys.* **71**, 4694–4702.
Borisov, A. Y. (1978). *In* "The Photosynthetic Bacteria" (R. K. Clayton and W. R. Sistram, eds.), pp. 323–340. Plenum, New York.
Breton, J., and Geacintov, N. E. (1980). *Biochim. Biophys. Acta* **594**, 1–32.
Breton, J., Geacintov, N. E., and Swenberg, C. E. (1979). *Biochim. Biophys. Acta* **548**, 616–635.
Burke, J. J., Ditto, C. L., and Arntzen, C. J. (1978). *Arch. Biochim. Biophys.* **187**, 252–263.
Butler, W. L. (1980). *Proc. Natl. Acad. Sci. U.S.A.* **77**, 4697–4701.
Butler, W. L., Tredwell, C. J., Malkin, R., and Barber, J. (1979). *Biochim. Biophys. Acta* **545**, 309–315.
Campillo, A. J., Kollman, V. H., and Shapiro, S. L. (1976a). *Science* **193**, 227–229.
Campillo, A. J., Shapiro, S. L., Kollman, V. H., Winn, K. R., and Hyer, R. C. (1976b). *Biophys. J.* **16**, 93–97.
Campillo, A. J., Hyer, R. C., Monger, T. G., Parson, W. W., and Shapiro, S. L. (1977). *Proc. Natl. Acad. Sci. U.S.A.* **74**, 1997–2001.
Dewey, T. G., and Hammer, G. G. (1980). *Biophys. J.* **32**, 1023–1036.
Dexter, D. L. (1953). *J. Chem. Phys.* **21**, 836–850.
Forster, T. W. (1965). *In* "Modern Quantum Chemistry" (O. Sinanoglu, ed.), pp. 187–253. Academic Press, New York.
Franck, J., and Teller, E. (1938). *J. Chem. Phys.* **6**, 861–872.
Gantt, E. (1975). *BioScience* **25**, 781–788.
Gantt, E., Lipchultz, C. A., and Zilinskas, B. (1976). *Biochim. Biophys. Acta* **430**, 375–388.
Geacintov, N. E., Breton, J., Swenberg, C., Campillo, A. J., Hyer, R. C., and Shapiro, S. L. (1977a). *Biochim. Biophys. Acta* **461**, 306–312.
Geacintov, N. E., Breton, J., Swenberg, C. E., and Paillotin, G. (1977b). *Photochem. Photobiol.* **26**, 629–638.
Govindjee, R. ed. (1975). "Bioenergetics of Photosynthesis." Academic Press, New York.
Grabowski, J., and Gantt, E. (1978a). *Photochem. Photobiol.* **28**, 47–54.
Grabowski, J., and Gantt, E. (1978b). *Photochem. Photobiol.* **28**, 39–45.
Hervo, G., Paillotin, G., Thiery, J., and Breuze, G. (1975). *J. Chim. Phys. Physiochem. Biol.* **6**, 761–766.
Huber, D. L. (1979). *Phys. Rev. B* **20**, 2307–2314.
Huber, D. L. (1980). *Phys. Rev. B* **22**, 1714–1721.
Kenkre, V. M., and Knox, R. S. (1974). *Phys. Rev. B* **9**, 5279–5286.
Kenkre, V. M., and Knox, R. S. (1976). *J. Lumin.* **12/13**, 187–194.
Knox, R. S. (1975). *In* "Bioenergetics of Photosynthesis" (Govindjee, ed.), pp. 183–221. Academic Press, New York.
Knox, R. S. (1977). *In* "Primary Processes of Photosynthesis" (J. Barber, ed.), Vol. 2, pp. 55–97. Elsevier, Amsterdam.
Mauzerall, D. (1976a). *J. Phys. Chem.* **80**, 2306–2309.
Mauzerall, D. (1976b). *Biophys. J.* **16**, 87–91.
Mauzerall, D. (1978). *Photochem. Photobiol.* **28**, 991–998.

Monger, T. G., and Parson, W. W. (1977). *Biochim. Biophys. Acta* **460**, 393–407.
Nordlund, T. M., and Knox, W. H. (1981). *Biophys. J.* **36**, 193–201.
Paillotin, G. (1976a). *J. Theor. Biol.* **58**, 219–235.
Paillotin, G. (1976b). *J. Theor. Biol.* **58**, 237–252.
Paillotin, G., and Swenberg, C. E. (1979). *Ciba Found. Symp.* **61**, 201–215.
Paillotin, G., Swenberg, C. E., Breton, J., and Geacintov, N. E. (1979). *Biophys. J.* **23**, 513–533.
Pearlstein, R. M. (1967). *Brookhaven Symp. Biol.* No. 19, pp. 8–15.
Pearlstein, R. M., (1982). In "Integrated Approach to Plant and Bacterial Photosynthesis" (Govindjee, ed.). Academic Press, New York. In press.
Pellegrino, F., Wong, D., Alfano, R. R., and Zilinskas, B. A. (1982). *Photochem. Photobiol.* **34**, 691–696.
Pope, M., and Swenberg, C. E. (1982). "Electronic Processes in Organic Crystals." Oxford Univ. Press, London and New York.
Porter, G., Tredwell, C. J., Searle, G. F. W., and Barber, J. (1978). *Biochim. Biophys. Acta* **501**, 232–245.
Robinson, G. W. (1967). *Brookhaven Symp. Biol.* No. 19, pp. 16–48.
Sarai, A. (1979). *Chem. Phys. Lett.* **63**, 360–366.
Sarai, A., and Yomosa, S. (1980). *Photochem. Photobiol.* **31**, 575–583.
Searle, G. F. W., Barber, J., Harris, L., Porter, G., and Tredwell, C. J. (1977). *Biochim. Biophys. Acta* **459**, 390–401.
Searle, G. F. W., Barber, J., Porter, G., and Tredwell, C. J. (1978). *Biochim. Biophys. Acta* **501**, 246–256.
Searle, G. F. W., Tredwell, C. J., Barber, J., and Porter, G. (1979). *Biochim. Biophys. Acta* **545**, 496–507.
Shipman, L. (1980). *Photochem. Photobiol.* **31**, 157–167.
Soules, T. F., and Duke, C. B. (1971). *Phys. Rev. B* **3**, 262–274.
Suna, A. (1970). *Phys. Rev.* **81**, 1716–1739.
Swenberg, C. E., and Geacintov, N. E. (1973). In "Organic Molecular Photophysics" (J. B. Birks, ed.), Vol. 1, pp. 489–564. Wiley, New York.
Swenberg, C. E., Geacintov, N. E., and Pope M. (1976). *Biophys. J.* **16**, 1447–1452.
Swenberg, C. E., Geacintov, N. E., and Breton, J. (1978). *Photochem. Photobiol.* **28**, 999–1006.
Teramoto, E., and Shigesada, N. (1967). *Prog. Theor. Phys.* **37**, 29–51.
Thornber, J. P., Markwell, J. P., and Reinman, S. (1979). *Photochem. Photobiol.* **29**, 1205–1216.
Wolber, P. K., and Hudson, B. S. (1979). *Biophys. J.* **28**, 197–210.
Wong, D., Pellegrino, F., Alfano, R. R., and Zilinskas, B. A. (1981). *Photochem. Photobiol.* **33**, 651–662.
Yu, W., Pellegrino, F., and Alfano, R. R. (1977). *Biochim. Biophys. Acta* **460**, 171–181.

CHAPTER 8

Statistical Theory of the Effect of Mutiple Excitation in Photosynthetic Systems

David Mauzerall

The Rockefeller University
New York, New York

I.	INTRODUCTION	215
II.	STATISTICS	216
	A. *Poisson Statistics*	216
	B. *Semiannihilation and Total Annihilation*	218
	C. *Multitrapped Units*	219
	D. *Escape at Traps*	220
	E. *Complications*	224
	F. *Quenchers*	224
III.	KINETICS	225
	A. *Annihilation*	225
	B. *Resonances*	227
IV.	APPLICATION	228
	A. *Cross Section of the Photosynthetic Unit*	228
	B. *Number of Traps per Unit*	228
	C. *Total Annihilation in System II*	230
V.	COMPARISON WITH OTHER TREATMENTS	230
VI.	CONCLUSIONS	232
	APPENDIX I: CALCULATION OF YIELD WITH ESCAPE AT TRAPS	233
	A. *Semiannihilation*	234
	B. *Total Annihilation*	234
	APPENDIX II	234
	REFERENCES	235

I. Introduction

The availability of lasers has created new and fruitful areas of research in many fields. The time range of nanoseconds to picoseconds has been opened to the study of energy and electron transfer processes in photosynthesis by the use of short pulses of light. The combination of submicrosecond pulses and monochromatic light of variable wavelength has allowed the absolute optical cross sections of photosynthetic units (PSUs) to be determined. However, the high photon fluxes of such light pulses inevitably lead to multiple excitation. Pulse energies of 1 mJ/cm^2 and nanosecond duration are sufficient to excite molecules multiply, leading to photoionization and other reactions. At about this energy direct multiphoton effects also occur, and the field of nonlinear spectroscopy has become very active. The effect of multiple excitations is aggravated in photosynthetic systems where ~ 300 molecules may act as antennae for a single reaction center. Thus a pulse of a few microjoules can show a nonlinear quantum yield of fluorescence. These effects are reinforced with picosecond pulses, since all absorbed photons arrive well within the fluorescence lifetime. The multiple excitations in the antennae of photosynthetic systems differ from those in molecules and some crystals. This is because the excitation in the molecular array appears to be very localized. Because of the finite lifetime of this localized excitation in the array, and in the trap where charge transfer occurs, more than one excitation will be present at once whenever the frequency of photon absorption by the array is comparable to the reciprocal of the lifetime. The interaction of these excitations either in the antenna array or at the trap can produce a variety of effects, ranging from simple quenching, through formation of triplet states, to electron transfer and formation of ions. In general these processes will compete with spontaneous emission, and thus the observed quantum yield of fluorescence will decrease in the multiple-excitation region. An intriguing effect in photosynthetic systems is the increased yield of fluorescence under conditions of saturated illumination, i.e., when the traps are filled. Depending on the time scale, excitations following the first may show this increased fluorescence yield, followed by the quenching processes when again the frequency of excitation exceeds the reciprocal of the lifetime which is itself often longer than in the open trap condition. It is the interplay of these various effects with those of the structure and topology of the photosynthetic array that produces the observed changes in fluorescence yield. This chapter will analyze the effects of multiple excitations on yields of fluorescence and photochemistry in photosynthetic systems. It will stress the statistical effects but will include an approximate treatment of kinetic effects as compared with other kinetic treatments.

Reviews of measurements in photosynthetic systems with short pulses of laser light, in addition to those in this book, are available (Holton and Windsor, 1978; Breton and Gaecintov, 1981; Govindjee, 1978).

II. Statistics

A. Poisson Statistics

Since even a microscopic sample of a photosynthetic system contains tens of thousands of PSUs, the distribution of hits n to the units following uniform

8. STATISTICS OF MULTIPLE EXCITATIONS

illumination is given by the Poisson distribution:

$$P_n = x^n e^{-x}/n!, \quad x = \sigma \int_0^T I\, dt = \sigma E, \qquad (1)$$

where x is the number of average hits, I is the intensity of the light, the integral is over the light pulse length, and E is the photon flux in photons per square angstrom. A hit is defined as absorption of a photon by a PSU which leads to an observable effect, be it oxygen formation, a change in fluorescence yield, or whatever. For simplicity and in agreement with most observations it is assumed that the quantum yield of the inherent photochemistry at the trap is unity. It can be shown that, if this is not so, then the measured optical cross section σ is $\phi'\sigma'$, where σ' is the true cross section and ϕ' is the actual quantum yield. The optical cross section is the probability of a PSU absorbing photons from the flux E. It is simply proportional to the conventional decadic molar absorption coefficient ε measured in solution. For randomly oriented samples, σ (in square angstroms) $= 3.82 \times 10^5 \varepsilon$. The average number of photons absorbed, or hits, per unit x, is then given by σE, while the specific number of hits per unit, $n = 0, 1, 2, \ldots$, is given by the Poisson distribution [Eq. (1)]. In what follows we will often refer to σ, which is obtained from x by fit of the data.

The simplest effect of multiple hits is nothing. That is, the *same* effect is observed for units hit one or more times. The fraction of these units is unity minus the number missed. The latter is $P_0 = e^{-x}$. This fraction is called the cumulative one-hit Poisson distribution, and thus the observed yield of oxygen, the increase in fluorescence yield $\Delta\phi$, etc., is

$$Y = \phi_0(1 - e^{-x}), \qquad (2)$$

where ϕ_0 is the maximum or saturated yield measured as $x \to \infty$ and e^{-x} are simply those units not hit [Eq. (1), $n = 0$]. In direct measurements of fluorescence, the quantum yield of fluorescence is more useful, since fluorescence is proportional (at least at very small x) to x. The quantum yield is obtained by simply dividing Eq. (2) by the average number of hits:

$$\phi = Y/x = \phi_0(1 - e^{-x})/x. \qquad (3)$$

As $x \to 0$, $\phi \to \phi_0$, showing that ϕ_0 can be considered the quantum yield of a linear system: molecules or quanta produced per quanta absorbed.

In general the yield is given by

$$Y = \sum_0^\infty \phi_n P_n, \qquad (4)$$

and ϕ_n is determined by fitting the data. A model can be fit to the ϕ_n by induction. In general, it is simpler to propose a model, calculate Y, and test for a fit to the data.

B. Semiannihilation and Total Annihilation

The fact that the quantum yield of fluorescence excited by short (<10 nsec) pulses of light decreases at energies comparable to that necessary to saturate photosynthesis indicates that some annihilation of the multiple excitations can occur. There is good evidence that the first excitation always fills a trap in a unit, i.e., an excitation in the antenna, P*, on encountering a trap, T, causes charge transfer $T^{\pm}$ to occur:

$$P^* + T \to P + T^{\pm}. \tag{5}$$

This evidence is the high quantum yield of the primary photoreactions. The reversibility of reaction (5) will be discussed in Section II,D. Second and further excitations may be destroyed in (a) semiannihilation or (b) total annihilation. In the former one excitation survives, and in the latter none survive. The products can be triplet states, ionization, or simply heat. An example of semiannihilation is the combination of two singlet states to form a doubly excited state: $P^* + P^* \to P^{**} + P$. In most molecules such higher excited states decay to the lowest excited state in subpicosecond times, emitting heat: $P^{**} \to P^* + \Delta$. The overall result is the loss of one excitation. An example of total annihilation is ionization and recombination to the ground state plus heat: $P^* + P^* \to P^+ + P^- \to 2P + \Delta$. Our experiments have shown that the ionization reaction occurs at the encounter limit between triplet states of porphyrins. The simplest case of saturation, Eq. (2), can be explained by semiannihilation. In this case the excitations annihilate themselves sequentially, either in the antenna or in the trap, depending on the time scale of the light pulse, and the remaining excitation fills the trap. In the spirit of statistics, it does not matter which excitation it is that survives. The important thing is that some effect is observed (one filled trap), no matter how many excitations have been absorbed.

The case of total annihilation is intriguing, for it represents a parity detector: Odd hits have an effect, and even hits undo the effect. The first hit can fill a trap, but the second hit obliterates the first either in the antenna or in the trap depending on the time scale. The third hit again can fill a trap, but the fourth hit demolishes the third, etc. The yield in this case (b) is given by the sum over the odd hits:

$$Y = Y_0 \sum_0^\infty \frac{x^{2n+1} e^{-x}}{(2n+1)!} = \frac{Y_0}{2}(1 - e^{-2x}). \tag{6}$$

A simple proof due to Schreiber is given in Mauzerall (1979). The rather surprising result is that the form of the saturation curve is precisely the same as in case (a). However, only half the traps are filled by a saturating flash, and the optical cross section for these events is twice the true value.

Intuitively one expects a half-yield, since as $x \to \infty$, the numbers of odd and even hits become identical. The $2\sigma(2x = 2\sigma E)$ arises because, as $x \to 0$, the quantum yield Y/x is Y_0, not $Y_0/2$. The first hit cannot possibly know it will be followed (or not followed) by a second hit. This simple fact is often obscured in "kinetic" treatments of these phenomena. Yet the saturation can only rise to half-maximum, and the odd Poissonians neatly mesh into an exponential of twice the argument. The half-yield allows a test for total annihilation. If a second saturating flash is given within a suitable time, it will excite half the unfilled traps, i.e., a yield of one-fourth; a third flash would produce a yield of one-eighth, etc. It is possible that observations on photosynthetic bacteria and system I may be interpretable in this way (Mauzerall, 1979). Evidence for the occurrence of total annihilation in the oxygen system will be presented in Section IV,C.

C. Multitrapped Units

It is implicit in the foregoing that there is only one trap per PSU. A trap is defined as the site where the conversion of a hit to an observable effect, be it oxygen formation, an increase in fluorescence yield, etc., occurs. Multiple-trap and single-trap PSUs are referred to as the "lake model" (many, even an infinite number of traps) and the "puddle model" (single traps). It is often thought that Poisson statistics do not apply to the former, but this is not so. Our solution to the problem requires only two assumptions: (1) a non-discrimination clause and (2) an operating rule. (1). An excitation has equal a priori probability of encountering open or filled traps. (2) On encountering an empty trap the excitation fills it. On encountering a filled trap, annihilation occurs which may be (a) semi, or (b) total.

1. Semiannihilation

In this case the first excitation always fills one trap. If there are two traps, the second hit fills one-half of the new trap, the third hit one-fourth, etc. For two traps the filling is two-thirds, four-ninths, etc. Thus the probability of filling new traps on the jth hit is $(t - 1/t)^{j-1}$, where t is the number of traps per unit. These probabilities can be summed for n hits, weighted by Poissons, and summed over all hits (for details, see Mauzerall, 1980).

$$Y = \phi_0 \sum_1^\infty P_n \sum_1^n \left(\frac{t-1}{t}\right)^{j-1} = \phi_0 t(1 - e^{-x/t}). \tag{7}$$

Thus a PSU with any number of traps shows a simple Poisson saturation curve. The cross section is the average per trap, not per unit. This is more clearly seen by writing the quantum yield for the process

$$\phi = Y/x = \phi_0(1 - e^{-x/t})/x/t \tag{8}$$

2. Total Annihilation

As before, the first hit always fills one trap. However, for two traps, the second and all succeeding hits do nothing. The second hit fills one-half of an empty trap and obliterates one-half of a previously filled trap, and so on *ad infinitum*. For three traps the second hit fills two-thirds of a trap and empties one-third, for a net gain of one-third, the third hit fills one-ninth of a trap, etc. The jth hit fills $(t - 2/t)^{j-1}$ new traps. Proceeding as before (Mauzerall, 1980):

$$Y = \phi_0 \sum_1^\infty P_n \sum_1^n \left(\frac{t-2}{t}\right)^{j-1} = \frac{\phi_0 t}{2}(1 - e^{-2x/t}). \qquad (9)$$

This equation is the generalization of Eq. (6). It predicts that only one-half of the traps will be filled by a saturating flash, but with twice the true optical cross section per trap. Note that this equation is magic for $t = 2$. The result is precisely the simple saturation of Eq. (3). System II may be of this kind.

D. Escape at Traps

1. Escape at Filled Traps

It has been proposed that excitations flow to other traps in system II units (Joliot and Joliot, 1964) and also system I units (Myers and Graham, 1963; Butler, 1968), when first traps are filled. This requires a finite escape from or reflection at a filled trap. We use the term "escape" to bring out the similarity of this process to that of the imperfect sink in diffusion problems. Collins and Kimball (1949) have shown the inadequacy of representing the process as a mere reflection. It is possible to provide an alternative to escape: Filled traps may be uncoupled from the antenna in a time of 35 nsec (Mauzerall, 1972), and a recoupling to unfilled traps would give the same effect as that observed. However, here let us only consider escape. Call the escape coefficient A. The case of $A = 0$ has been treated above. The case of finite A requires a computer calculation. An algorithm for accomplishing this is given in Appendix I. The case of $A = 1$ has a closed solution. A unit hit 1 to t times produces n effects, but only t effects for greater than t hits:

$$Y = \phi_0\left(\sum_1^t nP_n + \sum_{t+1}^\infty tP_n\right) = \phi_0\left[t - \sum_0^t (t-n)P_n\right] \qquad (10)$$

(for details, see Mauzerall, 1980). This equation has $t + 1$ terms, and so the yield is readily calculated for finite t. The analogous equation for the steady state has been proposed to explain the supralinear light saturation curves of photosynthesis often obtained in continuous light (Herron and Mauzerall, 1972). The rate of photosynthesis is $NK\alpha = N\sigma I(1 - X^t)$, where N is the total

[Figure: Y/t vs X/t curves for t = 1, 2, 4, and ∞]

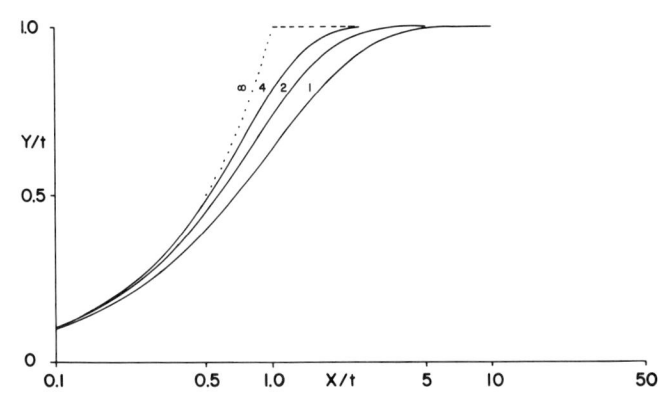

FIG. 1. The normalized yield Y/t is plotted versus the normalized hits x/t for the case of total escape ($A = 1$) at filled traps and for $t = 1, 2, 4$, and ∞ (dotted curve). All the yield curves for various t collapse onto the $t = 1$ curve as $A \to 0$ when normalized. Total reflection: $A = 1$ and $Y = t - \sum_0^t (t - n) P_n$. (From the *Proceedings of the 5th International Photosynthesis Congress.* International Science Services, Jerusalem.)

number of traps, X is the fraction of filled traps in a unit, and $1/\alpha$ is the turnover time. The saturation curve changes from hyperbolic ($t = 1$) to linear ($t = \infty$). Amusingly enough, it was also concluded from these studies that a PSU contained at least three or four traps (Section IV, B). Figure 1 shows the yield curves for various t. Several points can be made. (1) As $x \to 0$, all curves, irrespective of a or t, extrapolate to the same quantum yield ϕ_0. (2) It is often assumed that, for $A = 1$, the yield curve versus x is a straight line, with a break at ϕ_0. Figure 1 shows that this is true only for $t = \infty$. For finite t, the distributions of Mr. Poisson insist on a curvature. (3) For a given t, the yield is a nonlinear function of A, changing a small amount for growing A and reaching the $A = 1$ values very slowly (Fig. 2). For $A = 0.5$, the yield at the $1/e$ point increases only 15%. (4) The yields at finite A are sharper functions of x than Poissonian, and they are *not* a function of average hits x/t. This allows separation of x and t in principle. In fact, the effect is small, and values of $A = 0$–0.5 and $t = 2$–4 fit available data (Fig. 3). Larger values of A are improbable, and $A \sim 1$ can be rejected with present data. Within error none of our data on optical cross sections for oxygen and $\Delta \phi$ fit curves sharper than Poissonian curves. The computer calculation directly demonstrates the unlikelihood of $A \sim 1$. For large A, the program running time is directly proportional to the lifetime of the excitation in a unit. For efficient transfer the lifetime required becomes prohibitively long. We conclude that, although reflection at closed traps allows a separation of X and t in principle, in fact high-precision ($\sim \pm 1\%$ error) data would be required for $A \sim 0.5$,

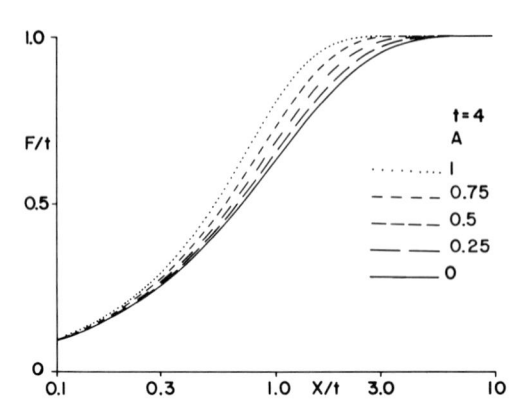

FIG. 2. The normalized yield Y/t is plotted versus the normalized hits x/t as a function of the escape fraction A at $t = 4$. The effect of A is nonlinear, approaching the $A = 1$ curve very slowly.

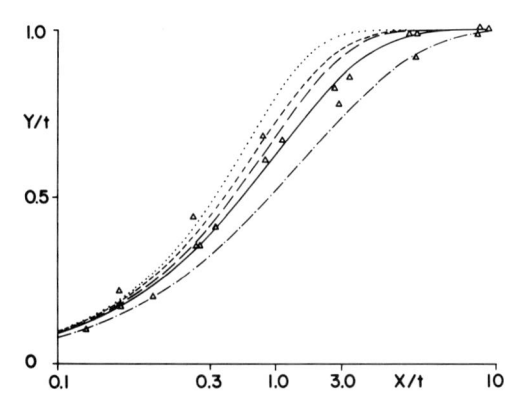

FIG. 3. The normalized yield Y/t values plotted versus the normalized hits x/t. The triangles are the increase in the quantum yield of fluorescence measured with a dim pulse (<0.3 hit) 30 μsec after a 7-nsec pulse of varying intensity. The curves are theoretical with varying escape parameters A or B and number of traps. From left to right the curves are: dots, $A = 1, t = 4$; dashes, $A = 1, t = 2$; long dashes, $A = 0.5, t = 4$; solid line, $A = B = 0$; long dashes and dots, $B = 0.5, t = 4$.

and the requirement becomes more severe as $A \to 0$, where t is indeterminate. Luckily, there is more than one way to skin a cat, and a method to determine t is described in Section IV,B.

2. *Escape at Open Traps*

For small values of escape ($B < 0.5$), the effect of escape at open traps almost mirrors the effect of escape at closed traps (Fig. 4). It is larger, since the yield limit at $B = 1$ is simply 0. The yields are calculated by the method given in Appendix I. The yield curve is now broadened. If the same reflectivity

8. STATISTICS OF MULTIPLE EXCITATIONS 223

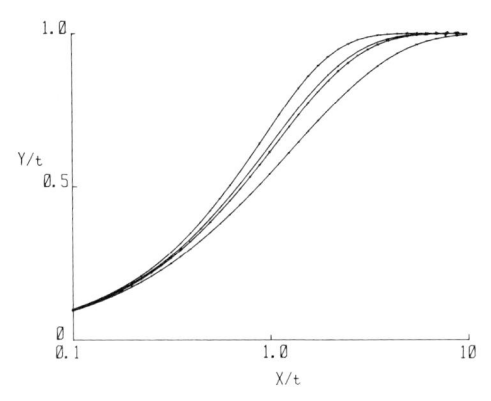

FIG. 4. The normalized yield Y/t is plotted versus the normalized hits x/t as a function of escape from filled traps A and from empty traps B for $t = 4$ and semiannihilation. The curves are: (a) $A = 0.5$, $B = 0$; (b) $A = 0$, $B = 0$; (c) $A = 0.5$, $B = 0.5$; (d) $A = 0$, $B = 0.5$, Note that at equal probability of escape (c) the Poisson curve (b) is almost recovered.

is applied to both traps, the Poisson curve is almost regained (Fig. 4). The real difference is the much increased time required for the excitations to fill a trap or be annihilated. This time, as measured by the cycling time of the calculation, increases at least 10-fold for a reflectivity of 0.5, and at least 100-fold for an A or B value of 0.9. Since it probably takes an excitation a few tens of picoseconds to find a trap, and the observed lifetime is only a few hundred picoseconds, it is clear that a reflectivity >0.5 is incompatible with present data. A large value of B (with A small) also makes the system inefficient, i.e., the observed value of σ is small. Moreover, the highly broadened curve does not fit the oxygen or $\Delta\phi$ yield data. Thus a large escape at empty traps is ruled out. Shipman (1980) has obtained good trapping and lifetimes compatible with experiment using a model involving much energy exchange between antenna and traps in system II. Our arguments show that there is a severe limit on the extent of this exchange based on yield data at energies where multiple excitations occur.

3. *Escape and Total Annihilation*

Escape allows a decrease in annihilation at filled traps but exposes excitations at empty traps to ensuing destruction. This is seen in the yield curves for $t = 4$ of Fig. 5. The no-escape curve is Eq. (9), showing a maximum yield of 0.5 and $\sigma = 2$ times true σ. With $A = 0.5$ (filled traps) the yield rises to two-thirds and σ decreases to ~ 1.5, while with $B = 0.5$ (open traps) the yield decreases to one-third and σ increases to ~ 3. If both escapes are present, one again almost regains Eq. (9). On increasing A, the maximum yield rises as $(2 - A)^{-1}$ and $\sigma \to 1$ times true σ, while on increasing B, the yield is no longer monotonic but the high-energy yield decreases as $(1 - B)/(2 - B)$.

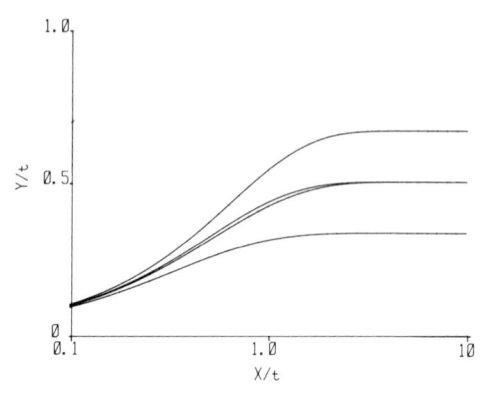

FIG. 5. The normalized yield Y/t is plotted versus the normalized hits x/t as a function of the escape parameters A (filled traps) and B (empty traps) for $t = 4$ and total annihilation. The curves are: (a) $A = 0.5$ $B = 0$; (b) $A = 0$, $B = 0$; (c) $A = 0.5$, $B = 0.5$; (d) $A = 0$, $B = 0.5$. Again the Poisson curve (b) is almost recovered at equal probability (c). Note now that in this case the final saturated yield is a function of A and B.

E. Complications

Before proceeding we must consider two possible complications. First, a distribution of σ's will broaden the yield curve. It is imaginable that nature could be so unkind as to form such a distribution of σ so as to cancel the sharpness of a finite reflectivity A. I will only summarize the results of our calculations on distributions. For narrow, monophasic distributions ($\sigma_{max} \leq 4\sigma_{min}$) the broadening of the yield curve is within present error ($\sim \pm 3\%$). The shape or the distribution (square, triangular, Gaussian) has little effect on the shape of the curve but can affect the average σ because of the exponential weighting. A wide distribution ($\sigma_{max} > 10\sigma_{min}$) broadens the yield curve significantly. Such distributions are unlikely (Mauzerall, 1980), and we conclude that nature has not been unkind.

A biphasic distribution for σ's significantly broadens the yield curve for $\sigma_2 \geq 4\sigma_1$. This brings up the second complication, an experimental artifact. If the illumination is inhomogeneous, either the light itself or, by absorption and scatter in the sample, the higher intensity part of the curve is drawn out. This can unfortunately be well fit by a minor component of smaller σ. At wavelengths of strong absorption this effect can be seen if the sample appears colored to the eye. Our measurements at 590 nm on less than a monolayer of green *Chlorella* are free of this artifact.

F. Quenchers

One of the products of the annihilation process may be a triplet state. This state becomes an "annihilation sink" because its lifetime is in general far longer than that of the singlets. Evidence exists that the triplet state of carotenoids is formed from that of chlorophyll (Mathis, 1977; Witt, 1971). It has often been suggested (Mauzerall, 1972; Butler *et al.*, 1973) that the primary charge transfer process is a quencher of further excitation, and in fact this is the basis of the annihilation at filled traps (Section II,B and C). On the other hand, if these triplets are formed in low yield, or if the lifetime of these quenching states is much less than the duration of the light pulse, the fluorescence yield may very well increase after the first hit (untrapped state). These are systems in which the fluorescence yield is determined only by the number of previous hits to the system. They can be described by Poisson statistics:

$$\phi = \frac{1}{x}[P_1\phi_0 + P_2(\phi_0 + \phi_1) + P_3(\phi_0 + \phi_1 + \phi_2) + \cdots]$$

$$= \frac{1}{x}\phi_0 \sum_1^\infty P_n + \phi_1 \sum_2^\infty P_n + \phi_2 \sum_3^\infty P_n + \cdots$$

$$= \frac{1}{x}(\phi_0 P_{1c} + \phi_1 P_{2c} + \phi_2 P_{3c} + \cdots), \qquad (11)$$

where P_{nc} is the nth cumulative Poisson distribution. The case of $\phi_0 = \phi_0$ and $\phi_1 = \phi_1$, and $\phi_2 = \phi_n$, $n \geq 2$, is of interest because of its similarity to that of kinetic annihilation (Section III,A). The second hit has a different (higher) yield than the first, but all following hits have a yield of ϕ_2, which may be zero. The yield can be immediately written from Eq. (11):

$$\phi = (\phi_0 + \phi_1 + \phi_2)(1 - e^{-x})/x - (\phi_1 + \phi_2)e^{-x} - \tfrac{1}{2}\phi_2 x e^{-x} \qquad (12)$$

In general these saturation curves are sharper as a function of x than the corresponding kinetic- or lifetime-determined curves (Section III,A). They do not appear to fit data obtained with short pulses.

III. Kinetics

A. Annihilation

The previous discussion has been concerned with time-independent probabilities. These probabilities depended only on the state of the traps which underwent instantaneous changes. For slow changes, i.e., changes on a time scale greater than T of the pulse, one simply time-scans with a delayed pulse and reads the kinetics directly. This was done in determining changes in fluorescence quantum yield over the range 10 nsec to 0.1 sec (Mauzerall, 1972). The effects of changes on the time scale of the pulse T are more complicated. The fluorescence yield as a function of pulse energy is an example of this problem. The quenching process has been described by simple second-order kinetic (Swenberg et. al., 1976), by a generalized kinetic equation (Paillotin et al., 1979), and by a chemical kinetics treatment (Sonneveld et al., 1979). I will compare these approaches with mine in Section V. The general problem is complex because of the random occurrence of the excitations during the pulse. But since there are a large number of traps and units in any experiment, the average frequency of excitation is well defined. However, the jump to simple chemical kinetics misses the essential statistical character of the events when $x \sim 1$.

In an attempt to avoid these complexities and simplicities, an approach based on stochastic considerations was adopted (Mauzerall, 1976a,b, 1978). The absorption of a photon by a unit converts it from state a to state b. The quantum yield of fluorescence of a hit when in state a is ϕ_0. The fluorescence yield of hits when in state b is zero, or more generally ϕ_1. The n hits during a pulse T are distributed with a mean frequency n/T, $n > 2$. Similarly the stochastic events comprising lifetimes of the excited state b, or any first-order process, are distributed with a mean frequency k_1. Thus the probability of being in either state is simply the frequency of its formation normalized to the sum of the frequencies:

$$b = \frac{n/T}{n/T + k_1} = \frac{n}{n + k_1 T} \quad \text{and} \quad a = 1 - b = \frac{k_1 T}{n + k_1 T}. \tag{13}$$

The total fluorescence is given by the fraction in each state times its yield weighted by a Poisson factor for the unit hit n times. Also, by simple annihilation, one excitation will always remain, giving a yield ϕ_0:

$$F = \phi_0 \sum_{1}^{\infty} P_n + \sum_{2}^{\infty} (n-1)P_n(\phi_0 a + \phi_1 b)$$

$$= \phi_0 \sum_{1}^{\infty} P_n + \phi_0 \sum_{2}^{\infty} (n-1)P_n + \sum_{2}^{\infty} (\phi_1 - \phi_0)(n-1)P_n b$$

$$= \phi_0 \sum_{1}^{\infty} nP_n + (\phi_1 - \phi_0) \sum_{2}^{\infty} (n-1)P_n b, \tag{14}$$

$$\phi = \frac{F}{x'} = \phi_0 + \frac{\phi_1 - \phi_0}{x'} \sum_{2}^{\infty} (n-1)P_n \frac{n}{(n + k_1 T)}. \tag{15}$$

Where x' is the average number of hits per *unit*, $x' = \sigma' E = t\sigma E$. The distinction between units and traps is crucial. Units determine fluorescence yield, and traps determine photochemical yield. The decrease in fluorescence yield ($\phi_1 < \phi_0$) described by this equation changes smoothly from exponential $\phi = [\phi_1 x' + (\phi_0 - \phi_1)(1 - e^{-x'})]/x'$ to hyperbolic $\phi = (\phi_0 k_1 T + \phi_1 x')/(x' + k_1 T)$ as $k_1 T$ goes from a small to a large number (>10). The derivation does not hold for $T < 1/k_1 = \tau$, since we are concerned with distributions of hits and lifetimes during the pulse. The lifetime of the excited state in photosynthetic systems is ~ 500 psec, thus the analysis of fluorescence excitation by picosecond pulses requires the more general treatment given by Paillotin et al. (1979). The earlier rejection of Eq. (15) because of incompatability with the results of picosecond experiments (Mauzerall, 1976a,b, 1978) was a mistake. The quantum yield of fluorescence during a saturating pulse depends on when the annihilations occur with respect to the lifetime [Eq. (15), unit size], while the yield of the effect of filling traps (oxygen, $\Delta\phi$) depends only on whether or not annihilations takeplace [Eq. (7), trap size]. The use of Eqs. (13) and (15) in identifying the number of traps is shown in Section IV,B.

B. Resonances

It has been observed that the quantum yield of fluorescence of some systems increases and then decreases as a function of increasing pulse energy. This will occur when the risetime for the fluorescence increase ($\Delta\phi$) is similar to or less than the duration of the pulse and is a natural consequence of our description of the excitation process. When the mean hits are about two (per trap), the yield will rise. But as the mean frequency of hits becomes greater than the lifetime, the yield will decrease. The lifetime of this high yield state will in general be longer than that of the original state, and thus annihilation will occur at lower energies than when the second state is not formed, e.g., in a short pulse, $T < \tau$. The yield will be

$$\phi = \frac{1}{x} \{P_1\phi_0 + P_2[\phi_0 + \phi_2 f(1-b) + \phi_0(1-f) + \phi_1 fb]$$

$$+ P_3[\phi_0 + 2\phi_2 f(1-b) + 2\phi_0(1-f) + 2\phi_1 fb] + \cdots\}$$

$$= \frac{1}{x}\left\{\phi_0 \sum_1^\infty P_n + \phi_0 \sum_2^\infty (n-1)P_n + \sum_2^\infty (n-1)P_n[f(\phi_2 - \phi_0) + fb(\phi_1 - \phi_2)]\right\}$$

$$= \frac{1}{x}\left\{\phi_0 \sum_1^\infty nP_n + \sum_2^\infty (n-1)P_n[f(\phi_2 - \phi_0) + fb(\phi_1 - \phi_2)]\right\}$$

$$= \phi_0 + \frac{1}{x}\sum_2^\infty (n-1)P_n[f(\phi_2 - \phi_0) + fb(\phi_1 - \phi_2)], \quad (16)$$

where P_n are the distributions of hits per trap, x represents the average hits per trap, and $b = nt/(nt + k_2 T)$. The number of traps per unit enters, since the lifetime refers to the excitation in a unit. The factor f measures the probability of being in the high-yield state during the flash. It is approximately $1 + (e^{-k'T} - 1)/k'T$. The exact factor is a weak function of n and is given in Appendix II. We now fit the fluorescence yield of Chlorella excited by a ~400-nsec dye laser pulse to Eq. (16) (Ley and Mauzerall, 1981). Since x has been measured by the saturation curve for the increased fluorescence rise, and k' by direct measurement of this risetime ($1/k' = 35$ nsec., Mauzerall, 1972), and k_2 is known, the fit of the data is a priori. This in itself is a strong justification of the present approach. The fluorescence yield data of the Leiden group (Sonneveld et al., 1979), using long flashes, can be similarly explained. I have also fit the data of Monger et al. (1976) on the similar fluorescence yield of photosynthetic bacteria. The value for σ was obtained from their saturation curve of the B870 signal, and the lifetime calculated (~ 1 nsec) is the same as that measured in a trapless mutant. The fit implies that there is *one* trap per unit. It is also inferred that the risetime of the fluorescence yield increase is ≤ 1 nsec, in strong contrast to that observed (35 nsec) in the oxygen-producing system. The changes in σ for various mutants is also deduced. Thus this analysis is of very general applicability and produces useful information (Mauzerall, 1978).

IV. Application

A. Cross Section of the Photosynthetic Unit

The fit of the Poissonian curves described in Section II allows the determination of x/t, the average number of hits per trap. If the photon flux of the incident pulse is known and the system is optically thin, the optical cross section per trap is obtained, $x/t = \sigma E/t$. Arthur Ley has carried out such measurements and has obtained $\sigma = 90$ Å^2 at 596 nm for normal *Chlorella* (Ley and Mauzerall, 1981). This cross section is identical for both oxygen and $\Delta\phi$ measurements, showing that both respond to the same photochemistry. Ley has also shown that σ varies with the light intensity under which the *Chlorella* are grown and has analyzed the changes as a constant number of oxygen units per cell containing a fixed core of chlorophyll a, to which is added a fixed ratio of chlorophyll a and b antennae (Ley, 1981). By determining the absolute cross section of chlorophyll *in vivo*, one can reduce this cross section to the average number of chlorophylls per trap, ~ 300. This in turn can be combined with the Emerson–Arnold unit (~ 3000 chlorophylls per oxygen) to give the quantum requirement for oxygen production, 10. Thus one has completed the cycle of 50 years of experimentation and arguments in photosynthesis.

B. Number of Traps per Unit

We are now in a position to determine the number of traps in a PSU. The cross section per trap is obtained by a fit of oxygen or $\Delta\phi$ yield data to Eq. (7); e.g., see Fig. 3. The decrease in fluorescence yield during the 7-nsec

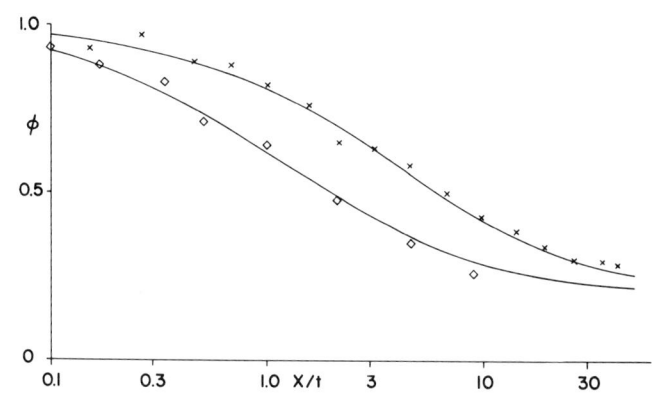

FIG. 6. The normalized yield of fluorescence ϕ at the first 7-nsec flash ($\times$) and 30 μsec later ($\diamond$) is the plotted versus hits per trap. The hits are calibrated by the saturation curve for the fluorescence yield increase, and the data are the average of several experiments. (For experimental details, see Mauzerall 1976a,b, 1978.) (From the *Proceedings of the 5th International Photosynthesis Congress*, International Science Services, Jerusalem.)

pulse of light is then determined and is plotted on the same scale of hits per trap in Fig. 6. Since the lifetime of the fluorescent state in these dark-adapted algae is known to be about 0.45 nsec (Breton and Geacintov, 1981), the kT value or hits per unit is ~16. Since a value of 4 is measured on the hits-per-trap scale, the number of traps per unit is 4. A confirmation of this is obtained by repeating this experiment with the algae preilluminated by a similar saturating flash. The data are plotted as the second curve in Fig. 6. In the "light-adapted" state, the lifetime is known to be about 1.7 nsec (Breton and Geacintov, 1981). Thus, kT is now expected to be ~4. A value of 1 is measured, and again the number of traps per unit is 4. Finally, it is obvious that the ratio of these two curves (aside from small changes in shape as kT changes) gives directly the number of traps per unit, namely, 4. One can in fact use this number to deduce the lifetime of the two states of the system.

Similar data were presented earlier (Mauzerall, 1976a,b, 1978), but the argument was confused by opposing the lifetime and the traps calculations and rejecting the former because of incompatability with picosecond measurements. It is now clear that both calculations are correct in their own domains. The lifetime argument in particular does not apply to sublifetimes (i.e., picosecond) pulses. In this domain it is best to use the general treatment of Paillotin *et al.* (1979) for closed traps, or a kinetic simplification thereof (Ley and Mauzerall, 1981). The case of the open traps and picosecond pulses is still open.

C. Total Annihilation in System II

Arthur Ley and I have simultaneously determined the optical cross sections (per trap) of oxygen formation and of the fluorescence yield rise in *Chlorella* (Ley and Mauzerall, 1981). The simultaneous measurement in one sample allows absolute proof (within experimental error) that these cross sections are identical. Thus the photochemistry in the trap as measured by oxygen yield is precisely the same as that measured by fluorescence under the specified conditions: very weak test flash, delayed by 30 μsec. On extending these measurements to very high intensities, a reversible decrease in yield of both oxygen and fluorescence is observed. Again the decrease is quantitatively identical for both observables. The reversability shows that trivial photochemical destruction is not involved. The traps are really missed on increasing the photon flux. This is just as expected from the previous discussion of total annihilation (Section II,B), where two excitations demolish one another. The simplest possibility predicts that the yield will level at 0.5 maximum, and the data are consistent with this assumption. Photochemical damage at higher fluxes complicates any attempt to prove this point directly. The half-saturation point is at $\sim$ 2000 hits per unit, corresponding to a lifetime of $\sim$ 80 psec. This short lifetime, about one-fifth of the dark-adapted fluorescent state lifetime, suggests that total annihilation occurs between excitations in the antenna, before they reach a trap. Thus annihilation between an excitation and a closed trap in the oxygen system may be semi, i.e., the trap remains closed, while annihilation outside the trap is total. This is consistent with the previous data on the filling of traps in the oxygen system and may also explain at least part of the decrease in fluorescence yield seen with short (picosecond) pulses. The observation of a decreased yield (oxygen or $\Delta\phi$) with picosecond pulses at moderate energies of a few hits per unit would verify this interpretation. However, the alternative that total annihilation occurs only rarely, in $\sim 0.1\%$ of hits, and thus requires a large number of hits to be observed, has been supported by our recent experiments with 5 nsec pulses of light that showed no decrease in fluorescent yield at equivalent excitation rates.

V. Comparison with Other Treatments

Other approaches to the effects of multiple excitation have stressed a kinetic description of the events occurring under these conditions. I also began with this approach but then set it aside for two reasons. First, the yield of fluorescence or oxygen is determined by ratios of rate constants, and thus a kinetic description contains excess parameters. Second, by itself the kinetic approach misses the essential statistics of these events, which are particularly important near $x = 1$. A complete description combines both approaches, as does that of Paillotin et al. (1979) or that given in Section III.

The simplest kinetic approach is to add a second-order annihilation term to the first-order decay of the excitation (Swenberg et al., 1976). This is inadequate at both ranges of excitation. At the low end, $x \sim 1$, it ignores the statistics and thus cannot be correct, especially for the case of open traps. At the high end, second-order chemical kinetics are inadequate when the excitations are an appreciable fraction of the sites. As implied before (Mauzerall, 1976a), the fit of an equation to broad, monotonic, and scattered data in no way justifies the assumptions. It is interesting that my equation [Eq. (15)] contains the factor $n(n - 1)$ for $n < kT$. This factor becomes $n - 1$ as $n > kT$ and saturation occurs. In kinetic terms the statistical theory gives the correct "second-order" term at a low excitation density and smoothly becomes linear as the excitations pile up on one another at a high excitation density.

The approach of Paillotin et al. (1979) uses a more correct stochastic equation for the annihilation and brings in the necessary Poisson statistics. Unfortunately it is only applicable to the closed trap, and thus the $t = 1$ case, at limiting values of the ratio of decay to annihilation, r. As $r \to \infty$, the second-order equation of Swenberg et al. (1976) is obtained, and as $r \to 0$, may equation [Eq. (3)], i.e., $t = 1$, is obtained. The parameter r is claimed to be dimensionless and thus contains much more than the two rate constants. In fact the annihilation rate constant is normalized to some concentration of molecules per unit. Thus, notwithstanding the impressive mathematical edifice, the case considered is essentially trivial. The important parameter t is bypassed, and the ambiguous parameter r is buried in the erudite mathematical haze.

To be more specific, the attempt by Paillotin et al. (1979) to assign a lower limit to t rests on their *assumption* that the unit size (per trap) is 300 chlorophylls. This is putting the cart before the horse. Our statistical method directly determines the absolute size of the unit per trap, and we infer t from experiments with open and filled (by photons alone, no poisons) traps (Section IV,A and B). The size of the unit per trap in *Chlorella* varies from 100 to 400 chlorophyllis depending on the light growth conditions (Ley and Mauzerall, 1981). The question of annihilation kinetics involves the mysterious r parameter. It is defined as twice the first-order rate constant for loss of excitation divided by the second-order rate constant for annihilation. Paillotin et al. (1979) hold that the data are fit by $r \sim 5$. The direct comparison of first- and second-order rate constants has plagued and confused kineticists since their inception. A concentration-normalizing factor is needed to correct the units to render r dimensionless and thus harmless. Obviously it can always be chosen to make $r > 1$. If one chooses this factor as the excitation density of one per unit (their domain), as the authors imply, then many excitations per unit (domain), in fact t, are needed for the annihilation rate to exceed the first-order rate. This implies that the excitations are quite invisible to one another, i.e., the annihilation cross section is very small. All experience is to the contrary. In contrast, our kinetic model essentially states that annihilation is encounter-limited and is simply determined by the excitation lifetime relative to the excitation frequency.

A straightforward chemical kinetic approach to the variation in fluorescence yield during flashes (15–350 nsec) has been taken by the Leiden group (Sonneveld et al., 1979). The model is overparametrized: Eight simultaneous equations and 12 variables seem excessive in the light of Gauss's (possibly apocryphal) statement about 5 variables and the elephant. It is far from obvious that the chosen parameters are unique, and error figures are not quoted. The yield-versus-flash energy data of their Fig. 2 (Sonneveld et al., 1979) is an example of the resonance type described in Section III,B and requires only 3 parameters to fit. All these can be (or have been) separately measured: the optical cross section, the risetime of the fluorescence increase, and the lifetime of the high-yield state.

VI. Conclusions

The statistical approach as formulated here leads to directly measurable parameters which can be correlated with other properties of the photosynthetic system. The stress on hits in the system leads to identification of the optical cross section of a PSU and then to a connection between the classic Emerson–Arnold unit and the quantum requirement for photosynthesis (Section IV,C). A distinction must be made between the cross section of a trap, where photochemistry occurs (oxygen, $\Delta\phi$), and that of a unit, which determines the immediate fluorescence yield. From this distinction the number of traps per unit and the lifetime of the excited states can be obtained by a specific double-pulse experiment (Section IV,B). Information on the annihilation process is also obtained by distinguishing traps and units. In measuring photochemistry, the first hit is the most important, and only the fact of annihilation is needed to obtain simple Poisson curves. Whether the annihilation occurs at the trap or outside will be simply determined by the lifetime in the antenna versus that in the trap. That it will occur is the simple assumption. In measuring immediate fluorescence yield the time of occurrence of the annihilation during the lifetime of the excitation is important. My data, and those of others are most consistent with the following model. Most photons are captured by the antenna, and the transfer to connected funnels or the bed or core of chlorophyll a to which traps are coupled occurs rapidly (~ 50 psec). This rapid transfer is made possible by the packaging of antenna pigment (chlorophyll *a* and *b* protein in green systems, phycobiliproteins in red and blue-green algae) wherein energy transfer is very fast if not delocalized. These packages may be arranged to allow essentially one-dimensional energy transfer by linear and energy level gradient stacking, as in phycobilisomes (Rosinski *et al.*, 1980). The trapping in system II centers may well be reversible, thus equalizing the traps and the funnel or bed or core chlorophyll *a*. Our evidence suggests that total annihilation may occur in the antenna, possibly through photoionization and recombination. Semiannihilation occurs in the core or trap on the time scale of excitation lifetime, i.e., the "deep" trapping time, ~500 psec. My calculation of the escape of excitation from traps (Section II,D) shows that the effect is small on yields of photochemistry and negligible if escape from open and closed traps is equal (rather unlikely). Thus the annihilation process may depend on the time and energy scale of the light pulse. We predict that picosecond pulses would miss a large fraction of the traps. This would be measured as a decreased yield of oxygen or $\Delta\phi$.

Lavorel (1979) has attempted to explain the miss parameter in the phenomenological equations describing the Kok cycle by nonlinear photochemical effects. He assumes that annihilations in the antenna can be distinguished from those in traps by the ability to fill all traps by oversaturating the former, while the latter ("more difficult to imagine") could not be so saturated. The analysis presented in this chapter shows that there is no theoretical justification of his assumption. Moreover, the oxygen photosystem (Section IV,C) may behave in just the opposite way.

Measurements of the fluorescence lifetime as a function of increasing excitation energy have shown the expected decrease (Breton and Gaecintov, 1980). A quantitative fit of these data, including both the open and closed trap units, will go far toward amplifying our concept of PSUs. It has taken 50 years to begin to close this chapter of the book of photosynthesis so dramatically opened by the experiment of Emerson and Arnold.

Appendix I: Calculation of Yield with Escape at Traps

Let A and B be the probability of escape of an excitation at closed and open traps, respectively. These probabilities are to be interpreted following Collins and Kimball's (1949) discussion of encounter-limited reactions to achieve continuity between the strict encounter limit and the reversible encounter leading to usual second-order reactions. Escape is defined as movement of the excitation a sufficient distance from the trap that the chances of its encountering a new trap or the previous trap are equal. The first hit always fills a trap, and therefore the yield is simply P_1, the single-hit Poissonian for a given x. The following algorithm allows calculation of the yields for successive hits.

A. SEMIANNIHILATION

(1) $E(1 - B)(T - C)/T \to H$,
(2) $AEC/T + BE(T - C)/T \to E$,
(3) $H + C \to C$.

The first line calculates the new filled traps H as the excitation E times the probability of filling $(1 - B)$ times the fraction of open traps $(T - C)/T$, where T is the total number of traps per unit and C is the number of closed traps. The second line calculates the residual excitations: the number of excitations E times the probability of escape A times the fraction of closed traps plus those reflected in the open traps, $BE(T - C)/T$. The initial conditions are $C = 1$ (from the first hit) and $E = 1$ (the nth hit). This calculation is cycled until E is small (e.g., 10^{-4}). C is then multiplied by P_n and summed to a yield, P is incremented, and a new cycle begins. When the contributions from P_n and C are negligible, the yield at a given x is available.

B. TOTAL ANNIHILATION

One simply adds the opening of previously closed traps to the above algorithm: $E(1 - A)C/T \to D$ is inserted between lines (1) and (2); and line (3) is replaced by $H + C - D \to C$, which removes these newly opened traps from the closed pool. The cycling time increases as A or $B \to 1$, reflecting the longer time taken by an excitation to stick in a trap or be annihilated.

Appendix II

If a slow change in a parameter, e.g., the fluorescence yield, occurs during the light pulse T, the average of the random excitations into either state is required. An approximation is simply the average over the whole pulse length:

$$f = \int_0^T (1 - e^{-k'T}) dt \bigg/ \int_0^T dt = 1 + \frac{e^{-k'T} - 1}{k'T}, \tag{A1}$$

where k' is the rate constant for the fluorescence yield increase. However, because of the random occurrence of hits during the pulse, the zero time of the rise, which begins at the first hit, is a random function of the number of hits during the pulse. On the average, the first hit will occur at $T/n + 1$.

Thus the factor is

$$f = \int_{T/n+1}^{T} (1 - e^{-k'T}) dt \bigg/ \int_{T/n+1}^{T} dt = 1 + \frac{e^{-k'T} - e^{-k'T/n+1}}{k'Tn/n+1}. \quad (A2)$$

For large $k'T$ (>10), f is 1 and decreases to $1 - (1/k'T)$ as $x \to \infty$. For smaller $k'T$, a larger change occurs, up to 50%, but the magnitude of f decreases, the limit being $kT(n+2)/2(n+1)$ as $kT \to 0$.

REFERENCES

Breton, J., and Geacintov, N. E. (1981). *Biophys. Biochim. Acta* **594**, 1–32.
Butler, W. L. (1968). *Ann. Rev. Plant Phys.* **29**, 345–378.
Butler, W. L., Visser, J. W. M., and Simons, H. L. (1973). *Biochim. Biophys. Acta* **292**, 140–151.
Collins, F. C., and Kimball, G. E. (1949). *J. Colloid Sci.* **4**, 425–437.
Govindjee, R. (1978) *Photochem. Photobiol.* **28**, 935–1038.
Herron, H., and Mauzerall, D. (1972). *Plant Physiol.* **50**, 141–148.
Holten, D., and Windsor, M. W. (1978). *Annu. Rev. Biophys. Bioeng.* **7**, 189–227.
Joliot, A., and Joliot, P. (1964). *C. R. Hebd. Seances Acad. Sci.* **258**, 4617–4625.
Lavorel, J. (1979). *Symp. Honor M. Kamen, San Diego, Calif.*
Ley, A. C. (1981). In preparation.
Ley, A. C., and Mauzerall, D. C. (1982). *Biophys. J.* **33**, 262a, abstract T-PM-P. 102; *Biochim. Biophys. Acta.* In press.
Mathis, P. (1977). *In* "Primary Processes of Photosynthesis" (J. Barber, ed.), pp. 269–302. Elsevier/North-Holland, Amsterdam.
Mauzerall, D. (1972). *Proc. Natl. Acad. Sci. U.S.A.* **69**, 1358–1362.
Mauzerall, D. (1976a). *Biophys. J.* **16**, 87–91.
Mauzerall, D. (1976b). *J. Phys. Chem.* **80**, 2306–2309.
Mauzerall, D. (1978). *Photochem. Photobiol.* **28**, 991–998.
Mauzerall, D. (1979). *Photochem. Photobiol.* **29**, 169–170.
Mauzerall, D. (1980). *Proc. Int. Photosynth. Congr. 5th*, Haldiki, Greece. Jerusalem (in press).
Monger, T. G., Cogdell, R. J., and Parsons, W. W. (1976). *Biochim. Biophys. Acta* **449**, 136–153.
Myers, J., and Graham, J. R. (1963). *Plant. Physiol.* **38**, 105–116.
Paillotin, G., Swenberg, C. E., Breton, J., and Gaecintov, N. E. (1979). *Biophys. J.* **25**, 513–533.
Rosinski, J., Hainfeld, J. F., Rigbi, M., and Siegelman, H. W. (1980). *Ann. Bot.* **47**, 1–12.
Shipman, L. L. (1980). *Photochem. Photobiol.* **31**, 157–167.
Sonneveld, A., Rademaker, H., and Duysens, L. N. M. (1979). *Biochim. Biophys. Acta* **548**, 536–551.
Swenberg, C. E., Gaecintov, N. E., and Pope, M. (1976). *Biophys. J.* **16**, 1447–1452.
Witt, H. T. (1971). *Q. Rev. Biophys.* **4**, 365–477.

PART II
Vision

BIOLOGICAL EVENTS PROBED BY ULTRAFAST LASER SPECTROSCOPY

CHAPTER 9

An Introduction to Visual Pigments and Purple Membranes and Their Primary Processes

Robert Callender

Department of Physics
The City College of The City University of New York
New York, New York

I.	Introduction	239
II.	The Free Chromophore	241
	A. Structure	241
	B. Absorption Spectra	242
	C. Photochemistry	243
III.	Chromophore Binding and Color	243
	A. Rhodopsin	243
	B. Bacteriorhodopsin	245
IV.	Light and Dark Reactions	246
	A. Rhodopsin	246
	B. Bacteriorhodopsin	248
V.	The Primary Photochemical Event	250
	A. Isomerization	250
	B. Kinetics	252
	C. Energy Storage	252
	D. Yields	253
	E. Spectral Red Shift	253
	F. Models	253
	References	254

I. Introduction

Retinal, the aldehyde of vitamin A, is the chromophore of two important and interesting pigment systems: visual pigments (Wald, 1968) and purple membranes (Oesterhelt and Stoeckenius, 1971). Rhodopsin, the name now applied to all visual pigments, and bacteriorhodopsin, the name often given to the protein of purple membranes because of its similarity to rhodopsin, consist of the chromophore covalently bound by a protonated Schiff base linkage to the apoprotein, opsin and bacterioopsin, respectively. The role of both pigment systems is one of energy transduction. In visual pigments, the protein pigment is embedded in the membranes of specialized photoreceptor cells, the rods and cones of vertebrates and the rhabdomeres of invertebrates. The energy of the absorbed photon is eventually converted to a change in electrical potential across the photoreceptor cells that is then transmitted to the nervous system through standard synaptic processes. The primary role of visual pigments is then light absorption followed by activation of an as yet poorly defined mechanism that eventually leads to the change in membrane potential.

Bacteriorhodopsin, the cell membrane protein of *Halobacterium halobium*, uses light energy directly to transport protons across the cell membrane against their electrochemical gradient. Normally, *H. halobium* uses oxygen in respiration to synthesize ATP. However, the clever bacteria manufacture the purple protein bacteriorhodopsin during oxygen-poor conditions. Bacteriorhodopsin offers a separate energy source to the bacteria by converting light energy to a proton gradient, pumping protons from inside to outside the cell. This proton gradient is then available as an energy source to drive other chemical reactions in a chemiosmotic mechanism leading to the production of ATP. This energy transduction mechanism is initiated with the absorption of a photon by the retinal chromophore of bacteriorhodopsin.

In any discussion involving primary processes, it is convenient to discuss rhodopsin and bacteriorhodopsin together since both are involved in the photochemistry of retinal. Also, many other properties of these two systems are similar, and both have been extensively studied by physical chemistry techniques, experimental and theoretical. In fact it is the particular success of a number of physical techniques, including the ability to obtain moderate resolution of the three-dimensional structure of bacteriorhodopsin with the promise of higher resolution in the near future, that makes these biological systems so interesting even apart from their intrinsic interest. It appears quite likely that a detailed molecular understanding will be obtained within a few years. Indeed, a quite sound semiquantitative understanding appears to be emerging.

This chapter serves as a review of the relevant properties of visual pigments and bacteriorhodopsin necessary in understanding their primary photochemistry, and as an introduction to Chapters 10–13. As such, we will not discuss the physiological and biochemical aspects of these systems. There are a number of detailed accounts of these subjects for both rhodopsin (Hagins, 1972; Ebrey and Honig, 1975; Knowles and Dartnall, 1977; O'Brien, 1978; Hubbell and Bownds, 1979; Montal, 1979; Ostroy, 1977) and bacteriorhodopsin (Henderson, 1977; Stoeckenius *et al.*, 1979).

Section II of this chapter discusses the properties of the free chromophore with regard to its structure and electronic behavior, Section III examines the

9. INTRODUCTION TO VISUAL PIGMENTS AND PURPLE MEMBRANES 241

properties of the protein-bound chromophore and its photochemistry, and Section IV discusses the primary photochemistry of the two pigment systems and attempts to set forth the general and specific characteristics of the primary event. More detailed reviews on various aspects of those topics are available and may be consulted (Honig and Ebrey, 1974; Stone and Dratz, 1977; Honig, 1978; Ottolenghi, 1980).

II. The Free Chromophore

A. STRUCTURE

Figure 1 shows various isomers of retinal. Retinal consists of a β-ionylidene ring bonded to a five double-bonded polyene chain. For retinal, the terminal end group is a carbonyl oxygen which is replaced by a nitrogen for the Schiff base. The molecule has many sterioisomers with cis and trans configurations about the double bonds. Hence 13-*cis*-retinal is cis with respect to the 13,14-double bond, etc. Steric hindrance in places precludes precise planarity of the module. The 1-CH_3 groups and 7-H and 5-CH_3 and 8-H force the ring out of the plane of the polyene chain in all isomers, and there is a similar departure from chain planarity due to the 13-CH_3 and 11-H steric hindrance in 11-*cis*-retinal.

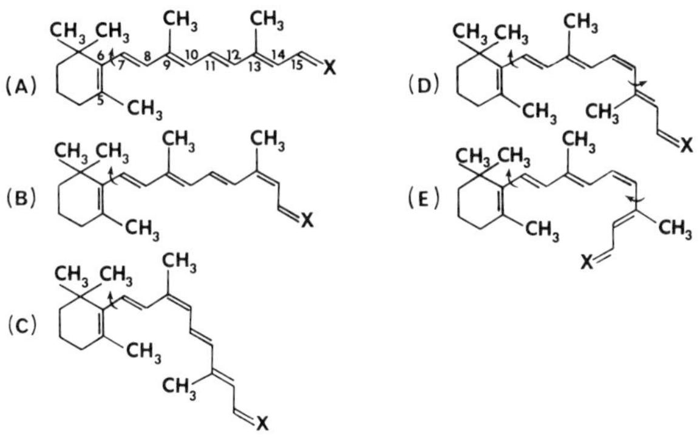

FIG. 1. Configurations of various isomers of retinal (X = O), its Schiff base (X = NR), and its protonated Schiff base (X = NH^+R). (a) All-trans; (b) 13-cis; (c) 9-cis; (d) 11-cis, 12-s-trans; (e) 11-cis, 12-s-cis. Bonds with a nonplanar conformation are denoted by arrows. R is usually an alkyl group.

The torsional potential about the 6,7-single bond is quite broad, and angles of 30–120° are possible (Honig and Karplus, 1971). A twisted s-cis conformation is preferred. Honig and Karplus (1971) predicted a slightly favored 12-s-cis conformation for 11-*cis*-retinal, and this was the conformation found in crystals (Gilardi *et al.*, 1971; Hamanaka *et al.*, 1972). In solution, 11-*cis*-retinal exists as a mixture of 12-s-cis and 12-s-trans, the composition depending on temperature and solvent (Rowan *et al.*, 1974; Birge *et al.*, 1976; Ebrey *et al.*, 1975; Becker *et al.*, 1974).

B. Absorption Spectra

The main absorption band in retinals lies near 380 nm, for Schiff bases near 360 nm, and for protonated Schiff bases near 450 nm. See Fig. 2 for some selected absorption data. The transition is $\pi-\pi^*$ excitation and corresponds, using the symmetry labels of C_{2h} associated with linear polyenes, to B_{1u}. Two smaller blue-shifted bands are observed lying near 280 and 250 nm in retinals. While the assignment of these bands is a matter of some disagreement, recent work appears to indicate that the 280-nm band is a "cis peak" and that the 250-nm band has not been adequately interpreted (Honig, 1978).

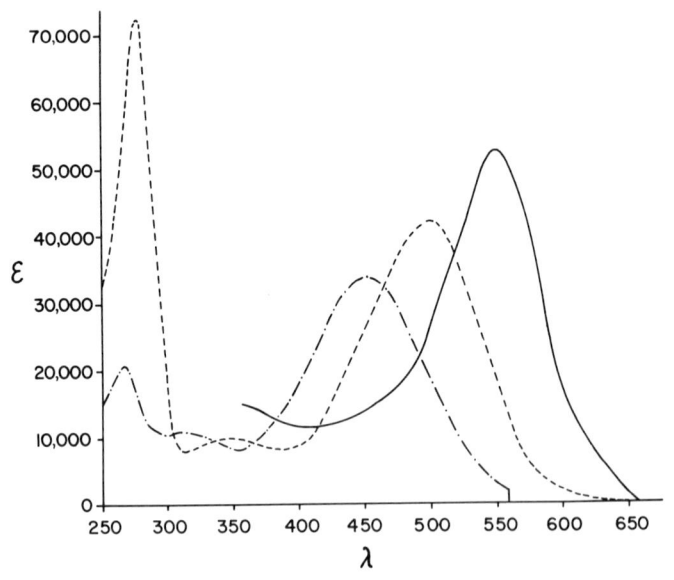

FIG. 2. Typical absorption curves of the protonated retinal Schiff bases (PRSB), the primary pigment rhodopsin, and the primary photoproduct of rhodopsin, bathorhodopsin. ———, bathorhodopsin; ———, rhodopsin; —·—, PRSB. In this case the absorption data of the pigments are those of the cattle system. [Data for rhodopsin taken from Ebrey and Honig (1975) and for bathorhodopsin from Yoshizawa and Horiuchi (1973).] The PRSB is 11-cis-retinal–*n*-butylamine HCl dissolved in methanol (Callender, unpublished data).

9. INTRODUCTION TO VISUAL PIGMENTS AND PURPLE MEMBRANES 243

The red shift in absorption maximum from 360 nm for Schiff bases of retinal to 450 nm for protonated Schiff bases arises (Fig. 2) from π-electron delocalization induced by the presence of the proton. Both ground and excited electronic states are downward shifted in energy, but the effect is larger for the excited state. Upon excitation, there is a net transfer of positive charge from the Schiff base to the ionone ring, which is stabilized by the addition of H^+ at the Schiff base. (See Chapter 12 for more details).

C. PHOTOCHEMISTRY

Excitation of retinals results in photoisomerization. The yields depend on both the end group and the solvent and vary with wavelength (Rosenfeld *et al.*, 1976). The quantum yield for isomerization of retinals is about 0.1 (Rosenfeld *et al.*, 1974; Waddell and Hopkins, 1977) and less than 0.1 and strongly wavelength-dependent for protonated Schiff bases (Rosenfeld *et al.*, 1977). As we will discuss below, these values are much lower than those observed for the protein-bound chromophore; and, further, the quantum yields of protein-bound chromophores are wavelength-independent. While tripled sensitized isomerization yields are considerably larger for protonated Schiff bases (Fisher and Weiss, 1974; Alchalel *et al.*, 1975), this mechanism is not believed to be responsible for the difference between chromophores in solution and *in situ* because of the long isomerization times for the chromophore in solution compared to the very short isomerization times for the protein-bound chromophore (Rosenfeld *et al.*, 1977).

III. Chromophore Binding and Color

A. RHODOPSIN

Visual pigments are formed when an appropriate chromophore is combined with the apoprotein opsin linked via a protonated Schiff base between the chromophore and the ε-amino group of a lysine. Many retinal isomers (Fig. 1) appear to form pigments. The fact that the absorption structure of rhodopsin matched that of opsin combined with 11-*cis*-retinal was the first line of evidence identifying the isomer of the native chromophore (see Wald, 1968). The 9-cis isomer also readily binds and forms an artificial pigment called isorhodopsin. The 13-cis and all-trans isomers do not bind, apparently because they are too long (Matsumoto and Yoshizawa, 1978). Many isomers and synthetically modified retinals form artificial pigments (Crouch *et al.*, 1975; DeGrip *et al.*, 1976; Ebrey *et al.*, 1975; Blatz *et al.*, 1970; Gawinowicz *et al.*, 1977; Honig *et al.*, 1979b), and a specific ionone-binding site has been proposed (Matsumoto and Yoshizawa, 1975).

Resonance Raman spectroscopy has proved to be a very powerful technique for probing the *in situ* chromophore structure of visual pigments and bacteriorhodopsin (for reviews, see Callender and Honig, 1977; Mathies, 1979). These measurements are complicated by the fact that the samples are highly photolabile, so that later photoproducts can result in the measurement beam. However, special techniques have been developed to control and eliminate this problem (Oseroff and Callender, 1974; Mathies *et al.*, 1976; Callender *et al.*, 1976; Terner *et al.*, 1977; Campion *et al.*, 1977; Marcus and Lewis, 1977, 1978). Oseroff and Callender (1974) provided the first definitive proof that the Schiff base of visual pigments was protonated in their low-temperature study. This feature of the binding properties is a key to understanding many of the properties of visual pigments, as we will discuss below. These experiments can also determine chromophore isomeric structure by an analysis of the "fingerprint region" between 1100 and 1400 cm^{-1}. Mathies *et al.* (1977) showed that the resonance Raman spectrum of the 11-cis protonated Schiff base of retinal and the corresponding 9-cis isomer were remarkably similar, respectively, to the spectra of rhodopsin and isorhodopsin. They thus concluded that the configuration of the protein-bound chromophores was very similar to that of model chromophores free in solution. The differences between the model and pigment spectra presumably reflect some degree of protein–chromophore interaction.

A number of experiments point toward a single-bond 12-s-trans conformation of the 11-*cis*-retinal in rhodopsin. First, 12-s-cis conformations should have relatively weak absorption bands, whereas the absorption band of rhodopsin is relatively strong (Burke *et al.*, 1973). Second, the retinal analog 14-methylretinal, which should exist as a 12-s-trans conformer from steric considerations, forms a pigment with essentially the same absorption, circular dichroism, and quantum yield properties as rhodopsin (Ebrey *et al.*, 1975). Third, the resonance Raman spectrum of rhodopsin has been interpreted to show the 12-s-trans conformation (Callender *et al.*, 1976; Cookingham and Lewis, 1978).

The fact that visual pigments absorb in the visible ranging from about 430 to 600 nm, whereas Schiff bases of retinal absorb in the ultraviolet, and the evident color regulation is of great interest (Fig. 2). This variation must be due to the opsin, since the chromophore is always 11-*cis*-retinal. The Raman evidence showing that the Schiff retinal–opsin linkage is protonated explains part of the red shift, since protonated Schiff bases absorb near 440 nm. Several theoretical models have proposed that charged groups (or highly polarized groups) are strategically situated near retinal in the opsin-binding site and control color (Kropf and Hubbard, 1958; Suzuki *et al.*, 1974; Honig *et al.*, 1976; Kliger *et al.*, 1977; Walch and Ingraham, 1973; Irving *et al.*, 1970). It seems almost certain that a negatively charged counterion is associated with the positively charged protonated Schiff base group, and it is fairly clear that at least one more charge is necessary to regulate color properly (Honig et al., 1976). Recently, Honig, Nakanishi, and their co-workers have presented evidence that this second charge is located along the polyene chain near the the 11,12-double bond (Honig et al., 1979b). The location is based on a theoretical analysis of the absorption maxima obtained from a series of saturated retinal analogs in solution and bound to bovine opsin. The 11,12-dihydroretinal showed quite distinctive behavior. Presumably, the various opsins regulate color by placing this second charge at key positions. (This work will be discussed more fully in Chapter 12.)

B. BACTERIORHODOPSIN

Bacteriorhodopsin exists as two distinct forms, the so-called dark-adapted and light-adapted. Only the light-adapted form appears to be capable of pumping protons (Ohno et al., 1977; Loizer et al., 1978). For samples in the dark and at thermal equilibrium, i.e., the dark-adapted bR (bacteriorhodopsin) 558, chromophore extraction experiments indicate an approximately 50:50 mixture of two isomers, 13-cis and all-trans (Oesterhelt et al., 1973; Pettei et al., 1977). There is also absorption evidence that bR558 represents two separate species, bR568 corresponding to the trans isomer and bR548 corresponding to the cis isomer (Aton et al., 1979). Exposure to light drives the 13-cis isomer to the all-trans form, and this pigment is called light-adapted bacteriorhodopsin, bR568. Extraction of the chromophore of bR568 results in an all-trans isomer (Oesterhelt et al., 1973; Pettei et al., 1977). Retinal–apoprotein binding studies have shown that all-trans and 13-cis isomers will form pigments and that the absorption structure of the all-trans bound isomer is identical to that of bR568 (Oesterhelt and Schuhmann, 1974). The fact that 9-cis and 11-cis isomers will not bind (Oesterhelt and Schuhmann, 1974) and that the C22 retinal will bind (Tokunaga et al., 1977) with bacterioopsin, while the reverse is the case with opsin, suggests that the bacteriorhodopsin-binding site is more extended than that of rhodopsin.

Resonance Raman studies have indicated that the Schiff base of both bR568 and bR558 is protonated (Lewis et al., 1974; Aton et al., 1977, 1979). However, the *in situ* chromophore isomer identity has not been established from this technique, since the Raman fingerprint regions of bR558 and bR568 do not uniquely match any isomer spectra of protonated Schiff bases of retinal in solution (Aton et al., 1977; Terner et al., 1977). This may be because the all-trans and 13-cis model chromophore spectra are unhappily so similar and minor protein effects on the chromophore make the configurational analysis difficult. It is also possible that the protein cavity has a more major effect and that the chromophore configuration is one not found in solution. Nevertheless, the Raman spectra of the two bacteriorhodopsin forms are quite unlike those of 9-cis and 11-cis and resemble the all-trans and 13-cis model spectra; the dark-adapted Raman spectrum is closer to the 13-cis model spectrum than is the light-adapted spectrum. In this sense, the Raman studies are in accord with the extraction studies.

A study paralleling the rhodopsin work which examined the absorption maxima of dihydroretinals combined with bacterioopsin has strongly indicated that a second negative charge (the first being the counterion associated with the protonated Schiff base) controlling color in bacteriorhodopsin is located near the β-ionine ring of the bound chromophore (Nakanishi et al., 1981). In this case, the difference in absorption maxima between the bound retinal analog and its solution value was largest for 5,6-dihydroretinal and decreased for 7,8-dihydroretinal, 9,10-dihydroretinal, etc. Thus, the basic mechanism for color regulation appears to be the same as for visual pigments, although the exact placement of charge is different. (See Chapter 12 for a more detailed discussion.)

IV. Light and Dark Reactions

A. Rhodopsin

Absorption of a photon by rhodopsin produces a red-shifted pigment called bathorhodopsin. Bathorhodopsin thermally decays to a series of spectrally distinct pigment forms, terminating in vertebrates with the dissociation of all-*trans*-retinal from opsin (see, e.g., Hubbard et al., 1965). This is called bleaching, since rhodopsin absorbs in the visible while retinals absorb in the ultraviolet near 380 nm. The various stages in the sequence were first identified through low-temperature absorption studies where it is possible to stop the process at a particular intermediate. More recently, flash photolysis studies have verified the bleaching sequence at room temperature as given in Fig. 3a and have characterized their kinetic lifetimes (see, e.g., Goldschmidt et al., 1976). The sequence in invertebrates, e.g., squid rhodopsin, is somewhat different in that retinal does not dissociate from opsin; rather the pigment thermally reverts to the parent rhodopsin pigment.

All pigments studied and bacteriorhodopsin are characterized by the formation of a bathorhodopsin-like red-shifted pigment as a result of light absorption (Fig. 2). For obvious reasons, this is called the primary photochemical event and has been studied using picosecond techniques. All subsequent reactions are thermal, light playing no further biological role. The known characteristics of the rhodopsin–bathorhodopsin transformation, and

9. INTRODUCTION TO VISUAL PIGMENTS AND PURPLE MEMBRANES   247

10^{-12} sec → BATHORHODOPSIN$_{543}$ ← 10^{-12} sec

↓ 10^{-7} sec

LUMIRHODOPSIN$_{497}$

↓ 10^{-5} sec

METARHODOPSIN I$_{478}$

↕ 10^{-3} sec

METARHODOPSIN II$_{380}$

↓ 10^2 sec

ALL-TRANS RETINAL$_{380}$ + OPSIN

RHODOPSIN$_{498}$
(11-cis)

ISORHODOPSIN$_{485}$
(9-cis)

(a)

10^{-12} sec → K$_{630}$

↓ 10^{-6} sec

L$_{550}$

↓ 10^{-4} sec

M$_{412}$

↕ 10^{-3} sec

M*$_{412}$

↓ 10^{-3} sec

N$_{520}$

↓

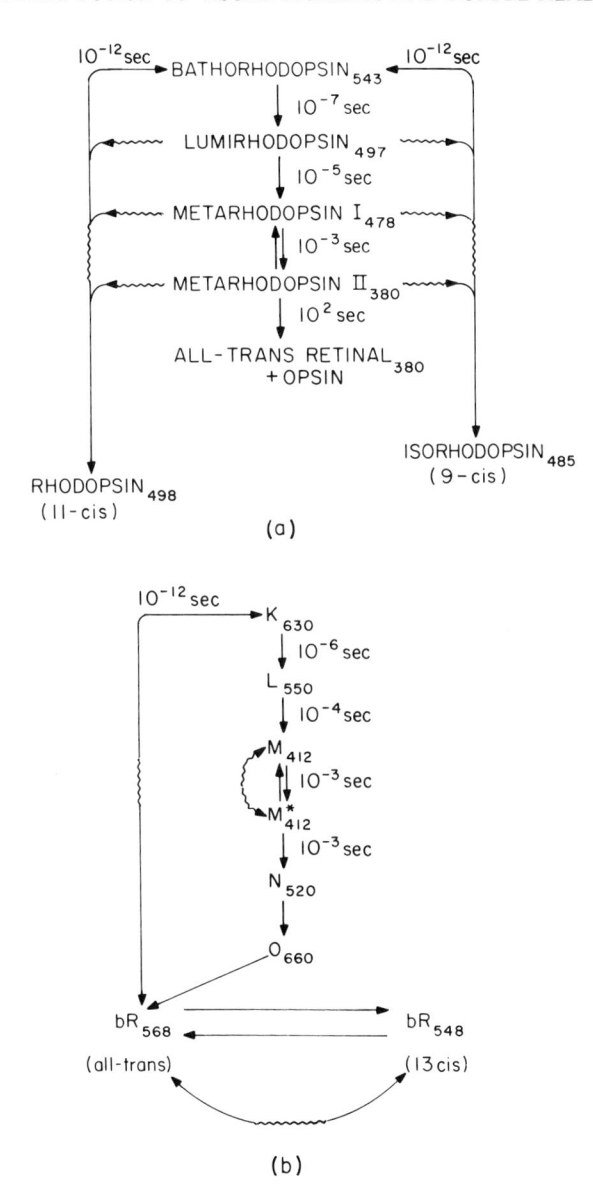

O$_{660}$

bR$_{568}$ ⇌ bR$_{548}$
(all-trans)   (13 cis)

(b)

FIG. 3. (a) The bleaching sequence of rhodopsin. (b) The photochemical cycle of bacteriorhodopsin. Relative free energies are represented by vertical position in the figure. [Reproduced, with permission, from the *Annual Review of Physical Chemistry*, Volume **29**. © 1978 by Annual Reviews Inc.]

its clear role in the light-to-chemical energy conversion step, will be discussed in the next section.

Resonance Raman experiments have shown that both the chromophores of metarhodopsin I and II are close to an all-trans configuration, with the Schiff base linkage protonated in metarhodopsin I and unprotonated in metarhodopsin II (Doukas *et al.*, 1978). There is good evidence that metarhodopsin I in invertebrate rhodopsin is also an all-trans protonated Schiff base, although the analysis was somewhat complicated by the mixed composition of the sample (Sulkes *et al.*, 1977).

There is a net uptake of one proton by the protein and a loss of the chromophore Schiff base proton to the protein in the metarhodopsin I-metarhodopsin II transformation (Matthews *et al.*, 1963). It has been suggested that there are protonated and unprotonated forms of metarhodopsin II (Emrich and Reich, 1974). These data also tend to indicate the presence of one or more donor–acceptor groups at the binding site. Heat of protonation data (Cooper and Converse, 1976), kinetic and thermodynamic evidence (Henselman and Cusanovich, 1976), and the fact that metarhodopsin II can convert to a now protonated intermediate, metarhodopsin III (not shown in Fig. 3a), also suggest the one or more donor–acceptor groups in the binding site.

There appear to be protein changes which accompany bleaching. Circular dichroism measurements of detergent-solubilized rhodopsin indicate changes in the helical content of rhodopsin; these changes, however, appear not to take place in intact membranes (Shichi, 1971). Thus, it is possible that these results for the bleaching sequence may only reflect differences in protein stability. More directly, the exposure of new SH groups accompanies bleaching (McDowell and Williams, 1976), tryptophan fluorescence properties change (Ebrey, 1972), and hydrogen tritium exchange experiments show that hydrogen-bonded peptides are more exposed following light absorption (Downer and Englander, 1975). Further, there are large volume increases (Lamola *et al.*, 1974) in the lipid microenvironment in the metarhodopsin I-metarhodopsin II transition. Whether or not these changes are intimately involved in the transduction process is, at present, conjecture. Important functional properties of proteins often only require minor structural change.

B. BACTERIORHODOPSIN

Like rhodopsin, the absorption of a photon by the primary pigments of bacteriorhodopsin initiates a series of dark reactions (see, e.g., Henderson, 1977; Lozier *et al.*, 1975, 1978; Oesterhelt, 1976) that are very similar to those observed in the case of rhodopsin (Fig. 3b). Here, however, the photochemistry is cyclic and the chromophore does not detach from the protein.

The cycle of light-adapted bacteriorhodopsin is complete in about 10 msec. As in visual pigments, the photochemical cycle of bacteriorhodopsin has been examined in low-temperature absorption as well as flash photolysis studies. However, the bacteriorhodopsin cycle is not as well defined and appears to be somewhat more complicated than in visual pigments.

Proton pumping has been found to accompany the cycle. Chance *et al.* (1975) found that the kinetics of proton release were the same as the formation of M412 at $-40°C$ but somewhat different at higher temperatures. Lozier *et al.* (1976) found that M412 formation was somewhat faster than flash-induced acidification of the medium but somewhat slower than M412 decay. Proton rebinding occurs within milliseconds and appears to correspond to photocycle recovery. The most recent evidence suggests that up to two protons are released per photocycle (Hess and Kuschmitz, 1978; Ort and Parson, 1979) and that up to two protons may actually be pumped (Bogomoli *et al.*, 1979; Govindjee *et al.*, 1980), depending on ionic strength.

Resonance Raman studies have established that bR568 (Lewis *et al.*, 1974; Aton *et al.*, 1977), K630 (Terner *et al.*, 1979a), and L550 (Terner *et al.*, 1979b; Marcus and Lewis, 1978) contain protonated Schiff bases of the retinal chromophore, whereas M412 is unprotonated (Lewis *et al.*, 1974; Aton *et al.*, 1977). These findings agree with the red-shifted absorption maxima of the former pigments and the short-wavelength absorption of M412. A key question is whether or not the Schiff base proton itself is pumped across the protein, since the chromophore is deprotonated and reprotonated with kinetic behavior similar to that of the pumping mechanism. This, however, has not been shown, and more complicated mechanisms would have to be considered should more than one pump proton per cycle be firmly established.

There is now clear evidence of chromophore isomerization during the photochemical cycle. From photochemical arguments, Rosenfeld *et al.* (1977) concluded that photoisomerization was the primary event in bacteriorhodopsin. There have been a number of experiments showing that isomerization is involved in the photochemical cycle. Chromophore extraction work has yielded all-*trans*-retinal for bR568 (Oesterhelt *et al.*, 1973; Pettei et al., 1977), a 50:50 mixture of 13-cis and all-trans for M412 (Pettei *et al.*, 1977) and, more recently, pure 13-cis-retinal for the M412 and L pigments Tsuda *et al.*, 1980a). Using mass-labeled chromophores, Braiman and Mathies (1980) have convincingly argued that the chromophore of M412 has the 13-cis configuration from resonance Raman results. In addition, the Raman fingerprint region of K630 is substantially different from that of bR568 (Pande *et al.*, 1981). These latter results at this early stage do not define the type of isomerization but strongly indicate that a chromophore isomerization occurs in the primary event. The simplest explanation of these data is that a double-bond isomerization, perhaps all-trans to 13-cis, occurs in the bR568–K630 transformation.

V. The Primary Photochemical Event

The primary photochemical event is generally taken to be the formation of bathorhodopsin in visual pigments and the function of K630 in bacteriorhodopsin (Fig. 3). There is, however, a problem in definition of the primary event for visual pigments. At liquid helium temperatures, a pigment called hypsorhodopsin, absorbing near 430 nm, is formed photochemically from rhodopsin; hysorhodopsin can thermally decay to bathorhodopsin at temperatures somewhat below 77 K (Yoshizawa and Horiuchi, 1973). This has prompted the suggestion that hypsorhodopsin is a precursor of bathorhodopsin. However, the experimental data are both in short supply and conflicting. Some picosecond studies have indicated that hypsorhodopsin is a precursor of bathorhodopsin (Shichida et al., 1978; Kobayashi, 1979, 1980), while others observe only the fast (less than 6 psec) formation of bathorhodopsin (Busch et al., 1972; Sundstrom et al., 1977; Peters et al., 1977; Huppert et al., 1977; Monger et al., 1979, Doukas et al., 1980), the studies being performed using the same pigment systems and under similar conditions. Some yield studies have indicated that hypsorhodopsin is formed parallel to bathorhodopsin with lower yields (Sarai et al., 1980; Tsuda et al., 1980b), suggesting hypsorhodopsin formation is a low-probability side reaction. There is a clear need for further study, and the identity of hypsorhodopsin is of major interest. We assume below that bathorhodopsin is the primary photoproduct. Our discussion is largely unaffected by the characteristics of hypsorhodopsin; it will need further elaboration once the real role of hypsorhodopsin has been established.

We attempt here to list key characteristics of the primary photochemical event. A great deal is known, and any fully developed mechanistic model must confront the issues presented below. A detailed quantum mechanical picture is the goal; and, in this author's view, this exciting goal is likely to be achieved in the next few years. Toward this end, lively discussions and various models have been presented in the last few years. We discuss the various models in terms of what is known to date concerning the primary event.

A. Isomerization

That chromophore isomerization plays a fundamental role in bathorhodopsin formation is on very firm grounds. The original argument (Hubbard and Kropf, 1958; Yoshizawa and Wald, 1963) for photochemical isomerization arises from the observation that bathorhodopsin is a common and interconvertible photointermediate between two cis isomers, rhodopsin (11-cis chromophore) and isorhodopsin (9-cis chromophore). This strongly suggests that the configuration of bathorhodopsin is all-trans. Further, the argument is reinforced by the fact that the chromophore of metarhodopsin I is all-trans and that this pigment also photoreverts to rhodopsin and isorhodopsin with a quantum yield similar to that of bathorhodopsin (Doukas *et al.*, 1978).

More recently, resonance Raman measurements have shown that a major geometrical change has taken place in the rhodopsin–bathorhodopsin transformation, since the fingerprint spectral regions of the two pigments are so different (Oseroff and Callender, 1974; Aton *et al.*, 1980; Erying and Mathies, 1979). The fingerprint region of rhodopsin's Raman spectrum is very close to that of 11-cis protonated Schiff bases of retinal in solution (Mathies *et al.*, 1977) and quite different from other isomer spectra. Bathorhodopsin's fingerprint Raman spectrum is quite close to that of all-trans model spectra (Aton *et al.*, 1980). There are, however, strong lower-frequency lines present in the bathorhodopsin spectrum that are absent in the model studies. This has prompted the suggestion that the configuration of bathorhodopsin is close to but not identical to trans (transoid), likely twisted about single bonds (Aton *et al.*, 1978, 1980). In a recent study, using artificial pigments with isotopically modified retinals, it has been suggested that the low-frequency lines arise from out-of-plane hydrogen motions whose Raman cross sections have been enhanced because of a twisted nonplanar trans geometry of the chromophore of bathorhodopsin (Eyring *et al.*, 1980). Also, studies have removed doubts concerning the feasibility of isomerization occurring at low temperatures by showing that isorhodopsin with its 9-cis chromophore is formed photochemically from rhodopsin (11-cis chromophore) even at 5.5 K (Yoshizawa and Horiuchi, 1973; Aton *et al.*, 1978).

It should finally be mentioned that an 11-cis-retinal analog where 11-cis-to-trans isomerization is prevented by bridging C-10 to C-12 with a four-carbon chain exhibits *no* photochemistry when regenerated with opsin (Mao *et al.*, 1981).

The evidence for isomerization occurring in the bR568–K630 transformation is also very convincing. Although there is no pigment analogous to isorhodopsin, the close photochemical similarities between bacteriorhodopsin and rhodopsin strongly suggest an isomerization (Rosenfeld *et al.*, 1977; Hurley *et al.*, 1977). Also, as noted above, chromophore extraction studies and Raman studies show that a trans-to-13-cis isomerization occurs in the formation of M412 from bR568. Recently, Raman studies have shown that the fingerprint regions of bR568 and K630 are quite different and (Pande *et al.*, 1981).

B. Kinetics

The picosecond behavior of the primary event are discussed and reviewed in Chapters 10–13 and will not be discussed here (see also Ottolenghi, 1980, for review). Suffice it to say that both bathorhodopsin and K630 are formed on the picosecond time scale (less than 100 psec in all studies and at all temperatures), that K630 is likely formed from a red-shifted (to K630) precursor, and that there is some evidence that bathorhodopsin is also formed from a precursor.

A major result from these studies is that the rate of bathorhodopsin formation is substantially altered by deuterating the exchangeable protons of rhodopsin and bathorhodopsin, strongly suggesting that proton movements are involved in the primary step (Peters et al., 1977). Based on these data, it was suggested that the Schiff base proton, the only exchangeable proton of the chromophore, tunneled closer to the Schiff base in the primary step (Peters et al., 1977). However, subsequent Raman studies probed this point and concluded that the Schiff base of rhodopsin is fully protonated and the degree of protonation is unchanged in the rhodopsin–bathorhodopsin transformation (Aton et al., 1980; Narva and Callender, 1980). This conclusion was made based on the observation that the Schiff base stretching frequency lies near 1655 cm^{-1} and shifts downward about 25 cm^{-1} when the Schiff base proton is exchanged for a deuteron for both rhodopsin and bathorhodopsin. This could only occur if the proton is covalently linked to the nitrogen of the Schiff base, allowing a significant interaction between C=N stretching and C=N—H bending motions. Experiments show that the bacteriorhodopsin system behaves similarly (Terner et al., 1979a; Pande et al., 1981). It seems reasonable to conclude, then, that any proton movement participating in the primary event would be thus associated with the apoprotein opsin.

C. Energy Storage

It has been recently shown that over one half of the energy of the absorbed photon (~ 55 kcal/mole) is converted to chemical energy in the rhodopsin–bathorhodopsin transition (Honig et al., 1979a; Cooper, 1979). Although a similar value is not available for bacteriorhodopsin, it is likely that here too significant energy storage takes place, since proton pumping requires energy input. That so much photon energy is converted to chemical energy is remarkable and places a severe constraint upon models. This, as well as the other properties of the primary event, namely, the large photochemical quantum yields and the red spectral shifts observed in the primary process, are discussed in detail in Chapters 12 and 13.

D. Yields

The forward quantum yields are 0.67 (Dartnall, 1972; Hurley *et al.*, 1977) for the rhodopsin–bathorhodopsin transition and 0.3 for bR568 to K630 (Hurley *et al.*, 1977). The reverse quantum yields are 0.5 (Suzuki and Callender, 1981) and 0.7 (Hurley *et al.*, 1977), respectively. The wavelength and temperature independence of these values, as well as their large magnitude, are important in understanding the *in situ* energy levels of the chromophore and the dynamical properties of the primary process (see Chapters 12 and 13).

E. Spectral Red Shift

In all rhodopsins studied and in bacteriorhodopsin, the absorption of a photon by the pigment leads to a red-shifted photoproduct, bathorhodopsin and K630. This leads one to conclude that spectral red shift is a general important property of the primary event which must be taken into account by any model.

F. Models

We discuss very briefly the important features of various models for the primary event in visual pigments and purple membranes described in the above discussion. More detailed information can be found in Chapters 10 and 11, and particularly Chapters 12 and 13. See also the recent review of Ottolenghi (1980). While there is no quantitative model for the primary event for visual pigments and purple membranes, three classes of qualitative and semiquantitative models have been proposed over the last few years.

In the first group (Peters *et al.*, 1977; Applebury *et al.*, 1978; Fransen *et al.*, 1976; van der Meer *et al.*, 1976; Harosi *et al.*, 1978; Favrot *et al.*, 1979), the action of light upon the chromophore causes the Schiff base proton to translocate toward the Schiff base in the parent pigment–primary photoproduct transitions. The observed fast kinetics times are quite consistent with these models, as only the movement of a single proton is involved; and the deuteration-dependent kinetic results and spectral red shifts discussed above are natural consequences of this view. There are, however, serious problems with this mechanism. The need for isomerization in the primary event for both visual pigments and purple membranes as described above is ignored or viewed as a subsequent thermal transition. The recent Raman results showing that the Schiff bases of the parent pigment and primary photoproduct are both fully protonated and the lack of any change in protonation before and after light absorption in both pigment systems are in direct conflict. Moreover, it is difficult to understand how significant energy storage can come about by simple proton translocation.

A second class of models (Warshel, 1978; Warshel and Deakyne, 1978; Lewis, 1978) proposes that the polarized excited state of the polyene chromophore induces proton translocation of the apoprotein which persist after relaxation to the ground state, forming a stable batho product. The two approaches differ in the defined geometry of the batho photoproduct. Lewis (1978) suggests an undefined change in polyene conformation which does not involve isomerization. Warshel (1978; Warshel and Deakyne, 1978) does not exclude isomerization. Both views hold that proton translocations occurs for apoprotein protons, avoiding problems with the Raman data on this matter, and that the formation of bathorhodopsin is mainly due to charge stabilization by a protein conformational change without necessarily requiring chromophore isomerization. Lewis' model does not effectively deal with the large body of evidence of an isomerized batho product, however. The main difficulty with the model of Warshel is the very small barrier between rhodopsin and bathorhodopsin (~ 6 kcal/mole) which would lead to relatively large rhodopsin thermal bleaching, which is not observed (Honig et al., 1979a). Also, neither model accounts for the observed substantial energy storage in the rhodopsin–bathorhodopsin transition.

The last class (Wald, 1968; Rosenfeld et al., 1977; Hurley et al., 1977; Honig et al., 1979a) treats isomerization as the fundamental event, as opposed to the above models. Proton transfer is a secondary step, taking place in the protein after and caused by chromophore isomerization to a thermalized ground state photoproduct. Energy storage is accomplished by the separation of charge that occurs when the postively change protonated Schiff base changes position upon isomerization. This can lead to significant photon to pigment energy conversion in the order to that observed. Spectral red shifts are also natural consequences of the separation of the protonated Schiff base and its assumed counterion.

Acknowledgments

This work supported in part by a PSC-BHE Faculty Research award grant, in part by grants from the National Institution of Health (EYO 2515 and EYO 3142), and in part by a grant from the National Science Foundation (PCM 79-02683).

References

Alchalel, A., Honig, B., Ottolenghi, M., Rosenfeld, T. (1975). *J. Am. Chem. Soc.* **97**, 2161–2166.
Applebury, M. L., Peters, K. S., and Rentzepis, P. (1978). *Biophys. J.* **23**, 375–382.
Aton, B., Doukas, A. G., Callender, R. H., Becher, B., and Ebrey, T. G. (1977). *Biochemistry* **16**, 2995–2999.
Aton, B., Callender, R. H., and Honig, B. (1978). *Nature (London)* **273**, 784–786.

Aton, B., Doukas, A. G., Callender, R. H., Becher, B., and Ebrey, T. G. (1979). *Biochim. Biophys. Acta* **576**, 424–428.
Aton, B., Doukas, A. G., Narva, D., Callender, R. H., Dinur, U., and Honig, B. (1980). *Biophys. J.* **29**, 79–94.
Becker, R. S., Berger, S., Dalling, D. K., Grant, D. M., and Pugmire, R. J. (1974). *J. Am. Chem. Soc.* **96**, 708–7014.
Birge, R. R., Sullivan, M. J., and Kohler, B. E. (1976). *J. Am. Chem. Soc.* **98**, 358–367.
Blatz, P. E., Dewhurst, P. B., Balasubramaniyan, V., Balasubramaniyan, P., and Lin, M. (1970). *Photochem. Photobiol.* **11**, 1–15.
Bogomoli, R. A., Lozier, R. H., Sivorinovsky, G., and Stoeckenius, W. (1979). *Biophys. J.* **25**, 318a.
Braiman, M., and Mathies, R. (1980). *Biochemistry* **19**, 5421–5428.
Burke, D. C., Faulkner, T. R., and Moscowitz, A. (1973). *Exp. Eye Res.* **17**, 557–572.
Busch, G. E., Applebury, M. L., Lamola, A., and Rentzepis, P. M. (1972). *Proc. Natl. Acad. Sci. U.S.A.* **69**, 2802–2806.
Callender, R., and Honig, B. (1977). *Annu. Rev. Biophys. Bioeng.* **6**, 33–55.
Callender, R. H., Doukas, A., Crouch, R., and Nakanishi, K. (1976). *Biochemistry* **15**, 1621–1629.
Campion, A., El-Sayed, M. A., and Terner, J. (1977). *Biophys. J.* **20**, 369–375.
Chance, B., Porte, M., Hess, B., and Oesterhelt, D. (1975). *Biophys. J.* **45**, 913–917.
Cookingham, R., and Lewis, A. (1978). *J. Mol. Biol.* **119**, 569–577.
Cooper, A. (1979). *Nature (London)* **282**, 531–533.
Cooper, A., and Converse, C. A. (1976). *Biochemistry* **15**, 2970–2978.
Crouch, R., Purvin, V., Nakanishi, K., and Ebrey, T. (1975). *Proc. Natl. Acad. Sci. U.S.A.* **72**, 1538–1542.
Dartnall, H. J. A. (1972). *Handb. Sens. Physiol.* **7**, Part 1, 122–145.
DeGrip, W. J., Liu, R. S. H., Ramamurthy, V., and Asato, A. (1976). *Nature (London)* **262**, 416–418.
Doukas, A. G., Aton, B., Callender, R. H., and Ebrey, T. G. (1978). *Biochemistry* **17**, 2430–2435.
Doukas, A. G., Stefanic, V., Suzuki, T., Callender, R. H., and Alfano, R. R. (1980). *Photochem. Photobiol.* **1**, 305–308.
Downer, N. W., and Englander, S. W. (1975). *J. Biol. Chem.* **252**, 8092–8104.
Ebrey, T. G. (1972). *Photochem. Photobiol.* **15**, 585–588.
Ebrey, T. G., and Honig, B. (1975). *Q. Rev. Biophys.* **8**, 124–184.
Ebrey, T., Govindjee, R., Honig, B., Pollock, E., Chan, W., Crouch, R., Yudd, A., and Nakanishi, K. (1975). *Biochemistry* **14**, 3933–3441.
Emrich, H. M., and Reich, R. (1974). *Z. Naturforsch., Teil B* **29**, 577–591.
Erying, G., and Mathies, R. (1979). *Proc. Natl. Acad. Sci. U.S.A.* **76**, 33–38.
Erying, G., Curry, B., Mathies, R., Fransen, R., Palings, I., and Lutenberg, J. (1980). *J. Am. Chem. Soc.* **19**, 2410–2418.
Favrot, J., Leclerq. J. M., Roberge, R., Sandorfy, C., and Vocelle, D. (1979). *Photochem. Photobiol.* **29**, 99–108.
Fisher, M., and Weiss, K. (1974). *Photochem. Photobiol.* **20**, 422–432.
Fransen, M. R., Luyten, W. C. M. M., van Thuijl, J., Lutenberg, J., Jansen, P. D. A., van Brengel, P. G. J., and Daeman, F. J. M. (1976). *Nature (London)* **260**, 726–727.
Gawinowicz, M. A., Balogh-Nair, V., Sabol, J. S., and Nakanishi, K. (1977). *J. Am. Chem. Soc.* **99**, 7720–7721.
Gilardi, R. D., Karle, I. L., Karle, J., and Sperling, W. (1971). *Nature (London)* **232**, 187–188.
Goldschmidt, C. R., Ottolenghi, M., and Rosenfeld, T. (1976). *Nature (London)* **263**, 169–171.
Govindjee, R., Ebrey, T., and Crofts, A. R. (1980). *Biophys, J.* **30**, 321–342.

Hagins, W. A. (1972). *Annu. Rev. Biophys. Bioeng.* **1**, 131–158.
Hamanaka, T., Mitsui, T., Ashida, T., and Kakudo, M. (1972). *Acta Crystallogr., Sect. B* **28**, 214–222.
Harosi, F. I., Favrot, J., Leclercq, J. M., Vocelle, D., and Sandorfy, C. (1978). *Rev. Can. Biol.* **37**, 257–271.
Henderson, R. (1977). *Annu. Rev. Biophys. Bioeng.* **6**, 87–109.
Henselman, R. A., and Cusanovich, M. A. (1976). *Biochemistry* **15**, 5321–5325.
Hess, B., and Kuschmitz, D. (1978). *In* "Frontiers in Biological Energetics" (P. L. Dutton, J. S. Leigh, and A. Scarpa, eds.), Vol. 1, pp. 257–264. Academic Press, New York.
Honig, B. (1978). *Annu. Rev. Phys. Chem.* **29**, 31–57.
Honig, B., and Ebrey, T. G. (1974). *Annu. Rev. Biophys. Bioeng.* **3**, 151–177.
Honig, B., and Karplus, M. (1971). *Nature (London)* **229**, 558–560.
Honig, B., Greenberg, A. D., Dinur, U., and Ebrey, T. G. (1976). *Biochemistry* **15**, 4593–4599.
Honig, B., Ebrey, T. G., Callender, R. H., Dinur, U., and Ottolenghi, M. (1979a). *Proc. Natl. Acad. Sci. U. j.A.* **76**, 2503–2507.
Honig, B., Dinur, U., Nakanishi, K., Balogh-Nair, V., Gawinowicz, M. A., Arnaboldi, M., and Motto, M. G. (1979b). *J. Am. Chem. Soc.* **101**, 7084–7086.
Hubbard, R., and Kropf, A. (1958). *Proc. Natl. Acad. Sci. U.S.A.* **44**, 130–139.
Hubbard, R., Bownds, D., and Yoshizawa, T. (1965). *Cold Spring Harbor Symp. Quant. Biol.* **30**, 301–315.
Hubbell, W., and Bownds, M. D. (1979). *Annu. Rev. Neurosci.* **2**, 17–34.
Huppert, D., Rentzepis, P. M., and Kliger, D. (1977). *Photochem. Photobiol.* **25**, 193–197.
Hurley, J. B., Ebrey, T. G., Honig, B., and Ottolenghi, M. (1977). *Nature (London)* **270**, 540–542.
Irving, C. S., Byers, G. W., and Leermakers, P. A. (1970). *Biochemistry* **9**, 858–864.
Kliger, D., Milder, S. J., and Dratz, E. A. (1977). *Photochem. Photobiol.* **25**, 277–286.
Knowles, A., and Dartnall, H. J. A. (1977). *In* "The Eye, Vol. 2B, The Photobiology of Vision" (H. Davson, ed.), 2nd ed., pp. 478–501. Academic Press, New York.
Kobayashi, T. (1979). *FEBS Lett.* **106**, 313–316.
Kobayashi, T. (1980). *Photochem. Photobiol.* **32**, 207–215.
Kropf, A., and Hubbard, R. (1958). *Annu. N.Y. Acad. Sci.* **74**, 266–280.
Lamola, A. A., Yamane, T., and Zipp, A. (1974). *Biochemistry* **13**, 738–745.
Lewis, A. (1978). *Proc. Natl. Acad. Sci. U.S.A.* **75**, 549–554.
Lewis, A., Spoonhower, J., Bogomolni, R. A., Lozier, R. H., and Stoeckenius, W. (1974). *Proc. Natl. Acad. Sci. U.S.A.* **71**, 4462–4466.
Lozier, R. H., Bogomolni, R. A., and Stoeckenius, W. (1975). *Biophys. J.* **15**, 955–963.
Lozier, R. H., Niederberger, W., Bogomolni, R. A., Hwang, S., and Stoeckenius, W. (1976). *Biochim. Biophys. Acta* **440**, 545–556.
Lozier, R. H., Niederberg, W., Ottolenghi, M., Sivorinovsky, G., and Stoeckenius, W. (1978). *In* "Energetics and Structure of Halophilic Micro-organisms" (S. R. Caplan and M. Ginzburg, eds.), pp. 123–141. Elsevier/North-Holland, New York.
McDowell, J. H., and Williams, T. P. (1976). *Vision Res.* **16**, 643–646.
Mao, B., Tsuda, M., Ebrey, T., Akita, H., Balogh-Nair, V., and Nakanishi, K. (1981). *Biophys. J.* **35**, 543–546.
Marcus, M. A., and Lewis, A. (1977). *Science* **195**, 1328–1330.
Marcus, M. A., and Lewis, A. (1978). *Biochemistry* **17**, 4722–4735.
Mathies, R. (1979). *In* "Chemical and Biochemical Applications of Lasers" (C. B. Moore, ed.), Vol. 4, pp. 55–100. Academic Press, New York.
Mathies, R., Oseroff, A. R., and Stryer, L. (1976). *Proc. Natl. Acad. Sci. U.S.A.* **73**, 1–5.
Mathies, R., Freedman, T. B., and Stryer, L. (1977). *J. Mol. Biol.* **109**, 367–372.
Matsumoto, H., and Yoshizawa, T. (1975). *Nature (London)* **258**, 523–526.

9. INTRODUCTION TO VISUAL PIGMENTS AND PURPLE MEMBRANES

Matsumoto, H., and Yoshizawa, T. (1978). *Vision Res.* **18**, 607–609.
Matthews, R. G., Hubbard, R., Brown, P. K., and Wald, G. (1963). *J. Gen. Physiol.* **47**, 215–240.
Monger, T. G., Alfano, R. R., and Callender, R. H. (1979). *Biophys. J.* **27**, 105–116.
Montal, M. (1979). *Biochim. Biophys. Acta* **559**, 231–257.
Nakanishi, K., Balogh-Nair, V., Arnaboldi, M., Tsujimoto, K., and Honig, B. (1981). *J. Am. Chem. Soc.* **102**, 7945–7947.
Narva, D., and Callender, R. H. (1980). *Photochem. Photobiol.* **32**, 273–277.
O'Brien, P. J. (1978). *In* "Receptors and Recognition" (P. Cuatrecasas and M. F. Greaves, eds.), Vol. 6, pp. 109–150. Chapman & Hall, London.
Oesterhelt, D. (1976). *Ciba Found. Symp. Energy Transform. Biol. Syst.* **34**, 147–167.
Oesterhelt, D., and Schuhmann, L. (1974). *FEBS Lett.* **44**, 262–265.
Oesterhelt, D., and Stoeckenius, W. (1971). *Nature (London), New Biol.* **233**, 149–152.
Oesterhelt, D., Meentzen, M., and Schuhmann, L. (1973). *Eur. J. Biochem.* **40**, 453–463.
Ohno, K., Takeuchi, Y., and Yoshida, M. (1977). *Biochim. Biophys. Acta* **462**, 575–580.
Ort, D. R., and Parson, W. W. (1979). *Biophys. J.* **25**, 341–50.
Oseroff, A. R., and Callender, R. (1974). *Biochemistry* **13**, 4243–4248.
Ostroy, S. (1977). *Biochim. Biophys. Acta* **463**, 91–125.
Ottolenghi, M. (1980). *Adv. Photochem.* **12**, 97–200.
Pande, J., Callender, R. H., and Ebrey, T. G. (1981). *Proc. Natl. Acad. Sci. U.S.A.* (in press).
Peters, K., Applebury, M. L., and Rentzepis, P. M. (1977). *Proc. Natl. Acad. Sci. U.S.A.* **74**, 3119–3123.
Pettei, M. J., Yudd, A. P., Nakanshi, K., Henselman, R., and Stoeckenius, W. (1977). *Biochemistry* **16**, 1955–1959.
Rosenfeld, T., Alchalel, A., and Ottolenghi, M. (1974). *J. Phys. Chem.* **78**, 336–340.
Rosenfeld, T., Alchalel, A., and Ottolenghi, M. (1976). *Proc. Lisbon Conf. Excited States Biol. Mol.* pp. 540–554.
Rosenfeld, T., Honig, B., Ottolenghi, M., Hurley, J., and Ebrey, T. G. (1977). *Pure Appl. Chem.* **49**, 341–351.
Rowan, R., Warshel, A., Sykes, B. D., and Karplus, M. (1974). *Biochemistry* **13**, 970–981.
Sarai, A., Kakitani, T., Shichida, Y., Tokunaga, F., and Yoshizawa, T. (1980). *Photochem. Photobiol.* **32**, 199–206.
Shichi, H. (1971). *Photochem. Photobiol.* **13**, 499–502.
Shichida, Y., Kobayashi, T., Ohtani, H., Yoshizawa, T., and Nagakura, S. (1978). *Photochem. Photobiol.* **27**, 335–341.
Stoeckenius, W. R., Lozier, R. H., and Bogomolni, R. A. (1979). *Biochim. Biophys. Acta* **505**, 215–278.
Stone, W. L., and Dratz, E. A. (1977). *Photochem. Photobiol.* **26**, 79–85.
Sulkes, M., Lewis, A., Lemley, A. T., and Cookingham, R. (1977). *Proc. Natl. Acad. Sci. U.S.A.* **73**, 4266–4270.
Sundstrom, V., Rentzepis, P. M., Peters, K., and Applebury, M. L. (1977). *Nature (London)* **276**, 645–646.
Suzuki, H., Komatsu, T., and Kitajima, H. (1974). *J. Phys. Soc. Jpn.* **37**, 177–185.
Suzuki, T., and Callender, R. H. (1981). *Biophys. J.* **34**, 261–265.
Terner, J., Campion, A., and El-Sayed, M. A. (1977). *Proc. Natl. Acad. Sci. U.S.A.* **74**, 5212–5216.
Terner, J., Hsieh, C.-L., Burns, A. R., and El-Sayed, M. (1979a). *Proc. Natl. Acad. Sci. U.S.A.* **76**, 3046–3050.
Terner, J., Hsieh, C.-L., and El-Sayed, M. (1979b). *Biophys. J.* **26**, 527–541.
Tokunaga, F., Govindjee, R., Ebrey, T. G., and Crouch, R. (1977). *Biophys. J.* **19**, 191–198.
Tsuda, M., Glaccum, M., Nelson, B., and Ebrey, T. (1980a). *Nature (London)* **287**, 351–353.

Tsuda, M., Tokunaga, F., Ebrey, T., Yue, K., Marque, J., and Eisenstein, L. (1980b). *Nature (London)* **287**, 461–462.
van der Meer, K., Mulder, J. J. C., and Lutenberg, J. (1976). *Photochem. Photobiol.* **24**, 363–367.
Waddell, W. H., and Hopkins, D. L. (1977). *J. Am. Chem. Soc.* **99**, 6457–6459.
Walch, A., and Ingraham, L. L. (1973). *Arch. Biochem. Biophys.* **156**, 261–266.
Wald, G. (1968). *Science* **162**, 230–239.
Warshel, A. (1978). *Proc. Natl. Acad. Sci. U.S.A.* **75**, 2558–2562.
Warshel, A., and Deakyne, C. (1978). *Chem. Phys. Lett.* **55**, 459–463.
Yoshizawa, T., and Horiuchi, S. (1973). *In* "Biochemistry and Physiology of Visual Pigments" (H. Langer, ed.), pp. 69–81. Springer-Verlag, Berlin and New York.
Yoshizawa, T., and Wald, G. (1963). *Nature (London)* **197**, 1279–1286.

CHAPTER 10

Dynamics of the Primary Events in Vision

K. S. Peters and N. Leontis

Department of Chemistry
Harvard University
Cambridge, Massachusetts

I.	Rhodopsin	259
II.	Isorhodopsin	266
III.	Hypsorhodopsin	267
IV.	Concluding Remarks	267
	References	268

Busch *et al.* (1972) published, in the *Proceedings of the National Academy of Sciences*, the first picosecond laser study on the kinetics of the primary event in vision; their results raised many questions concerning the Hubbard–Kropf isomerization hypothesis and served as the impetus for further picosecond studies on the visual pigments for the following eight years. This chapter is an account of these many studies. The emphasis will be placed on the contribution of picosecond experiments in establishing the sequence as well as the dynamics of the events initiating the visual process in response to the absorption of light, and on the structural information that can be discerned from these studies. Most of the discussion will be set against the background developed in Chapter 9, and the results of the picosecond studies as they relate to current theories of visual initiation will be more fully discussed in Chapter 12.

I. Rhodopsin

It has long been realized that before any molecular mechanism for initiation of the visual process can be discussed it is first necessary to establish the sequence of the intermediates and their subsequent kinetics. The initial studies on the intermediates encountered in the bleaching process can be traced back to the experiments of Boll (1877) and Kühne (1878) in 1877, with their observation of color changes in visual pigments caused by exposure to light. The more systematic approach that has been the dominant methodology employed in sequence elucidation is low-temperature matrix isolation of the intermediates and their characterization by visible absorption spectroscopy. With this technique Wald (1968) and his many co-workers have established that the first observed intermediate upon photolysis of bovine rhodopsin at 77 K is bathorhodopsin, whose absorption spectrum λ_{max} of 543 nm is red-shifted with respect to rhodopsin's spectrum, λ_{max} 498 nm. The configuration of rhodopsin's retinal chromophore, which is attached to the protein opsin through a Schiff base with the ε-amino group of a lysine residue, is cis about the 11,12-bond, whereas when the photolyzed sample at 77 K is warmed to room temperature and the chromophore is isolated, retinal is found in the all-trans configuration. The overall quantum yield for structural isomerization is 0.67 (Dartnall et al., 1936). Since it was known that isomerization about a double bond could be induced by a photon, Hubbard and Kropf proposed that the action of light induced isomerization and consequently that the first observed photoproduct at 77 K, bathorhodopsin, contained the chromophore in the all-trans configuration.

The challenge facing Rentzepis and co-workers (1972) was the demonstration that, at physiological temperatures, bathorhodopsin represents a discrete intermediate in the bleaching sequence, as matrix isolation studies are always subject to criticism of perturbation effects. Prior to their study, the rate of formation of bathorhodopsin was not known and the room temperature rate of decay of less than 1 μsec had to be extrapolated (Busch et al., 1972) from low-temperature experiments between -50 and $-67°C$. Their approach to the study of the photodynamics of rhodopsin at physiological temperatures was through the recently developed technique of picosecond laser absorption spectroscopy. To put this work in perspective, it should be recalled that only a very limited number of picosecond kinetic studies on chemical systems had been undertaken and that this was the first on a protein. The apparatus employed in this initial experiment utilized a neodymium–glass laser, 6-psec time resolution, and 530 nm for photolysis and 561 nm from benzene Stokes scattering for interrogation. The data were recorded photographically. Upon photolysis of rhodopsin which had been purified and solubilized in Ammonyx LO, an absorption increase appearing in less than 6 psec was observed at 561 nm and decayed in 30 nsec. From the time dependence of decay kinetics, they determined that a single transient intermediate was being observed, which was assigned to bathorhodopsin as it is the only intermediate whose absorption spectrum is red-shifted with respect to that of rhodopsin. Thus bathorhodopsin is indeed a discrete intermediate in the bleaching process. The surprising aspect of these experiments is the formation time of bathorhodopsin. The authors stated, "Although our kinetic data do not lead to structural information, it seems to us that the extreme rapidity of the formation of prelumirhodopsin (bathorhodopsin) could scarcely allow a major structural change between rhodopsin and prelumirhodopsin, e.g., complete geometric isomerization of the retinyl group from a truly 11-cis isomer to an all trans isomer with concomitant accommodating change in the opsin structure," The immediate impact of this work was a critical reevaluation of the isomerization hypothesis.

In the initial picosecond study on rhodopsin, only one wavelength was employed in monitoring the formation kinetics of bathorhodopsin, which immediately raised the question as to whether the observed transient was bathorhodopsin. During the next five years several technical innovations (Netzel and Rentzepis, 1974; Alfano and Shapiro, 1970), including the continuum cell for interrogation and the vidicon tube for detection, were developed, so that by 1977 it was possible to measure absorption changes throughout the visible spectrum. To establish bathorhodopsin firmly as the first observed intermediate at room temperature, Sundstrom *et al.* (1977) extended the initial picosecond studies to examination of the kinetics of the changes in the absorption spectrum from 400 to 700 nm. They observed (Fig. 1) within 6-psec of photolysis of detergent-solubilized rhodopsin a bleaching of the ground state rhodopsin, revealed as a negative change in the absorption spectrum from 400 to 510 nm, and concomitant formation of bathorhodopsin with a positive absorption change from 540 to 680 nm, firmly establishing the primacy of bathorhodopsin.

There is one striking difference between the room temperature picosecond absorption spectrum and the spectrum observed (Applebury *et al.*, 1974)

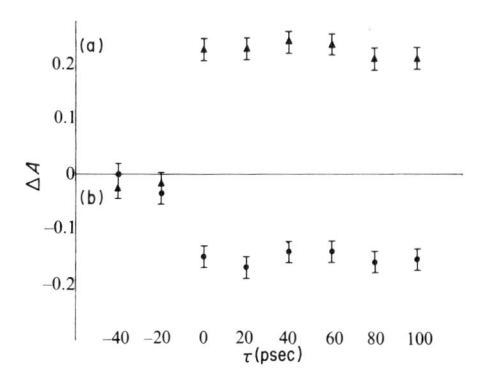

FIG. 1. (a) Formation of the prelumirhodopsin band with a maximum at 580 nm induced by a 530-nm 6-psec pulse. ▲, 580 nm; ●, 480 nm. (b) Depletion rate of the rhodopsin band at 480 nm after excitation with the same picosecond pulse. Similar kinetics are observed over the entire spectrum. Both figures represent an average of five kinetic records. The variance is given by the error bars.

in low-temperature glasses (Fig. 2). The ratio of the maximum (580 nm) to minimum (490 nm) change in absorption ΔA in the picosecond difference spectrum at room temperature is 1:1. At low temperatures (77 K) under photostationary state conditions the ratio of 580 to 490 nm is 3:1. This dramatic temperature effect is not an artifact of the picosecond experiment as later illustrated by Peters et al., (1977), for when the picosecond absorption experiment is carried out at 77 K, the ratio increases to 3:1. There are at least two possible explanations for this temperature dependence; either the absorption profile of rhodopsin and/or bathorhodopsin is temperature-dependent, or the quantum yield for the production of bathorhodopsin increases with decreasing temperature. The absorption spectra of retinals are known to be temperature-dependent (Sperling and Rafferty, 1969). When solutions of all-*trans*-retinal are cooled from room temperature to 77 K, the first optical transition at 380 nm red-shifts and increases its

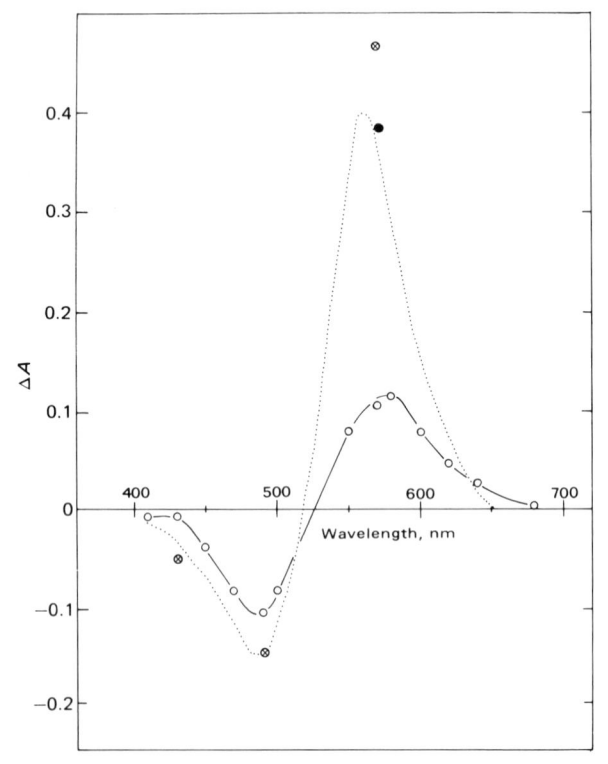

FIG. 2. Bathorhodopsin difference spectrum, recorded at 298 K (○), at 77 K (●), and at 4 K (⊗), 60 psec after excitation; ....., difference spectrum generated by photostationary state studies in low-temperature glasses at 77 K. (Data from Peters et al., 1977.)

intensity of λ_{max} by 10%. A similar effect is observed with 11-*cis*-retinal, except that there is a more pronounced temperature effect as the λ_{max} intensity increases by a factor of 2.0. In light of this observation, the results observed for rhodopsin–bathorhodopsin are in direct contrast with expectations. As rhodopsin contains the chromophore in the 11-cis configuration and bathorhodopsin in the supposed all-trans configuration, then as the temperature is lowered from room temperature to 77 K, the ΔA ratio of 580:490 should change from 1:1 to 1:2, the opposite of what is observed. A second possible explanation is a temperature-dependent quantum yield for bathorhodopsin formation, which can be accommodated by invoking a second intermediate that can be populated in parallel with bathorhodopsin at higher temperatures. To be consistent with the experimental data, however, this new intermediate cannot have a visible absorption spectrum. In this scheme, at 77 K the excited state of rhodopsin decays, forming bathorhodopsin with a bathorhodopsin/rhodopsin λ_{max} ratio of 3:1. When the photolysis is carried out at higher temperatures, a second channel is open to the decay of rhodopsin's excited state, thereby reducing the yield of bathorhodopsin, manifested by a reduction in the ratio of the λ_{max} values. Clearly further studies are in order for elucidation of this anomalous temperature effect.

Following the report of Sundstrom *et al.* (1977), Monger *et al.* (1979) published their picosecond studies, employing a neodymium–glass laser, on Ammonyx-solubilized rhodopsin. Though the formation kinetics for bathorhodopsin were consistent with previous reports, the observed spectrum was somewhat different in that two isobestic points were observed at 460 and 510 nm, whereas Sundstrom *et al.* (1977) observed only one isobestic point at 525 nm. The second isobestic point indicates that a species absorbing to the blue of rhodopsin is formed parallel with bathorhodopsin and persists beyond 100 psec. There have been previous reports of blue absorptions paralleling bathorhodopsin in the nanosecond studies of Benasson *et al.* (1977), the nature of which is yet to be explained, though it appears to be power-dependent. Because of the kinetic characteristics of the blue intermediate, Monger *et al.* (1979), do not attribute it to hypsorhodopsin.

The question that arose subsequently concerned just how fast bathrhodopsin formed at room temperature, as this is an important parameter in any molecular dynamic model. From a technological standpoint in 1977, it was not possible to resolve the risetime of bathorhodopsin. The alternate approach, initiated by Peters *et al.* (1977), was to attempt to slow down the process, providing it was thermally activated, by examining the picosecond kinetics at lower temperatures from which room temperature kinetics could be extrapolated. When the kinetics of bathorhodopsin, monitored at 570 nm, were measured at 77 K, the risetime of the intermediate was found to be within the time resolution of the laser pulse, that is, within 6 psec. Not until 20 K could the actual rate of formation of bathorhodopsin be resolved, determined to be 9 psec. The rate of formation decreased as the temperature was further lowered, so that at 10 K the rate was 20 psec, and at 4 K, 36 psec (Fig. 3). The kinetics of two other wavelengths, 440 and 490 nm, which characterize the bleaching of the ground state of rhodopsin, were monitored as a function of temperature and were found to be temperature-independent.

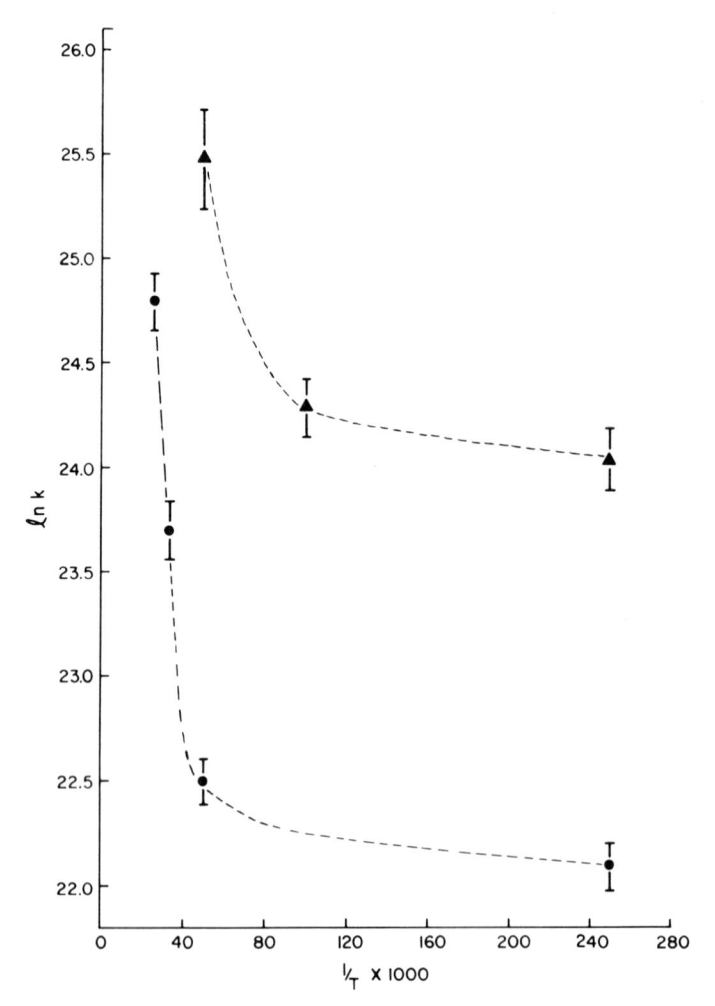

FIG. 3. An Arrhenius plot of ln k for the formation of prelumirhodopsin versus $1/T$ (Kelvin) × 10^3. ▲, rhodopsin; ●, D-rhodopsin.

From an Arrhenius analysis of the 570-nm kinetic data, they noted that the process giving rise to bathorhodopsin was virtually temperature-independent from 10 to 4 K, thus exhibiting nonclassic kinetic behavior. Since a temperature-independent process is characteristic of a tunneling phenomenon, and as protons had previously been proposed to be involved in the primary event in vision, a deuterium isotope effect upon the formation kinetics of bathorhodopsin was examined. By suspending rod outer segments in deuterium oxide all exchangeable protons were replaced by deuterons. A dramatic deuterium isotope effect was subsequently observed upon the formation kinetics, with the half-life increasing to 250 psec at 4 K, a sevenfold decrease in the rate constant (Fig. 3).

As revealed by resonance Raman spectroscopy (Oseroff and Callender (1974), the only exchangeable proton associated with the retinal chromophore under the conditions employed is that of the Schiff base, so that Peters *et al.* (1977) has proposed that the primary event following light absorption is a transfer of proton toward the Schiff base nitrogen. However, this proposal is not consistent with recent findings of resonance Raman studies (Eyring and Mathies, 1979), which show that the degree of Schiff base protonation is the same for both rhodopsin and bathorhodopsin.

A further interesting aspect of these low-temperature studies is the absence of changes in the kinetics of the transients monitored at 440 and 490 nm. Taken in conjunction with quantum yield studies, it had previously been assumed (Rosenfeld *et al.*, 1977) that the excited state decay of rhodopsin had a branch point with one path leading directly to bathorhodopsin and the other back to ground state rhodopsin, with respective quantum yields of 0.67 and 0.33. If this model holds, it is then anticipated that, as bathorhodopsin is being formed, there must be a concomitant repopulation of the ground state which would be revealed at 4 K by an initial bleach at 490 nm followed by an increase in absorption with a half-life of 36 psec. However, this is not observed, and therefore the kinetic scheme must be modified. Two are consistent with the kinetic data. If bathorhodopsin is to be formed directly from the excited state of rhodopsin, this must then occur with a quantum yield of 1 and at some point beyond bathorhodopsin in the bleaching sequence there is a decay back to rhodopsin with a quantum yield of 0.33, so that the the overall quantum yield for bleaching is still 0.67. A more attractive hypothesis is that there is another intermediate between rhodopsin's excited state and bathorhodopsin, so that the excited state partitions its decay between ground state rhodopsin and the new intermediate, all in less than 6 psec at 4 K, and subsequently the new intermediate decays into bathorhodopsin in 36 psec at 4 K, with total efficiency. Honig *et al.* (1979) employ this latter scheme in their recent model of the initial events in vision where bathorhodopsin's precursor has the all-trans configuration.

II. Isorhodopsin

The strongest support for the isomerization hypothesis was the observation of Yoshizawa and Wald (1967) that rhodopsin at 77 K could be photoconverted to isorhodopsin, whose retinal chromophore is in the 9-cis configuration. As previously discussed, when rhodopsin is photolyzed at 77 K with blue light, bathorhodopsin is formed. In turn, when bathorhodopsin is irradiated with green light, not only is rhodopsin regenerated but also a new species is formed, isorhodopsin:

$$\text{Rhodopsin} \rightleftharpoons \text{bathorhodopsin} \rightleftharpoons \text{isorhodopsin}$$
$$(11\text{-}cis) \qquad (\text{all-}trans) \qquad (9\text{-}cis)$$

Similarly, isorhodopsin photoreverts to rhodopsin, proceeding through bathorhodopsin. Thus the most probable configuration of a chromophore that can be photoconverted to either 11-cis or 9-cis is all-trans. If bathorhodopsin was the common intermediate in the interconversion of rhodopsin and isorhodopsin, it remained to be shown that the initial photoproducts of rhodopsin and isorhodopsin were one in the same, bathorhodopsin.

Green *et al.* (1977) were the first to undertake a picosecond study of the photochemistry of isorhodopsin, prepared by the isomerization of rhodopsin's chromophore at 77 K through irradiation at 568 nm and warming to room temperature, producing samples containing better than 95% isorhodopsin. The experimental approach was similar to the initial work of Busch and co-workers (1972), where the sample was photolyzed at 530 nm, produced from a neodymium–glass laser, and interrogated with a single frequency at 561 nm. Within 3 psec, a transient was observed which was attributed to bathorhodopsin. As found in the initial study on the primary photoproduct of isorhodopsin, a complete spectrum was necessary. Thus Monger *et al.* (1979) extended these initial studies to include wavelengths of interrogation from 460 to 700 nm. They observed within 3 psec of excitation the bleaching of the ground state of isorhodopsin with the simultaneous formation of bathorhodopsin. That the same bathorhodopsin is formed from rhodopsin and isorhodopsin was argued based upon isobestic point analysis. The observed isobestic points in the picosecond absorption difference spectrum of rhodopsin and isorhodopsin are 510 and 500 nm, respectively. Since the absorption spectrum of isorhodopsin is blue-shifted approximately 10 nm with respect to that of rhodopsin, and if the bathorhodopsin from both precursors is the same, then a 10-nm shift in the isobestic point would be anticipated, which is in accord with their observation. These studies strongly support the isomerization hypothesis.

III. Hypsorhodopsin

Since the discovery of hypsorhodopsin by Yoshizawa (1972) and coworkers a dilemma has existed as to its role in the visual process. When digitonin-solubilized rhodopsins from cattle, frog, chicken, and squid are irradiated at 4 K, along with bathorhodopsin, a new intermediate with λ_{max} 440 nm is formed which has been termed hypsorhodopsin. When the sample temperature is subsequently warmed to 35 K, hypsorhodopsin decays into bathorhodopsin. The question raised by these low-temperature studies is whether hypsorhodopsin is the direct precursor of bathorhodopsin or whether it is formed parallel with bathorhodopsin. Schichida *et al.* (1977, 1978) investigated the role of hypsorhodopsin produced from squid rhodopsin solubilized in digitonin in the visual sequence at physiological temperatures with picosecond absorption spectroscopy. The laser system employed in these studies was a mode-locked ruby with a 20-psec time resolution and an excitation wavelength of 347 nm. When monitoring at 430 nm, a transient absorption attributed to hypsorhodopsin was observed, formed within the time profile of the excitation pulse and decaying in 50 psec. With this decay was an increase in absorption at 550 nm, assigned to bathorhodopsin. Consequently, they concluded that hypsorhodopsin was the immediate precursor of bathorhodopsin. As these observations were contrary to those of Sundstrom *et al.* (1977) and Monger *et al.* (1979) for cattle rhodopsin solubilized in Ammonyx, they noted that the difference might be attributed to the different wavelengths for excitation, the detergents employed, or the intrinsic difference in the opsin structure.

As the kinetics of formation of hypsorhodopsin could not be time-resolved at room temperature, Schichida *et al.* (1978) sought to slow down this process by examining the picosecond kinetics at liquid nitrogen temperatures. By monitoring at 421 nm hypsorhodopsin was found to be formed in 70 psec at 77 K. Paralleling this formation was a decay at 541 nm, which was attributed to the hypsorhodopsin precursor, presumably an excited state whose absorption at 541 nm is identical to that of rhodopsin. During the 190 psec of interrogation hypsorhobopsin was observed not to decay and, furthermore, there was no indication of bathorhodopsin formation during this time interval.

IV. Concluding Remarks

Rhodopsin's photochemistry has fascinated its researchers for over one hundred years, and though many facets of its photochemistry have been clarified we are still far from a comprehensive understanding. The sequence of early photochemical events is yet to be established, and it is clear that many more experiments will be necessary before the roles of hypsorhodopsin and bathorhodopsin are elucidated. Perhaps the most promising experimental approach to the sequence problem is time-resolved picosecond resonance Raman spectroscopy. A difficulty with absorption spectroscopy is that it does not provide detailed structural information. Also, on the short time scales changes in absorption spectra do not necessarily reflect chromophore alterations, as heat relaxation and protein rearrangements can be manifested as spectral shifts (Green et al., 1979). Picosecond resonance Raman spectroscopy can be used to surmount these difficulties.

One of the most intriguing aspects to emerge from the picosecond studies on rhodopsin is the difference in the dynamics of the photochemistry of the retinal Schiff base in solution and in protein. In solution cis–trans isomerization of n-butylamine retinal Schiff base occurs in 11 nsec (Huppert et al., 1977), while in the protein isomerization apparently occurs in less than 6 psec. The quantum yield for isomerization of the free chromophore at 77 K is zero (Waddell et al., 1973), while in the protein it occurs with a high quantum yield even at 4 K. Thus the challenge that awaits us is elucidation of the nature of the interaction of the protein with the chromophore that so drastically modifies the dynamics of its photochemistry.

References

Alfano, R. A., and Shapiro, S. L. (1970). *Phys. Rev. Lett.* **24**, 584–486.
Applebury, M. L., Zuckerman, D., Lamola, A. A., and Jovin, T. (1974). *Biochemistry* **13**, 3448–3458.
Bensasson, R., Land, E. J., and Truscott, T. G. (1977). *Photochem. Photobiol.* **26**, 601–605.
Boll, F. (1877). *Arch. Anat. Physiol. Abt.* **1**, 4.
Busch, G. E., Applebury, M. L., Lamola, A. A., and Rentzepis, P. M. (1972). *Proc. Natl. Acad. U.S.A.* **69**, 2802–2806.
Dartnall, H. J. A., Goodeve, C., and Lythgoe, R. (1936). *Proc. R. Soc. London, Ser. A* **156**, 158–170.
Eyring, G., and Mathies, B. (1979). *Proc. Natl. Acad. Sci. U.S.A.* **76**, 33–37.
Green, B. H., Monger, T. G., Aton, B., and Callender, R. H. (1977). *Nature (London)* **269**, 179–180.
Green, B. I., Hochstrasser, R. M., and Weisman, R. B. (1979). *J. Chem. Phys.* **70**, 1247–1259.
Honig, B., Ebrey, T., Callender, R., Dinur, U., and Ottolenghi, M. (1979). *Proc. Natl. Acad. Sci. U.S.A.* **76**, 2503–2507.
Huppert, D., Rentzepis, P. M., and Kliger, D. S. (1977). *Photochem. Photobiol.* **25**, 193–197.
Kühne, W. (1878). *In* "On the Photochemistry of the Retina and on Visual Purple" (M. Foster, ed.), pp. 248–290. Macmillan, London.
Monger, T. G., Alfano, R. R., and Callender, R. H. (1979). *Biophys. J.* **27**, 105–115.
Netzel, T. L., and Rentzepis, P. M. (1974). *Chem. Phys. Lett.* **29**, 337–342.
Oseroff, A., and Callender, R. (1974). *Biochemistry* **13**, 4243–4248.
Peters, K. S., Applebury, M. L., and Rentzepis, P. M. (1977). *Proc. Natl. Acad. Sci. U.S.A.* **74**, 3119–3132.

Rosenfeld, T., Honig, B., Ottolenghi, M., Hurley, J., and Ebrey, T. G. (1977). *Pure Appl. Chem.* **49**, 341–351.
Schichida, Y., Yoshizawa, T., Kobayashi, T., Ohtani, H., and Nagakura, S. (1977). *FEBS Lett.* **80**, 214–216.
Schichida, Y., Kobayashi, T., Ohtani, H., Yoshizawa, T., and Nagakura, S. (1978). *Photochem. Photobiol.* **27**, 335–341.
Sperling, W., and Rafferty, C. N. (1969). *Nature (London)* **224**, 591–594.
Sundstrom, V., Rentzepis, P. M., Peters, K. S., and Applebury, M. L. (1977). *Nature (London)* **267**, 645–646.
Waddell, W. H., Schaffer, A. M., and Becker, R. S. (1973). *J. Chem. Soc.* **95**, 8223–8224.
Wald, G. (1968). *Nature (London)* **219**, 800–807.
Yoshizawa, T. (1972). *Handb. Sens. Physiol.* **7**, Part 1, 146–179. Springer-Verlag,
Yoshizawa, T., and Wald, G. (1967). *Nature (London)* **214**, 566–572.

CHAPTER 11

Primary Events in Bacteriorhodopsin

Thomas G. Ebrey

Department of Physiology and Biophysics
University of Illinois at Urbana-Champaign
Urbana, Illinois

I.	INTRODUCTION	271
II.	ABSORPTION MEASUREMENTS	273
	A. *Early Measurements*	273
	B. *Low Temperature*	273
	C. *Ultrafast Measurements*	274
	D. *Hypsochromically Shifted Early Intermediates*	276
	E. *Kinetics of the Back Reaction from K to Bacteriorhodopsin*	276
	F. *Primary Photoproduct of the 13-Cis Form of Bacteriorhodopsin*	277
III.	FLUORESCENCE MEASUREMENTS	277
	A. *Early Measurements*	277
	B. *Fluorescence from Intermediates*	278
	C. *Ultrafast Measurements*	278
	D. *Relationship of the Fluorescence to the Photochemistry*	278
	REFERENCES	280

I. Introduction

Bacteriorhodopsin (bR), the only protein found in the purple membrane of *Halobacterium halobium*, contains retinal as its chromophore. This is the most obvious of a number of striking similarities to rhodopsin (see reviews in Stoeckenius *et al.*, 1979; Honig, 1979). For example, the chromophore appears to be bound to the protein in both pigments in an identical manner, via a protonated Schiff base. In both cases the action of light creates a high free energy intermediate which then drives the subsequent physiological events. Moreover, as in the case of rhodopsin, the action of light on bR is extremely rapid, so that the photoproduct is formed within picoseconds. Based on both flash photolysis and low-temperature spectroscopy, the general outline of the photochemistry of bR can be sketched. It has two stable or metastable states—the dark-adapted form (bR^{DA}) and the light-adapted (bR^{LA}) form. Chromophore extraction shows that the light-adapted form is a homogeneous pigment, all of whose chromophores have the all-trans conformation, while the dark-adapted pigment is a heterogeneous mixture of pigments containing both all-trans and 13-cis chromophores. The simplest and most generally accepted model is that the all-trans pigment found in bR^{DA} is identical with the all-trans pigment found in bR^{LA}; in the dark-adapted pigment the 13-cis and all-trans forms are in equilibrium. Most work has concentrated on bR^{LA} because this form can use light to pump protons across the cell membrane. However, some low-temperature and flash photolysis work has been done with the dark-adapted pigment.

The intermediates in the bR cycle are shown in Fig. 1. The first five intermediates have been given the names K′, K, L, M, and O. K′ can only be observed as a transient, so K is the first intermediate that can be isolated as a stable species at low temperatures. There are a number of similarities between the K intermediate and the first photoproduct of rhodopsin, called bathorhodopsin, or prelumirhodopsin; some workers call the K intermediate the bathointermediate or the batho product of bR. These similarities include both photoproducts being bathochromically shifted from their parent pigments and both being stable at liquid nitrogen temperatures. The L intermediate has some similarities to both lumirhodopsin and metarhodopsin I (all are slightly blue-shifted from the parent pigment), but these may be just superficial. The M intermediate is similar to metarhodopsin II in that both are unprotonated Schiff bases, hence absorb in the near-ultraviolet or blue region of the spectrum.

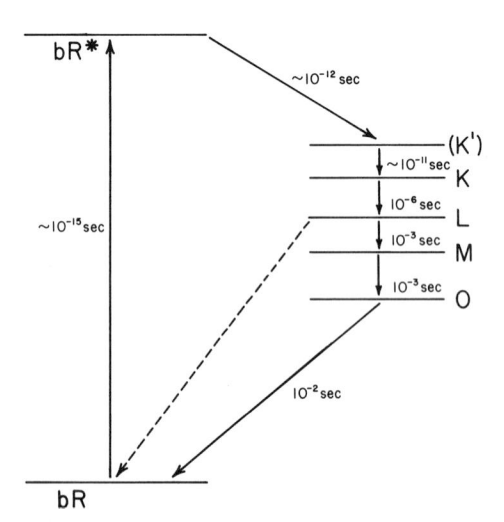

FIG. 1. Photochemical and thermal reactions of bR. The dashed line represents a branch pathway from L back down to bR.

In principle, there are two types of ultrafast kinetic measurements that can be made—absorption and fluorescence. There is a weak fluorescence from bR. More recently, Doukas et al. (1981) have observed the fluorescence kinetics from rhodopsin. Unfortunately, unlike the case of chlorophyll in photosynthesis, so far observations on bR fluorescence have not been able to be closely correlated with the absorption measurements. This may be due to the unusual electronic structure of the chromophore of bR.

II. Absorption Measurements

A. Early Measurements

Lozier et al. (1975) made the first kinetic absorption measurements for bR^{LA} using microsecond time scale flash photolysis techniques. The first photoproduct they observed, K, was a bathochromically shifted species whose formation time was faster than 1 μsec and which decayed to the next intermediate with a half-life of 2 μsec at room temperature. Kung et al. (1975) and Dencher and Wilms (1975) also used flash photolysis over a similar time scale and found results similar to those of Lozier et al. Kung et al. tentatively proposed an intermediate preceding K but later suggested it was probably artifactual.

B. Low Temperature

At low temperatures (77 K) the K intermediate is stabilized or "frozen in" and so, by placing the bR sample in a low-temperature Dewar, the intermediate can be studied using conventional recording spectrophotometers. At 77 K, light seems to produce a photo-steady state containing just two species, bR and K. If the proportion of K in this mixture can be determined, then its absorption spectrum can be calculated. Hurley and Ebrey (1978) tried to determine the amount of K in the photo-steady state by preparing a bR sample at pH 10, in 25% NaCl, forming a photo-steady state containing K at 77 K, and then (rapidly) transferring the sample to a Dewar at $-60°$C. At this temperature K decays to the M intermediate, and the latter is stable. Since M absorbs at 412 nm, far from the absorption maximum of bR, the amount of bR left in the photo-steady state at 77 K can be determined if one assumes that all the K goes to M upon warming. This assumption may not be correct, for others have shown that, under conditions different from those that Hurley et al. used, there is a branching from the L intermediate back to bR, so some K might not be converted to M. Later Tokunaga et al. (1981) determined the amount of K in a photo-steady state at 77 K with another technique which avoids this assumption. Hurley et al. calculated that the λ_{max} of K at 77 K is at 630 nm, $\varepsilon = 78,000$, while Tokunaga et al. determined the λ_{max} to be at 628 nm, $\varepsilon = 70,000$. Since upon cooling the λ_{max} of bR^{LA} shifts approximately 10 nm to the red (from 568 to 578 nm), one can reasonably guess that the λ_{max} of K also shifts about 10 nm upon cooling to 77 K, so its room temperature λ_{max} is estimated to be 620 nm.

C. Ultrafast Measurements

Kaufmann et al. (1976) made the first picosecond time scale measurements of the light-induced absorption changes in the spectrum of bR. They showed that the primary photochemical intermediate of bR, K, was formed in less than about 10 psec at room temperature; it did not decay measurably for at least 300 psec. The difference spectrum seen between bR and K 10 psec after excitation resembled both that seen on the microsecond time scale and that seen with low-temperature spectroscopy. Kaufmann et al. also suggested that there may be a transient before the formation of the batho product; this observation was followed up in a more detailed study by Applebury et al. (1978).

A certain amount of controversy and confusion has surrounded the interpretation of the transient absorption changes and the identification of the transients seen in various studies on rapid absorbance changes. The sequence of the light-induced intermediates of bR that can be observed very soon after a picosecond or subpicosecond flash can be described by $bR^{LA} \to X \to Y$. I believe that the simplest and most straightforward explanation of the data of the various groups is that both X and Y are ground state species. Y is identical with the batho product K characterized at low temperatures, while X is a transient ground state precursor of the batho product, closely resembling it in most properties. Let me now discuss the data and their interpretation in more detail.

Working at temperatures lower than those of the earlier studies of Kaufmann et al. (1976), Applebury et al. (1978) found that they could resolve the formation of the metastable intermediate Y. The formation time was temperature-dependent, ranging from 11 psec at room temperature to ~ 50 psec at liquid helium temperatures (1.8K).

Kryukov et al. (1981) have recently determined a value for the formation time of Y (at 13 K) of 80 psec; this value does not coincide with that expected from Fig. 4 of Applebury et al. (1978), about 30 psec. The origin of this discrepancy is unknown. The rate of formation of X, the precursor of Y, could not be resolved. Applebury et al. (1978) have suggested that X is an excited state of bR and Y is the batho product, K. However, since the photoconversion from bR to K is not complete, with almost 70% of the excited bRs returning to their ground state, if X is an excited state, the ground state of bR should be repopulated with the same kinetics as X disappears; in contrast, the ground state of bR is repopulated much faster than X disappears. This suggests that, at room temperature, X and Y are both ground state species (Honig et al., 1979).

Moreover, Ippen et al. (1978), using subpicosecond laser spectroscopy, have time-resolved this first photoproduct, X, appearing after light absorption in bR, and have found that its formation time is about 1 psec, consistent with the less than 10-psec time of Kaufmann et al. and Applebury et al. The 1-psec formation time also suggests that X is a ground, not an excited, state. Ippen et al. could not observe formation of the Y species.

Applebury et al. (1978) made a number of interesting observations about the properties of the species X and Y. They have quite similar absorption spectra, they have approximately the same extinction coefficient, and their λ_{max} values are shifted from one another by about 10–20 nm. Hurley et al. (1977) and Honig et al. (1979) have suggested on the basis of these resemblances that X and Y are very similar forms of the chromoprotein, the difference between them probably being due to some relaxation event in the apoprotein following formation of the photochemical product. Since the primary species which can be stabilized at 77 K is called K, and Applebury et al.'s Y is also stable at this temperature, Y must correspond to K. Its transient ground state precursor, X, since it relaxes to K, has been denoted K'. This nomenclature shall be used for the rest of this chapter.

An immediate question is, Why did Applebury et al. see two species—K and K'—while Ippen et al. reported only one? The latter authors probably did not see the transition from K' to K because they used 615 nm for the probe wavelength. This wavelength is almost exactly at the isosbestic point between K and K', hence although Ippen et al. measured the formation time of K', they could not see the transformation from K' to K (Dinur et al., 1981). How the rate of formation of K' compares to the rate of deexcitation in the excited state as measured by fluorescence lifetime will be discussed in a later section of this chapter.

Two quite interesting observations have been made concerning the rate of decay of K' and the rate of formation of K (Applebury et al., 1978). One is that at very low temperatures the rate starts to become temperature-independent. Second, when the pigment has been placed in D_2O, which should allow for the exchange of many of its protons for deuterons, the rate of transformation from K' to K at a given temperature slows down by a factor of about 2. The chromophore has only one exchangeable proton, the one which protonates the Schiff base. The protein itself has numerous potentially exchangeable groups, roughly one per amino acid or about 300 in all. However, some recent work has suggested that about half the protons of rhodopsin which are expected to be exchangeable are not (Osborne and Nabedryk-Viala, 1977); a similar situation may apply for bR.

The origin of the temperature independence and deuterium dependence referred to above is unknown. Peters *et al.* (1977) saw a similar phenomena in rhodopsin, with a larger deuterium effect, approximately sixfold, and attributed it to the movement through a tunneling mechanism of the proton on the Schiff base, so that either rhodopsin or its batho product would be an unprotonated Schiff base. Applebury *et al.* made a similar suggestion for bR; however, a protonation–deprotonation of the Schiff base is at variance with a wide variety of spectral evidence indicating both bR and K are protonated Schiff bases (see Chapter 9). Hurley *et al.* suggested that a conformational relaxation of the protein following chromophore isomerization might be responsible for the proton movements. Goldanskii (1979) has pointed out that the size of the deuterium effect for rhodopsin is too small (hence for bR even more so) for proton tunneling; he has suggested that some substantial part of the chromophore itself may have to tunnel through a barrier during isomerization–relaxation.

D. Hypsochromically Shifted Early Intermediates

For some visual pigments, there is a blue-shifted intermediate, hypsorhodopsin, which is stable near liquid helium temperatures and converts to bathorhodopsin upon warming to liquid nitrogen temperatures (Yoshizawa, 1972). It has been proposed that this intermediate might be a precursor of the batho intermediate of visual pigments, but recent evidence suggests it probably is not (Tsuda *et al.*, 1980). For bR the batho product is the only stable photo intermediate detected at liquid helium temperatures (Iwasa *et al.*, 1979). Thus, if there is an analog in bR for the hypso-type intermediate, it must be unstable even at ~ 10 K.

E. Kinetics of the Back Reaction from K to Bacteriorhodopsin

Kryukov *et al.* (1981) measured the rate of the reaction from the batho product K back to bR. A photo-steady state was prepared by irradiation of bR at 13 K. It was not clear whether the bR was light-adapted or not. The rate at which the K was photoconverted back to bR could not be resolved but was less than 30 psec. The authors have concluded that, since in their interpretation of the data the rate of the bR → K reaction is significantly slower than the K → bR reaction (80 psec versus less than 30 psec), these two decay processes do not occur from the same excited state. However, this conclusion is clouded by the identification of the forward process they measured *not* as the decay of the excited state, but the ground state transition

K′ → K. The measurements of Ippen et al. (1978) suggest that the excited state decays in about 1 psec, so the back reaction should also take about 1 psec. Thus, Kryukov's value of less than 30 psec for the K → bR reaction is consistent with the predicted value of 1 psec for the decay of the excited state of K back to bR. However, it still is not understood exactly how the back reaction from K to bR proceeds and in particular if there is an intermediate similar or identical to K′. The measurements of Kryukov et al. are an important step in beginning the elucidation of the K → bR photoreaction.

F. PRIMARY PHOTOPRODUCT OF THE 13-CIS FORM OF BACTERIORHODOPSIN

Tokunaga et al. (1977) showed that at liquid nitrogen temperatures light produced a bathochromically shifted photoproduct for that portion of dark-adapted bR with a 13-cis chromophore. Using flash photolysis, Sperling et al. (1977) also found a bathochromically shifted photoproduct that was rapidly formed from the 13-cis pigment, which they designated C^{610}. They also inferred the existence of a precursor of C^{610}, C^x, but could not obtain information as to its spectral properties. Iwasa et al. (1981) could not detect this species as a stable intermediate at liquid nitrogen or liquid helium temperatures.

III. Fluorescence Measurements

A. EARLY MEASUREMENTS

A second photochemical process has been studied in bR, the fluorescence from the chromophore. This fluorescence has a very small yield at room temperature, about 0.02%, but upon cooling at liquid nitrogen temperatures the yield increases approximately 20-fold and therefore the fluorescence has often been studied at this temperature (Lewis et al., 1976; Alfano et al., 1976). The identification of the fluorescent species at liquid nitrogen temperatures as bR and not one of its photoproducts, or contaminants of the purple membrane preparation such as carotenoids, has been made by us and others (Alfano et al., 1976; Govindjee et al., 1979; Lewis et al., 1976). It is generally assumed that the fluorescence seen at room temperature is also from bR, although the criteria used in this identification are not as rigorous as for the fluorescence observed at liquid nitrogen temperatures. Indeed, there are significant changes in the fluorescence emission spectra upon cooling from room temperature to 77 K (Shapiro et al., 1978). It may be that these changes are due to the fluorescence from the photocycle intermediates, which can be present in significant quantities at some of the intermediate temperatures that have been studied. As for the emission spectrum at 77 K, we (Alfano et al.,

1976; Govindjee et al., 1979) and others (Grillbro and Kriebel, 1977) found peaks at 670, 710, and 725 nm. Lewis and co-workers (1976) and Sineshchekov and Litvin (1977) found emission peaks at 670, 720 (which may be our 710 and 725 combined), and in addition a third peak at 790 nm, which we did not observe. It would be useful if other workers would carefully measure the emission spectrum of bR under a variety of conditions.

B. Fluorescence from Intermediates

Fluorescence has been looked for in studies on the photocycle intermediates K, M, and O. K does not appear to fluoresce, although its fluorescence could have gone undetected if it had a yield less than $\sim 15\%$ that of bR (Alfano et al., 1976; Lewis et al., 1976; Govindjee et al., 1979). The M and O intermediates have been shown by Gillbro and Kriebel (1977) to fluoresce. L and any other intervening intermediates have not yet been studied.

C. Ultrafast Measurements

Alfano et al. (1976) made the first measurements of the fluorescence lifetime of bR. At 90 K, they found a lifetime of 45 ± 5 psec using a train of picosecond pulses about 7 nsec apart. In the case of bR, the K intermediate, which is formed as a result of light absorption with a quantum efficiency of about 0.3, has been shown to have very little fluorescence and so would not interfere at this temperature. Shapiro and colleagues (1978) later measured bR fluorescence at 77 K using both the train of pulses and the single-pulse technique to see whether the train of pulses had an effect on the fluorescence lifetime. The single-pulse measurements gave essentially the same lifetime as the measurements using the train of picosecond pulses. The lifetime they obtained was fairly close to that of Alfano et al. (The difference in the temperatures of the two experiments—90 and 77 K—have been shown by Shapiro et al. to cause about a twofold difference in lifetime.) At room temperature, Alfano et al. and Shapiro et al. could not detect any fluorescence with a lifetime of longer than about 10 psec. In contrast to these results, Hirsch et al. (1976), using an interesting new technique, determined a value for the room temperature fluorescence lifetime of about 15 psec. This contradiction has not been resolved.

D. Relationship of the Fluorescence to the Photochemistry

Alfano et al. showed that, if one calculates the intrinsic radiative lifetimes from the fluorescence yield, the fluorescence lifetime at 77 K, and the change in yield with temperature, then the predicted radiative lifetime is much longer (25–125 nsec) than that obtained from integrating the oscillator strength of the lowest excited state (6 nsec). This suggested to them that the fluorescing state seen at 90 K was a forbidden state. This may not be correct (see below). As noted previously, Kaufmann et al. (1978) and Applebury et al. (1978) measured the photochemical lifetime of bR at 77 K. They found that formation of the stable primary photoproduct, K, at 77 K required ~20 psec, much less than the 40–60 psec found for the fluorescence lifetime. Moreover, as noted above, the actual lifetime of the excited state, bR* → K', is considerably shorter than the lifetime for formation of K (K' → K). Thus, at liquid nitrogen temperatures the lifetime of the bR state which fluoresces is much longer than the lifetime of the state which photochemically forms the K' intermediate of bR. Moreover, the fluorescence and the photochemistry have very different temperature dependencies. This suggests that these two processes occur from different excited states (Govindjee et al., 1979). Lewis (1979) has pointed out that an interesting consequence of the photochemical and fluorescent states being different is that the inference may be incorrect that the fluorescent state is forbidden because of its long radiative lifetime. Rather, Lewis suggested that the unexpectedly low yield of fluorescence may be because most of the light absorbed does not reach the fluorescent excited state but does populate the photochemical excited state.

If the fluorescence at room temperature was directly related to the fluorescence observed at 77 K, then one could calculate a fluorescence lifetime at room temperature from (1) the lifetime measurement at 77 K, and (2) the 20- to 30-fold increase in yield upon lowering the temperature. Alfano et al. estimated the fluorescence lifetime at room temperature to be about 2.5 psec. Shapiro et al. (1978), using somewhat more accurate values for the temperature dependence of the yields, calculated a lifetime of about 1.5 psec. This is quite close to the photochemical lifetime measured by Ippen et al. of about 1 psec. Thus, at room temperature it may be that the fluorescent state and the photochemical state are the same.

In summary, the sequence of events in the photochemical transformation of light-adapted bR seems relatively clear. There is no evidence for a hypso-type intermediate, the first species being an unstable species, K', which goes on to K; K is stable at liquid nitrogen temperatures. However, it appears that the other methods available for monitoring the excited state, fluorescence, do not produce a simple unified picture, especially at low temperatures.

ACKNOWLEDGMENTS

I would like to thank my colleagues Rajni Govindjee, Barry Honig, and Ken Kaufmann for reading the manuscript for this chapter. This work was supported by NSF PCM 79-11772.

References

Alfano, R. R., Yu, W., Govindjee, R., Becher, B., and Ebrey, T. G. (1976). *Biophys. J.* **16**, 541–545.
Applebury, M. L., Peters, G. S., and Rentzepis, P. M. (1978). *Biophys. J.* **23**, 375–382.
Dencher, N., and Wilms, M. (1975). *Biophys. Struct. Mech.* **1**, 259–271.
Dinur, U., Honig, B., and Ottolenghi, M. (1981). *Photochem. Photobiol.* **33**, 523–528.
Doukas, A. G., Lu, F. Y., and Alfano, R. R. (1981). *Biophys. J.* **35**, 505–508.
Gillbro, T., and Kriebel, A. (1977). *FEBS Lett.* **79**, 29–32.
Goldanskii, V. I. (1979). *In* "Tunneling in Biological Systems" (B. Chance, D. C. DeVault, H. Frauenfelder, R. A. Marcus, J. R. Schrieffer, and N. Sutin, eds.), pp. 663–714. Academic Press, New York.
Govindjee, R., Becher, B., and Ebrey, T. G. (1978). *Biophys. J.* **22**, 67–77.
Hirsch, M. D., Marcus, M. A., Lewis, A., Mahr, H., and Frigo, N. (1976). *Biophys. J.* **16**, 1399–1409.
Honig, B. (1979). *Ann. Rev. Phys. Chem.* **29**, 31–57.
Honig, B., Ebrey, T. G., Callender, R. H., Dinur, U., and Ottolenghi, M. (1979). *Proc. Natl. Acad. U.S.A.* **76**, 2503–2507.
Hurley, J. B., and Ebrey, T. G. (1978). *Biophys. J.* **22**, 49–66.
Hurley, J. B., Ebrey, T. G., Honig, B., and Ottolenghi, M. (1977). *Nature (London)* **270**, 540–542.
Ippen, F. P., Shank, G. V., Lewis, A., and Marcus, M. (1978). *Science* **200**, 1279–1281.
Iwasa, T., Tokunaga, F., Yoshizawa, T., and Ebrey, T. (1979). *Photochem. Photobiol.* **31**, 83–85.
Kaufmann, K. J., Rentzepis, P. M., Stoeckenius, W., and Lewis, A. (1976). *Biochem. Biophys. Res. Commun.* **68**, 1109–1115.
Kaufmann, K. J., Sundstrom, V., Yamane, T., and Rentzepis, P. M. (1978). *Biophys. J.* **22**, 121–123.
Kryukov, P. G., Lazarev, Y. A., Matveetz, Y. A., Sharkov, A. V., and Terpugov, E. L. (1981). In press.
Kung, M. C., DeVault, D., Hess, B., and Oesterhelt, D. (1975). *Biophys. J.* **45**, 907–911.
Lewis, A. (1979). *Biophys. J.* **25**, 79a.
Lewis, A., Spoonhower, J. P., and Perreault, G. (1976). *Nature (London)* **260**, 675–678.
Lozier, R. H., Bogomolni, R. A., and Stoeckenius, W. (1975). *Biophys. J.* **15**, 955–963.
Osborne, H. B., and Nabedryk-Viala, E. (1977). *FEBS Lett.* **84**, 217–220.
Peters, K., Applebury, M., and Rentzepis, P. M. (1977). *Proc. Natl. Acad. Sci. U.S.A.* **74**, 3119–3123.
Shapiro, S. L., Campillo, A. J., Lewis, A., Perreault, G. J., Spoonhower, J. P., Clayton, R. K., and Stoeckenius, W. (1978). *Biophys. J.* **23**, 383–393.
Sineshchekov. V. A., and Litvin, F. F. (1977). *Biochim. Biophys. Acta* **462**, 450–466.
Sperling, W., Carl, P., Rafferty, C., and Dencher, N. (1977). *Biophys. Struct. Mech.* **3**, 79–94.
Stoeckenius, W., Lozier, R. H., and Bogomolni, R. A. (1979). *Biochim. Biophys. Acta* **505**, 215–278.
Tokunaga, F., and Yoshizawa, T. (1981). In press.
Tokunaga, F., Iwasa, T., and Yoshizawa, T. (1976). *FEBS Lett.* **72**, 33–38.
Tsuda, M., Tokunaga, F., Ebrey, T., Yue, K. T., Margue, J., and Eisenstein, L. (1980). *Nature* **287**, 461–462.
Yoshizawa, T. (1972). *In* "Handbook of Sensory Physiology, VII/2, Photochemistry of Vision" (H. J. A. Dartnall, ed.), pp. 146–179. Springer-Verlag, Berlin and New York.

CHAPTER 12

Theoretical Aspects of Photoisomerization in Visual Pigments and Bacteriorhodopsin

Barry Honig[*]

Department of Physiology and Biophysics
University of Illinois at Urbana-Champaign
Urbana, Illinois

I.	Introduction	281
II.	Absorption Spectra	283
	A. *Polyenes*	283
	B. *Pigments*	284
III.	The Primary Photochemical Event	286
	A. *Ground State Energetics*	286
	B. *Photoisomerization and Charge Separation*	288
IV.	Excited State Processes	290
	A. *Potential Energy Curves*	290
	B. *Isomerization Dynamics*	292
	C. *Proton Transfer Processes*	294
V.	Concluding Remarks	295
	References	296

I. Introduction

Understanding the mechanism by which a photon is converted to chemical energy in visual pigments and bacteriorhodopsin requires that the nature of the primary photochemical event in both pigments be described in molecular detail. That is, it is necessary to characterize the changes that take place in the chromophore and the protein when rhodopsin is converted to bathorhodopsin or when bacteriorhodopsin is converted to its primary photoproduct. These changes involve dynamical processes on excited state potential energy surfaces, which may in turn induce additional ground state transformations. A large body of data, the product of extensive experimental studies over a period of more than 20 yr, has brought us to the point where the nature of these events seems well understood, at least in qualitative terms (for reviews, see Honig, 1978; Ottolenghi, 1980). However, the increase in our

[*] Present address: Department of Biochemistry, Columbia University, New York, New York 10032.

knowledge has generated new, highly intricate questions, many of which still remain unresolved. The purpose of this chapter is to discuss the present status of our theoretical understanding of the photochemical processes that occur in visual pigments and bacteriorhodopsin and to describe the variety of complex problems currently under active theoretical and experimental investigation.

A characterization of the spectroscopic properties of visual pigments is an obvious first step in the study of their photochemistry. Indeed, the problem of accounting for the absorption spectra of visual pigments has been a subject of great interest for many years (see, e.g., Wald, 1968). The pigments have absorption maxima ranging from 430 to 580 nm, and this variation must be interpreted in terms of opsin variability since the chromophore is apparently always 11-*cis*-retinal, which itself absorbs in the ultraviolet ($\sim$ 380 nm). It is known from Raman studies that the chromophore is bound to the protein via a protonated Schiff base linkage to the ε-amino group of a lysine (Oseroff and Callender, 1974). This accounts for part of the wavelength shift, since protonated Schiff bases of retinal absorb at about 450 nm when isolated in solution. However, it does not explain an absorption maximum near 500 nm for most rod pigments and maxima ranging out to 580 nm for cone pigments.

An identical problem arises in trying to account for the absorption maximum of the purple membrane protein of *Halobacterium halobium*, which is frequently referred to as bacteriorhodopsin (for review, see Stoeckenius *et al.*, 1979). Bacteriorhodopsin also uses retinal (all-*trans*) as a chromophore, it is also bound to the protein via a protonated Schiff base linkage to the ε-amino group of a lysine (Lewis *et al.*, 1974), and it is red-shifted to 570 nm (in the range of the reddest visual pigments). It seems clear that similar mechanisms for shifting absorption maxima are operating in both systems. These will be explored in the next section.

There has been considerable controversy concerning the nature of the excited state process which initiates the visual response. As discussed elsewhere in this volume, there now seems little doubt that the primary event involves isomerization of the 11-cis chromophore to a transoid conformation, however, the exact nature of the isomerization is still not completely understood. For example, the quantum yield for the process is much higher than that of model compounds (Rosenfeld *et al.*, 1977), and the actual conformation of the chromophore of the bathorhodopsin photoproduct is not known. Moreover, other events such as proton transfer also appear to be important components of the primary event (Peters *et al.*, 1977). As is the case for absorption spectra, the protein appears to play a major role in regulating the photophysical properties of the chromophore.

In order to understand the chromophore–protein interactions that initiate the visual response, it is essential to obtain a reasonable representation of the ground and excited state potential energy surfaces. Section III shows how an approximate ground state curve for rhodopsin may be extracted from available experimental data. When the experimentally derived curve is compared to that expected on the basis of the properties of the isolated chromophore, significant differences are found. These must be due to the specific effects of the protein on the energetics of the chromophore. A model which accounts for the energetics as well as the spectroscopic changes associated with the primary photochemical event is introduced in Section III.

Section IV contains an analogous discussion for excited states. An experimentally derived potential energy curve is presented, and this is compared to that expected on the basis of quantum mechanical calculations. As is the case for the ground state, it is necessary to invoke specific chromophore–protein interactions to account for the dramatically different photophysical behavior of opsin-bound and free chromophore in solution. Recent theoretical simulations of the picosecond isomerization processes in visual pigments and bacteriorhodopsin are also described in Section IV.

II. Absorption Spectra

A. POLYENES

The factors that determine the spectroscopic properties of linear polyenes, such as retinal, are well understood (see, e.g., Salem, 1966). It appears to be a general principle that long-wavelength absorption is correlated with increased π-electron delocalization and decreased single–double bond alternation. This behavior can be seen when comparing cyanine dyes, which have an odd number of atoms, to polyenes, which have an even number of atoms. Cyanines, whose π electrons are largely delocalized, have a long wavelength absorption maximum equal to about 1000 Å times the number of double bonds and have approximately equal bond lengths; that is, single and double bonds are equivalent. Polyenes tend to absorb at much shorter wavelengths than cyanines of the same length and exhibit considerable bond alternation. The spectroscopic properties of conjugated hydrocarbon chains are related to the extent that their π electrons are delocalized (Labhart, 1957). Any mechanism that increases delocalization and reduces bond alternation, even in a polyene with an even number of atoms, should induce bathochromic shifts in the absorption maxima.

When the Schiff base of retinal is protonated, a positive charge is partially delocalized throughout the π system. Resonance structures such as those of Fig. 1b–d will contribute, hence increase π-electron delocalization and decrease bond alternation. The latter effect can be measured in Raman experiments from the frequency downshift of the C=C stretching vibration upon protonation (for a review of Raman measurements, see Callender and

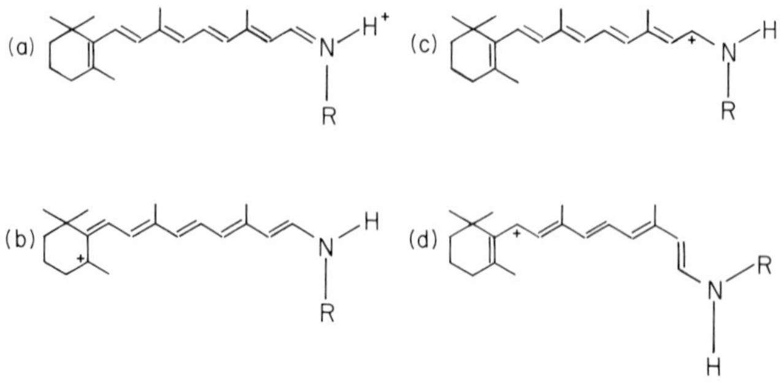

FIG. 1. Resonance structures of the protonated Schiff base of retinal. (a) The classic polyene-like structure; (b–d) structures which contribute to enhanced π-electron delocalization.

Honig, 1977). The bathochromic shift that occurs upon protonation can also be qualitatively understood in terms of increased π-electron delocalization. However, the magnitude of the shift is difficult to determine experimentally since the protonated Schiff base cation does not exist in an isolated form and can only be measured in solution where significant environmental effects are to be expected (Honig, 1978).

Theoretical calculations (Suzuki et al., 1974; Honig et al., 1976) confirm the original prediction of Blatz (see, e.g., Blatz and Mohler, 1975) that the isolated protonated Schiff base of retinal should absorb near 600 nm. However, solutions of protonated Schiff bases absorb at ∼450 nm, much to the blue of this theoretical value. This is due in large part to the state of association of the counterion. In nonpolar solvents, where the salt is not dissociated, the effect of a counterion is to pull the positive charge to the nitrogen, increase the contribution of the structure in Fig. 1a, increase bond alternation, and thus induce a blue shift (Blatz and Mohler, 1975; Honig et al., 1976).

B. PIGMENTS

The discussion of the previous section suggests that it is possible to view pigment spectra as being blue-shifted from the ∼600 nm expected for the isolated protonated Schiff base, or red-shifted from the corresponding solution value ∼450 nm which includes the effect of the counterion. The solution value seems more appropriate for the stable form of a particular pigment, since the positively charged Schiff base is almost certainly balanced in the different proteins by a negatively charged amino acid (Honig et al., 1976). (The generalization that buried net charges are not found in proteins should be particularly valid in the low dielectric environment provided by membranes.) However, this constraint need not be applied to the unstable transient photoproducts of both visual pigments and bacteriorhodopsin.

It is possible to use the resonance structures of Fig. 1 to consider the types of effects that can influence pigment spectra. The effect of a counterion has already been rationalized in these terms. If we assume a fixed counterion, variability in absorption maxima can arise from the positioning of additional charged groups around the retinal chromophore. For example, a negative charge located near the β-ionone ring will increase the contributions of the structures in Fig. 1b and d and thus induce a red shift, while a positive charge at the same location will induce a blue shift. It is not always possible, however, to rely completely on resonance diagrams to predict the effect of a particular environmental perturbation. In general, external charges will interact with many positions along the polyene chain, thus stabilizing some resonance structures while destabilizing others. In order to predict the effect of a charge at a particular location, it is, in general, necessary to carry out a quantum mechanical calculation of transition energies. Semiempirical schemes (and corresponding computer programs) for this purpose are widely available but must be used with extreme caution (Dinur et al., 1980).

While the concept of charged groups influencing pigment absorption maxiam has been widely accepted, there has been no direct evidence for the existence of such groups or their location with respect to the chromophore. However, the theoretical considerations discussed above led to the synthesis of a series of compounds which now make it possible to identify groups on the opsin, which play the major role in determining the absorption maxima of various pigments.

The approach is based upon the fact that saturation of even one double bond in the polyene chain completely disrupts the π-electron system. Thus, a series of retinal-like molecules with different bonds saturated (dihydroretinals; Arnaboldi et al., 1979) provide compounds that are, sterically, nearly identical but whose light absorption properties are localized at different positions along the polyene chain. An absorption spectrum is measured for the protonated Schiff base of each compound in solution and compared to that of the corresponding pigment formed upon binding of the chromophore to an opsin. Since the magnitude of the shifts induced by external charges depends on their distance from the π system, by observing the spectral shift for each compound it is possible to locate the position of the groups on the protein that determine the absorption maximum of the chromophore (Honig et al., 1979a).

Experimental and theoretical studies have been carried out so far for two pigments, bovine rhodopsin and bacteriorhodopsin. For bovine rhodopsin, the results strongly imply the existence of a negatively charged amino acid in the vicinity of the 11,12-double bond (Fig. 2a). Moreover, there are unlikely to be spectroscopically important electrostatic interactions near the β-ionone ring. In contrast, the results for bacteriorhodopsin (Nakanishi et al.,

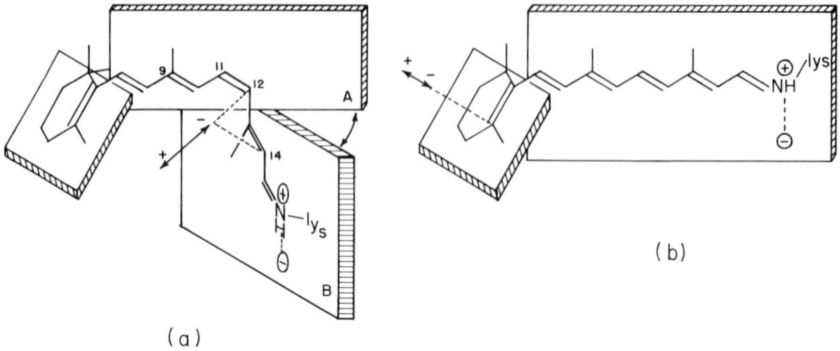

FIG. 2. External point charge models for (a) bovine rhodopsin and (b) bacteriorhodopsin. Dashed lines indicate approximately a 3-Å separation from a negative charge to a designated atom.

1981) indicate that a negative charge is located quite close to the β-ionone ring in that pigment (Fig. 2b).

III. The Primary Photochemical Event

A. Ground State Energetics

As mentioned in Section I, the photochemical event that triggers visual excitation corresponds to a photoisomerization of the 11-cis chromophore to a transoid conformation. This conformational change in the chromophore must in some way activate the protein so as to initiate the series of biochemical events that give rise to the receptor potential of the cell. One way of characterizing the activation mechanism is to consider the energy changes associated with the primary event. A number of years ago it was shown that a significant fraction of the photon's energy was stored in the rhodopsin–bathorhodopsin transformation (Rosenfeld et al., 1977). A lower limit of 20 kcal/mole was estimated for the extent of energy storage, which should be compared to the approximately 50 kcal/mole provided by the photon (Honig et al., 1979b). A recent photocalorimetric study has established that bathorhodopsin is in fact 35 kcal/mole higher in energy than rhodopsin itself (Cooper, 1979). Since conformational energy differences between cis and trans isomers of retinal-like molecules rarely exceed a few kilocalories per mole, it is clear that the pigment has found a way to alter the energetics of isomerization significantly.

Give an energy difference of 35 kcal/mole, the activation energy between rhodopsin and bathorhodopsin must be somewhat higher (above 40 kcal/mole to account for the stability of bathorhodopsin at 77 K). This value is close to that estimated from electrophysiological studies on dogfish bipolar cells and is somewhat higher than the lower limit determined from psychophysical estimates of the level of thermal noise in photoreceptors (see discussion in Honig et al., 1979b). The large barrier to the thermal population of the bathorhodopsin photoproduct is necessary for rhodopsin to function as a highly sensitive photon detector and provides a useful criterion in the evaluation of various models.

The barrier to thermal isomerization in retinal isomers is about 25 kcal/mole (Hubbard, 1966), but similar numbers are not available for the various protonated Schiff bases. These are expected to be somewhat smaller, since the increased π-electron delocalization in these systems somewhat weakens their double bonds. It seems that photoisomerization about a double bond was chosen by evolution in part because it occurs so rarely in the ground state, thus ensuring a low level of thermal noise. However, the rate of thermal isomerization in retinals is still about two orders of magnitude higher than the upper limit for the rate of thermal isomerization of visual pigments. Thus, the interaction with the opsin must enhance the inherent stability of the ground state conformation of the retinal chromophore. A schematic potential energy curve based on the evidence discussed in this section is given in Fig. 3.

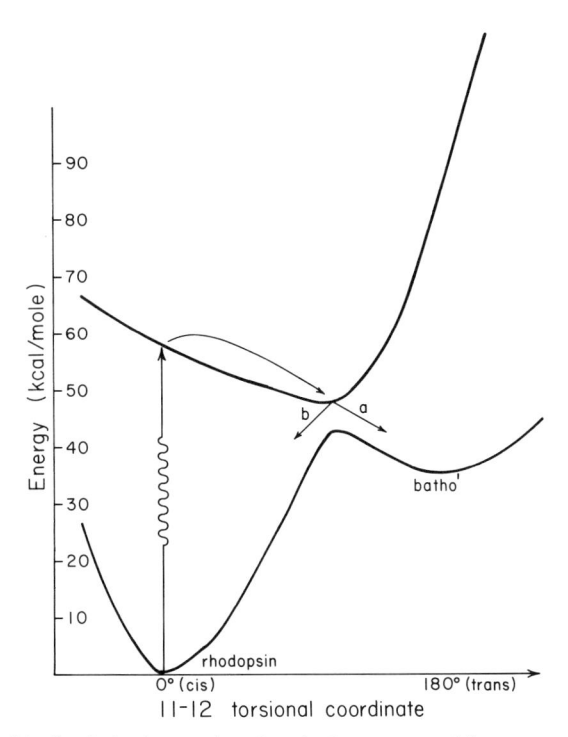

FIG. 3. Empirically derived ground and excited state potential energy curve for bovine rhodopsin (see text).

B. Photoisomerization and Charge Separation

As discussed in Section III,A, the energetics of cis-trans isomerization at the opsin-binding site are very different from what is expected on the basis of the solution properties of the chromophore alone. The large spectral red shift seen in the formation of the batho photoproduct is also atypical and, together with the ground state energy changes, needs to be explained in terms of an isomerization within the highly specific environment of the opsin matrix.

In Section II it was shown that variability in the absorption maxima of visual pigments is due to the positioning of charged amino acids on the opsin at different locations around the chromophore. The immediate environment of the Schiff base nitrogen is assumed to be similar to that in nonpolar solvents; that is, a negatively charged amino acid forms a salt bridge with the positively charged Schiff base. This group has little or no role in wavelength regulation in the primary pigments but appears to assume major importance as a photochemical determinant.

A fundamental difference between the counterion in the protein and one in solution is that the latter is mobile and the former is not. In the opsin matrix, the counterion would not be expected to follow the Schiff base nitrogen during the course of an isomerization. Thus, as shown in the model for the primary event, an ionic bond has to be broken before an isomerization can occur, and this is energetically unfavorable in a low-dielectric medium (Honig et al., 1979b). It is, unfortunately, difficult to estimate quantitatively the increase in energy associated with charge separation in a protein, especially in the absence of a three-dimensional structure. Based on precedents from other proteins, it is likely that the Schiff base-counterion linkage is "solvated" by additional protein dipoles. A photoisomerization could well lead to a configuration where the Schiff base finds itself in an environment in which it undergoes unfavorable electrostatic interactions with the protein or, alternatively, in a hydrophobic environment where no favorable solvation occurs. In either case, a net increase in electrostatic energy will result from isomerization, and the order of magnitude involved can be estimated by assuming a simple coulombic interaction.

With reasonable values for the dielectric constant it is possible to account for the 35 kcal/mole energy difference between rhodopsin and bathorhodopsin by assuming a charge separation of a few angstroms (Honig et al., 1979b). Since a photoisomerization will increase the nitrogen counterion distance from 3 to ~ 7 Å, a large energy increase is in fact a natural consequence of a photoisomerization within the protein matrix. (In addition, strain energy resulting from twisting about different bonds in both the chromophore and the protein could account for some smaller component of the total energy differential.) A recent theoretical study on photoisomerization within the protein matrix shows quite nicely how conformation distortion and charge separation can combine to account for most of the observed energy increase accompanying the rhodopsin–bathorhodopsin transformation (Birge and Hubbard, 1980). The large thermal barrier to the rhodopsin–bathorhodopsin transformation can also be understood in these terms, since

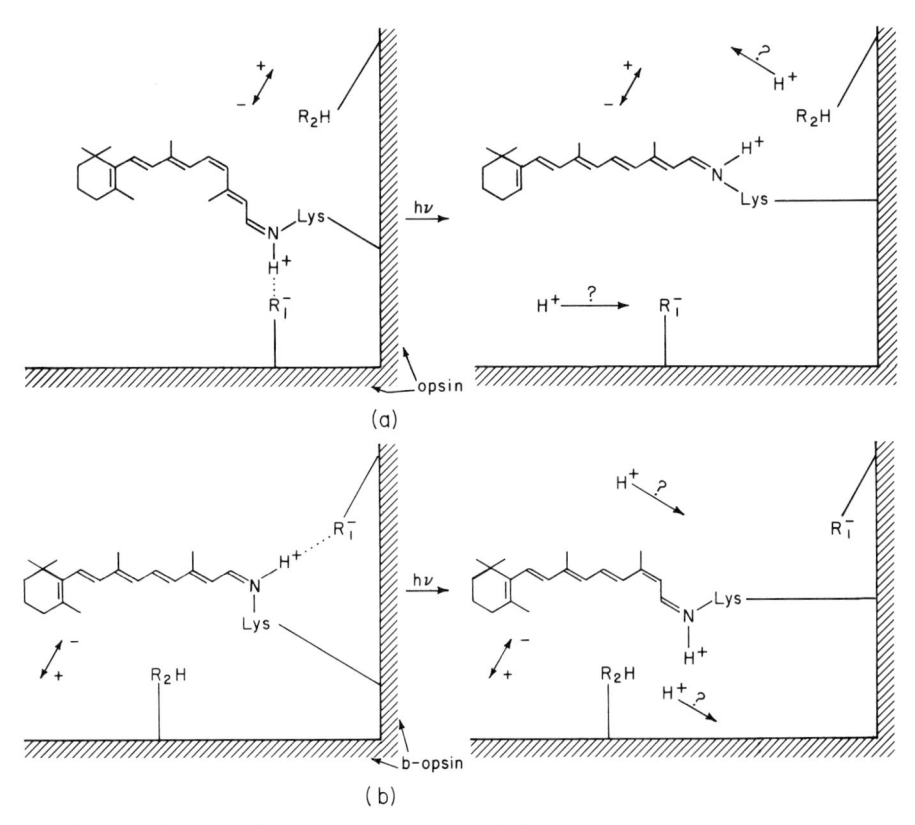

FIG. 4. Models for the primary events in visual pigments and bacteriorhodopsin. (a) The 11-cis chromophore of rhodopsin is depicted with its Schiff base forming a salt bridge with a negative counterion. The additional charge pair near the 11,12-double bond represents the group that regulates the absorption maxima of different pigments. The photochemical event is an isomerization about the 11,12-double bond forming the species referred to as batho' in the text. The pK values of the Schiff base and those of other groups on the protein, such as R_1 and R_2, are strongly affected by photoisomerization because a salt bridge has been broken, a positive charge has moved near R_2, and R_1 is now a bare negative charge. The batho' → bathorhodopsin transition involves ground state protein relaxation apparently involving the movement of protons. Possible proton transfer steps resulting from charge separation are depicted. (b) For bacteriorhodopsin, the isomerization is trans–cis (probably about the 13,14-double bond) rather than cis–trans, but all other events are assumed to be equivalent.

an ionic bond has to be broken before isomerization can occur. In this way, the inherently large activation energy for cis–trans isomerization is increased by an electrostatic term that is not present in solution (Honig *et al.*, 1979b; Birge and Hubbard, 1980).

A final and satisfying consequence of the charge separation model is that it provides a natural explanation of the red shift that results from the primary photochemistry. As discussed in Section II, the presence of a counterion blue-shifts absorption maxima so that the charge separation induced by isomerization will always lead to a red-shifted primary photoproduct. It is important to emphasize in this regard that this shift in the transient photoproduct is in addition to the shifts induced by the other groups that regulate wavelength in the stable forms of the pigments.

These consequences of isomerization are not limited to a particular cis or trans isomer. Any isomerization can induce a charge separation of the protonated Schiff base from its counterion so that the mechanism can work equally well for the 11-cis-to-all-trans isomerization in visual pigments (Fig. 4a) and for a trans–cis isomerization in light-adapted bacteriorhodopsin (Fig. 4b). In fact, the general mechanism is applicable to any pigment with a flexible and charged chromophore, even though the specific bond that rotates and the direction of isomerization may vary.

IV. Excited State Processes

A. POTENTIAL ENERGY CURVES

The photochemical properties of molecules related to the visual chromophore have been studied in detail under the assumption that they might simulate the behavior of the pigments (see recent review in Ottolenghi, 1980, for a detailed discussion). These properties turned out to be quite complex depending on solvent, isomer, and end group. Moreover, the photoisomerization quantum yields for the model compounds are relatively small (in general < 0.1) and exhibit a marked wavelength and temperature dependence. In contrast the quantum yield for photoisomerization in visual pigments is quite large (~ 0.67) and, remarkably, exhibits no wavelength or temperature dependence from room temperature down to 77 K. It is evident that the photochemical properties of visual pigments are significantly different from those of model compounds. Thus, the situation encountered for spectroscopic properties and ground state energetics is repeated for photochemical properties; the protein must play a central role in modifying the behavior of the isolated chromophore.

The approach to understanding the role of the protein that was described in the previous section involved accounting for differences in spectroscopic transition energies (or ground state energies) between the free and opsin-bound chromophore, in terms of specific chromophore–protein interactions. This is more difficult to accomplish when treating photochemical properties, because the complex behavior of retinal-like molecules in solution has not been adequately explained. Indeed, the nature of the excited state potential energy surfaces of even a simple polyene such as butadiene is still a subject of active research efforts. In contrast, it has been possible to provide what appears to be a reliable description of the excited state potential for cis–trans isomerization in visual pigments, primarily because their photochemistry is much simpler than that of model compounds.

A qualitative description of the torsional potential about the 11,12-double bond has been extracted from experimental data (Rosenfeld et al., 1977). It has been pointed out that the wavelength and temperature independence of the primary photochemistry can most easily be interpreted in terms of a barrierless excited state potential energy surface connecting the 11-cis and all-trans isomers. That the back photoreaction from bathorhodopsin to rhodopsin is also wavelength- and temperature-independent implies that the photochemistry in either direction involves the population of a common energy minimum from which the repopulation of either ground state occurs. A potential energy curve of this type (Fig. 3) will give rise to the highly efficient photoisomerization characteristic of visual pigments. The question arises however as to why photoisomerization of model compounds is a far less efficient process.

Quantum mechanical calculations of potential energy curves do not give an unambiguous answer. Recent calculations report a barrierless potential for cis–trans isomerization of the isolated protonated Schiff base and, as discussed in the next section, allow a simulation of the dynamics of isomerization in good agreement with experimental studies on bovine rhodopsin (Birge and Hubbard, 1980). However, the low quantum yield for isomerization in a protein-free environment is not accounted for by the calculation alone. Other calculations (U. Dinur and B. Honig, unpublished data; Ottolenghi, 1980, Fig. 4) yield a small barrier for isomerization in the isolated molecule and thus cannot by themselves account for the photochemistry of the pigments. (The shapes of the calculated potentials in both studies are quite similar, with the exception of the very small barrier found in the latter.)

If the potential for the isolated protonated Schiff base indeed has no barrier, it may be possible that the low quantum yield for isomerization in solution is due to solvation effects, particularly in polar solvents (Birge and Hubbard, 1980). A study of quantum yields in nonpolar solvents would be of great interest in this regard. If the low quantum yield in solution is due to a small intrinsic barrier, it is possible that this barrier is removed by specific interactions with the opsin (Rosenfeld et al., 1977). Interaction with the counterion may produce small effects in the right direction, but these do not seem to be very large (Birge and Hubbard, 1980). Another possibility is that the same charges that influence pigment spectra play an important role in photochemical catalysis. For example, in the perpendicular excited state (twisted by 90° from the planar configuration) a large dipole is calculated across the 11,12-double bond with the positive end localized on C-12 (Birge and Hubbard, 1980; U. Dinur and B. Honig, unpublished data). Thus, the negative amino acid located near this atom would be expected to stabilize twisted configurations and possibly remove any barriers or, in the absence of a barrier, make the potential much steeper. Future theoretical and experimental studies may allow us to clarify these issues.

B. ISOMERIZATION DYNAMICS

The important discovery that bathorhodospin is formed in less than 6 psec (Busch *et al.*, 1972; Green *et al.*, 1977) has generated considerable interest in the problem of describing the dynamics of cis–trans isomerization in polyenes related to the visual chromophore. The fact that the potential energy curve for isomerization about the 11,12-bond of the chromophore of rhodopsin has no barrier in the excited state implies that picosecond or even subpicosecond isomerization times are possible (Honig *et al.*, 1979b). Moreover, the absence of a barrier implies that the process should have little or no temperature dependence. Consistent with this expectation are the observations that the quantum yield for the rhodopsin–bathorhodopsin transformation is unchanged between room temperature and 77 K (Hurley *et al.*, 1978) and that bathorhodopsin is formed quite readily even at liquid helium temperatures (Yoshizawa, 1973; Aton *et al.*, 1978). (The quantum yield at liquid helium temperatures has not been determined.)

Rosenfeld *et al.* (1977) suggested that the photochemical behavior of cattle rhodopsin could best be rationalized in terms of a model in which excitation of either rhodopsin or bathorhodopsin generated a thermally relaxed population of states in the common excited state minimum along the C-11–C-12 torsional coordinate. It was argued that repopulation of either ground state occurred exclusively from this region of the potential surface with a branching ratio of about 2:3 to 1:3 for bathorhodopsin and rhodopsin, respectively. The basis of the model was the observation that the quantum yields for the forward and back reactions summed to unity, implying that the same product distribution was obtained independent of the initial state of the pigment.

The recent report (Suzuki and Callender, 1981) that the quantum yields sum to a value (~ 1.2) slightly greater than unity requires that the available date be interpreted in terms of a more general description of the isomerization process (Miller and George, 1971; Warshel and Karplus, 1975; Birge and Hubbard, 1980). Following excitation to the excited state surface the chromophore twists to the common minimum (corresponding to the perpendicular configuration along the torsional coordinate) and vibrates in this potential well. There is a probability of crossing to the ground state for each pass through the minimum, but the crossing occurs so as to preserve angular momentum. That is, each pass originating from the cis side of the minimum can only produce a ground state trans conformer (with probability a in Fig. 3) and crossing from the trans side produce cis conformers with probability b (Fig. 3).

The model of Rosenfeld et al. (1977), in which thermal relaxation occurs in the excited state, implies that the crossing probability is quite small, so that many oscillations occur before a significant ground state population is accumulated (see, e.g., analysis in Suzuki and Callender, 1981). In this dynamical picture, the molecule has no "memory" of whether it was originally excited in a cis or a trans conformation. If, as it now appears (Suzuki and Callender, 1981), the quantum yields sum to greater than unity, the crossing probability must be relatively large; that is, the excited state must be significantly depleted following each pass through the minimum. Suzuki and Callender (1981) have analyzed the available data and find that values of $a = 0.38$ and $b = 0.30$ can reproduce the experimental results. Thus a relatively small number of torsional oscillations are made before the photochemistry is complete. The theoretical simulations (Fig. 5) of Birge and

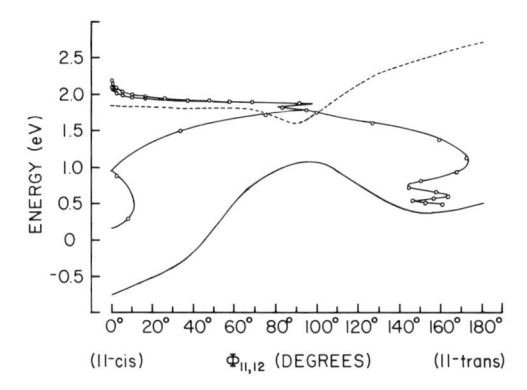

FIG. 5. Theoretical simulation of the molecular dynamics of cis–trans isomerization in rhodopsin. Two trajectories are shown in which the molecule reaches the activated complex in ~ 1 psec and oscillates with a torsional frequency of ~ 150 cm^{-1}. The lower right trajectory leads to a distorted transoid product (bathorhodopsin) in approximately the same time. Trajectories such as these are responsible for depletion of the S_1 surface, leaving a fraction of less than e^{-1} of the molecules in the excited state. The solid circles indicate trajectory increments of 0.1 psec. (From Birge and Hubbard, 1980, Fig. 11. Adapted with permission from *Journal of the American Chemical Society* **102**, 2195. Copyright 1980 American Chemical Society.) The potential energy curves (a solid line for the ground state and a dashed line for the excited state) are based on quantum mechanical calculations; note, however, the similarity to the curves deduced from experiments that are shown in Fig. 2.

Hubbard (1980) are in excellent agreement with this empirically derived result. These authors have obtained ~2-psec isomerization times which they believe to be overestimates. It appears then that the motion along a barrierless excited state potential energy surface provides a reasonable explanation of the observed photophysical properties of visual pigments. (See Chapter 13 for a detailed discussion of isomerization dynamics.)

C. PROTON TRANSFER PROCESSES

After excitation of either bovine rhodopsin at low temperature or bacteriorhodopsin at room temperature, Rentzepis and co-workers (Peters et al., 1977; Applebury et al., 1978) detected a transient absorption that appeared in less than the 6-psec resolution time of their apparatus. The transient state decayed to the bathorhodopsin (or K610) ground state via a temperature-dependent process that was resolvable below 20 K and was slowed down by deuterium exchange. It was suggested that the transient corresponded to the first singlet excited state, S_1, of the parent pigment, so that the measured absorption would have to arise from a transition to a higher singlet ($S_1 \rightarrow S_n$).

Recently it has been argued that the transient is not an excited state species but rather corresponds to a ground state precursor of bathorhodopsin (Aton et al., 1978; Honig et al., 1979b). About 33% of the excited rhodopsin molecules fail to form bathorhodopsin and simply repopulate their own ground states (Hurley et al., 1978). The picosecond data (Peters et al., 1977; Applebury et al., 1978; Monger et al., 1979) suggest that this process is complete in less than 6 psec at all temperatures, because no transient absorption changes corresponding to ground state repopulation have been detected at longer times. Thus, the observed transient (which, for example, decays at 10 K with a half-life of 29 psec) could not be the first singlet because the apparent decay kinetics of the first singlet (ground state repopulation in less than 6 psec) do not match those of the transient.

Further evidence suggesting that the transient absorbance change is due to a ground state process is based on the finding discussed above—that ground state repopulation occurs from the common minimum along the coordinate connecting rhodopsin and bathorhodopsin. Repopulation of both states from this common excited state requires that they appear in identical times, corresponding to the lifetime of the excited state. Thus, if the rhodopsin ground state is repopulated in less than 6 psec, it follows that the transient is also a ground state species (which has been called batho'; Honig et al., 1979b). Similar arguments may also be applied to bacteriorhodopsin where the appearance of a precursor (J625) which rises in ~1 psec (Ippen et al., 1978) and the relatively slow appearance of K610 (~11 psec at room temperature (Applebury et al., 1980) allow a somewhat more detailed analysis to be made (Dinur et al., 1981).

A sequence of events consistent with the available picosecond data is

$$\text{Rhodopsin} \xrightarrow{h\nu} \text{rhodopsin*} \rightarrow \text{batho}' \rightarrow \text{bathorhodopsin},$$

where the photochemical process is a cis–trans isomerization of the chromophore and the batho' → bathorhodopsin step is a deuterium-sensitive process involving a ground state relaxation of the protein induced by isomerization and may take from less than 6 psec to 36 psec, depending on the temperature. It is this relaxation which gives rise to the transient. [In bacteriorhodopsin, the corresponding sequence is bacteriorhodopsin → bacteriorhodopsin* → J625 → K610 (Dinur et al., 1981).] It should be pointed out that ground state proton transfer is a plausible consequence of the photoisomerization depicted in Fig. 4, because the charge separation should significantly alter pK values in the vicinity of the chromophore. Because Raman studies (Eyring and Mathies, 1979; Aton et al., 1980) show that the Schiff base proton does not change its state of protonation as a result of the primary photochemistry, another proton must be involved. The proton transfer processes depicted in Fig. 4 are meant to represent possible modes of ground state relaxation.

V. Concluding Remarks

Studies on pigment spectra have demonstrated that appropriately positioned charged or polar amino acids are responsible for producing the colors of visual pigments and bacteriorhodopsin. Protein-mediated electrostatic interactions also appear to play a central role in the light transduction mechanism in both pigment systems. While the possibility of electrostatic interactions playing a central role in mediating biological processes has long been recognized, there has been little direct evidence as to the relative importance and magnitude of the effects involved. The results described in this chapter demonstrate that they can be extremely large and, in fact, can dominate other interactions.

A second conclusion that may be derived from recent work on visual pigments is that proteins are capable of carrying out quite remarkable physical chemistry. This comes as no surprise, but it is perhaps useful to be reminded of this point, hence cautioned against the dangers of taking results from model systems as representative of proteins. Protonated Schiff bases of retinal in solution have not been made to absorb near 570 nm, and polyenes do not isomerize efficiently in their singlet manifold, particularly at low temperatures. Nevertheless, the placement of what is apparently a single charged group near the chromophore has been shown to produce enormous red shifts, while simple electrostatic factors also appear to produce a photochemical mechanism with no precedent in nonbiological systems. Understanding these and related phenomena in other systems constitute an extremely interesting area of biophysical research.

Acknowledgements

This work was supported in part by the NSF PCM 81-18088 and the NIH GM 30518.

References

Applebury, M. L., Peters, K. S., and Rentzepis, P. (1978). *Biophys. J.* **23**, 375–382.
Arnaboldi, M., Mottom., Tsujimoto, K., Balogh-Nair, U., and Nakanishi, K. (1979). *J. Am. Chem. Soc.* **101**, 7082–7084.
Aton, B., Callender, R. H., and Honig, B. (1978). *Nature (London)* **273**, 784–786.
Aton, B., Doukas, A. G., Callender, R. H., Dinur, U., and Honig, B. (1980). *Biophys. J.* **29**, 79–94.
Birge, R. R., and Hubbard, L. (1980). *J. Am. Chem. Soc.* **102**, 2195–2205.
Blatz, P., and Mohler, J. (1975). *Biochemistry* **14**, 2304–2309.
Busch, G. E., Applebury, M. L., Lamola, A., and Rentzepis, P. M. (1972). *Proc. Natl. Acad. Sci. U.S.A.* **69**, 2802–2806.
Callender, R., and Honig, B. (1977). *Annu. Rev. Biophys. Bioeng.* **6**, 33–55.
Cooper, A. (1979). *Nature (London)* **282**, 531–533.
Dinur, U., Honig, B., and Schulten, K. (1980). *Chem. Phys. Lett.* **72**, 493–497.
Dinur, U., Honig, B., and Ottolenghi, M. (1981). *Photochem. Photobiol.* **33**, 523–527.
Erying, G., and Mathies, R. (1979). *Proc. Natl. Acad. Sci. U.S.A.* **76**, 33–38.
Green, B., Monger, T., Alfano, R., Aton, B., and Callender, R. (1977). *Nature (London)* **264**, 1979–180.
Honig, B. (1978). *Annu. Rev. Phys. Chem.* **29**, 31–57.
Honig, B., Greenberg, A. D., Dinur, U., and Ebrey, T. G. (1976). *Biochemistry* **15**, 4593–4599.
Honig, B., Ebrey, T. G., Callender, R. H., Dinur, U., and Ottolenghi, M. (1979a). *Proc. Natl. Acad. Sci. U.S.A.* **76**, 2505–2507.
Honig, B., Dinur, U., Nakanishi, K., Balogh-Nair, V., Gawinowicz, M., Arnaboldi, M., and Motto, M. (1979b). *J. Am. Chem. Soc.* **101**, 7084–7086.
Hubbard, R. (1966). *J. Biol. Chem.* **241**, 1814–1818.
Hurley, J. B., Ebrey, T. G., Honig, B., and Ottolenghi, M. (1978). *Nature (London)* **270**, 540–542.
Ippen, E. P., Shank, C. V., Lewis, A., and Marcus, M. (1978). *Science* **200**, 1279–1281.
Labhart, H. (1957). *J. Chem. Phys.* **27**, 957–963.
Lewis, A., Spoonhower, J., Bogomolni, R. A., Lozier, R. H., and Stoeckenius, W. (1974). *Proc. Natl. Acad. Sci. U.S.A.* **71**, 4462–4466.
Miller, W., and George, T. (1971). *J. Chem. Phys.* **56**, 5637–5645.
Monger, T. G., Alfano, R. R., and Callender, R. H. (1979). *Biophys. J.* **27**, 105–116.
Nakanishi, K., Balogh-Nair, V., Arnaboldi, M., Tsujimoto, R., and Honig, B. (1980). *J. Am. Chem. Soc.* **102**, 7945–7947.
Oseroff, A. R., and Callender, R. (1974). *Biochemistry* **13**, 4243–4248.
Ottolenghi, M. (1980). *Adv. Photochem.* **12**, 97–200.
Peters, K., Applebury, M. L., and Rentzepis, P. M. (1977). *Proc. Natl. Acad. Sci. U.S.A.* **74**, 3119–3123.

Rosenfeld, T., Honig, B., Ottolenghi, M., Hurley, J., and Ebrey, T. G. (1977). *Pure Appl. Chem.* **49**, 341–351.
Salem, L. (1966). "Molecular Orbital Theory of Conjugated Systems." Benjamin, New York.
Stoeckenius, W. R., Lozier, R. H., and Bogomolni, R. A. (1979). *Biochim. Biophys. Acta* **505**, 215–278.
Suzuki, H., Komatsu, T., and Kitajima, H. (1974). *J. Phys. Soc. Jpn.* **37**, 177–185.
Suzuki, T., and Callender, R. H. (1981). *Biophys. J.* (in press).
Wald, G. (1968). *Science* **162**, 230–239.
Warshel, A., and Karplus, M. (1975). *Chem. Phys. Lett.* **32**, 11–14.
Yoshizawa, T. (1973). *In* "Biochemistry and Physiology of Visual Pigments" (H. Langer, ed.), pp. 169–181. Springer-Verlag, Berlin and New York.

CHAPTER 13

Simulation of the Primary Event in Rhodopsin Photochemistry Using Semiempirical Molecular Dynamics Theory

Robert R. Birge

Department of Chemistry
University of California
Riverside, California

I. Introduction	299
II. Theoretical Approaches in the Simulation of Intramolecular Dynamics	300
III. Semiempirical Molecular Dynamics Theory and Rhodopsin Photochemistry	302
A. Potential Surfaces of Isomerization	303
B. The Molecular Dynamics Formalism	306
C. Analysis of Photoisomerization Trajectories and Quantum Yields	309
IV. Comments and Conclusions	315
References	316

I. Introduction

The primary event of visual transduction is a cis → trans photoisomerization (for recent reviews, see Chapter 12 and Ottolenghi, 1980; Birge, 1981). One experimental observation that might appear to conflict with the concept of a photoisomerization is the finding that bathorhodopsin is formed at ambient temperatures in less than 6 psec following the absorption of a photon by rhodopsin (Busch et al., 1972). More recent measurements indicate that bathorhodopsin is formed in ∼3 psec (Green et al., 1977). This formation time is considered by some to be too short to accommodate a cis → trans isomerization of a molecule as large as retinal; in particular, experiments indicate that the protonated Schiff base of 11-*cis*-retinal exhibits nanosecond isomerization times in solution (Huppert et al., 1977). Furthermore, the quantum yield for bathorhodopsin formation ($\phi = 0.67$; Dartnall, 1972) is uncharacteristically large for polyene cis → trans photoisomerization quantum yields (Ottolenghi, 1980).

Recent semiempirical molecular dynamics calculations, however, theoretically predict that a one-bond 11-cis → 11-trans photoisomerization can occur in the opsin binding site in ~2 psec with a quantum yield of 0.62 (Birge and Hubbard; 1980; 1981). These calculations are based on hypothetical assumptions concerning the rhodopsin-binding site and simplifications of the formalistic treatment of the isomerization dynamics. A critical examination of these approximations is important before one accepts the predictions of our model as worthy of serious consideration.

This chapter reviews the theoretical procedures used in our semiempirical molecular dynamics formalism. We will first briefly introduce the reader to the general theory of molecular dynamics to provide some perspective on the computational methods. Subsequent sections discuss our semiempirical molecular dynamics procedures with a goal of critically examining some of the approximations inherent in our method. We will attempt to demonstrate what effect the principal approximations have on the calculated quantum yields and isomerization kinetics of the rhodopsin → bathorhodopsin photochemical transformation.

II. Theoretical Approaches in the Simulation of Intramolecular Dynamics

The theoretical treatment of intramolecular dynamic processes in molecules can be arbitrarily divided into two basic approaches. One approach generates the potential surface in a separate set of calculations and then uses the resulting surface to predict classically the motion of the nuclei along a specified reaction path. [We will use the term "reaction path" as a generic identification of the intramolecular processes (e.g., isomerization about a given bond).] The other approach calculates the forces on the nuclei at a given starting geometry by direct evaluation of the molecular Hamiltonian and incrementally updates the positions of the nuclei in response to the calculated forces, thereby following the trajectory until the reaction (e.g., isomerization) is completed. Accordingly, the first approach uses a static potential surface, and the latter a dynamic potential surface, to simulate the reaction trajectory.

The static approach has the advantage that a potential surface need be calculated only once. Furthermore, the number of potential surface calculations can be judiciously selected by assuming that the reaction will follow the path of steepest descent (Miller et al., 1980). This reaction path is obtained by starting at some initial position on the potential surface (typically a saddle point) and evaluating the gradient vector $\mathbf{u}$ defined for atom i as,

$$\mathbf{u}_i = \frac{-c}{\sqrt{m_i}} \left(\frac{\partial V}{\partial x_i} \hat{i} + \frac{\partial V}{\partial y_i} \hat{j} + \frac{\partial V}{\partial z_i} \hat{k} \right) = \frac{-c}{\sqrt{m_i}} \nabla_i V, \qquad (1)$$

where c is a constant which normalizes the $3N$-dimensional vector $\mathbf{u}$ to unity, m_i is the mass of atom i, and V is the potential energy at positions x_i, y_i, and z_i.

The rationale for representing the reaction path as the path of steepest descent can be understood by evaluating Newton's second law which states that the motion of the ith atom obeys the equation

$$\mathbf{F}_i = m_i \mathbf{a} = m_i \frac{\partial^2 \mathbf{r}_i}{\partial t^2}, \tag{2}$$

where $\mathbf{F}_i$ is the force acting on the ith atom and $\mathbf{a}$ is the acceleration of the atom (the second derivative of the position vector $\mathbf{r}_i$ as a function of time). If the forces on the atoms are conservative, then we can assume that the forces are given by the potential gradient,

$$\mathbf{F}_i = -\nabla_i V. \tag{3}$$

Equations (2) and (3) combine to yield a set of $3N$ second-order differential equations (Porter and Raff, 1976),

$$\frac{d}{dt}\left(\frac{\partial T}{\partial x_i}\right) + \frac{\partial V}{\partial x_i} = 0, \tag{4a}$$

$$\frac{d}{dt}\left(\frac{\partial T}{\partial y_i}\right) + \frac{\partial V}{\partial y_i} = 0, \tag{4b}$$

$$\frac{d}{dt}\left(\frac{\partial T}{\partial z_i}\right) + \frac{\partial V}{\partial z_i} = 0, \tag{4c}$$

where T, the kinetic energy of the system, is given by

$$T = \frac{1}{2} \sum_{i=1}^{N} m_i \left[\left(\frac{\partial x_i}{\partial t}\right)^2 + \left(\frac{\partial y_i}{\partial t}\right)^2 + \left(\frac{\partial z_i}{\partial t}\right)^2\right]. \tag{5}$$

Solution of the second-order differential Eqs. (4a)–(4c) generates the trajectories, and if the molecule is started from a rest position, the trajectories will seek a minimum energy along the path of steepest descent.

One problem inherent in the use of the static approach is that changes in initial conditions will frequently require the calculation of a new reaction path. Furthermore, if the molecule starts out with a large amount of kinetic energy, a priori prediction of the reaction path is often computationally difficult. Wang and Karplus (1973) and Warshel (1976) have shown that the dynamic approach can provide insights into reaction mechanisms that are not obvious from evaluation of static reaction paths.

The above discussion suggests that the dynamic approach is preferable for simulating intramolecular dynamics because of its inherent generality. In particular, one is never required to specify a reaction path because the calculation generates its own reaction path in the course of determining the trajectory. However, the computer time required to carry out a simulation of polyatomic intramolecular dynamics is prohibitive if one is using a sophisticated SCF-MO quantum mechanical Hamiltonian. The following example serves to illustrate the problem.

The accurate simulation of dynamic processes typically requires the use of very small time increments (on the order of 10^{-15} sec). At each time increment, the eigenvalues and eigenvectors of the SCF-MO Hamiltonian must be calculated and the gradient matrix evaluated. One such calculation for an unsubstituted hexaene polyene using an INDO-CISD semiempirical MO formalism (see below) takes >25 min on a Cray computer (the fastest commercially available computer) and involves $>10^{12}$ floating point operations. The simulation of a 2-psec photoisomerization would require about a month of CPU time on a Cray computer and would cost $\sim \$1,200,000$. It would be difficult to convince a granting agency to fund a calculation of this magnitude.

In contrast, the static reaction path approach discussed in the following section requires only ~ 6 hrs of Cray CPU time to generate INDO-CISD ground and excited state hexaene isomerization surfaces and ~ 15 min of CPU time per photoisomerization trajectory. Although our approach is less accurate than the dynamic surface approach, it is computationally tractable and can provide useful insights into the dynamics of photoisomerization.

III. Semiempirical Molecular Dynamics Theory and Rhodopsin Photochemistry

Our molecular dynamics procedures are based on the static reaction path approach. Our formalism is semiempirical because we use semiempirical MO theory and consistent force field methods to determine the ground and excited state potential energy surfaces. A number of semiempirical approximations are also introduced in the molecular dynamics procedures.

Determination of the kinetics and quantum yields of photoisomerization is divided into two separate calculations. The first part involves determination of the potential energy surface for isomerization in terms of a single reaction path variable. The primary event of rhodopsin photochemistry is a 11-cis → 11-trans photoisomerization of the 11-cis protonated Schiff base chromophore. Accordingly, the reaction path variable is the 11,12-dihedral angle ($\phi_{11,12}$), although other selected geometric variables are allowed to change in response to changes in the $\phi_{11,12}$ coordinate (see below). The second part of the procedure calculates a classical trajectory of isomerization along the reaction path determined in the previous step.

A. POTENTIAL SURFACES OF ISOMERIZATION

The retinal chromophore in rhodopsin is covalently bound to the opsin protein via a protonated Schiff base linkage to a lysine residue (Oseroff and Callender, 1974). If we assume that the β-ionylidene ring is trapped in a hydrophobic cleft and therefore does not move significantly during the photoisomerization, the lysine residue must change its conformation to accommodate the isomerization. Our hypothetical model of the active site is shown at the top of Fig. 1. The specific details of our model of the active site will be summarized below. It is clear from analysis of Fig. 1 that our reaction path for isomerization must include contributions from the conformational energy of the lysine residue, as well as the conformational energy of the polyene system, as a function of the 11,12-dihedral angle.

We calculated the conformational energy of the polyene system by using INDO-CISD all-valence electron MO theory. The details and rationale of our procedures have been discussed previously (Birge and Hubbard, 1980). One important aspect of our MO procedures is the inclusion of both single and double excitation configuration interaction (CI) (typically ~ 130 single and ~ 350 double excitations). The importance of including double CI for calculating the excited state properties of the visual chromophores cannot be overemphasized (Birge et al., 1975, 1978; Birge and Pierce, 1979; Birge and Hubbard, 1980). If double CI were not included, the calculated excited state potential surface would not be barrierless, and it is this barrierless excited state surface which is directly responsible for the picosecond isomerization kinetics (see below).

Since a calculation of this type would be computationally intractable if the entire molecular framework of the retinyl chromophore were included, we approximate the PSB of retinal using a $C_{11}NH_{14}$ moiety. A standard geometry was assumed for all the calculations [$R_{C=C} = 1.35$ Å, $R_{C-C} = 1.46$ Å, $R_{C-H} = 1.08$ Å, $R_{C=N} = 1.296$ Å, $R_{N-H} = 1.20$ Å, all bond angles 120°, all dihedral angles planar trans (180°) except $\phi_{6,7}$, which is s-cis (45°), and $\phi_{11,12}$, which is variable]. (It should be noted that the trajectory calculations include the inertial effects of the complete molecular framework of the chromophore, including the β-ionylidene ring as well as the hydrocarbon chain of the lysine residue.)

Accordingly, the conformational energy of the polyene portion has not been adiabatically minimized along the $\phi_{11,12}$ reaction path. This observation is very important and suggests that we will overestimate isomerization times because we will underestimate the downward slope of the reaction path toward $\phi_{11,12} = 90°$. We have carried out INDO-CISD calculations which indicate that a partial adiabatic mapping of the orthogonal system will drop the energy of the lowest excited singlet to within ~ 0.1 eV of the ground state potential surface at $\phi_{11,12} = 90°$ (R. R. Birge, B. M. Pierce, and

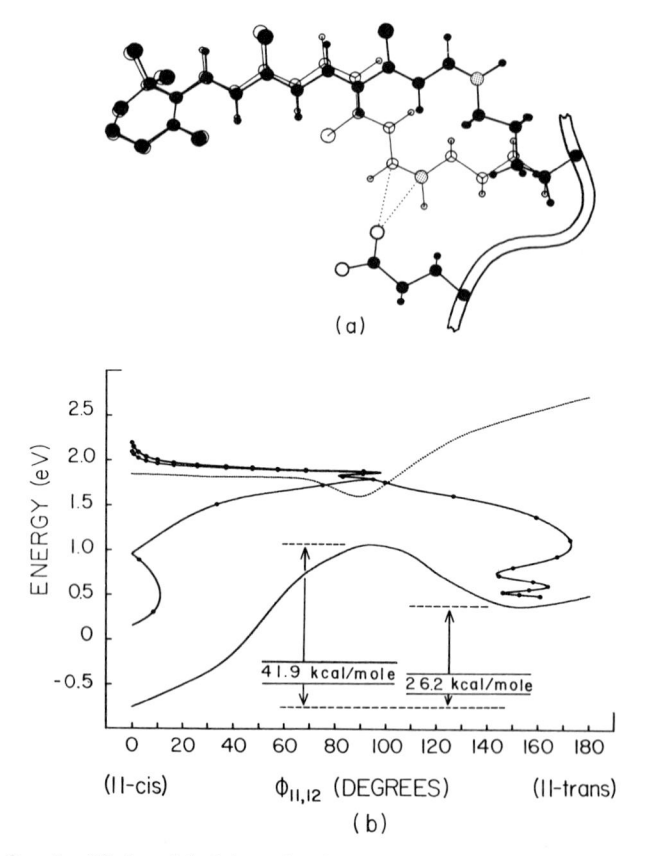

FIG. 1. Our simplified model of the active site of rhodopsin (a) and the resultant potential surfaces in the ground and first excited singlet states as a function of the C-11–C-12 dihedral angle (b). Trajectories associated with cis → trans isomerization in rhodopsin are shown in (b). The excited state trajectory enters the activated complex in ~1.1 psec and oscillates with an average frequency of torsional motion of ~4.5 × 10^{12} Hz (~150 cm^{-1}). The lower-left trajectory leads to the starting geometry (rhodopsin) in 1.8 psec; the lower-right trajectory leads to the isomerized product (bathorhodopsin) in 2.2 psec. These trajectories, along with the two torsional oscillations through $\phi_{11,12} = 90°$ which precede them, are responsible for depleting the $S_1(\pi\pi^*)$ surface, leaving a fraction of less than e^{-1} (0.37) of the molecules in the excited state. The solid circles indicate trajectory increments of 0.1 psec. The two different excited state trajectories differ in excitation energy by 0.1 eV but coalesce before reaching the activated complex to produce wavelength-independent isomerization times and quantum yields. The lower-left trajectory actually continues beyond the range of the horizontal axis to $\phi_{11,12} = -11°$ but is shown reflected back toward positive dihedral angles for convenience. The cis → trans isomerization is accomplished without significantly disturbing the β-ionylidene ring. The counterion consists of the carboxylate group of a glutamic acid residue. One of the oxygen atoms of the carboxylate group is placed 3 Å (dotted lines) from the C-15 carbon and the N-16 imino nitrogen atoms. The ribbon connecting the lysine and glutamic acid residues indicates schematically that both residues are attached to the same protein backbone. (From Birge and Hubbard, 1980. Adapted with permission from *Journal of the American Chemical Society* **102**, 2195, Copyright 1980 American Chemical Society.)

L. M. Hubbard, unpublished data). This observation has important implications for calculating the quantum yields of photoisomerization (see below).

The conformational energy of the lysine residue was calculated using the empirical potential functions proposed by Lifson and Warshel (1968, 1970). The conformation of the lysine residue in rhodopsin was arbitrarily assumed to be that which yielded the minimum force field free energy with no external constraints (see top of Fig. 1). The conformation of the lysine residue as a function of 11,12-dihedral angle was determined by force field minimization with the constraint that only the first four carbon atoms, and their associated hydrogens, were allowed to move in response to motion of the polyene moiety. (See Fig. 1 and note that the lysine carbon atoms are counted starting at the polyene end.) The fact that a "no constraint" minimization of the lysine residue was performed for the planar 11-*cis*-retinyl geometry will probably bias the calculation to overestimate the "true" conformational energy of the lysine residue in bathorhodopsin. This bias will also tend to increase the calculated isomerization times slightly.

We summarize our binding site assumptions below.

(1) The retinal chromophore is covalently bound to the opsin active site via a protonated Schiff base linkage to a lysine residue of the protein (see Fig. 1).

(2) The β-ionylidene ring is trapped in a hydrophobic cleft. During the isomerization, no atom of the β-ionylidene ring can move by more than 0.05 Å and the center of mass must remain fixed.

(3) The photochemical isomerization from the 11-cis to the 11-trans conformation is accomplished entirely as a one-bond rotation about the 11,12-bond. All other internal degrees of freedom of the chromophore are fixed at the original conformation of the chromophore [6-s-cis ($\phi_{6,7} = 45°$), 12-s-trans].

(4) The first four carbon atoms of the hydrocarbon portion of the lysine residue, and the hydrogens bonded to these carbon atoms, are allowed to seek their minimal energy conformation during the isomerization process. No other distortions of the protein are allowed.

(5) A single counterion consisting of the carboxylate group of a glutamic acid residue is placed near the N=C-15 group as shown in Fig. 1. The counterion is held fixed in space during isomerization, and the electrostatic stabilization lost due to charge separation is not corrected for protein dielectric effects. These assumptions will tend to overestimate the contribution of charge separation to the calculated bathochromic shift. The calculated bathochromic shift is, in fact, overestimated ($\Delta\lambda_{calc} = 110$ nm, $\Delta\lambda_{obs} = 43$ nm). An additional source of error in our treatment of the active site is associated with our neglect of the second counterion (Arnaboldi *et al.*, 1979;

Honig et al., 1979a,b). The second counterion is believed to be above the plane of the polyene in the vicinity of atoms C-12 to C-14 (see Chapter 12, Fig. 2). A counterion in this location may stabilize the orthogonal ($\phi_{11,12} = 90°$) conformation more than the 11-cis ($\phi_{11,12} = 0°$) conformation because electronic reorganization during isomerization shifts electron density out of this region of the molecule [q(C-12) + q(C-13) + q(C-14) = -0.011 (S_1, $\phi_{11,12} = 0°$), $+0.385$ ($S_1, \phi_{11,12} = 90°$); see Birge and Hubbard, 1980, Fig. 1)]. However, the precise location of this counterion relative to the motion of the chromophore during isomerization is not known, and the second counterion could increase or decrease the excited state barrier to isomerization.

While the above assumptions are for the most part hypothetical, the resulting potential surfaces (Fig. 1) agree reasonably well with the available experimental data. A comparison of various observed parameters with corresponding values calculated based on the potential surfaces of Fig. 1 indicates that all the spectroscopic and thermodynamic trends are correctly predicted. In particular, the calculated increases in λ_{max} and oscillator strength in going from rhodopsin to bathorhodopsin are in reasonable agreement with experiment [λ_{max}(rhod) = 475 nm (498 obs), f(rhod) = 0.87 (0.8 obs), λ_{max}(batho) = 585 nm (540 obs), f(batho) = 0.98 (0.9 obs)]. The potential energy difference between bathorhodopsin and rhodopsin is calculated to be 26 kcal/mole (Fig. 1), which is approximately 9 kcal/mole too small based on recent calorimetric measurements (Cooper, 1979).

B. The Molecular Dynamics Formalism

The intramolecular dynamics of photoisomerization are classically determined by solving Newton's equation of motion [Eqs. (4) and (5)] using the potential energy surfaces shown in Fig. 1 to calculate the gradients in Eq. (4). The inertial effects of the entire molecular system shown at the top of Fig. 1 are included (note, however, that the counterion remains fixed during isomerization). As previously discussed, our potential surfaces are not adiabatic and therefore the true barrier is probably more negative than that depicted in Fig. 1. Therefore, the gradient of the potential energy driving the isomerization will be underestimated. We have chosen to parametrize the other equations in our molecular dynamics formalism to overestimate the trajectory time also. Accordingly, our procedures will probably calculate an upper limit to the "true" isomerization time, where we define the true isomerization time in reference to that which would be calculated using a formalism including all the degrees of freedom.

The formalism places restraints on the energy available to the torsional motion by limiting the sources of the torsional kinetic energy to that provided by the potential surface. Accordingly, "excess" vibrational energy associated with the other vibrational degrees of freedom is prevented from partitioning into the torsional kinetic energy. In contrast, an efficient pathway is provided for the transfer of torsional kinetic energy into "nonproductive" vibrational modes based on a vibrational continuum approximation given by

$$(\Delta E_{kin}/\Delta t)_{t=\tau} = -C_m E_{kin}^2(\tau) \quad (E_{kin} > h\nu_1), \tag{6}$$

where $(\Delta E_{kin}/\Delta t)_{t=\tau}$ is the rate of loss of torsional kinetic energy at trajectory time τ, $E_{kin}(\tau)$ is the torsional kinetic energy at $t = \tau$, and $h\nu_1$ is the energy of the lowest vibrational mode capable of scavenging torsional kinetic energy (see below). C_m is a semiempirical constant which is equal to the vibrational coupling efficiency. An approximate (classic) upper limit to C_m can be obtained by assuming that a single vibrational quantum transition is present in the molecule, which is exactly equal in energy to the instantaneous torsional kinetic energy, and that this vibrational mode can scavenge all the torsional kinetic energy at a rate equal to the frequency of the vibration. This assumption yields an upper limit of

$$C_m \leq h^{-1}, \tag{7}$$

where h is Planck's constant. Molecular dynamics calculations including explicit vibrational force fields have been used to calibrate C_m. Equation (7), in conjunction with Eq. (6), is found to overestimate the magnitude of vibrational scavenging by factors of 2–10. However, the continuum model embodied in Eq. (6) will invariably overestimate coupling when the torsional kinetic energy is lower than the energy of the lowest mode available for vibrational scavenging. Accordingly, $(\Delta E_{kin}/\Delta t)$ is set equal to zero whenever E_{kin} is less than the energy of this lowest mode.

The trajectory calculations presented here are designed to err in the direction of overestimating trajectory times. Accordingly, we have chosen a relatively large value for C_m of $\frac{1}{2}$ hr^{-1} and a relatively small $h\nu_1$ of 50 cm^{-1}. We are confident that these parameters will overestimate vibrational scavenging of the torsional kinetic energy, but calculations including all degrees of freedom would be required to test this point. Furthermore, the possibility of reverse coupling whereby vibrational energy is transferred back to the torsional degree of freedom is ignored in our trajectory calculation.

The experimental observation of a wavelength-independent quantum yield for rhodopsin isomerization indicates that excess vibrational energy of the chromophore (E_{vib}) is rapidly transferred to the protein matrix. We simulate this process using a density of states approximation,

$$\left(\frac{\Delta E_{vib}}{\Delta t}\right)_{t=\tau} = -\left|\frac{\Delta E_{vib}}{\Delta t}\right|_{avg} (1 - e^{-\rho[E_{vib}(\tau) - kT]}) - \left(\frac{\Delta E_{kin}}{\Delta t}\right)_{t=\tau}, \tag{8}$$

where $(\Delta E_{\text{vib}}/\Delta t)_{t=\tau}$ is the rate of vibrational relaxation at trajectory time τ, $|\Delta E_{\text{vib}}/\Delta t|_{\text{avg}}$ is the absolute (exponential) average vibrational relaxation rate, ρ is the density of states factor, $E_{\text{vib}}(\tau)$ is the vibrational energy at trajectory time τ, k is the Boltzmann constant, T is the temperature (300 K), and $(\Delta E_{\text{kin}}/\Delta t)_{t=\tau}$ is calculated using Eq. (6). Equation (8) indicates that vibrational energy is lost through relaxation processes but gained through the transfer of torsional kinetic energy into the vibrational manifold. The salient parameters were adjusted to produce a wavelength-independent quantum yield for isomerization yielding $|\Delta E_{\text{vib}}/\Delta t|_{\text{avg}} = 2$ eV/psec and $\rho = 1$ eV^{-1} (Birge and Hubbard, 1980).

A theoretical prediction of quantum yields for isomerization based on our trajectory analysis can be obtained by calculating the probability of crossing into the ground state $a_0^2(\tau)$ as a function of the trajectory time τ. A time-dependent quantum mechanical treatment (Miller and George, 1972; Warshel and Karplus, 1975; Weiss and Warshel, 1979) yields

$$a_0^2(\tau) = \left\{ \int_0^\tau \left[\left\langle \psi_0 \left| \frac{\partial \psi_1}{\partial t} \right\rangle a_1(t) \exp\left(-\frac{i}{\hbar} \int_0^t \Delta E_{10} dt' \right) \right] dt \right\}^2, \qquad (9)$$

where ψ_0 and ψ_1 are the electronic wave functions of the ground and excited states and ΔE_{10} ($= E_1 - E_0$) is the time-dependent potential energy difference between the excited state and the ground state. Unfortunately, solution of the nonadiabatic coupling function $\langle \psi_0 | \partial \psi_1/\partial t \rangle$ is not possible within the confines of our semiempirical trajectory formalism without the introduction of serious *ad hoc* assumptions. Accordingly, we calculate $a_0^2(\tau)$ using a semiclassical solution to Eq. (9):

$$a_0^2(\tau) = \exp -\left\{ \left[\frac{4 \Delta W(\tau)}{3\hbar} \right] \left[\frac{2 \Delta W(\tau)}{(\partial^2 \Delta E_{10}/\partial t^2)_{t=\tau}} \right]^{1/2} \right\}, \qquad (10)$$

where $\Delta W(\tau)$ is the adiabatic potential energy difference between the ground and excited states adjusted with respect to the Born–Oppenheimer local minimum (Birge and Hubbard, 1980):

$$\Delta W(\tau) = \Delta E_{10}(\tau) - \Delta E_{10}(lm) + \tfrac{1}{4} E_{\text{vib}}(\tau), \qquad (11)$$

where $\Delta E_{10}(lm)$ is the difference in the excited state and ground state potential energy surfaces at their minimum energy separation ($\phi_{11,12} = 90°$). The purpose of using $\Delta W(\tau)$ rather than $\Delta E_{10}(\tau)$ in our calculation of ground state crossing probabilities is to introduce coordinate relaxation processes into our model semiempirically. Note that the potential surfaces shown in Fig. 1 were calculated assuming only one degree of freedom for the polyene coordinates ($\phi_{11,12}$), and that all other internal degrees of freedom of the polyene were held fixed at the original (ground state) values. It is reasonable to assume, however, that equilibration of the other degrees of freedom of the polyene residue will occur following excitation and that this process will be roughly proportional in time to the rate of vibrational energy loss to the protein matrix. In other words, as the polyene distributes the excess vibrational energy to the protein, the polyene simultaneously relaxes to a minimum excited state energy so that the difference in energy between the ground and excited state potential surfaces decreases. Equation (11) naively predicts that the S_1 and S_0 surfaces will converge when $E_{vib} = 0$, but in actuality the vibrational energy is calculated to decrease to no less than 0.07 eV (~ 560 cm^{-1}) above the zero-point energy prior to complete ($>99.9\%$) internal conversion to the ground state. Recent calculations of the first excited singlet state energy of the protonated Schiff base of retinal using a partial gradient minimization of the excited state geometry predict that the S_1 and S_0 surfaces will approach an energy difference of ~ 0.1 eV (R. R. Birge, B. M. Pierce, and L. M. Hubbard, unpublished data). Accordingly, the assumptions inherent in the application of Eq. (11) are justified. However, our assumption that the S_0 and S_1 potential surface separation is proportional to $\frac{1}{4}E_{vib}$ is empirical and is based on the expectation that coordinate relaxation is roughly four times slower than radiationless decay for the present system.

The calculations reported in this chapter were carried out using trajectory increments of 10^{-15} sec. At each time increment, the classic (Newtonian) equations of motion are solved in double precision (16 significant digits) using five-point Lagrangian interpolation formulas to calculate the energy of the potential surface at $\phi_{11,12}(\tau)$ and the gradients of force and the temporal derivatives that appear in Eqs. (4)–(6), (8), and (10) (Birge and Hubbard, 1981). Test calculations were performed using smaller (10^{-16} sec) and larger (10^{-14} sec) increments in time to determine the effect of changes in this arbitrary parameter on the calculated properties. Calculations using increments of 10^{-15} and 10^{-16} sec produced results which agreed to within four significant digits. Although an increment of 10^{-14} sec appears to be adequate for calculations of trajectory times, this interval is too coarse to calculate $a_0^2(\tau)$ accurately because of calculational error associated with evaluation of the $\partial^2 \Delta E_{10}/\partial t^2$ term in Eq. (10).

C. ANALYSIS OF PHOTOISOMERIZATION TRAJECTORIES AND QUANTUM YIELDS

Isomerization trajectories following excitation into the first excited singlet states of rhodopsin and bathorhodopsin are shown in Figs. 1 and 2. The molecules are promoted into the excited state from a rest position with an excess vibrational energy (E_{vib}) of 0.25 or 0.35 eV. This excess energy is

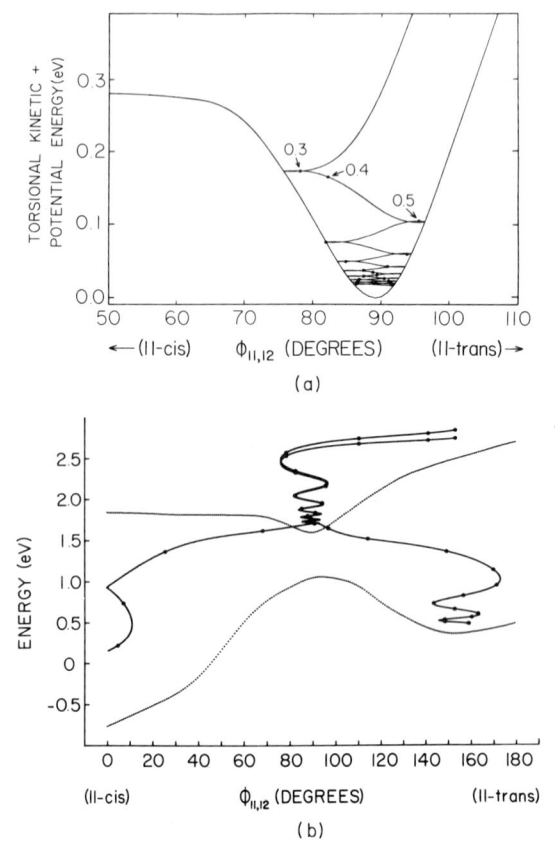

FIG. 2. Molecular dynamics of trans → cis isomerization in bathorhodopsin. (a) Total energy associated with the torsional degree of freedom as a function of dihedral angle ($\phi_{11,12}$) in the vicinity of the excited state activated complex. The solid circles represent trajectory increments of 0.1 psec, and trajectory times of 0.3, 0.4, and 0.5 psec following excitation are indicated. Note that the first trajectory passing through $\phi_{11,12} = 90°$ is trapped in the activated complex because the torsional kinetic energy is rapidly transferred to other vibrational degrees of freedom and the remaining kinetic energy is insufficient to override the small barrier associated with the activated complex. (b) Molecular dynamics of trans → cis isomerization in bathorhodopsin. The excited state trajectory enters the activated complex in ~0.3 psec and oscillates with an average frequency of torsional motion of ~4.6 × 10^{12} Hz (~155 cm^{-1}). The lower-left trajectory leads to the isomerized product (rhodopsin) in 1.8 psec; the lower-right trajectory leads to the starting geometry (bathorhodopsin) in 2.2 psec. These trajectories, along with the nine torsional oscillations through $\phi_{11,12} = 90°$ which precede them, are responsible for depleting the $S_1(\pi\pi^*)$ surface, leaving a fraction of less than $e^{-1}(0.37)$ of the molecules in the excited state. (From Birge and Hubbard, 1981.)

rapidly dissipated so that the molecules enter the potential well at $\phi_{11,12} = 90°$ with virtually identical vibrational energy regardless of initial energy. Because the dynamics of internal conversion to the ground state are associated entirely with the torsional behavior in the potential well, the calculated quantum yields of isomerization are very insensitive to the amount of initial excess vibrational energy. The calculations therefore reproduce the experimentally observed wavelength independence of the quantum yields. It should be noted that we adjusted the parameterization of Eq. (8) to produce this wavelength independence, so the above observations are not indicative of successful theoretical prediction but rather successful parameterization.

An important limitation of our theoretical treatment of bathorhodopsin is our neglect of the 9,10-torsional coordinate. Accordingly, excitation of bathorhodopsin is falsely predicted to produce only rhodopsin (11-*cis*) or bathorhodopsin (11-*trans*) but no isorhodopsin (9-*cis*). This limitation is not a serious defect in simulating cattle pigment photochemistry, because the observed quantum yield of the formation of isorhodopsin from bathorhodopsin is only 0.054 (Suzuki and Callender, 1981).

Figures 1 and 2 show only two of the many trajectories associated with depopulation of the excited state manifold and repopulation of the ground state. Analyses of ground state repopulation dynamics for the photoisomerizations of rhodopsin and bathorhodopsin are presented in Fig. 3. We define the "isomerization time" as the time required to repopulate the "relaxed" ground state leaving a fraction of less than $e^{-1}(0.37)$ molecules in the excited state. Examination of Fig. 3 indicates that only four torsional oscillations (in the activated complex) are necessary to return $>63\% (1 - e^{-1})$ of the excited molecules to the ground state during 11-cis → 11-trans isomerization of rhodopsin. The isomerization time is calculated to be 2.19 psec. In contrast, 11 torsional oscillations are necessary to return 63% of the excited molecules to the ground state during 11-trans → 11-cis isomerization of bathorhodopsin (Fig. 3). The reason for this significant difference is analyzed below, and it is surprising to observe that the net isomerization time is calculated to decrease slightly to 1.82 psec. In summary, therefore, our calculations predict that isomerization of rhodopsin and bathorhodopsin occurs in ~2 psec following excitation.

The quantum yield of isomerization is calculated by statistically analyzing the probabilities of trajectory splitting into the ground state manifold. As shown in Fig. 3, graphs of the probability of crossing into the ground state as a function of time display sharply peaked maxima. These maxima correspond to trajectory passes through the $\phi_{11,12} = 90°$ dihedral angle coordinate along the potential surfaces of Figs. 1 and 2. The difference in the ground and excited state potential energy is minimized at $\phi_{11,12} = 90°$ (ΔW is small), and the surface curvature and velocity are maximized at $\phi_{11,12} = 90°$

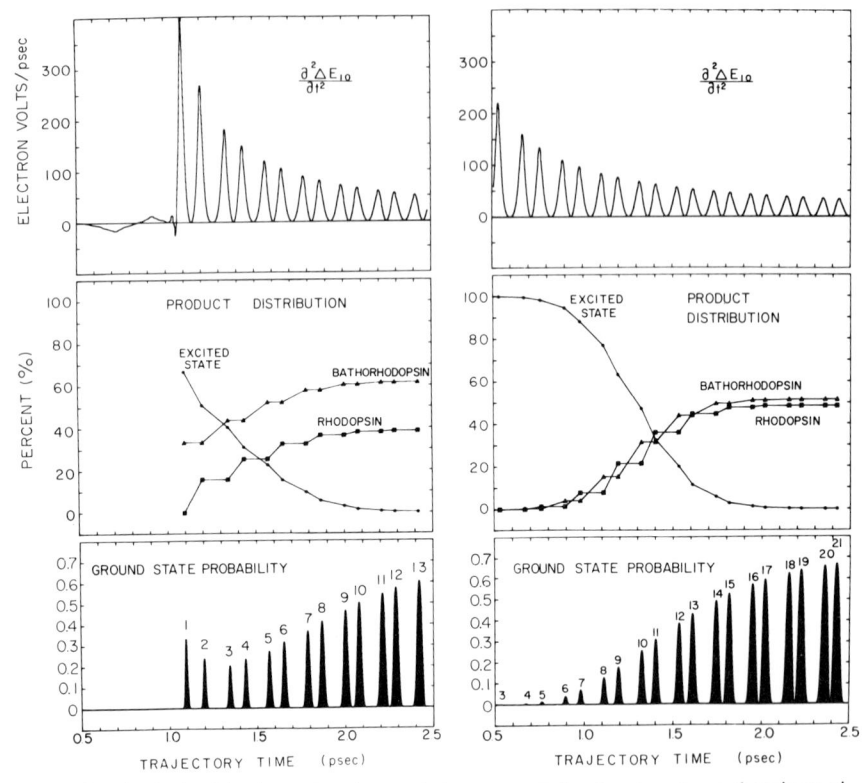

FIG. 3. Analysis of the dynamics of ground state repopulation for cis → trans photoisomerization in rhodopsin (left three figures) and trans → cis photoisomerization in bathorhodopsin (right three figures). The ground state product distribution as a function of time following excitation is shown in the plots marked "product distribution." Quantum yields of photoisomerization calculated by reference to these plots yield rhodopsin → bathorhodopsin ($\phi = 0.62$, left middle) and bathorhodopsin → rhodopsin ($\phi = 0.48$, right middle). The probability of crossing into the ground state $a_0^2(\tau)$ and the second derivative of the potential energy difference between the ground and excited states are plotted as a function of trajectory time in the bottom two and top two figures, respectively. (From Birge and Hubbard, 1981.)

($\partial^2 E_{10}/\partial t^2$ is maximum; see Fig. 3). Evaluation of Eq. (10) indicates that a small ΔW and a large $\partial^2 E_{10}/\partial t^2$ produce a maximum probability of crossing into the ground state. The sharpness of the probability function a_0^2 at $\phi_{11,12} = 90°$ is primarily due to the high curvature of the excited state potential well at the orthogonal dihedral angle.

A calculation of the quantum yield of rhodopsin photoisomerization proceeds as follows. The first trajectory following excitation enters the "activated complex" in approximately 1 psec and arrives at $\phi_{11,12} = 90°$ in

1.098 psec (Fig. 3). The probability of crossing into the ground state during this first pass is 0.333, which means that ~33% of the molecules cross into the ground state to produce isomerized products in ~1.9 psec. The remaining 67% of the molecules continue on the excited state trajectory which is now "trapped" in the activated complex in the low-frequency (~150 cm^{-1}) torsional mode of the 11,12-dihedral angle. The observation that the molecular dynamics calculations predict a low-lying torsional vibrational mode of ~150 cm^{-1} can be used to verify the accuracy of our computational procedures. We calculated the vibrational energy levels of the excited state torsional potential surface by fitting the potential surface to a 144-point Fourier series and then expanding the Hamiltonian in a basis set of a free internal rotor. The results are shown in Fig. 4 and indicate that the lowest-lying eigenfunctions are separated by 163 cm^{-1} (1 ← 0), 157 cm^{-1} (2 ← 1), 151 cm^{-1} (3 ← 2), and 145.9 cm^{-1} (4 ← 3). After two passes through the 90° region, our molecular dynamics procedures predict that the torsional energy $[E_{kin}(\tau) + E_s(\phi)]$ is ~400 cm^{-1} which corresponds approximately to the third vibrational eigenfunction ($v = 2$, Fig. 4) and a vibrational frequency of 151 cm^{-1} in good agreement with the frequency classically predicted. The second pass through the 90° region occurs at 1.2 psec, and the probability of crossing into the ground state is 0.238. Accordingly, ~24% of the remaining excited state molecules undergo internal conversion to produce ~16% unisomerized product in ~1.6 psec. Torsional oscillations in the activated complex continue to pass through the 90° region transferring molecules to the ground state. After 14 torsional oscillations, >99.9% of the excited state molecules have crossed into the ground state to produce 61.5% isomerized product (bathorhodopsin) and 38.5% unisomerized starting material (rhodopsin). The above statistical analysis predicts a quantum yield of isomerization of 0.62 which is in reasonable agreement with the observed value of 0.67 (Dartnall, 1972).

Analysis of the quantum yield of bathorhodopsin photoisomerization is also presented in Fig. 3. Following excitation the molecule reaches the activated complex in less than 0.3 psec because of the steep slope of the 11,12-torsional potential surface. [Our simplified potential surface neglects partitioning torsional relaxation into distortion of the 9,10-double bond, and therefore rhodopsin (11-*cis*) is the only photochemical product theoretically predicted.] The molecule enters the activated complex with a considerable excess of vibrational energy (Fig. 2), and this excess energy prevents the first few trajectory passes through the 90° region from efficiently crossing into the ground state (ΔW is large). A total of 11 passes through the 90° dihedral angle are required before >63% of the excited state molecules internally convert to the ground state manifold to produce product in ~1.8 psec (Fig. 3). After 18 passes, 99.8% of the molecules are in the ground state, 48.4%

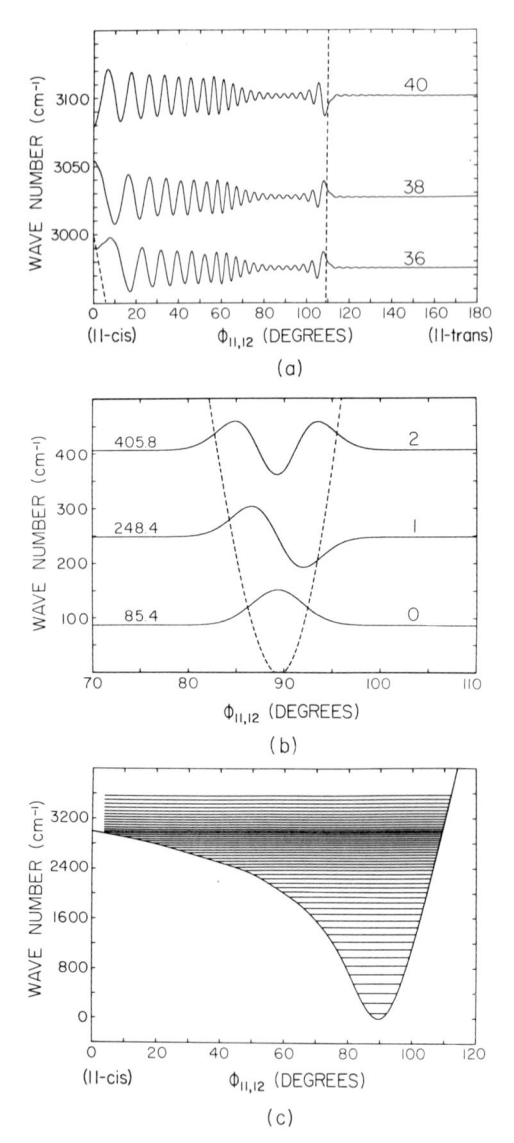

FIG. 4. The vibrational energy levels of the excited state $\phi_{11,12}$ torsional mode of the polyene–lysine system calculated by fitting the excited state potential surface shown in Fig. 1 to a 144-point Fourier series and then expanding the Hamiltonian in a 25-term free rotor basis set. The potential surface was smoothed slightly to minimize the number of terms that had to be included in the basis set. Consequently, the surfaces shown here do not correspond exactly to those used in the molecular dynamics calculations (Figs. 1 and 2). Independent calculations of the vibrational energy spacings provide a useful test of the molecular dynamics calculations and serve to indicate whether or not the classical equations of motion are correctly treated in the computer simulation (see text). We are currently developing a semiempirical molecular dynamics formalism that explicitly includes the quantization of the torsional eigenfunctions and the effect of Franck–Condon factors on the dynamics of internal conversion to the ground state. (From Birge et al., 1982.)

in the 11-cis conformation (rhodopsin), and 51.4% in the original 11-transoid conformation (bathorhodopsin). The calculated quantum yield is therefore 0.48, in good agreement with the observed value of 0.5 (Suzuki and Callender, 1981).

The trajectory associated with the bathorhodopsin → rhodopsin transformation appears somewhat anomalous in that the first pass through $\phi_{11,12} = 90°$ is trapped in the activated complex region even though it appears that the trajectory "energy" is significantly higher than the barrier height of the local potential well. The confusion is resolved by recognizing that the vertical axis in the bottom graph of Fig. 2 represents the total energy of the chromophore–lysine system. The total energy is a sum of three components: $E_s(\phi)$, the energy of the torsional potential surface; $E_{kin}(\tau)$, the kinetic energy of the torsional motion; and $E_{vib}(\tau)$, the vibrational energy associated with the remaining modes of the polyene–lysine moiety. The first and all subsequent trajectory elements are trapped in the activated complex because the kinetic energy of torsional motion is efficiently transferred to other "nonproductive" vibrational modes of the polyene–lysine system [see discussion concerning Eq. (6)]. Although the first trajectory pass is a relatively high-velocity trajectory reaching a kinetic energy maximum of 0.23 eV (5.3 kcal/mole), "reflection" of the first trajectory at $\phi_{11,12} \cong 75°$ back into the region of the activated complex occurs because the torsional kinetic energy is insufficient to override the local barrier associated with the activated complex potential well (see top of Fig. 2).

IV. Comments and Conclusions

The analysis presented in the previous section leads to the following conclusion. A one-bond cis → trans isomerization in the active site of rhodopsin can occur with reasonable quantum efficiency on a picosecond time scale. A comparison of our calculated results with experimental observation is presented below:

Rhodopsin $\xrightarrow{h\nu}$ bathorhodopsin $\quad \Delta t_{calc} = 2.2$ psec (obs = 3 psec),
$\quad\quad\quad\quad\quad\quad\quad\quad\quad\quad\quad\quad\quad\quad \phi_{calc} = 0.62 \quad$ (obs = 0.67),

Bathorhodopsin $\xrightarrow{h\nu}$ rhodopsin $\quad \Delta t_{calc} = 1.8$ psec (obs = ?),
$\quad\quad\quad\quad\quad\quad\quad\quad\quad\quad\quad\quad\quad\quad \phi_{calc} = 0.48 \quad$ (obs = 0.5),

where the observed formation time of 3 psec is from Green *et al.* (1977) and the observed quantum yields are from Dartnall (1972) (rhodopsin) and Suzuki and Callender (1981) (bathorhodopsin).

Our calculated rhodopsin → bathorhodopsin isomerization time is believed to be overestimated, and yet the above comparison suggests we are underestimating the isomerization time. As discussed in Chapter 12, however, the formation time to generate photochemically bathorhodopsin is actually a sum of two components. The first component is the isomerization time which we calculate to be ~2.2 psec. The second component is the proton translocation time which is a ground state process and probably involves the transfer of a proton from one counterion to another counterion in the vicinity of the chromophore (see Chapter 12; Honig et al., 1979a,b). Although at low temperatures the rate of this process is observed to be dominated by tunneling (Peters et al., 1979), at ambient temperatures the rate can, to a first approximation, be described using classical transition state theory,

$$\tau_H = 1/k_H = [(kT/h)\exp(-\Delta E_0^\ddagger/RT)]^{-1}, \qquad (12)$$

where $\triangle E_0^\ddagger$ is the activation energy. Based on the low-temperature (4 K) kinetics of bathorhodopsin transformation, where the observed rate constant (2.8×10^{11} sec^{-1}; Peters et al., 1977) is associated with proton tunneling, the barrier to this process is calculated (using the formalism of Löwdin, 1965) to be in the range 1.0–1.6 kcal/mole (based on a translocation distance of ~1 Å). At ambient temperature, this barrier yields a proton transfer time of 0.9–2.4 psec [based on Eq. (12)]. The observed formation time of bathorhodopsin at ambient temperature is ~3 psec (Green et al., 1977). Accordingly, the isomerization time is 0.6–2.1 psec. Although this prediction is crude and subject to considerable error, the analysis suggests our molecular dynamics calculations overestimate the isomerization time as originally predicted.

References

Arnaboldi, M., Motto, M., Tsujimoto, K., Balogh-Nair, V., and Nakanishi, K. (1979). *J. Am. Chem. Soc.* **101**, 7082–7084.
Birge, R. R. (1981). *Annu. Rev. Biophys. Bioeng.* **10**, 315–354.
Birge, R. R., and Hubbard, L. M. (1980). *J. Am. Chem. Soc.* **102**, 2195–2205.
Birge, R. R., and Hubbard, L. M. (1981). *Biophys. J.* **34**, 517–534.
Birge, R. R., and Pierce, B. M. (1979). *J. Chem. Phys.* **70**, 165–178.
Birge, R. R., Schulten, K., and Karplus, M. (1975). *Chem. Phys. Lett.* **31**, 451–455.
Birge, R. R., Bennett, J. A., Pierce, B. M., and Thomas, T. M. (1978). *J. Am. Chem. Soc.* **100**, 1533–1539.
Birgue, R. R., Bocian, D. F., and Hubbard, L. M. (1982). *J. Am. Chem. Soc.* In press.
Busch, G. E., Applebury, M. L., Lamola, A., and Rentzepis, P. M. (1972). *Proc. Natl. Acad. Sci. U.S.A.* **69**, 2802–2806.
Cooper, A. (1979). *Nature (London)* **282**, 531–533.
Dartnall, H. J. A. (1972). *Handb. Sens. Physiol.* **7**, Part 1, 122–145.

Green, B., Monger, T., Alfano, R., Aton, B., and Callender, R. (1977). *Nature (London)* **269**, 179–180.

Honig, B., Ebrey, T. G., Callender, R. H., Dinur, U., and Ottolenghi, M. (1979a). *Proc. Natl. Acad. Sci. U.S.A.* **76**, 2503–2507.

Honig, B., Dinur, U., Nakanishi, K., Balogh-Nair, V., Gawinowicz, M., Arnaboldi, M., and Motto, M. (1979b). *J. Am. Chem. Soc.* **101**, 7084–7086.

Huppert, D., Rentzepis, P. M., and Kliger, D. S. (1977). *Photochem. Photobiol.* **25**, 193–197.

Lifson, S., and Warshel, A. (1968). *J. Chem. Phys.* **49**, 5116–5129.

Lifson, S., and Warshel, A. (1970). *J. Chem. Phys.* **53**, 582–594.

Löwdin, P. O. (1965). *Adv. Quantum Chem.* **2**, 213–254.

Miller, W. H., and George, T. (1972). *J. Chem. Phys.* **56**, 5637–5645.

Miller, W. H., Handy, N. C., and Adams, J. E. (1980). *J. Chem. Phys.* **72**, 99–112.

Oseroff, A. R., and Callender, R. (1974). *Biochemistry* **13**, 4243–4248.

Ottolenghi M. (1980). *Adv. Photochem.* **12**, 97–200.

Peters, K., Appelbury, M. L., and Rentzepis, P. M. (1977). *Proc. Natl. Acad. Sci. U.S.A.* **74**, 3119–3123.

Porter, R. N., and Raff, L. M. (1976). *In* "Dynamics of Molecular Collisions," Part B (W. H. Miller, ed.), Modern Theoretical Chemistry, Vol. 2, pp. 1–52. Plenum, New York.

Suzuki, T., and Callender, R. H. (1981). *Biophys. J.* **34**, 261–270.

Wang, I. S. Y., and Karplus, M. (1973). *J. Am. Chem. Soc.* **95**, 8160–8162.

Warshel, A. (1976). *Nature (London)* **260**, 679–683.

Warshel, A., and Karplus, M. (1975). *Chem. Phys. Lett* **32**, 11–14.

Weiss, R. M. and Warshel, A., (1979). *J. Am. Chem. Soc.* **101**, 6131–6133.

PART III

Hemoproteins

CHAPTER 14

Introduction to Hemoproteins

Laura Eisenstein and Hans Frauenfelder

Department of Physics
University of Illinois at Urbana-Champaign
Urbana, Illinois

I.	The Importance of Hemoprotein Studies	321
II.	Structure and States	322
III.	Fast Elementary Processes	326
IV.	The Individual Steps	327
	A. Light Absorption and Energy Transfer	327
	B. Photodissociation	328
	C. Intermediate State(s)	328
	D. A Simple Scheme	329
	E. Quantum Yields	330
	F. The Transitions $B \to A$ and $B \to S$	330
	G. Overview	333
	H. Conformations	333
	I. Conformational Substates	334
	References	335

I. The Importance of Hemoprotein Studies

Most biological processes are well-controlled, precisely executed chemical reactions performed by proteins. The pathways, controls, and interrelations of these reactions form a nearly infinite set, and a complete phenomenological description may be impossible and also useless. A more promising approach is to select prototype systems and explore and understand these in depth. Hemoproteins are an excellent choice for such a task for the following reasons: (1) Hemoproteins occur in all aerobic and many anaerobic cells, and they perform a wide variety of activities, from oxygen storage (myoglobin, Mb) and transport (hemoglobin, Hb), to electron transfer (cytochromes) and catalysis (cytochrome P450). It may therefore be possible to understand how the protein modifies and controls one particular active center, the heme group, and adapts it to various functions. (2) Hemoproteins have been investigated in considerable detail by many techniques, and much is known about their structure and function. Ideas, models, and theories can consequently be tested quantitatively. (3) The isolated active center can be studied in model compounds, and the influence of the protein on elementary reaction steps can therefore be explored. In this chapter we restrict the discussion to a few hemoproteins, mainly Mb and Hb, and to two elementary processes, photodissociation and binding of small ligands. The treatment is brief and not meant as a complete review but as a first orientation.

II. Structure and States

Hemoproteins contain one or more heme groups embedded as active centers in the folded polypeptide chain. The protein modifies the reactivity of, and controls the access to, the heme. Mb, probably the hemoprotein studied in most detail, consists of 153 amino acids and has a molecular weight of 17,200. Hb is a tetramer built of four Mb-like subunits, two α and two β chains (Antonini and Brunori, 1971; Weissbluth, 1974). The heme group is formed from four pyrrole rings and a central iron atom (Smith, 1975; Dolphin, 1978). The active center of Mb and Hb, iron protoporphyrin IX, is shown in Fig. 1.

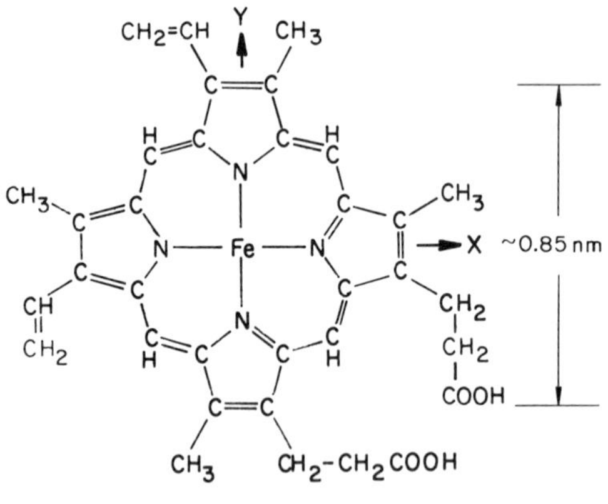

FIG. 1. Iron protoporphyrin IX.

The average spatial structure of Mb and Hb, as elucidated by x-ray diffraction (Kendrew et al., 1960; Kendrew, 1963; Perutz, 1963, 1968, 1970; Fermi, 1975; Takano, 1977a,b; Phillips, 1978, 1980), is the most basic information for an understanding of the working of these proteins. A cross section through Mb, drawn by computer graphics (Feldmann, 1976), is reproduced in Fig. 2. Additional structural data come from EXAFS (Eisenberger et al., 1976, 1978) and neutron diffraction experiments (Norvell et al.,

14. INTRODUCTION TO HEMOPROTEINS 323

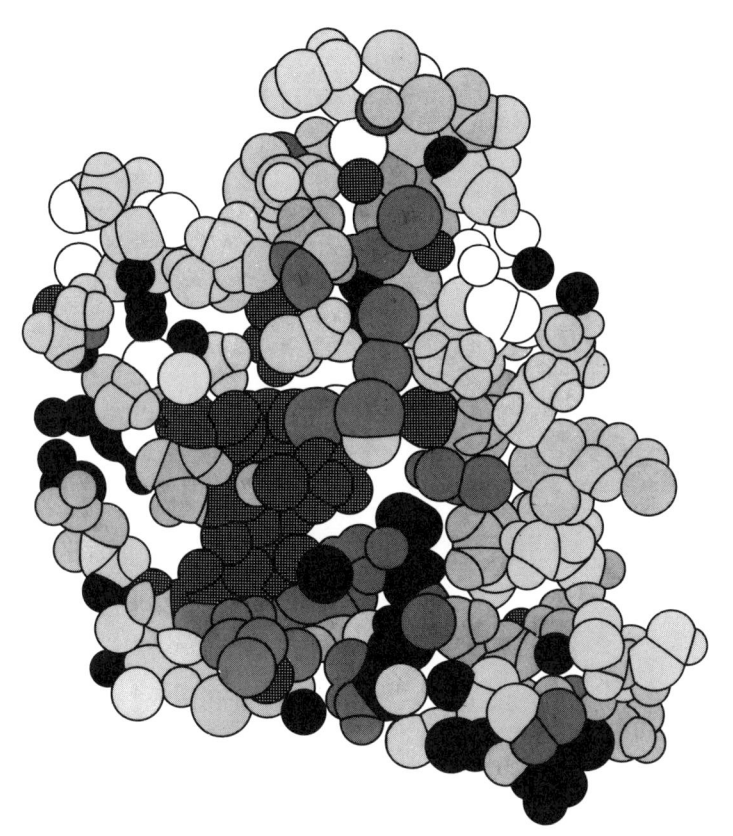

Fig. 2. Atomic mean square displacements deduced from x-ray diffraction of met Mb. A 5-Å slice perpendicular to the heme plane is shown, and $\langle x^2 \rangle$ values are indicated by shading, darker shades corresponding to smaller $\langle x^2 \rangle$. The heme plane is seen in cross section as a dark horizontal row of atoms at the middle left. Below the heme the bound water is visible as a light crescent in the tightly packed ligand pocket. The cavity of the proximal histidine, seen above the heme, is less tightly packed. The highly mobile atoms near the C-terminus are in the upper right (unshaded). (After computer graphics by R. J. Feldmann.)

1975). Mössbauer (Lang, 1970), Raman (Spiro and Strekas, 1972, 1974; Felton and Yu, 1978), and infrared (IR) (Alben and Caughey, 1968; Alben, 1978) spectroscopy, and magnetic susceptibility (Cerdonio *et al.*, 1977, 1978), yield further information on the ground state of the protein. The essential features of Mb without (deoxy Mb) and with bound ligand (MbO$_2$ or MbCO) are as follows.

In the ligand-free state (Mb), the heme is buckled and the iron has spin 2, is ferrous (nominally Fe^{2+}), and lies slightly out of the mean heme plane

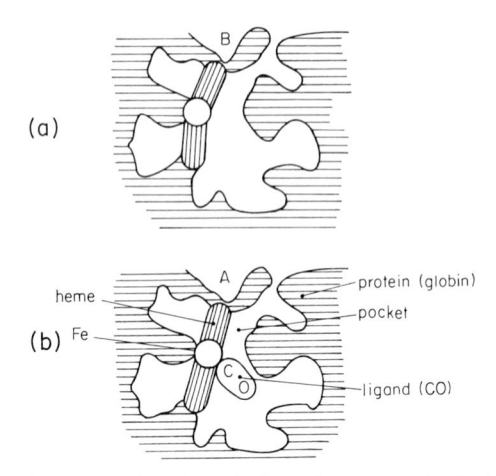

FIG. 3. A schematic cross section through the active center of Mb. (a) In the ligand-free state, B, the heme is buckled and the iron has spin 2 and has moved slightly out of the mean heme plane. (b) In the CO-bound state, A, the heme is planar and the iron has spin 0 and lies almost in the heme plane.

toward the proximal histidine, as shown in Fig. 3a. In the bound state, for instance MbCO (Fig. 3b), the heme is nearly planar; the iron has spin 0 and has almost moved into the heme plane. Optical, Raman, and IR absorptions differ in the ligand-bound and ligand-free states, and we show in Fig. 4 some relevant optical, resonance Raman, and IR spectra. The spectra can be interpreted in terms of the energy levels of the heme–iron system (Zerner and Gouterman, 1966; Zerner et al., 1966; Eaton et al., 1978; Case et al., 1979). The absorption spectra in the visible and near-ultraviolet (UV) region are due primarily to the allowed porphyrin $\pi \to \pi^*$ transitions which give rise to an intense peak in the near UV (Soret) and a weaker peak in the visible (α, β). The resonance Raman spectra correspond to heme vibrational modes. The IR lines shown in Fig. 4 are CO stretching vibrations in the ligand-bound and photodissociated states (Section III).

The structure of Mb is actually not static, as drawn in Fig. 2. A protein molecule in a given conformation can exist in a large number of structurally slightly different substates (Austin et al., 1975; Frauenfelder et al., 1979; Sternberg et al., 1979; Artymiuk et al., 1979; Frauenfelder and Petkso, 1980). At low temperatures, say below about 200 K, each molecule is frozen in a particular conformational substate. At higher temperatures, conformational motion occurs, and each protein molecule moves rapidly from one substate to another.

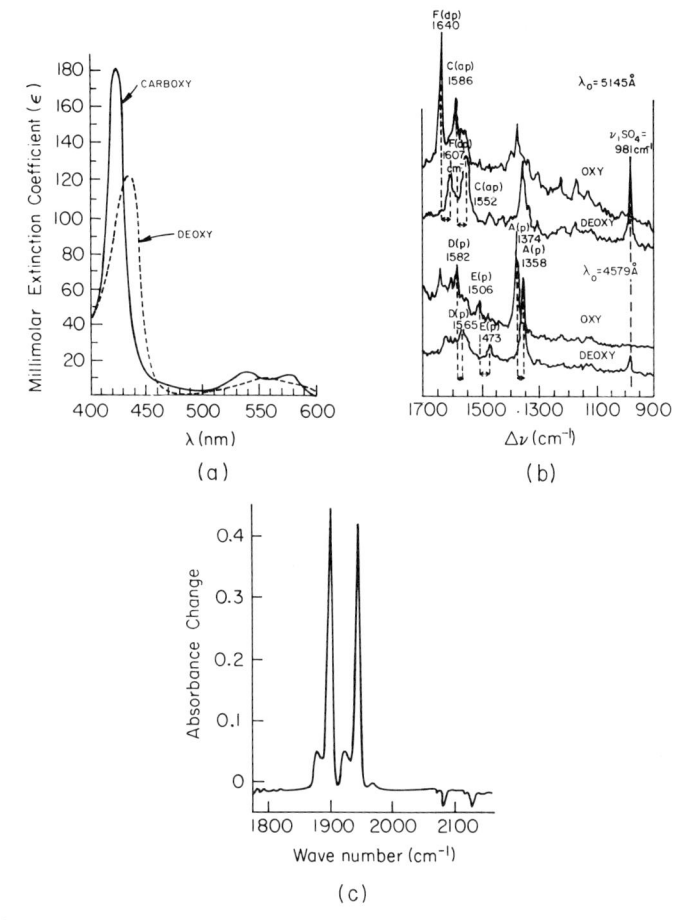

FIG. 4. (a) Optical absorption spectrum of sperm whale Mb in the deoxy and CO-bound states. (b) Resonance Raman spectra of oxy-Hb and deoxy-Hb in the visible ($\lambda_0 = 5145$ Å) and Soret ($\lambda_0 = 4579$ Å) scattering regions. The deoxy-Hb solution contained $(NH_4)_2SO_4$, the ν_1 (981 cm^{-1}) band of which is indicated. Frequencies which shift appreciably on oxygenation are labeled A, C, D, E, F, and their polarization: p, polarized; dp, depolarized; ap, anomalously polarized. (From Spiro and Strekas, 1974. Reprinted with permission from *Journal of the American Chemical Society* **96**, 338 (1974). Copyright 1974 American Chemical Society.) (c) Absorbance change $= \log[I_{eq}(\nu)/I_A(\nu)]$, as a function of wave number ν, at 12 K for Mb^{12}CO and Mb^{13}CO. $I_A(\nu)$ is the IR spectrum in the CO-bound state before illumination. $I_{eq}(\nu)$ is the IR spectrum after illumination. The bound states for both isotopes appear at about 1900 cm^{-1}; the weak lines of the photodissociated CO are seen at about 2100 cm^{-1}. (From Alben et al., 1982.)

III. Fast Elementary Processes

The final step in ligand binding and the initial step in dissociation occur at the heme group. The experimental exploration of these elementary processes and their theoretical interpretation may yield considerable insight into biomolecular reactions. Because the bound state MbL (Fig. 3b) can be photodissociated (Haldane and Lorraine-Smith, 1896), flash photolysis has become one of the most important tools for the investigation of ligand binding and dissociation.

The idea underlying studies on heme protein dynamics by flash photolysis is simple. Consider a heme protein H with bound ligand L, denoted by HL. Irradiation with light absorbed by HL breaks the bond; L dissociates from H and later rebinds. Photodissociation and recombination can be followed using the spectral differences between the ligand-free and -bound states given in Fig. 4. In this chapter we will review the dynamics of ligand binding to heme proteins as seen by three techniques. With optical absorption and resonance Raman spectroscopy, rebinding is sensed by the heme group; using IR spectroscopy the ligand is followed.

Some of the processes during and after the photodissociation can be described with the help of Fig. 5. This figure is speculative, and more work will be required before all transitions and states are identified, characterized, and understood, but the proposed scheme provides a framework for a discussion of the experiments. Photodissociation starts with HL in its ground state A (Fig. 3b). The photoflash, $\hbar\omega_0$, lifts the system to a highly excited state A** from which it decays to A*. Further transitions from A* are shown in Fig. 5. Tentatively we identify A* with an excited state of HL before the Fe—L bond is broken, and B with the state where the ligand is dissociated and has moved away from the heme group and iron, heme, and protein have assumed the deoxy configuration (Fig. 3a). B* is an intermediate state in

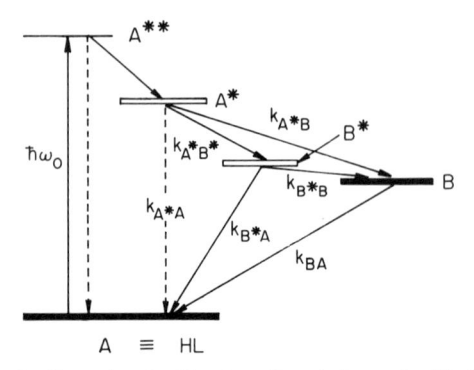

FIG. 5. States involved in photodissociation and rebinding.

which the Fe—L bond is broken but the system has not relaxed to the deoxy state. While we show only one state B* in Fig. 5, there will most likely be many such states corresponding to the different parts of the heme protein assuming the deoxy configuration on different time scales. In Fig. 5 the positions of the levels and the rates and branching ratios depend on the system under consideration and on the energy $\hbar\omega_0$.

IV. The Individual Steps

Figure 5 provides one possible scheme for discussing and interpreting experiments. Two types of features must be explored—general and specific ones. Which states and pathways occur in all heme proteins and which ones are specific for a particular protein? How do protein structure and heme composition control and change the various steps? How does the ligand affect the scheme? The many available different protein–ligand systems make answers to these questions feasible, but at the same time render the interpretation more difficult because the richness of choice has led to a dearth of overlapping and interlocking data. Much experimental work remains to be done.

Most or all of the steps in Fig. 5 are expected to be very fast at physiological temperatures. Their exploration consequently calls for fast techniques. Cooling to low temperatures slows down some of the elementary reactions and permits a more leisurely look, but some reactions are very fast even at 4 K. Combination of the two approaches, fast laser techniques and low temperatures, may be necessary for an elucidation of all transitions. In the following we discuss some of the transitions and states in more detail.

A. Light Absorption and Energy Transfer

The first two steps in photodissociation are absorption of light quanta in the protein and transfer of energy to the central iron atom. Early experiments (Bucher and Kaspers, 1947) showed that the quantum yield for photodissociation of CO from MbCO was unity for incident wavelengths between 280 and 546 nm. At 280 nm, the light is absorbed by the aromatic amino acids Trp and Tyr; above about 400 nm, absorption occurs mainly in the heme through $\pi \to \pi^*$ transitions (Fig. 4). Thus two steps must be discussed—the transfer from Trp and Tyr to the porphyrin, and the relaxation within the latter. The constancy of the quantum yield implies that the energy is transferred efficiently through the protein by the Förster mechanism (Weber and Teale, 1959). Indeed, fluorescence studies confirm an energy transfer approaching 100% from Trp and Tyr to the porphyrin (Sebban et al., 1980). Relaxation after the initial optical excitation is extremely fast. In fact, the transition to the state symbolically denoted by A* in Fig. 5 is so fast that direct measurements in the time domain are extremely difficult. Experiments in the frequency domain, however, yield some of the needed information. The simplest approach is to use the linewidth Γ and determine the lifetime as $\tau = \hbar/\Gamma$. The Soret lines in Fig. 4 have a width of $\Delta\lambda \sim 25$ nm; and the corresponding decay time is about 4 fsec. This argument is correct only if the line is homogeneous and the entire broadening is due to the natural linewidth. Inhomogeneous line broadening is expected in proteins (see Section IV,I), and additional methods are therefore required. One approach is optical hole burning (Hochstrasser, 1971; Shelby and McFarlane, 1979). Resonance Raman scattering (Adar et al., 1976; Friedman et al., 1977) and polarization spectroscopy (Andrews and Hochstrasser, 1980) promise to yield much of the needed information.

B. Photodissociation

Photodissociation implies breaking of the covalent bond between the iron and the ligand. Bond breakage occurs as a result of a radiationless transition from the excited electronic state of the porphyrin to a lower state with a d-d excitation of the iron atom (Zerner et al., 1966; Greene et al., 1978; Eisenstein et al., 1978; Chernoff et al., 1980). In the simplest case, the iron transition involves the promotion of an electron from a bonding state d_{xz} or d_{yz} to an antibonding d_{z^2} orbital. The resulting state B* (Fig. 5) may still differ from the stationary deoxy state B, and the characteristic time for bond breaking may be much shorter than that for the approach to the state B. Operationally we define photodissociation as disappearance of the prominent Soret peak at about 420 nm. With this criterion, photodissociation in MbCO and MbO_2 occurs within less than 5 psec (Reynolds et al., 1981). Observation at 615 nm shows that HbCO photodissociates within 0.5 psec (Shank et al., 1976). Changes in the resonance Raman spectrum also imply that HbCO dissociates within less than 30 psec (Terner et al., 1981). These experiments consequently all indicate that the transition A* → B* in Fig. 5 occurs on a picosecond time scale.

C. Intermediate State(s)

In Fig. 5 we show an intermediate state, B*, between the excited porphyrin state, A*, and the deoxy state, B. Such a state has been proposed by Phillipson et al. (1973), and evidence for one or more intermediates comes from a number of observations: Photodissociation of carboxymethylated cytochrome c (cm cyt c) leads to a state from which recombination is fast even at 4 K and which has an optical spectrum different from that of deoxy cm cyt c (Alberding et al., 1978a). In HbO_2, MbO_2, and MbCO, the dissociation rate is faster than the rate of appearance of the deoxy spectrum (Shank et al., 1976; Reynolds et al., 1981). In HbO_2, about 40% of the photodissociated oxygen recombines with a lifetime of about 200 psec at about 300 K (Chernoff et al., 1980), which is much faster than that ever measured for B → A. These observations imply the existence of intermediate states.

Little is known at present about the properties and number of intermediate states. One possible interpretation (Alberding et al., 1978a) is based on the molecular orbital calculations of Zerner et al. (1966) and postulates the assignments A (d_{xy}^2, d_π^4), B* $(d_{xy}^2, d_\pi^3, d_{z^2})$, and B $(d_{xy}^2, d_\pi^2, d_{z^2}, d_{x^2-y^2})$. A has spin 0 and lies in the heme plane, and the bond with CO is strong. B* would, by Hund's rule, have a spin of 1. The d_{z^2} in B* is antibonding, and the CO could dissociate. The $d_{x^2-y^2}$ orbital however, is still empty; since the occupation of this orbital makes the out-of-plane position favourable for the iron, Fe may still be in the plane and the rebinding of CO could be fast. In B, the iron has spin 2 and has moved out of the heme plane; rebinding now is slow because a spin change of 2 and movement into the heme plane are required. A great deal of work will be needed to elucidate all the properties of the intermediate state(s). One particular problem must be considered. Since B has, in general, a much longer lifetime than B*, the scheme shown in Fig. 5 leads to pumping (Frauenfelder, 1978): If the photolyzing pulse has a long duration compared to $1/k_{B*A}$ or a fast repetition rate compared to k_{BA}, the system will be preferentially moved to B. Signal averaging, under such circumstances, may cover up the existence of B*.

D. A SIMPLE SCHEME

Many features of the photodissociation and binding of ligands, particularly those slower than a few nanoseconds, can be discussed with a scheme that is simpler than Fig. 5 and contains only states A and B (Austin et al., 1975). The main features of the scheme are given as a potential energy plot in Fig. 6.

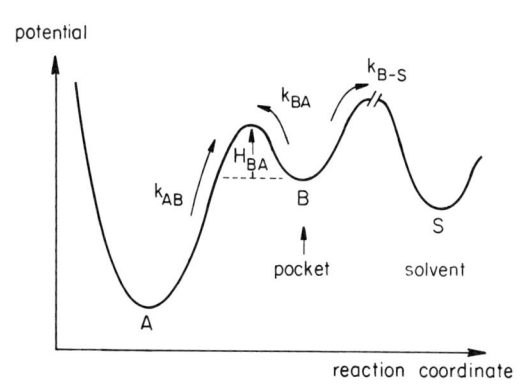

FIG. 6. Potential energy diagram for the binding of ligands to heme proteins. Details between the pocket and the solvent have been omitted.

States A and B are as in Fig. 5 and described above. The barrier, H_{BA}, between B and A is an effective barrier and may actually hide more than one step. Similarly, the barrier between state B in the pocket and the solvent S may in reality consist of a number of barriers. Photodissociation promotes the ligand from A to B. From B, the ligand can either rebind (B → A) or move into the solvent (B → ··· → S). We will not treat the rather complex situation obtained for complete rebinding (Austin et al., 1975; Alberding et al., 1978b,c; Hänggi, 1978) but will use Fig. 6 for a discussion of the fast processes at the active center.

E. Quantum Yields

Figures 5 and 6 imply that a quantum yield can only be defined unambiguously if the correct reaction scheme is known, and can only be measured properly with very short single pulses (Phillipson et al., 1973; Austin et al., 1975; Frauenfelder, 1978; Duddell et al., 1980b). We introduce three numbers here, Y_p, Y_B, and N^{out}. We define the *primary quantum yield* Y_p as the probability that an absorbed photon will dissociate the ligand. In terms of Fig. 5, Y_p is the probability that state B^* will be reached. Since the site of absorption and the transfer to the porphyrin depend on the wavelength of the incident photon Y_p may also depend on the wavelength. Indeed, picosecond experiments point to such a dependence (Reynolds et al., 1981). The *effective quantum yield* Y_B is defined as the probability that the system will move to state B, where iron, heme, and protein essentially have assumed the deoxy configuration (see, however, Section IV,H). N^{out} is defined as the fraction of the ligands photodissociated in the primary event that leave the protein and move into the solvent. N^{out} can be determined without too much trouble, but unambiguous measurements of Y_p and Y_B can be difficult. B^* may in many proteins be populated only for a short time, so that its observation can be very difficult. Even cooling to very low temperatures will not always slow k_{B^*A} sufficiently, because the transition can proceed by tunneling. Long or repeated pulses may pump the system into state B, and incorrect quantum yields may be inferred if this effect is not taken into account.

F. The Transitions B → A and B → S

Within the protein, a number of different geminate processes can occur. Of particular interest is the direct rebinding B → A which we call process I. The rate λ_I of I is approximately given by

$$\lambda_I \simeq k_{BA}[1 + (k_{BS}/k_{BA})], \tag{1}$$

where k_{BS} is the effective rate coefficient for the escape from the pocket (state B) into the solvent and k_{BA} is the effective rate coefficient for rebinding. The fraction of ligands that escape into the solvent is approximately given by

$$N^{out} \simeq [1 + (k_{BA}/k_{BS})]^{-1}. \tag{2}$$

Consequently, if $k_{BS} \gg k_{BA}$, nearly all the photodissociated ligands will move into the solvent and rebind very slowly from there. On the other hand, if $k_{BS} \ll k_{BA}$, essentially all the photodissociated ligands will rebind directly from the pocket and N^{out} will be very small. The ratio k_{BS}/k_{BA} determines whether the fast rebinding, B → A, or the slow multistep process, B → → S → → B → A, will dominate. The ratio depends on protein structure, ligand, temperature, and solvent properties such as viscosity and pH. Some general features, however, appear to be universal, and we will sketch these in the following.

Below about 200 K, the photodissociated ligands remain within the heme pocket (state B) (Chance et al., 1965) and process I can be studied in detail (Austin et al., 1975; Alberding et al., 1976, 1978c; Alben et al., 1980). The main results can be summarized as follows: Low-temperature rebinding is not exponential in time but follows approximately a power law. We will return to this feature in Section IV,I. Below about 60 K, rebinding is dominated by quantum mechanical tunneling. Above 60 K, rebinding proceeds mainly by over-the-barrier transitions, with an average rate $\lambda_I \simeq \bar{k}_{BA}$ given by the Arrhenius relation

$$\bar{k}_{BA} = A_{BA} \exp(-H^p_{BA}/RT), \tag{3}$$

where $R = 8.32$ J/mole is the gas constant. A few typical values of A_{BA} and H^p_{BA} are given in Table I. Straightforward extrapolation of the low-temperature parameters to 300 K gives the values of $\bar{k}_{BA}$ in the last column. By Eq. (1), process I should be faster than predicted by the extrapolation.

Process I has indeed been observed by optical (Alpert et al., 1974, 1979; Duddell et al. 1979, 1980a,b; Morris et al., 1982) and resonance Raman (Lyons et al., 1978; Friedman and Lyons, 1980; Lyons and Friedman, 1982)

TABLE I

TYPICAL VALUES OF A_{BA} AND H^p_{BA} FOR THE LOW-TEMPERATURE PROCESS I

System	log A_{BA} (sec^{-1})	H^p_{BA} (kJ/mole)	$\bar{k}_{BA}{}^a$ extrapolated to 300 K (sec^{-1})
MbO$_2$	8.5	8.4	9×10^6
MbCO	8.7	10.0	9×10^6
α-HbCO	9.4	4.3	5×10^8
β-HbCO	8.6	4.6	6×10^7

[a] Calculated by Eq. (3).

techniques at temperatures of ~ 300 K in MbO_2, HbCO, and HbO_2, but not in MbCO. The fractional intensity of I decreases with increasing temperature and increases with increasing solvent viscosity. These features can be understood by considering the physical nature of the transitions B → A and B → S.

In the transition B → A, the ligand binds covalently to the iron atom. During the transition, the iron moves toward the heme plane and changes spin, the heme becomes nearly planar, and the ligand approaches the iron. The activation enthalpy H_{BA}^p is small, the activation entropy large and negative. Since the process occurs deep inside the protein, we assume tentatively that it is not affected by the solvent viscosity and that $\bar{k}_{BA}$ is described by Eq. (3). The transition B → S involves two or more barriers. In a dynamic model, these correspond to the opening of channels within the protein (Case and Karplus, 1979; Beece et al., 1980, 1982). Opening and closing is influenced by the solvent viscosity; k_{BS}, therefore, cannot be described by an Arrhenius equation. Instead a modified Kramers equation must be used (Kramers, 1940; Skinner and Wolynes, 1978; Beece et al., 1980; 1982):

$$k_{BS} = (A_{BS}\eta^{-\kappa} + A_{BS}^0)\exp(-H_{BS}^*/RT). \qquad (4)$$

Here η is the solvent viscosity and κ is a parameter ($0 \leq \kappa \leq 1$) that describes the strength of the coupling of the relevant protein part to the solvent. A_{BS} and A_{BS}^0 are parameters that depend on the protein and the ligand; H_{BS}^* is typically on the order of 30 kJ/mole and thus is much larger than H_{BA}^p. With Eqs. (3) and (4), the crucial ratio $k_{BS}/\bar{k}_{BA}$ becomes

$$\frac{k_{BS}}{\bar{k}_{BA}} = \left(\frac{A_{BS}\eta^{-\kappa} + A_{BS}^0}{A_{BA}}\right)\exp\left[\frac{(H_{BA}^p - H_{BS}^*)}{RT}\right]. \qquad (5)$$

The observed features of process I and N^{out} can be understood with Eq. (5). Since $H_{BA}^p - H_{BS}^* \sim -25$ kJ/mole, the exponential in Eq. (5) will overpower the preexponential below a certain temperature and $k_{BS}/\bar{k}_{BA}$ will tend to zero. At low temperatures, the photodissociated ligand will remain in the pocket, and only process I will be observable. Above about 220 K, the preexponential in Eq. (5) will be the important factor, and the resulting ratio will depend crucially on protein structure. Some general trends, however, are independent of the details. With increasing viscosity, $k_{BS}/\bar{k}_{BA}$ will decrease, and the fractional intensity of I thus will increase, as observed (Morris et al., 1982). With increasing T, $k_{BS}/\bar{k}_{BA}$ will increase, N^{out} will increase, and the intensity of I will decrease. In MbO_2, HbO_2, $k_{BS}/\bar{k}_{BA} \sim 1$ at about 300 K and process I can still be observed. In MbCO, however, $k_{BS}/\bar{k}_{BA} \gg 1$, $N^{out} \sim 1$ above 300 K, and process I becomes very small so that it has not been observed. It can, however, not be completely absent; otherwise no rebinding to state A would occur.

The arguments given here explain the main observed features. Much work remains to be done, however. In particular, the various processes should be followed over the entire temperature range where they can be seen, and their dependence on protein structure should be explored and understood.

G. Overview

In Fig. 7, we schematically show the various binding processes discussed so far. We denote with I* the very fast rebinding from state B*, with I the rebinding from the pocket state B, with II rebinding from inside the protein, and with IV binding from the solvent. At normal ligand concentrations, I*, I, and II are independent of, and IV is proportional to, the ligand concentration in the solvent after photodissociation. Figure 7 is an idealization; it is unlikely that any protein at any temperature displays the various processes so neatly. While we expect the features to be general, the details, namely, the intensities and rates of each component, will be governed by the protein structure and by the ligand and will be functions of temperature and solvent viscosity. Some or all of the components may also be influenced by inhibitors.

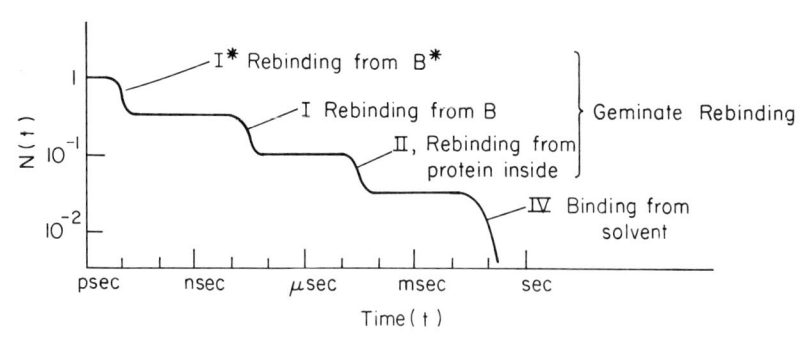

FIG. 7. An idealization of the rebinding kinetics. $N(t)$ is the fraction of heme protein molecules that have not yet rebound a ligand at time t after photodissociation. Processes corresponding to rebinding from states B* and B, inside the protein, and from the solvent are indicated.

H. Conformations

X-ray diffraction shows that liganded and unliganded heme proteins possess different static (Phillips, 1978, 1980) and dynamic (Frauenfelder and Petsko, 1980) structures. After photodissociation, the protein will try to reach the new equilibrium state. It is likely that the different components—iron, heme, and protein tertiary and quaternary structures—will approach equilibrium with vastly different rates. The structural change will affect the spectral properties, and disentangling structural relaxation and rebinding is not easy.

Even at very low temperatures, some changes, for instance, the spin change of the iron, proceed very rapidly. The heme and the protein may, however, remain frozen in the initial or in an intermediate state. Evidence for partial relaxation comes from IR (Yonetani et al., 1973a; Iizuka et al., 1974a), electron paramagnetic resonance (Yonetani et al., 1973b; Iizuka et al., 1974b), and Mössbauer (Spartalian et al., 1976) experiments.

The time dependence of structural relaxation after photodissociation at temperatures of about 300 K has been studied by optical (Sawicki and Gibson, 1976; Greene et al., 1978; Chernoff et al., 1980; Lindqvist et al., 1980) and resonance Raman (Friedman and Lyons, 1980; Lyons and Friedman, 1982; Terner et al., 1981) techniques. The optical studies show that a deoxy-like Soret line appears within a few picoseconds of photodissociation. However, the line is weaker, broader, and slightly red-shifted. In HbO_2, a first relaxation process appears within 100 psec, but in HbCO no change occurs in the first nanosecond. Further relaxation toward the deoxy spectrum occurs at about 1 μsec, characterized by an activation enthalpy of about 40 kJ/mole. These results are corroborated by resonance Raman investigations which show that the Raman spectrum of the photoproduct is similar to that of deoxy-Hb, but with frequencies 4–7 cm^{-1} lower. The lines shift toward the deoxy position at about 1 μsec; this relaxation is probably caused by a rearrangement of the tertiary structure. Quaternary effects are also seen, but they are about two orders of magnitude slower.

Nonequilibrium states, produced in metalloproteins by pulse radiolysis, have also been studied extensively by Blumenfeld and collaborators (Blumenfeld and Davidov, 1979).

I. CONFORMATIONAL SUBSTATES

In Section IV,F we remarked that rebinding at low temperatures was not exponential in time. This observation can be explained by assuming that a protein can exist in a very large number of somewhat different conformational substates, all with the same biological function. Different substates may have slightly different properties, however, and the the barrier heights H_{BA} in Fig. 6 may for instance change from substate to substate. At low temperature, say below 200 K, each protein remains frozen in a particular substate. Denote with $g(H) \, dH$ the probability of finding a barrier height between H and $H + dH$; the observed binding then is given by

$$N(t) = \int dH \, g(H) \exp(-kt), \qquad (6)$$

where k is related to H by an Arrhenius relation of the form of Eq. (3). Equation (6) can explain the observed nonexponential time dependence. From the observed $N(t)$, the probability density $g(H)$ can be found (Austin et al., 1975). Further evidence for conformational substates in Mb comes also from x-ray diffraction (Frauenfelder et al., 1979).

As the temperature is increased above about 200 K, relaxation among substates sets in and binding becomes exponential if the relaxation rate is faster than the binding rate. The relaxation rates are not yet well explored, and they probably differ from protein to protein and vary within each protein. Mössbauer studies (Keller and Debrunner, 1980; Parak et al., 1981) provide information about the relaxation time at the heme iron. The relaxation rate k_r can be approximated by an Arrhenius relation with $A_r \simeq 10^{13}$ sec^{-1} and $H_r \simeq 30$ kJ/mole which predicts $k_r \simeq 10^8$ sec^{-1} at 300 K. This number raises questions. Many of the laser experiments discussed here involve rates much faster than k_r. Should we expect the observed processes to be nonexponential in time? If they are exponential, how can we explain the relaxation? The exploration of the processes in heme proteins in the time range between picoseconds and nanoseconds is obviously still in an exploratory state and many surprises can still be expected.

REFERENCES

Adar F., Gouterman, M., and Aronowitz, S. (1976). *J. Phys. Chem.* **80**, 2184–2190.
Alben, J. O. (1978). *In* "The Porphyrins." (D. Dolphin, ed.), Vol. 3, pp. 323–345. Academic Press, New York.
Alben, J. O., and Caughey, W. S. (1968). *Biochemistry* **7**, 175–183.
Alben, J. O., Beece, D., Bowne, S. F., Eisenstein, L., Frauenfelder, H., Good, D., Marden, M. C., Moh, P. P., Reinisch, L., Reynolds, A. H., and Yue, K. T. (1980). *Phys. Rev. Lett.* **44**, 1157–1160.
Alben, J. O., Beece, D., Bowne, S. F., Doster, W., Eisenstein, L., Frauenfelder, H., Good, D., McDonald, J. D., Marden, M. C., Moh, P. P., Reinisch, L., Reynolds, A. H., and Yue, K. T. (1982). *Proc. Natl. Acad. Sci. U.S.A.* (submitted).
Alberding, N., Austin, R. H., Chan, S. S., Eisenstein, L., Frauenfelder, H., Gunsalus, I. C., and Nordlund, T. M. (1976). *J. Chem. Phys.* **11**, 4701–4711.
Alberding, N., Austin, R. H., Chan, S. S., Eisenstein, L., Frauenfelder, H., Good, D., Kaufmann, K., Marden, M., Nordlund, T. M., Reinisch, L., Reynolds, A. H., Sorensen, L. B., Wagner, G. C., and Yue, K. T. (1978a). *Biophys. J.* **24**, 319–329.
Alberding, N., Frauenfelder, H., and Hänggi, P. (1978b). *Proc. Natl. Acad. Sci. U.S.A.* **75**, 26–29.
Alberding, N., Chan, S. S., Eisenstein, L., Frauenfelder, H., Good D., Gunsalus, I. C., Nordlund, T. M., Perutz, M. F., Reynolds, A. H., and Sorensen, L. B. (1978c). *Biochemistry* **17**, 43–51.
Alpert, B., Banerjee, R., and Lindqvist, L. (1974). *Proc. Natl. Acad. Sci. U.S.A.* **71**, 558–562.
Alpert, B., El Mohsni, S., Lindqvist, L., and Tfibel, F. (1979). *Chem. Phys. Lett.* **64**, 11–16.
Andrews, J. R., and Hochstrasser, R. M. (1980). *Proc. Natl. Acad. Sci. U.S.A.* **77**, 3110–3114.
Antonini, E., and Brunori, M. (1971). "Hemoglobin and Myoglobin in Their Reactions With Ligands." North-Holland Publ., Amsterdam.
Artymiuk, P. J., Blake, C. C. F., Grace, D. E. P., Oatley, S. J., Phillips, D. C., and Sternberg, M. J. E. (1979). *Nature (London)* **280**, 563–568.
Austin, R. H., Beeson, K. W., Eisenstein, L., Frauenfelder, H., and Gunsalus, I. C. (1975). *Biochemistry* **14**, 5355–5373.

Beece, D., Eisenstein, L., Frauenfelder, H., Good, D., Marden, M. C., Reinisch, L., Reynolds, A. H., Sorensen, L. B., and Yue, K. T. (1980). *Biochemistry* **19**, 5147–5157.
Beece, D., Eisenstein, L., Frauenfelder, H., Good D., Marden, M. C., Reinisch, L., Reynolds, A. H., Sorensen, L. B., and Yue, K. T. (1982). *In* "Hemoglobin and Oxygen Binding" (C. Ho, ed.), pp. 363–369. Elsevier, New York.
Blumenfeld, L. A., and Davidov, R. M. (1979). *Biochim. Biophys. Acta* **549**, 255–280.
Bucher, T., and Kaspers, J. (1947). *Biochim. Biophys. Acta* **1**, 21–34.
Case, D. A., and Karplus, M. (1979). *J. Mol. Biol.* **132**, 343–368.
Case, D. A., Huynh, B. H., and Karplus, M. (1979). *J. Am. Chem. Soc.* **101**, 4433–4453.
Cerdonio, M., Congiu-Castellano, A., Mogno, F., Pispisa, B., Romani, C. L., and Vitale, S. (1977). *Proc. Natl. Acad. Sci. U.S.A.* **74**, 398–400.
Cerdonio, M., Congiu-Castellano, A., Calabrese, L., Morante, S., Pispisa, B., and Vitale, S. (1978). *Proc. Natl. Acad. Sci. U.S.A.* **75**, 4916–4919.
Chance, B., Schoener, B., and Yonetani, T. (1965). *In* "Oxidases and Related Redox Systems" (T. E. King, H. S. Mason, and M. Morrison, eds.), pp. 609–621. Wiley, New York.
Chernoff, D. A., Hochstrasser, R. M., and Steele, A. W. (1980). *Proc. Natl. Acad. Sci. U.S.A.* **77**, 5606–5610.
Dolphin, D., ed. (1978). "The Porphyrins." Academic Press, New York.
Duddell, D. A., Morris, R. J., and Richards, J. T. (1979). *J.C.S. Chem Commun.* pp. 75–76.
Duddell, D. A., Morris, R. J., Muttucumaru, N. J., and Richards, J. T. (1980a). *Photochem. Photobiol.* **31**, 479–484.
Duddell, D. A., Morris, R. J., and Richards, J. T. (1980b). *Biochim. Biophys. Acta* **621**, 1–8.
Eaton, W. A., Hanson, L. K., Stephens, P. J., Sutherland, J. C., and Dunn, J. B. R. (1978). *J. Am. Chem. Soc.* **100**, 4991–5003.
Eisenberger, P., Shulman, R. G., Brown, G. S., and Ogawa, S. (1976). *Proc. Natl. Acad. Sci. U.S.A.* **73**, 491–495.
Eisenberger, P., Shulman, R. G., Kincaid, B. M., Brown, G. S., and Ogawa, S. (1978). *Nature (London)* **274**, 30–34.
Eisenstein, L., Franceschetti, D. R., and Yip, K. L. (1978). *Theor. Chim. Acta* **49**, 349–359.
Feldmann, R. J. (1976). *Annu. Rev. Biophys. Bioeng.* **5**, 477–510.
Felton, R. H., and Yu, N. T. (1978). *In* "The Porphyrins" (D. Dolphin, ed.), Vol. 3, pp. 347–393. Academic Press, New York.
Fermi, G. (1975). *J. Mol. Biol.* **97**, 237–256.
Frauenfelder, H. (1978). *In* "Biomembranes, Part E, Biological Oxidations, Specialized Techniques" (S. Fleischer and L. Packer, eds.), Methods in Enzymology, Vol. 54, pp. 506–532. Academic Press, New York.
Frauenfelder, H., and Petsko, G. A. (1980). *Biophys. J.* **32**, 465–483.
Frauenfelder, H., Petsko, G. A., and Tsernoglou, D. (1979). *Nature (London)* **280**, 558–563.
Friedman, J. M., and Lyons, K. B. (1980). *Nature (London)* **284**, 570–573.
Friedman, J. M., Rousseau, D. L., and Adar, F. (1977). *Proc. Natl. Acad. Sci. U.S.A.* **74**, 2607–2611.
Greene, B. I., Hochstrasser, R. M., Weisman, R. B., and Eaton, W. A. (1978). *Proc. Natl. Acad. Sci. U.S.A.* **75**, 5255–5259.
Haldane, J., and Lorraine-Smith, J. (1896). *J. Physiol. (London)* **20**, 497–520.
Hänggi P. (1978). *J. Theor. Biol.* **74**, 337–359.
Hochstrasser, R. M. (1971). *In* "Probes of Structure and Function of Macromolecules and Membranes" (B. Chance, C. Lee, and J. Blaisie, eds.), Vol. I, pp. 57–64. Academic Press, New York.
Iizuka, T., Yamamoto, H., Kotani, M., and Yonetani, T. (1974a). *Biochim. Biophys. Acta* **371**, 126–139.

Iizuka, T., Yamamoto, H., Kotani, M., and Yonetani, T. (1974b). *Biochim. Biophys. Acta* **351**, 182–195.
Keller, H., and Debrunner, P. G. (1980). *Phys. Rev. Lett.* **45**, 68–71.
Kendrew, J. C. (1963). *Science* **139**, 1259–1266.
Kendrew, J. C., Dickerson, R. E., Strandberg, B. E., Hart, R. G., Davis, D. R., Phillips, D. C., and Shore, V. C. (1960). *Nature (London)* **185**, 422–427.
Kramers, H. A. (1940). *Physica (Amsterdam)* **7**, 284–304.
Lang, G. (1970). *Q. Rev. Biophys.* **3**, 1–60.
Lindqvist, L., El Mohsni, S., Tfibel, F., and Alpert, B. (1980). *Nature (London)* **288**, 729–730.
Lyons, K. B., and Friedman, J. M. (1982). In "Hemoglobin and Oxygen Binding" (C. Ho, ed.), pp. 333–338. Elsevier, New York.
Lyons, K. B., Friedman, J. M., and Fleury, P. A. (1978). *Nature (London)* **275**, 565–566.
Morris, R. J., Duddell, D. A., and Richards, J. T. (1982). In "Hemoglobin and Oxygen Binding" (C. Ho, ed.), pp. 339–343. Elsevier, New York.
Norvell, J. C., Nunes, A. C., and Schoenborn, B. P. (1975). *Science* **190**, 568–570.
Parak, F., Frolov, E. N., Mössbauer, R. L., and Goldanskii, V. I. (1981). *J. Mol. Biol.* **145**, 825–833.
Perutz, M. F. (1963). *Science* **140**, 863–869.
Perutz, M. F. (1968). *Proc. R. Soc. London, Ser. B* **173**, 113–140.
Perutz, M. F. (1970). *Nature (London)* **288**, 726–739.
Phillips, S. E. V. (1978). *Nature (London)* **273**, 247–248.
Phillips, S. E. V. (1980). *J. Mol. Biol.* **142**, 531–554.
Phillipson, P. E., Ackerson, B. J., and Wyman, J. (1973). *Proc. Natl. Acad. Sci. U.S.A.* **70**, 1550–1553.
Reynolds, A. H., Rand, S. D., and Rentzepis, P. M. (1981). *Proc. Natl. Acad. Sci. U.S.A.* **78**, 2292–2296.
Sawicki, C. A., and Gibson, Q. H. (1976). *J. Biol. Chem.* **254**, 1533–1542.
Sebban, P., Coppey, M., Alpert, B., Lindqvist, L., and Jameson, D. M. (1980). *Photochem. Photobiol.* **32**, 727–731.
Shank, C. V., Ippen, E. P., and Bersohn, R. (1976). *Science* **193**, 50–51.
Shelby, R. M., and McFarlane, R. M. (1979). *Chem. Phys. Lett.* **64**, 545–549.
Skinner, J. L., and Wolynes, P. (1978). *J. Chem. Phys.* **69**, 2143–2150.
Smith, K. M., ed. (1975). "Porphyrins and Metalloporphyrins." Elsevier, Amsterdam.
Spartalian, K., Lang, G., and Yonetani, T. (1976). *Biochim. Biophys. Acta* **428**, 281–290.
Spiro, T. G., and Strekas, T. C. (1972). *Proc. Natl. Acad. Sci. U.S.A.* **69**, 2622–2626.
Spiro, T. G., and Strekas, T. C. (1974). *J. Am. Chem. Soc.* **96**, 338–345.
Sternberg, M. J. E., Grace, D. E. P., and Phillips, D. C. (1979). *J. Mol. Biol.* **130**, 231–253.
Takano, T. (1977a). *J. Mol. Biol.* **110**, 537–568.
Takano, T. (1977b). *J. Mol. Biol.* **110**, 569–584.
Terner, J., Stong, J. P., Spiro, T. G., Nagumo, M., Nicol, M. F., and El-Sayed, M. A. (1981). *Proc. Natl. Acad. Sci. U.S.A.* **78**, 1313–1317.
Weber, G., and Teale, F. J. W. (1959). *Discuss. Faraday Soc.* **27**, 134–141.
Weissbluth, M. (1974). "Hemoglobin." Springer-Verlag, Berlin and New York.
Yonetani, T., Iizuka, T., Yamamoto, H., and Chance, B. (1973a). In "Oxidases and Related Redox Systems" (T. E. King, H. S. Mason, and M. Morrison, eds.), pp. 401–404. Univ. Park Press, Baltimore, Maryland.
Yonetani, T., Yamamoto, H., and Iizuka, T. (1973b). *J. Biol. Chem.* **249**, 2168–2174.
Zerner, M., and Gouterman, M. (1966). *Theor. Chim. Acta* **4**, 44–63.
Zerner, M., Gouterman, M., and Kobayashi, H. (1966). *Theor. Chim. Acta* **6**, 363–400.

CHAPTER 15

The Study of the Primary Events in the Photolysis of Hemoglobin and Myoglobin Using Picosecond Spectroscopy

Lewis J. Noe

Department of Chemistry
University of Wyoming
Laramie, Wyoming

I.	Introduction	339
II.	Review of Binding and Photodissociation of Molecular Oxygen and Carbon Monoxide in Heme Compounds	340
III.	Experimental Results	346
IV.	Picosecond Photodissociation Experiments, Related Experiments, and Results	348
V.	Conclusions	354
	References	355

I. Introduction

Research involving the photolysis of six-coordinate iron(II), CO, and O_2 complexes of hemoglobin (Hb) and myoglobin (Mb) has been of intense interest ever since Gibson's (1956a,b; Gibson and Ainsworth, 1957) initial description of the photoinitiated recombination kinetics of CO with deoxy-Hb and Mb. The primary reason for interest in this research area is derived from the ability to interrogate optically the dynamics of the religation of photolyzed heme in various time frames, ranging from picosecond to millisecond, in order to gain insight into the cooperative binding of O_2 to Hb. In the allosteric model, cooperativity is thought to be governed by a crucial R-T quaternary structural transformation. Certainly, a great deal of the present-day understanding of the spectroscopy and cooperative binding of O_2 in Hb must be credited to the x-ray-determined structure of Mb and Hb established by Kendrew *et al.* (1960) and Perutz *et al.* (1960). The reader may wish to consult, for example, the recent review by Parkhurst (1979) for a thorough discussion of the ligand kinetics and cooperativity in hemoglobin.

This chapter is devoted to a review of the dynamics of the initial or primary stages of Hb photolysis and religation that have been studied using picosecond spectroscopy. This will be done in three parts. In Section II, the binding in oxy-, carboxy-, and deoxy-Hb, Mb, and model chromophores will be discussed, with emphasis on the latest theoretically and experimentally determined electronic energy level information. In Section III, the picosecond absorption apparatus and methods will be discussed briefly. Finally, Section IV will review the picosecond absorption and resonance Raman experiments in terms of transient photoproduct spectra. Here, the emphasis will be on analysis of the results in terms of quantum yields of photodissociation and of possible radiationless pathways of photodissociation leading to ground state products and/or recombination. Related photolysis studies that have been analyzed using resonance Raman and transient absorption techniques in the nanosecond range will also be discussed briefly.

II. Review of Binding and Photodissociation of Molecular Oxygen and Carbon Monoxide in Heme Compounds

Crucial to any spectroscopically based interpretation of picosecond or nanosecond photodissociation experiments on oxy- and carboxy-Hb and Mb is a workable understanding of the disposition of the molecular orbitals and states in these or in model molecules. Thus far, the agreement between theoretical assignment and spectroscopic assignment of the electronic states for these molecules is not altogether satisfying. The origin of this variance lies in part in the electronic complexity of five- and six-coordinate iron(II) porphyrin complexes and the necessity for modeling calculations on a semiempirical basis or an *ab initio* basis limited to simplified iron porphyrin analogs. The source of the conjugate experimental problem lies in the difficult process involving identification of the numerous $Fe^{2+}(d^6)$ five- and six-coordinate metal–metal (dd), metal–ligand $(d\pi^*)$ charge transfer (CT), and ligand–metal (πd) CT states. Progress in both these areas has recently been made, permitting more accurate analyses of photodissociation experiments. Certain aspects of the binding and photodissociation of O_2 and CO in heme derivatives that are of fundamental interest in the interpretation of photodissociation experiments will be reviewed in this section.

The binding of O_2 to Hb, and Mb has been a subject of intensive investigation over the past 45 years. This work includes the Pauling and Coryell (1936; Pauling, 1977) 120° Fe—O—O bent model, the coplanar O—O—Fe ring model of Griffith (1956), the Weiss (1964) superoxide Fe^{3+}—O_2^- model, the two-electron oxidative addition model of Gray (1971), and the ozone model of Olafson and Goddard (1977). The question concerning the orientation of the O_2 molecule was resolved by Collman and associates (1974, 1975) who found a bent 126° Pauling-type geometry from their x-ray studies on a synthetic "picket fence" iron(II) porphyrin. Subsequent to this study Collman et al. (1976) suggested that the Fe—O—O geometry in HbO$_2$ was similar to that of the 126° geometry in the picket fence model compound, this conclusion being based on the closeness of the O$_2$ vibrational frequencies between these molecules. In contrast to these rather definitive structural studies, the geometry of CO in HbCO is not as well resolved, as shown in studies by Norvell et al. (1975), Heidner et al. (1976), Peng and Ibers (1976), and Case and Karplus (1978). For the purposes of theoretical calculations, however, most investigators assume a linear three-center Fe—C—O bond that is normal to the plane of the porphyrin.

A number of theoretical investigations, semiempirical and *ab initio*, have been carried out on model heme compounds with the intent of assigning specific orbital promotions and corresponding state descriptions and/or examining the stability of iron–porphyrin complexes as a function of various geometrical descriptions. Chronologically, these investigations include (1) the extended Hückel calculations of Zerner et al. (1966) using linear Fe—O—O geometry, (2) an *ab initio* valence bond calculation by Goddard and Olafson (1975; Olafson and Goddard, 1977) using NH$_2$ groups and NH$_3$ as representative of the porphyrin and imidazole moieties, respectively, (3) an *ab initio* calculation by Dedieu et al. (1976) using the picket fence geometry(ies) and NH$_3$ in place of the proximal imidazole, (4) an extended Hückel calculation by Kirchner and Loew (1977) using several different Fe—O—O geometries, (5) an extended Hückel calculation by Eaton et al. (1978) using the parameters of Zerner and associates, but done using a C_{2v} Fe—O—O symmetry restriction as an aid in assigning their experimental optical spectra, and finally (6) the Xα and PPP calculations of Case and Karplus (1979).

Recent advances in the optical assignments of divalent iron deoxy-, oxy-, and carboxy- Hb and Mb molecules are primarily the result of the analysis of polarized single-crystal absorption measurements and of circular dichroism (CD) and magnetic circular dichroism (MCD) measurements in solution. In terms of orbital promotions these spectral assignments have been made with reference to the theoretical predictions detailed in the references cited above. Included in the single-crystal studies are investigations carried out by (1) Eaton and Hochstrasser (1968) on Fe^{3+} Mb compounds, (2) Makinen and Eaton (1973) on HbCO and HbO$_2$, (3) Churg and Makinen (1978) on MbCN, MbO$_2$, MbCO, and Mb, and (4) Eaton et al. (1978) on HbO$_2$, HbCO, and Hb. The assignments made with the assistance of CD and MCD measurements in solution are the result of investigations by (1) Cheng et al. (1973) on high-spin Hb compounds in the near-infrared (IR) spectral region, (2) Stephens et al. (1976) on a review of CD and MCD measurements applied to Hb compounds, and (3) Eaton et al. (1978) on HbO$_2$, HbCO, and Hb.

Although the present catalog of ultraviolet (UV), visible, and near- IR spectral assignments for Hb derivatives is by no means definitive, it does serve as a common ground for a discussion of the photodissociation experiments described in the next section. The reader is referred to the articles by Eaton et al. (1978) and Case et al. (1979) for a summary of the optical assignments and for a comparison of the semiempirical and *ab initio* theoretical assignments to the observed spectral data. The reader may also wish to consult the text written by Weissbluth (1974) for an overall discussion of the structure, cooperativity, and electronic properties of Hb. The atomic and crystal field properties of divalent and trivalent iron are thoroughly discussed by Weissbluth. In the discussion which follows we will summarize the optical assignments in terms of the various classes of orbital promotions. Please note that in crystal field theory it is customary to choose the x- and y-axes as bisecting the pyrrole nitrogen atoms in the porphyrin plane, whereas in the recent articles dealing with MO theory these axes are chosen to bisect the methylene carbon atoms of the porphyrin moiety. Thus the d_{xy} orbital in the work of Zerner et al. (1966) is represented as the $d_{x^2-y^2}$ orbital by Case et al. (1979). Where necessary, we will use the crystal field axis convention.

The xy-polarized $\pi-\pi^*$ porphyrin transitions account for the well known Q, B, N, and L bands characteristic of O_2-, CO-, and deoxy-Hb and Mb as well as other metalloporphyrins. The basic features of these bands were first discussed theoretically by Kobayashi (1959a,b) and Weiss et al. (1965). In porphyrin, these bands are located at 555, 392, 317, and 263 nm, respectively. For reasons discussed by Case et al. (1979) the agreement between extended Hückel, Xα, and PPP average one-electron $\pi-\pi^*$ promotional energies and experimental band positions is reasonably consistent. There is no question as to the assignment of these $\pi-\pi^*$ bands.

Of particular interest in regard to photodissociation experiments are the Q- and B (Soret)-band systems because of their spectral positions in relation to second- and third-harmonic frequencies of neodymium–glass and neodymium–YAG pulsed lasers. Depending on the type of sixth axial ligand substituent, O_2, CO, or deoxy, the electronic origins of the Q and B bands lie at approximately 560 nm (oscillator strength, $f \sim 0.1$) and 415 nm ($f \sim 1$), respectively. Optically pumping the Q and B bands with 530- and 353-nm neodymium harmonics results in the photodissociation of HbCO, HbO$_2$, and the Mb analogs. Furthermore, it is common practice to follow the course of the photodissociation subsequent to the photolysis pulse by monitoring optical density changes in the Q and B bands. The energy and intensity of the B- and Q-band transitions are commonly accounted for by invoking configuration interaction (CI) between the nearly degenerate $\mathbf{D}_{4h}$ in-plane polarized porphyrin one-electron promotions $a_{2u}(\pi) \to e_g^*(\pi)$ and $a_{1u}(\pi) \to e_g^*(\pi)$. Theoretical results indicate the two-electron promotion $a_{2u}(\pi), a_{1u}(\pi) \to$

$e_g^*(\pi)$ assigned to the B- and Q-band systems contain little, if any, metal character. In addition to being less intense than the B band, the Q transition is split into the electronic origin, Q_0, and the first vibronic band, Q_v. Together, these band systems account for the gross recognizable features in Hb spectra. Of importance here is the fact that, in each of the derivatives of Hb, the Q and B transitions have significantly different shapes, intensities, and positions. The text by Antonini and Brunori (1971) discusses these differences. As a result of such differences one has at hand a very useful probe for following the temporal and spectral changes during and subsequent to the photodissociation of HbO_2 and HbCO. For example, optical density changes recorded as a function of time and wavelength in the Q- and B-spectral regions can be compared to the static Hb-HbO_2 and Hb-HbCO difference spectra. The analysis of this type of data can in principle reveal the extent of the photodissociation and/or recombination and also the existence of transient intermediates or electronic states whose lifetimes fall within the capabilities of the experimental apparatus.

A detailed discussion of the remaining CT transitions, $d \to \pi^*$, $\pi \to d$, and $\pi, d \to CO(\pi^*)$, metal–metal transitions, $d \to d$, and intraligand transitions on O_2 will not be undertaken in this chapter for several reasons. First, this is a fairly complex topic that has been reviewed in some detail both theoretically and experimentally in the papers by Eaton et al. (1978), Churg and Makinen (1978), and Case et al. (1979). Eaton and associates have labeled these experimentally observed transitions I–VII. The crystal field calculations of Eicher et al. (1976) and Huynh et al. (1974) should also be considered in the assignment of $d \to d$ transitions and states. Second, there are fairly sizable discrepancies between the calculated and experimental transition energies, greater than 2 eV in some instances, not to mention disagreement in regard to the assignment of metal states in HbO_2 between the crystal field approach and the ozone modeling of Goddard and Olafson (1975). Also, there are a great number of predicted CT transitions that have not as yet been observed. The most severe case is that of HbCO and MbCO where the only two observed transitions, other than porphyrin transitions, have been assigned to $d \to d$ promotions by Eaton et al. (1978) and Churg and Makinen (1978). Nevertheless, the developments summarized here represent a considerable advance in the understanding of the electronic states in Hb. Furthermore, although the absolute transition energies predicted by theory may be somewhat poor, there is reason to believe that the relative ordering of the electronic states is reasonable. This latter point is of importance in the assignment of mechanisms to the photodissociation of HbCO and HbO_2.

In order to facilitate the discussion of the metal and ligand orbitals, we have reproduced the energy level diagram by Greene et al. (1978) in Fig. 1. Although there are some changes in the energy levels indicated by the recent

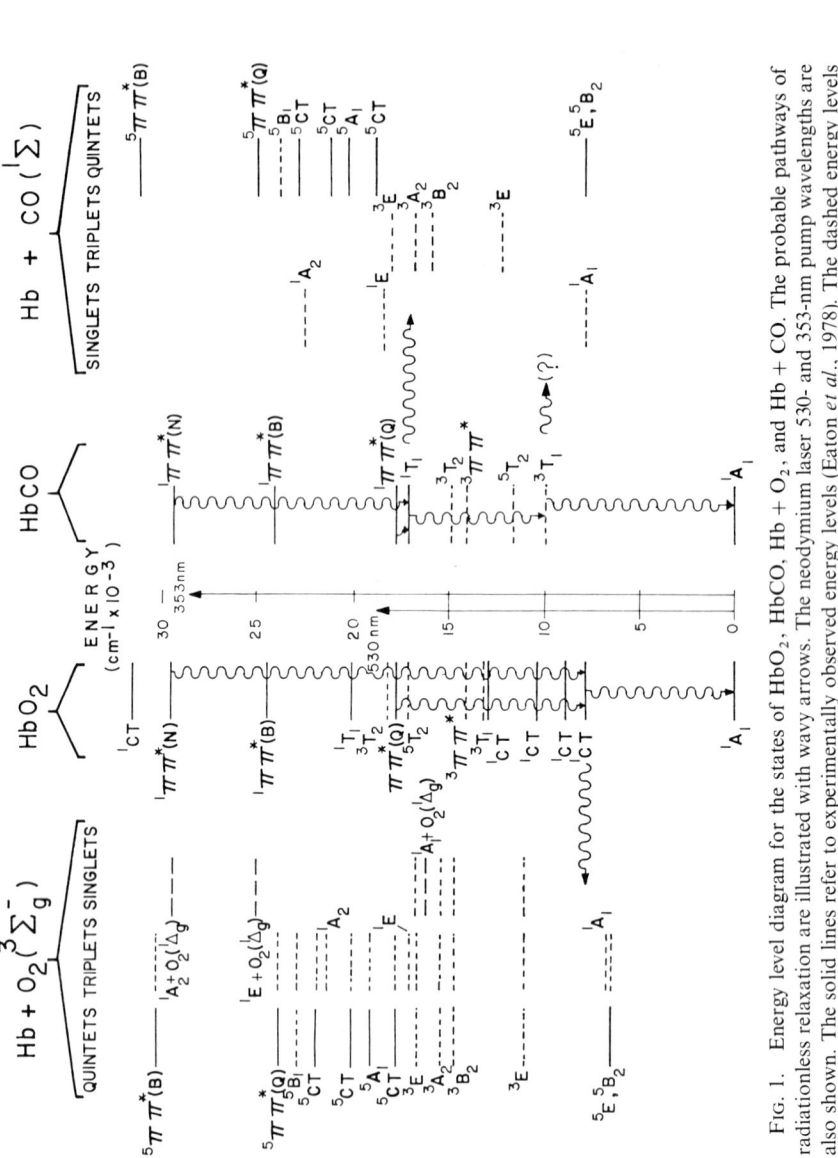

FIG. 1. Energy level diagram for the states of HbO_2, HbCO, Hb + O_2, and Hb + CO. The probable pathways of radiationless relaxation are illustrated with wavy arrows. The neodymium laser 530- and 353-nm pump wavelengths are also shown. The solid lines refer to experimentally observed energy levels (Eaton et al., 1978). The dashed energy levels refer to extended Hückel calculations of Eaton et al. Certain crystal field results of Eicher et al. (1976) were used in the diagram. (From Greene et al., 1978; reproduced with permission.)

PPP calculation of Case et al. (1979), Fig. 1 represents as it stands a summary of various types of spectroscopic assignments, both observed and calculated, using extended Hückel theory and the crystal field results of Eicher et al. (1976). The illustration brings to light the electronic complexity of the Hb chromophore.

In terms of octahedral crystal field notation, the ground states of HbCO and of HbO$_2$ are $^1A_1$, having a low spin d^6 configuration $d_{xy}^2 d_{xz}^2 d_{yz}^2$ (t_2^6), whereas according to Olafson and Goddard (1977), the ground state of Hb is a quintet $^5T_2$ having the high spin configuration $d_{xy}^1 d_{xz}^2 d_{yz}^1 d_{x^2-y^2}^1 d_{z^2}^1$ ($t_2^4 e^2$). Eicher et al. (1976) points out that one should represent the crystal field states of these compounds in tetragonal notation, such as C_{4v}, due to spin orbit splitting and large tetragonal distortion. This notation is used in Fig. 1 for the states of Hb; the $^5T_2$ ground state splits into the tetragonal ^5E and $^5B_2$ components, and the T_1 states split into A_2 and E components.

The highest crystal field $^1T_1$ state of HbCO is located slightly below the ^1Q porphyrin band at about 17,000 cm^{-1}. This state is derived from the observed 17,750 and 16,000 cm^{-1} transitions that have been assigned according to CD and single-crystal measurements to the tetragonally based transitions $^1A_1 \rightarrow {}^1E_1$ ($d_{xz}^2, d_{yz}^2 \rightarrow d_{z^2}, d_{x^2-y^2}$) and $^1A_1 \rightarrow {}^1A_2$ ($d_{xy}^2 \rightarrow d_{x^2-y^2}$), respectively. Case et al. (1979) suggest that these states should be ordered with E as the lowest. From this orbital description it can be seen that the $^1T_1$ state becomes dissociative upon population of the empty and antibonding d_{z^2} orbitals.

The indicated electronic energy level diagrams for HbO$_2$ and Hb are more complex than the one for HbCO. For HbO$_2$ there is an observed crystal field $^1T_1$ state, V, above the ^1Q porphyrin state, and a projected $^5T_2$ state just below it. Additionally, there are a number of projected CT bands located below an observed CT band, IV, located at 12,500 cm^{-1}. There is also a $^3T_1$ state projected to lie just above CT IV. Noteworthy is the fact that Olafson and Goddard (1977) view the binding in HbO$_2$ according to an ozone model, O (^3P) + O$_2$ ($^3\Sigma_g^-$), in which the low-spin S_0 ground state of HbO$_2$ is essentially the triplet oxygen O$_2$ ($^3\Sigma_g^-$) coupled to the triplet component of iron ^3E (or ^3T) having configuration $t_g^5 e^1$.

In Hb we find from Fig. 1 that there are a number of CT and $d \rightarrow d$ transitions located below the ^1Q band. Crystal field calculations place an $^1A_1$ state just above and almost degenerate with the $^5T_2$ ground state. These calculations also place a $^3T_1$ state a few thousand wave numbers above the $^1A_1$ state.

The spectroscopic work of Crosby (1975) on d^6 transition metal complexes suggests that the degree of spin orbit coupling present in iron(II) complexes does not preclude them from processing normal spin-based Kasha (1950) organic behavior. According to Crosby, this spin normality most likely does not exist for the d^6 complexes of ruthenium and osmium where the degree of spin orbit coupling is very much greater. Further evidence of normal behavior in this regard is suggested by the independence of photolysis yields of the Hb derivatives on excitation wavelength in the study by Hoffman and Gibson (1978) and by the resonance Raman work of Friedman and Hochstrasser (1974) on related molecules. Normal spin behavior in d^6 Hb derivatives is therefore assumed. In terms of the radiationless deactivation of excited state(s) populated via an optical pump of some type, this means that internal conversion is expected to occur to the lowest excited state in a given manifold having a fixed spin multiplicity. Further radiationless decay of this state is partitioned between direct internal conversion to the ground state if it has the same multiplicity and intersystem crossing to nearby states of different multiplicity via first or, less likely, second-order spin orbit mechanisms involving the ground state.

Thus, an important consideration in the analysis of HbCO and HbO$_2$ photodissociation experiments is a realization of the various radiationless pathways leading to ground state product and/or recombination. Certain possibilities concerning the photodissociation of HbCO and HbO$_2$ and radiationless relaxation are indicated in Fig. 1.

III. Experimental Results

The preparation of the various derivatives of Hb is adequately discussed by Antonini and Brunori (1971). Many spectroscopic and x-ray investigations have followed the standard preparations given by Perutz (1968) and Giuseppe et al. (1969). Horse and sperm whale Mb may also be purchased as a lyophilized powder in need of further purification. The procedure described by Bauer and Pacyna (1975) can be used to convert trivalent iron forms of Hb or Mb into corresponding divalent forms. The concentrated heme solutions may be stored in pellet form in liquid nitrogen, and samples prepared just before use with potassium phosphate or Tris [tris(hydroxymethyl)amino-methane] and bis-Tris [bis(2-hydroxymethyl)imino-tris(hydroxymethyl)methane] buffers. For Hb experiments, a suitably high Hb concentration is needed to ensure that the kinetics due to the photodissociation and recombination are directly related to cooperative tetramers rather than noncooperative dimers. This dimer–tetramer equilibrium is discussed by Gibson and Antonini (1967) and by Antonini et al. (1972). A procedure and device for saturating buffered solutions of HbCO or HbO$_2$ and Mb analogs with CO or O$_2$, respectively, is illustrated in Fig. 1 in the article by Noe et al. (1978). This apparatus is interconnected to the sample and reference cuvets and is used to exchange the solutions periodically in these cells when necessary.

Many of the experiments described in the next section have been done using a double-beam picosecond absorption spectrometer. A typical spectrometer of this type that is presently in use in the authors' laboratory is illustrated in Fig. 2. Noteworthy is the use of a 42-nm band-pass subtractive monochromator before the spectrograph that eliminates laser excitation light and a dual wide-area Reticon RL512SF 512-element, 2.5 mm high, intensified diode array detector. This system has a resolution of 0.82 Å per array element with an optical arrangement that permits one array to detect the sample beam, I, while the other array detects the reference beam, I_0.

A comprehensive analytical and experimental evaluation of transient absorption measurements using this type of apparatus has been made by

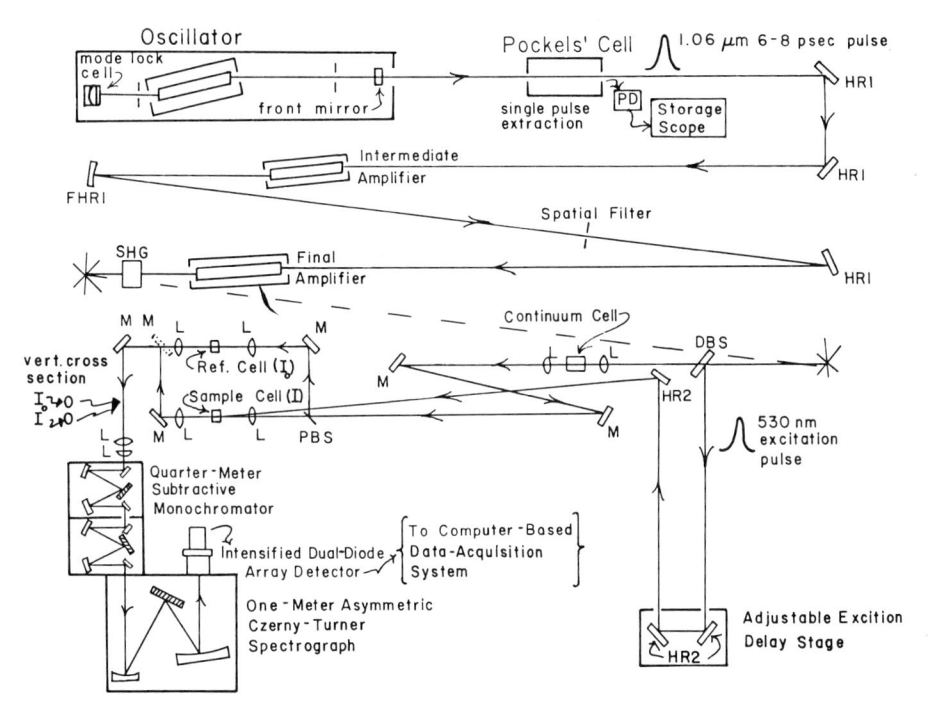

FIG. 2. Schematic diagram of a two-stage amplified, spatially filtered, Nd^{3+}-phosphate glass TEM_{00} laser, double-beam picosecond spectrometer. The diagram is not drawn to scale. The excitation (optical pump) is fixed for 530 nm, although other harmonics of the 1.06-μm fundamental can be used. The view of the spectrometer is from the top except for the vertical cross section of the I_0(reference)/I(sample) beams. PD, Photodiode; HR1, 1.06-μm high reflector; HR2, 530-nm high reflector; SHG, second harmonic generator; DBS, dichroic beam splitter; M, first surface aluminized mirror; L, lens; PBS, pellicle beam splitter. Typical characteristics of this laser system: (1) maximum repetition rate of four shots per minute; (2) average amplified 1.06-μm pulse of 40×10^{-3} J with a pulse duration of 6–8 psec.

Greene and associates (1979). For the purposes of this chapter a general description of the picosecond spectrometer will suffice. The mode-locked neodymium–laser permits one to pump optically or excite a sample, after single pulse extraction and amplification, at the fundamental wavelength 1.06 μm or at any of the harmonics 530, 353, or 265 nm, depending on the type of harmonic crystal generator(s) used. The unconverted fundamental light and/or harmonic light is then used to generate a continuum in the range 400–900 nm. The continuum light passes through the sample and is used for the purpose of interrogating, in a double-beam absorption mode, the time rate of change of the excited state population as a function of time after arrival of the excitation pulse. Since the continuum light has the same temporal half-width, 6–8 psec, as the excitation pulse, one can use this apparatus to measure lifetimes of electronic states or intermediates in the range of 10 psec to 5 nsec.

IV. Picosecond Photodissociation Experiments, Related Experiments, and Results

In this section we will summarize and briefly discuss the results of the various picosecond photodissociation investigations that have been carried out on the carboxy- and oxy- derivatives Hb and Mb. Additionally, a survey of related photodissociation experiments that have been analyzed using time-resolved resonance Raman techniques in the picosecond and nanosecond temporal range will be discussed.

The investigation by Shank et al. (1976) was the first to study the photolysis of HbO_2 and HbCO using subpicosecond pulses at 615 nm. The 615-nm pulses were generated from a passively mode-locked dye laser at a repetition rate of 10 kHz and used to excite and interrogate the Hb samples at times delayed from the excitation. Figure 3 shows the experimental results of Shank et al. for HbCO. After correction for the coherence effect between the pump and probe beams near zero delay, an induced absorption in HbCO was found to occur in 0.5 psec and remain at a steady state level with no apparent decay during the 20-psec observation period. For HbO_2, a similar rapid induced absorption occurred which was directly followed by a decay to the initial unphotolyzed level having a time constant of 2.5 psec. Experiments with Hb gave the same results as those for HbO_2. These results were interpreted by Shank et al. primarily on the basis of the quantum yield of photodissociation of HbCO and HbO_2. The work of Gibson and Ainsworth (1957), Suffran and Gibson (1977), and Alberding et al. (1978) has shown the quantum yield of photodissociation of HbCO to be 0.5. The

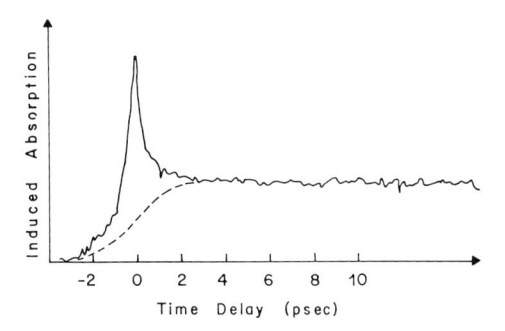

FIG. 3. Induced absorption at 615 nm in HbCO versus time delay between the excitation and probe pulses. The dashed line is the response corrected for a coherence artifact between the pump and probe beams. (From C.V. Shank, E.P. Ippen, and R. Bersohn. *Science* **193** pp. 50–51, No. 4247, 1976. Copyright 1976 by the American Association for the Advancement of Science.)

quantum yield of photodissociation of HbO_2 was set at 0.05 by Suffran and Gibson (1977). Using literature values for quantum yields similar to these, Shank *et al.* reasoned that the high yield for HbCO was due to a fast dissociation followed by a slow recombination. They also proposed that the low quantum yield for HbO_2 was due to a dissociation rate much slower than their observed 2.5-psec recovery, rather than a rapid recombination of photolyzed HbO_2 molecules.

Noe *et al.* (1978) studied the photodissociation of HbCO using 530-nm, 6- to 9-psec pulses from a neodymium–glass laser. The kinetics of the photodissociation were monitored by following absorbance changes from 0 to 300 psec in the Soret band at the single wavelength of 440 nm. Based on a kinetic calculation using observed optical density changes, they reported that the photodissociation took place in 11 psec with no evidence of recombination from 48 to 300 psec. At times delayed from the 530-nm excitation pulse starting at 50 μsec, they also studied the recombination kinetics by following absorption changes at 441.7 nm using a helium–cadmium laser, photomultiplier, and storage scope. These investigators proposed that the 11-psec lifetime, which is significantly longer than the 0.5-psec lifetime reported by Shank *et al.* (1976) exciting at 615 nm, was due to additional radiationless processes occurring in the HbCO-Hb manifold of electronic states that happen before reaching the $^5T_2$ ground state of Hb. Consideration of the relative position of the $^1T_2$ state of HbCO with reference to the 615- and 530-nm excitation wavelengths suggested by Fig. 1 in the previous section could be of use in clarification of these two studies. However, most of the energy level information detailed in Fig. 1 was not available at the time these studies were undertaken.

The photolysis of HbO_2 and HbCO has also been investigated by Greene and associates (1978) using 353- and 530-nm 8-psec excitation pulses generated with a neodymium–glass laser. The sample concentration was chosen to exhibit a maximum passive absorbance in the probing region while still transmitting a sufficient amount of probing light to facilitate accurate transient absorption measurements. The induced optical density changes were recorded in 50-nm-wide segments in both the Soret and Q-band spectral regions. By adding to the induced optical density difference spectra, which contain contributions from both photoproduct and bleached starting material, the appropriate amount of steady state unphotolyzed sample absorbance, these investigators were able to derive the transient absorbance spectra of photoproduct at certain fixed delay times. In order to do this it is necessary to know the fraction of the starting material photolyzed. From analysis of the transient absorbance spectra at 8, 20, and 680 psec, this group concluded that (1) the photoproduct of HbCO and HbO_2 appeared within 8 psec and each had nearly identical deoxy-Hb-like transient spectra but were considerably broadened from the steady state Hb spectrum, and (2) at 680 psec, recombination of CO and O_2 with the heme moiety was less than 10 and 20%, respectively. Based on the lack of evidence for an intermediate upon photolysis of Hb, and the fact that the transient spectra were consistently broadened throughout the 680-psec period of observation, Greene *et al.* suggested multiphoton processes as the probable cause of this broadening. They discussed this point on the basis of the energy level diagram given in Fig. 1 in terms of certain "bottleneck" states such as $^3T_1$ in HbCO and ^3CT in HbO_2, that could, for example, absorb additional photons pumping the heme to higher states.

Eisert and associates (1979) have followed the kinetics of the photodissociation of MbCO and MbO_2 with picosecond methods at various wavelengths in the Soret and visible Q-band systems. They reported that the photodissociation of these molecules took place in less than 8 psec and concluded that within the 450-psec observation period the absorbance changes in the Soret and Q-band regions corresponded closely with the steady state Mb-MbCO and Mb-MbO_2 difference spectra. They also found the initial MbCO photoproduct signal to undergo a $20 \pm 5\%$ decay, which was monitored at 440 nm and had a lifetime of 125 ± 50 psec. There was no evidence of such a decay in MbO_2. These results are shown in Fig. 4. They further concluded that at times greater than 300 psec after excitation that the MbCO photoproduct was the same as the MbO_2 photoproduct which was generated in 8 psec and that this photoproduct was probably deoxy-Mb or similar to it. Based on these findings, Eisert *et al.* proposed that the 125-psec relaxation observed for MbCO could be due to a pathway for the photodissociation that included a low-lying triplet state in the Mb manifold

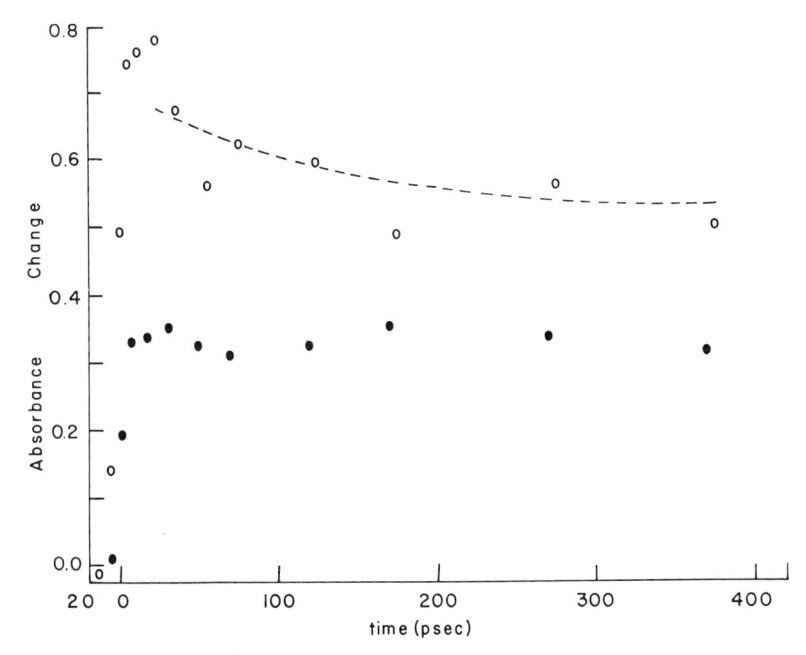

FIG. 4. Absorbance changes in MbCO and MbO$_2$ at 440 nm after single-pulse excitation at 530 nm. ○, MbCO; ●, MbO$_2$; ----, least-squares fit ($k = 7.6 \pm 2.7) \times 10^9$ sec^{-1}). (From Eisert et al., 1979.)

before relaxing via an intersystem radiationless process to the quintet ground state of Mb or that it could be related to differences in tertiary structural changes confined to the heme pocket.

The recent investigation by Chernoff and associates (1981) on geminate recombination of O$_2$ and Hb, using 530-nm pulses, has revealed some important new results in terms of heretofore undected rapid spectral changes in HbO$_2$ photoproduct, followed by a relatively slow 200 ± 70 psec recombination. In this study, which extended out to 1.2 nsec, photoproduct was observed to be promptly formed for both HbO$_2$ and HbCO, but no recombination of Hb plus CO was observed. Unlike HbO$_2$, there was no rapid spectral evolution observed for HbCO. Comparison of the induced difference spectra, Hb-HbCO and Hb-HbO$_2$, with corresponding steady state difference spectra revealed that the photoproduct spectra of both HbO$_2$ and HbCO were broader, weaker, and red-shifted in comparison to the fast-reacting form of deoxy-Hb, Hb*, discussed by Sawicki and Gibson (1976). This group has discussed these results in terms of specific proposals for radiationless pathways, for both dissociation and recombination, among the CT and crystal field states shown in Fig. 1. For HbCO, they suggest as an example the following dissociative pathway involving first-order spin orbit coupling as a possibility:

$$HbCO(^1A_1) \xrightarrow{530\,nm} HbCO(^1T_1) \longrightarrow HbCO(^3T_1) \longrightarrow HbCO(^5T_2) \longrightarrow Hb(^5T_2) + CO(^1\Sigma).$$

They also suggest that the recombination of Hb ($^5T_2$) with CO ($^1\Sigma$) to yield HbCO ($^1A_1$) is an energetically difficult process that should require substantial activation energy as well as second-order spin orbit processes. In the case of HbO$_2$, Chernoff and associates favor the proposal that the low-lying ^{3}CT states of HbO$_2$ are not strongly coupled to the Hb manifold. They may, however, couple through a first-order spin orbit path to the higher HbO$_2$ ($^5T_2$) state which is photodissociative, since it directly correlates to the Hb ($^5T_2$) ground state. This proposal would account for the small quantum yield of HbO$_2$ dissociation. They also suggest that, because of the involvement of certain first-order spin orbit processes in the recombination of O$_2$ ($^3\Sigma$) with Hb ($^5T_2$), the barrier to recombination may be significantly lower than in HbCO, thus increasing the possibility of early recombination in the HbO$_2$ system over that in the HbCO system.

Before ending this discussion on the detection and analysis of transient picosecond photoproducts of HbCO and HbO$_2$, the recent picosecond resonance Raman (RR) study of HbCO by Terner and associates (1980) should be described. Since the theory and applications of RR spectroscopy have been recently reviewed by Mathies (1979) and by Spiro and Stein (1977), we will limit the discussion to the results of this study which are shown in Fig. 5. The study was undertaken in an attempt to provide detailed vibrational-structural information on the picosecond iron–porphyrin photointermediate. Using 30-psec 576-nm pulses, these investigators observed three vibrational bands between 1540 and 1620 cm^{-1} which are distinct from the RR vibrational structure of HbCO and Hb in terms of band position. From previous work by Spiro and Burke (1976), these bands were identified as structure-sensitive porphyrin bands. In particular, the bands at 1552 and 1603 cm^{-1} have been correlated with the iron(II) prophyrin core size and their relative position with reference to the analogous vibrational bands in deoxy-Hb is a sensitive measure of the local heme geometry. These bands are 4–6 cm^{-1} lower in energy in the picosecond RR spectrum of HbCO than in the RR spectrum of deoxy-Hb. Terner and colleagues have associated this energy shift with an expanded porphyrin iron core size for the photoproduct of HbCO with the possibility of the iron atom residing in the porphyrin plane. They attribute these bands to a photointermediate that develops within 30 psec of the excitation. When their photolysis pulse is widened to 20 nsec, the RR spectrum remains unchanged and is in good agreement with the RR spectrum recorded by Lyons et al. (1978) using

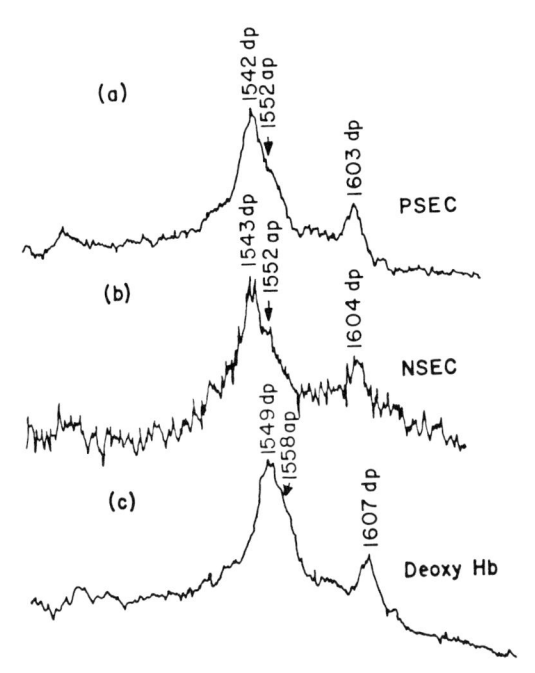

FIG. 5. (a) RR of the photoproduct of HbCO obtained by computer subtraction of partially photolyzed and nonphotolyzed flowing HbCO. Laser pulses were 30 psec in duration, 576 nm, 10 nJ, and applied at the rate of 0.8 MHz. (b) Identical to (a) with the exception of a 20-nsec pulse duration. (c) Steady-state RR of Hb at 5752 Å. (From Terner et al., 1980. Reprinted with permission from *Journal of the American Chemical Society* **102**, 3238–3239. Copyright 1980 American Chemical Society.)

10-nsec pulses. Terner and associates have proposed that this photointermediate whose lifetime extends into the nanosecond range corresponds to a molecular species having iron(II) in the quintet, $S = 2$, state confined to the heme plane. This species may in fact be some form of deoxy-Hb. They suggest that times longer than 20 nsec are required for the iron and proximal imidazole to move away from the heme plane.

We will conclude this section with a brief review of transient absorption and resonance Raman work that has been done in the nanosecond and longer time ranges. The recombination processes $Hb + O_2$ and $Hb + CO$ that take place in the subnanosecond range are undoubtedly geminate in nature and occur without competition from diffusion-controlled bimolecular processes. In the submicrosecond and longer time ranges, geminate and normal bimolecular processes should become competitive, and quaternary, R-T, structural transformations are also expected to take place. Recent transient absorption and RR studies on Hb and Mb in the nanosecond and longer time ranges have addressed this subject in terms of the analysis of characteristic recombination times, the effect of temperature on recombination times and barrier heights, and the quantum yields of dissociation. The dynamics of binding of CO or O_2 to Mb have also been examined theoretically by Case and Karplus (1979), with particular emphasis on the effect of the globin matrix on barriers along possible reaction paths. Chapter 14, on the binding kinetics and flash photolysis of heme proteins, should be consulted for a thorough review of this area.

For the concluding remarks in this section, only a brief survey of this research area is warranted. In the microsecond and longer time ranges, the following absorption investigations are representative: Sawicki and Gibson (1979), Alberding et al. (1978), and Austin et al. (1975). Transient absorption studies in the nanosecond time region have been reported by Alpert et al. (1974), Duddell et al. (1979), and Alpert et al. (1979). From a study on the temperature dependence of the recombination of CO with Hb, Duddell and co-workers suggested that the 100-nsec relaxation was geminate recombination rather than a tertiary structural change as originally postulated by Alpert et al. (1974). Alpert and associates (1979) also investigated the nanosecond photolysis of HbCO. They observed a rapid 200-nsec recombination and a slower millisecond recombination. Using two pulse excitation experiments and the fact that the nanosecond recombination was independent of CO pressure, this group concluded that the nanosecond recombination was indeed geminate. They were also able to achieve 100% dissociation of the HbCO complex, although the recombination kinetics varied with intensity and excitation conditions. This was also noticed by Noe et al. (1978) in their picosecond study. The RR study by Friedman and Lyons (1980) confirmed the fact that the photolysis of HbCO resulted in a quantum yield near unity followed by substantial recombination by 100 nsec. Interestingly, this RR investigation showed no significant recombination of photolyzed MbCO in times less than 100 nsec. Finally, the latest RR work by Lyons and Friedman (1981) has demonstrated the existence of a "trigger mechanism" for the R-T quaternary transformation in HbCO. This determination was made by the analysis of the position of the 1357 cm^{-1} Raman line as function of both excitation power and of time in the high end of the submicrosecond range.

V. Conclusions

The picosecond absorption studies and the R Raman study that have been reviewed in this chapter have addressed the subject of the photophysics of the dissociation processes for the oxy, deoxy, and carboxy-Hb derivatives and Mb in terms of the analysis of transient photoproduct spectra and quantum yields of dissociation. The analysis of this data has been discussed according to the dynamics of processes involving radiationless relaxation that are considered to be favourable as suggested by the latest experimental and theoretical energy level information.

Still, there is not a clear picture of the dynamics of ligand dissociation and religation in these molecules. In point of example, there are substantial differences between the picosecond transient photoproduct spectra of HbO_2 and MbO_2; the nanosecond RR photoproduct spectra of HbCO show substantial recombination in 100 nsec, while no significant recombination occurs in MbCO in this time; the early picosecond photoproduct of MbCO shows decay, while MbO_2 does not. These instances indicate the requirement of further spectroscopic and theoretical work on these and on model chromophores, particularly because of the apparently strong effect that equilibrium geometry, possibly quaternary, has on the differences in ordering of certain electronic levels between Hb and Mb. Besides theoretical calculations, this work should probably include studies on the effect of excitation wavelength and temperature on transient photoproduct spectra and quantum yields.

ACKNOWLEDGMENTS

I wish to thank C. V. Shank, R. M. Hochstrasser, P. M. Rentzepis, T. G. Spiro, and M. A. El-Sayed for permission to use their figures and/or sending me advance copies of their work. I also wish to thank the Research Corporation, the National Science Foundation, and the College of Arts and Sciences in The University of Wyoming for support of my work in this area of research.

REFERENCES

Alberding, N., Chan, S. S., Eisenstein, L., Frauenfelder, H., Good, D., Gunsalus, I. C., Nordlund, T. M., Perutz, M. F., Reynolds, A. H., and Sorensen, L. B. (1978). *Biochemistry* **17**, 43–51.
Alpert, B., Bannerjee, R., and Lindqvist, L. (1974). *Proc. Natl. Acad. Sci. U.S.A.* **71**, 558–562.
Alpert, B., El Mohsni, S., Lindqvist, L., and Tfibel, F. (1979). *Chem. Phys. Lett.* **64**, 11–16.
Antonini, E., and Brunori, M. (1971). *Front. Biol.* **21**, pp. 1–39.
Antonini, E., Anderson, N. M., and Brunori, M. (1972). *J. Biol. Chem.* **247**, 319–321.
Austin, R. H., Beeson, K. W., Eisenstein, L., Frauenfelder, H., and Gunsalus, I. C. (1975). *Biochemistry* **14**, 5355–5373.
Bauer, C., and Pacyna, B. (1975). *Anal. Biochem.* **65**, 445–448.
Case, D. A., and Karplus, M. (1978). *J. Mol. Biol.* **123**, 697–701.
Case, D. A., and Karplus, M. (1979). *J. Mol. Biol.* **132**, 343–368.
Case, D. A., Huynh, B. A., and Karplus, M. (1979). *J. Am. Chem. Soc.* **101**, 4433–4453.
Cheng, J. C., Osborne, G. A., Stephens, P. J., and Eaton, W. A. (1973). *Nature (London)* **241**, 193–194.
Chernoff, D. A., Hochstrasser, R. M., and Steele, A. W. (1981). *Proc. Natl. Acad. Sci. U.S.A.* **77**, 5606–5610.

Churg, A. K., and Makinen, M. W. (1978). *J. Chem. Phys.* **68**, 1913–1925.
Collman, J. P., Gagne, R. E., Reed, C. A., Robinson, W. T., and Rodley, G. S. (1974). *Proc. Natl. Acad. Sci. U.S.A.* **71**, 1326–1329.
Collman, J. P., Gagne, R. R., Reed, C. A., Halbert, T. R., Lang, G., and Robertson, W. T. (1975). *J. Am. Chem. Soc.* **97**, 1427–1439.
Collman, J. P., Brauman, J. I., Halbert, T. G., and Suslick, S. (1976). *Proc. Natl. Acad. Sci. U.S.A.* **73**, 3333–3337.
Crosby, G. A. (1975). *Acc. Chem. Res.* **8**, 231–238.
Dedieu, A., Rohmer, M.-M., Benard, M., and Veillard, A. (1976). *J. Am. Chem. Soc.* **98**, 3717–3718.
Duddell, D. A., Morris, R. J., and Richards, J. T. (1979). *J.C.S. Chem. Commun.* pp. 75–76.
Eaton, W. A., and Hochstrasser, R. M. (1968). *J. Chem. Phys.* **49**, 985–995.
Eaton, W. A., Hanson, L. K., Stephens, P. J., Sutherland, J. C., and Dunn, J. B. R. (1978). *J. Am. Chem. Soc.* **100**, 4991–5003.
Eicher, H., Bade, D., and Parak, F. (1976). *J. Chem. Phys.* **64**, 1446–1455.
Eisert, W. G., Degenkelb, E. O., Noe, L. J., and Rentzepis, P. M. (1979). *Biophys. J.* **25**, 455–464.
Friedman, J. M., and Hochstrasser, R. M. (1974). *Chem. Phys.* **6**, 155–165.
Friedman, J. M., and Lyons, K. B. (1980). *Nature (London)* **284**, 570–572.
Gibson, Q. H. (1956a). *J. Physiol. (London)* **134**, 112–122.
Gibson, Q. H. (1956b). *J. Physiol. (London)* **134**, 123–134.
Gibson, Q. H., and Ainsworth, S. (1957). *Nature (London)* **180**, 1416–1417.
Gibson, Q. H., and Antonini, W. (1967). *J. Biol. Chem.* **242**, 4678–4681.
Giuseppe, G., Parkhurst, L., and Gibson, Q. H. (1969). *J. Biol. Chem.* **244**, 4664–4667.
Goddard, W. A., III, and Olafson, B. D. (1975). *Proc. Natl. Acad. Sci. U.S.A.* **72**, 2335–2339.
Gray H. B. (1971). *Adv. Chem. Ser. No.* 100, pp. 365–389.
Greene, B. I., Hochstrasser, R. M., Weisman, R. B., and Eaton, W. A. (1978). *Proc. Natl. Acad. Sci. U.S.A.* **75**, 5255–5258.
Greene, B. I., Hochstrasser, R. M., and Weisman, R. B. (1979). *J. Chem. Phys.* **70**, 1247–1259.
Griffith, J. S. (1956). *Proc. R. Soc. London, Ser. A* **236**, 23–36.
Heidner, E. J., Ladner, R. M., and Perutz, M. F. (1976). *J. Mol. Biol.* **104**, 707–722.
Hoffman, B. M., and Gibson, Q. H. (1978). *Proc. Natl. Acad. Sci. U.S.A.* **75**, 21–25.
Huynh, B. H., Papaefthymiou, G. C., Yen, C. S., Groves, J. L., and Wu, C. S. (1974). *J. Chem. Phys.* **61**, 3750–3758.
Kasha, M. (1950). *Discuss. Faraday Soc.* No. 9, pp. 14–19.
Kendrew, J. C., Dickerson, R. B., Standberg, B. E., Hart, R. G., Phillips, D. C., and Shore, V. C. (1960). *Nature (London)* **185**, 422–427.
Kirchner, R. F., and Loew, G. H. (1977). *J. Am. Chem. Soc.* **99**, 4639–4647.
Kobayashi, H. (1959a). *J. Chem. Phys.* **30**, 1362–1363.
Kobayashi, H. (1959b). *J. Chem. Phys.* **30**, 1373–1374.
Lyons, K. B., and Friedman, J. M. (1981). *In* "Interactions Between Iron and Proteins in Oxygen and Electron Transport" (C. Ho, ed.). Elsevier, Amsterdam. In press.
Lyons, K. B., Friedman, J. M., and Fleury, P. A. (1978). *Nature (London)* **275**, 565–566.
Makinen, M. W., and Eaton, W. A. (1973). *Ann. N.Y. Acad. Sci.* **206**, 210–222.
Mathies, R. (1979). *In* "Chemical and Biological Applications of Lasers" (C. B. Moore, ed.), Vol. 4, pp. 55–99. Academic Press, New York.
Noe, L. J., Eisert, W. G., and Rentzepis, P. M. (1978). *Proc. Natl. Acad. Sci. U.S.A.* **75**, 573–577.
Norvell, J. C., Nunes, A. C., and Schoenborn, B. P. (1975). *Science* **190**, 568–570.
Olafson, B. D., and Goddard, W. A., III (1977). *Proc. Natl. Acad. Sci. U.S.A.* **74**, 1315–1319.
Parkhurst, L. J. (1979). *Annu. Rev. Phys. Chem.* **30**, 503–546.
Pauling, L. (1977). *Proc. Natl. Acad. Sci. U.S.A.* **74**, 2612–2613.

Pauling, L., and Coryell, C. D. (1936). *Proc. Natl. Acad. Sci. U.S.A.* **22**, 210–216.
Peng, S. M., and Ibers, J. I. (1976). *J. Am. Chem. Soc.* **98**, 8023–8026.
Perutz, M. F. (1968). *J. Cryst. Growth* **2**, 54–56.
Perutz, M. F., Rossmann, M. G., Cullis, A. F., Muirhead, H., Will, G., and North, A. C. T. (1960). *Nature (London)* **185**, 416–422.
Sawicki, C. A., and Gibson, Q. H. (1976). *J. Biol. Chem.* **251**, 1533–1542.
Sawicki, C. A., and Gibson, Q. H. (1979). *J. Biol. Chem.* **254**, 4058–4062.
Shank, C. V., Ippen, E. P., and Bersohn, R. (1976). *Science* **193**, 50–51.
Spiro, T. G., and Burke, J. M. (1976). *J. Am. Chem. Soc.* **98**, 5482–5489.
Spiro, T. G., and Stein, P. (1977). *Annu. Rev. Phys. Chem.* **28**, 501–520.
Stephens, P. J., Suterland, J. C., Cheng, J. C., and Eaton, W. A. (1976). *In* "The Excited States of Biological Molecules" (J. B. Birks, ed.), pp. 434–442. Wiley, New York.
Suffran, W. A., and Gibson, Q. H. (1977). *J. Biol. Chem.* **252**, 7955–7958.
Terner, J., Spiro, T. G., Nagumo, M., Nicol, M. F., and El-Sayed, M. (1980). *J. Chem. Soc.* **102**, 3238–3239.
Weiss, C., Kobayaski, H., and Gouterman, M. (1965). *J. Mol. Spectrosc.* **16**, 415–450.
Weiss, J. J. (1964). *Nature (London)* **202**, 83–84.
Weissbluth, M. (1974). *Mol. Biol. Biochem. Biophys.* **15**, 58–136.
Zerner, M., Gouterman, M., and Kobayaski, H. (1966). *Theor. Chim. Acta* **6**, 363–400.

PART IV
DNA

CHAPTER 16

Ultrafast Techniques Applied to DNA Studies

Stanley L. Shapiro

Molecular Spectroscopy Division
National Bureau of Standards
Washington, D.C.

I.	Introduction	361
II.	Nonlinear Optical Effects and Selective Action—Proposed Infrared Schemes	363
III.	Photochemical Reactions in Nucleic Acid Components Induced by Visible and Ultraviolet Pulses	365
IV.	Selectivity in a Mixture of Bases and in More Complex Nucleic Acid Components	369
V.	Selectivity in a RNA Component and Multiphoton Experiments with Viruses and Plasmids	371
VI.	Theory of Selectivity on Nucleic Acid Components	372
VII.	Selective Photodamage of Dye–Biomolecule Complexes	375
VIII.	Picosecond Depolarization Measurements and Torsional Motions in DNA	378
	References	382

I. Introduction

This chapter will describe the application of modern ultrafast temporal techniques to the study of DNA, DNA components, and DNA–dye complexes. It is not intended to be an exhaustive review, nor for that matter even a balanced representation of these types of studies today. I hope instead to direct attention to interesting recent developments by means of a few illustrative examples and to provide a cursory account sufficient to whet the appetite of the more intent reader.

What are subnanosecond techniques and why are they necessary? (see, e.g., Chapter 17 and Shapiro, 1977). For *any* earthly system of condensed matter, excited state interactions and relaxations can be observed on a nanosecond or picosecond time scale. All biological systems consist of liquid or solid matter, thus fulfilling the criterion of being composed of condensed matter. Large molecules like proteins possess so many degrees of freedom that energy relaxation can take place on a short time scale. For our purposes then, such large molecules can be thought of as miniature condensed-matter systems. The relaxation of electronic, vibrational, rotational, and librational energy, along with such important processes as electronic energy migration and charge transfer are just some of the many events that ordinarily take place on a nanosecond and picosecond time scale. Even vibrations of small molecules, such as methanol and ethanol in solution, relax in times on the order of 10–30 psec (von der Linde *et al.*, 1971; Alfano and Shapiro, 1972), whereas a slightly larger aromatic molecule such as rhodamine 6G can even relax on a subpicosecond time scale (Shank *et al.*, 1977).

There are two aspects of the use of ultrafast temporal techniques that will be described: (1) the use of ultrafast spectroscopy to probe the relaxation of a system and (2) the deliberate alteration of a system with ultrashort light pulses. The first aspect has already been alluded to—that is, the study of the dynamics of complex systems. Here the numerous processes normally encountered on the picosecond time scale are followed. Sometimes these processes are integral to the event itself, such as the beginning steps in the visual processes and in the energy migration and charge separation processes that occur in photosynthesis, as discussed elsewhere in this volume. At other times ultrafast techniques are used simply to probe systems and to learn about them—whether molecules relax independently of their environment or whether the surrounding interactions are crucial. By measuring the dynamics and comparing with theory, nanosecond and picosecond spectroscopies are employed to delineate and extract the meaning of complex interactions.

The second, and at this time more embryonic, aspect which we shall mainly discuss in this chapter, the selective alteration or even destruction of a complex molecule with a laser source, opens a modern branch of photochemistry–photobiology. After the discovery of intense laser sources, it was quickly recognized that new types of nonlinear optical effects were possible such as harmonic generation, stimulated Raman scattering, and parametric amplification, as well as other multiple-photon effects. Photochemists soon recognized the possibility of using these nonlinear optical effects induced by intense laser action to develop and study new types of reactions. Most prominent of these reactions for our purposes are those induced by multiphoton dissociation or ionization. By tuning powerful laser beams so as to excite molecular modes, specific reactions can often be efficiently selected by the multiphoton process. The separation of isotopes (Ambartzumian and Letokhov, 1972; Ambartzumian *et al.*, 1975) provides a prime example of the applicability of these new methods. The directionality of laser beams, the possibility of precision focusing, the ability to vary precisely the pulse duration, frequency, and intensity of a laser beam, as well as the ability to delay temporally laser pulses with respect to one another—these exceptional qualities have led to several serious proposals for the application of multiphoton chemistry in the biological area. More importantly, there have been recent experimental demonstrations. Let us therefore concentrate on a discussion of this exciting new area of research, an unexplored frontier.

II. Nonlinear Optical Effects and Selective Action— Proposed Infrared Schemes

The most intriguing proposals for selective action on biological molecules have involved the DNA molecule. The popularity of this molecule derives no doubt from the possibility of developing a new method of genetic manipulation. Because this molecule perhaps provides the most likely opportunity for directing biochemical processes by selective laser excitation, several groups have concentrated on the possibility of selective action on DNA and its nucleic acid components. Initially it was recognized that several major difficulties had to be overcome in order to observe selective action on DNA. The rapid electronic deactivation of DNA, indicated by the apparent absence of luminescence (Daniels and Hauswirth, 1971; Callis, 1979) at room temperature ($\phi \simeq 10^{-4}$–10^{-5}), signifies the extreme difficulty of monitoring the reactions by fluorescence techniques. Furthermore, there is the possibility that energy migration along the DNA chains can defeat the selectivity process (Shapiro et al., 1975). If vibrational selectivity is to be achieved, it must be done quickly, before the vibrational modes relax on the picosecond time scale. Letokhov (1975) first pointed out that, if selective action is to be obtained on DNA, picosecond techniques are essential. Because the absorption band in DNA is broad and featureless (although happily easily accessible by means of the fourth harmonic of the neodymium–YAG laser since the peak absorption is near 265 nm, whereas the fourth harmonic is at 266 nm), Letokhov originally proposed that selectivity could be achieved by means of vibrational processes. He immediately saw that picosecond techniques were necessary because vibrational relaxation in most molecules takes place in 10 psec or less in solution. Analogous to multiphoton dissociation schemes for such molecules as SF_6 and UF_6 in the gaseous phase (Ambartzumian et al., 1975), Letokhov (1975) proposed two vibrational methods. These schemes are shown schematically in Fig. 1. In the first infrared (IR)-/ultraviolet (UV) scheme an IR pulse selectively excites a vibrational mode, and then a visible or UV pulse dissociates the molecule; in the second IR scheme a very intense IR pulse dissociates the molecule by the absorption of numerous IR quanta, as many as 30 in SF_6. For the IR-UV scheme both pulses must be deposited synchronously with sufficient energy. Letokhov calculated that a power of several times 10^9 W/cm^2 was needed in each pulse to be

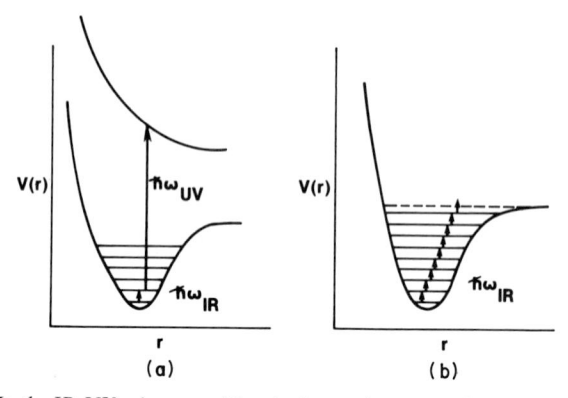

FIG. 1. (a) In the IR-UV scheme an IR pulse is tuned to a specific vibrational frequency. A synchronous UV pulse then raises the electronic energy above the dissociation level. (b) In the IR scheme, a very intense IR pulse is tuned to a specific vibrational frequency. A molecule dissociates by multiple photon absorption of IR radiation alone.

deposited within the depopulation time for the vibration. This high power requirement is a direct consequence of the fact that sufficient energy must be delivered in a time shorter than the vibrational relaxation time to prevent a loss of selectivity. The IR scheme requires even greater powers, since the energy must be deposited in a time comparable to the dephasing time of the molecular vibration, which is always shorter than the depopulation time for molecules in solution. Letokhov has further pointed out that the IR spectrum of thin films of thymine at low temperature is quite structured.

Letokhov has also suggested that the hydrogen bonding between the pair bases of adenine(A) and thymine(T), and guanine(G) and cytosine(C), might be selectively broken and might stimulate the untwisting of the DNA double helix. The AT pairs have a vibrational frequency mode at 1700 cm^{-1}, whereas CG pairs can be stimulated at 1720 cm^{-1}. The phosphate backbone may also be stimulated at 8.13 μm, because the PO_2 vibrational frequency is at 1230 cm^{-1}.

Certain difficulties would have to be overcome for the proposed IR schemes to become successful. Almost all biomolecules *in vivo* are surrounded by molecules which themselves absorb in the IR. The strong IR absorption bands of water are a prime example. There are at least two possible ways around this problem. For example, using D_2O as a solvent reduces the absorption problem (Kryukov *et al.*, 1978). Moreover, this difficulty can sometimes be overcome by exciting vibrational or electronic overtone bands. Such a scheme has been demonstrated for coumarin 6 in a CCl_4 solution (Laubereau *et al.*, 1975).

Even if these schemes are successful, at least one obscuring aspect appears. Because IR and UV pulses can theoretically only be focused down to a wavelength of light, selectivity would take place at all AT or CG pairs on the DNA chain within this focal diameter. Thus the DNA molecule could be broken at numerous locations, especially at longer IR wavelengths. So far, no IR experiments have been demonstrated. However, schemes with visible and UV pulses have been successful, and we shall now describe them.

III. Photochemical Reactions in Nucleic Acid Components Induced by Visible and Ultraviolet Pulses

At first glance, it also appears to be very difficult to induce selective reactions in the visible and UV with similar biological molecules. After all, absorption bands in large macromolecules are usually broad and featureless. However, with simultaneous illumination of pulses of different frequency, the multiphoton quantum spectrum of these molecules might vary, allowing selectivity. Intense short pulses would again be required both for the multiphoton process and also to improve the selectivity. Consequently, as a first step toward selective action on complex molecules, the research group at the Institute of Spectroscopy at Troitsk (Kryukov et al., 1979) investigated the interaction of intense visible and UV ultrashort pulses with components of nucleic acids rather than the more complex DNA molecule. They began by illuminating the samples with intense picosecond pulses and then using spectroscopic as well as chromatographic means to detect the products of the irreversible multiphoton reactions. In their initial experiments (Kryukov et al., 1979) dilute aqueous solutions of adenine and thymine were irradiated by 5-mJ 20-psec 266-nm pulses and 10-mJ 25-psec 532-nm pulses. These pulses were derived from the fourth and second harmonics of a neodymium–YAG laser, respectively. Illumination with 266-nm pulses alone produced an easily observable decrease in the optical density of the sample. The simultaneous addition of 532-nm pulses enhanced the observable change in optical density substantially. However, when a time delay of 100 psec was set between the 266 and 532-nm pulses, the enhancement disappeared, the result being the same as for the 266-nm pulses alone. Nor did illumination of the sample with 532-nm pulses alone cause any effect. The optical density change as a function of the irradiation dose for thymine in aqueous solution is plotted in Fig. 2a and b. The irradiation dose E is by definition given by $E \equiv M \int_0^\tau I \, dt$, where I is the intensity, M is the number of laser shots, and τ covers the pulse duration. From Fig. 2a, which shows the structure of thymine in the inset, we see that for 266-nm irradiation only, for a constant irradiation dose, the increase in optical density takes place at low intensities, but apparently saturation begins at higher intensities. The enhancement effect of simultaneous pumping with 266- and 532-nm pulses is shown in

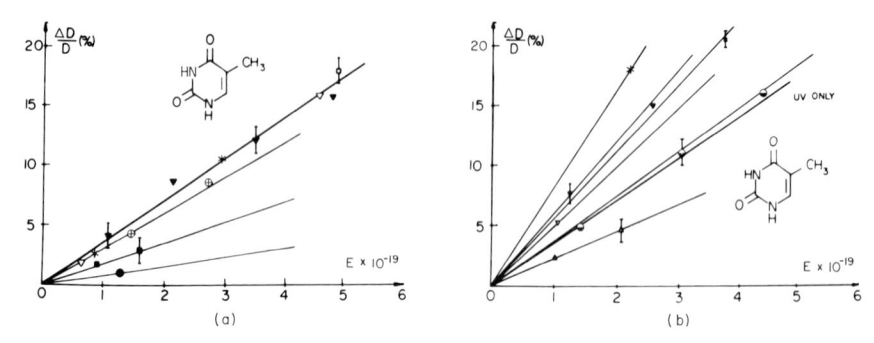

FIG. 2. The dependence of the relative change in optical density of thymine in aqueous solution on UV irradiation dose E. (a) Irradiation with $\lambda = 266$ nm. The irradiation intensity is in GW/cm^2: ●, 0.18; ■, 0.35; ⊕, 0.63; *, 1.52; □, 2.28; ▽, 2.53; ▼, 2.58. (b) Irradiation with $\lambda = 266$ nm and $\lambda = 532$ nm. The irradiation intensity at $\lambda = 266$ nm and $\lambda = 532$ nm is in GW/cm^2. ▽, 0.41 and 15.17; △, 0.52 and 1.78; ▽, 0.52 and 16.53; ⊖, 1.34 and 3.02; +, 4.91 and 7.00; X, 2.28 and 3.92. (After Kryukov et al., 1979.)

Fig. 2b. The greater change in density is evident at low intensities. Because starting from the ground state 532-nm pulses alone have no effect, it is apparent that there must be considerable absorption of the 532-nm pulses from the first excited singlet state to higher electronic states. The change in optical density per irradiation dose for all five of the DNA bases, adenine, thymine, cytosine, guanine, and uracil (Angelov et al., 1980a), is plotted versus optical intensity in Fig. 3. At pH 6.3 the change in optical density for uracil, for example, is far greater than that for adenine. Theory (Section VI) shows that the initial linear dependence of the change in optical density ΔD,

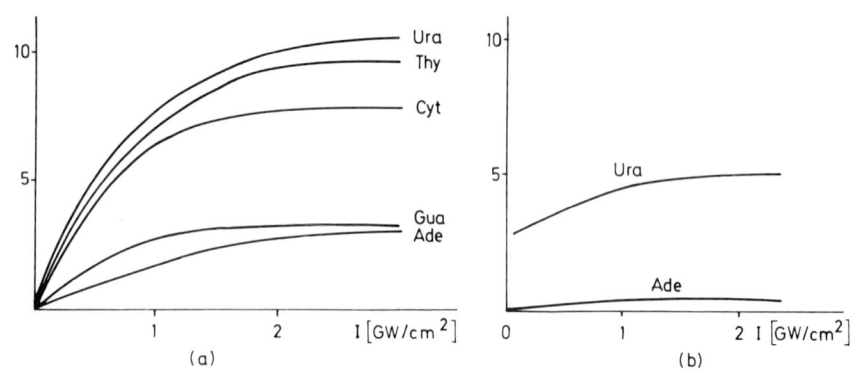

FIG. 3. The photoproduct yield versus irradiation intensity for all DNA bases at pH 6.3(a) and for uracil and adenine at pH 2.2(b). $\Delta D/DE \times 10^{19}$ photons^{-1} cm^2. (After Angelov et al., 1980a.)

16. ULTRAFAST TECHNIQUES APPLIED TO DNA STUDIES

when plotted in the form $\Delta D/DE$ versus I, is an indication of a two-step process. Intensities in the gigawatt-per-square centimeter range are required to observe this effect experimentally. The great difference in optical density changes produced upon the different bases immediately suggests a possibility of selectivity. There are at least two possible ways to monitor this selectivity. The first is to look at the difference spectra before and after illumination, and the second is to look at the end products of the irreversible photoreactions by means of chromatography. We shall return to this figure again later. Note, however, how much the selectivity is improved between uracil and adenine (Fig. 3b) at lower pH. Thus an important biological variable may be varied to produce a significant change in selectivity.

Before attempting the more complex experiment of a mixture of bases, the Russian group logically first observed the difference spectra and photoproducts formed with individual bases in aqueous solution (Kryukov et al., 1979). The differential spectrum obtained after irradiation of thymine is shown in Fig. 4. Analysis shows that there is a decrease in optical density for the bands at 221 and 266 nm, while new absorption bands appear at 195 and 302 nm. Similar difference spectra are obtained upon simultaneous illumination with 532- and 266-nm pulses, indicating the formation of similar if not identical photoproducts. According to Kryukov et al. (1979), the appearance of the transmission at 266 nm signifies the destruction of the pyrimidine ring of thymine, and the presence of the band at 195 nm confirms the splitting of the pyrimidine ring. The difference spectra formed for adenine under 266-nm

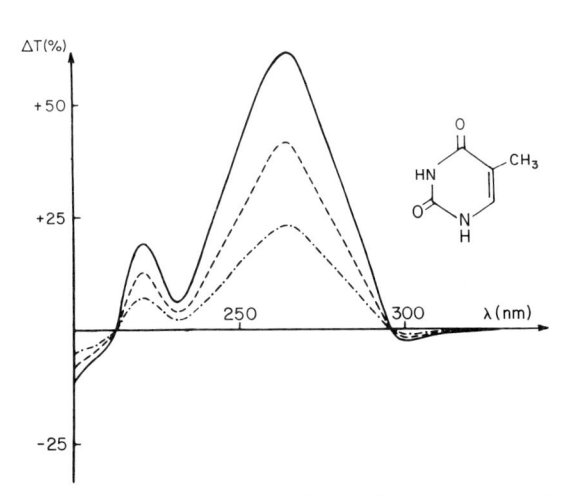

FIG. 4. Difference spectra obtained after irradiation of thymine solution. Dose of irradiation: $-\cdot-\cdot$, $E_{266} = 32.4$ J/cm^2, $E_{532} = 93.8$ J/cm^2; ---, $E_{266} = 103.4$ J/cm^2; ———, $E_{266} = 136.6$ J/cm^2, $E_{532} = 293$ J/cm^2. (After Kryukov et al., 1979.)

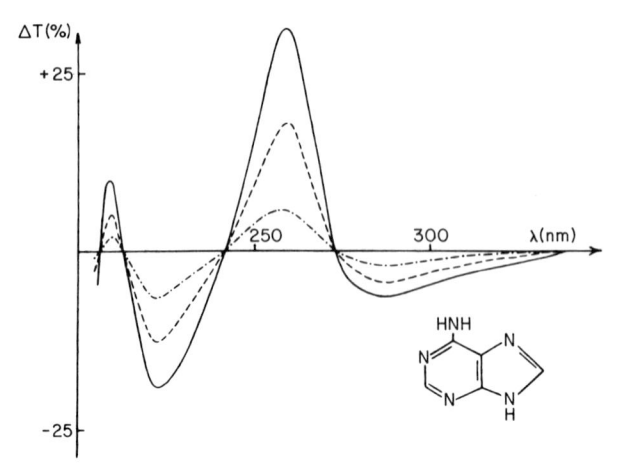

FIG. 5. Difference spectra after irradiation of an adenine solution. Dose of irradiation: —·—·, $E_{266} = 26.4$ J/cm^2, $E_{532} = 51.2$ J/cm^2; ----, $E_{266} = 66.8$ J/cm^2, $E_{532} = 131$ J/cm^2; ———, $E_{266} = 105.2$ J/cm^2, $E_{532} = 210.6$ J/cm^2.

irradiation are shown in Fig. 5. Although adenine is known to be among the most stable purine molecules, the figure shows that efficient multiphoton reactions lead to the appearance of absorption bands at 221 and 287 nm and also to an increasing transmission at 209 and 259 nm. According to Kryukov et al., this confirms violation of the purine nucleus system of conjugated bonds.

Chromatography was also used to look at the end products of the reaction. A typical chromatogram is shown in Fig. 6. The results are complicated, but they are consistent with the photoproducts being formed only by multiphoton chemistry. Although a detailed identification cannot be made for each species, the same photoproducts are not formed with low-intensity UV radiation. The appearance of new photoproducts is not surprising, because the absorption of two UV quanta means that a molecule acquires about 9.3 eV of energy, exceeding the ionization limits.

Later studies (Angelov et al., 1980b) have shown that the mechanisms of the two-step dissociation and the ensuing chemistry are somewhat more complicated than the picture described above. Research on the photoconductivity of these types of solutions shows that there is a photocurrent that rises quadratically with the intensity of the UV laser beam. This photocurrent can be traced back to the water solvent alone. Kinetic and spectral measurements show that the water itself is dissociated into radicals including a hydrated electron e_{aq}^-. The quantum yield for formation of these hydrated electrons is about 10%. In order to delineate the role of these hydrated electrons in the photochemistry of thymine in aqueous solution, the photo-

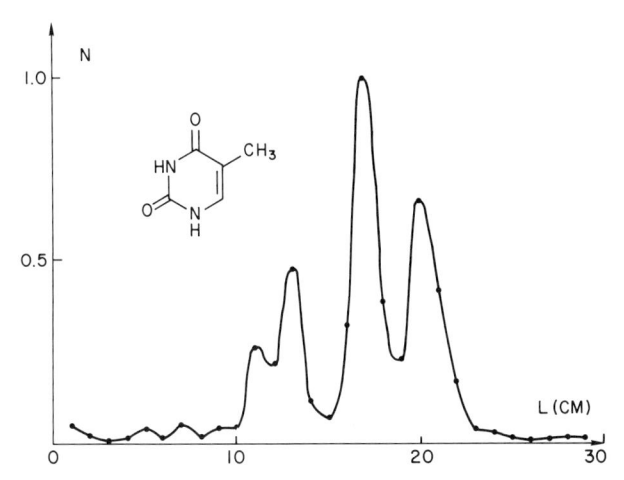

FIG. 6. Chromatogram obtained after irradiation of thymine aqueous solution by 266-nm pulses. The irradiation dose was 118 J/cm^2, and the irradiation intensity 2.4 GW/cm^2. (After Kryukov et al., 1979.)

decomposition efficiency was studied as a function of thymine concentration. These studies can be accomplished by varying the thymine concentration while keeping the optical density constant. In fact an increase in the photodecomposition frequency is observed at lower thymine concentrations, showing that the water radicals play an important role. However, it can be demonstrated by exciting a mixture of bases with 289-nm light that these radicals interact with excited molecules rather than with unexcited molecules. At 289 nm guanine mainly absorbs, whereas at 266 nm the pyrimidines thymine, cytosine, and uracil mainly absorb. Therefore if the water photolysis products interacted with the unexcited molecules, excitation at a wavelength of 289 nm would produce the same results as excitation at a wavelength of 266 nm—principally, the pyrimidine bases would decompose. But guanine decomposition does in fact dominate with 289-nm irradiation. Therefore, excited molecules are involved in the photochemistry. This experiment also illustrates how selectivity can be effected by varying the incident wavelength as well. We shall return to the subject when we describe the model later in this chapter.

IV. Selectivity in a Mixture of Bases and in More Complex Nucleic Acid Components

As shown previously in Fig. 3, the change in the optical density for different bases can be quite different for the same optical intensity. Plotting the change in optical density as a function of irradiation intensity yields curves for different bases that can be compared, allowing the experimentalist to choose an intensity where the selectivity coefficient is high. In Fig. 3b for uracil and adenine in an equimolar solution at pH 2.2, the selectivity defined by

$$S = (\Delta D_{Ura}/D_{Ura})/(\Delta D_{Ade}/D_{Ade}) \qquad (1)$$

can be greater than 20. Because of this high ratio, selectivity experiments were attempted with mixtures of adenine and uracil (Angelov *et al.*, 1980a). The difference spectra obtained for each base separately in aqueous solution are displayed along with the difference spectrum obtained when both bases are mixed together. The dissociation of uracil relative to adenine is so efficient that, at an irradiation intensity of 1.5 GW/cm^2, the photoproducts of adenine finally can be spotted in the difference spectrum of the mixture. The measured experimental selectivity was about 11, in good agreement with the selectivity predicted by the curves in Fig. 3, as it should be.

Selectivity experiments have also been performed with more complex nucleic acid components. In one set of experiments, in a fashion analogous to the selectivity experiments with uracil and adenine, uridine and adenosine photoproducts were detected (Fig. 7b) and selectivity demonstrated. For a

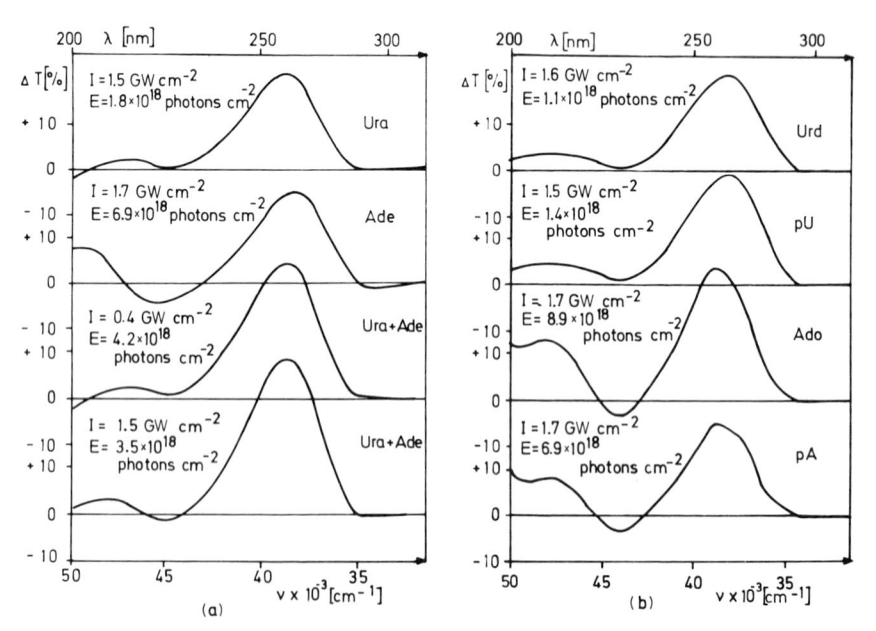

FIG. 7. Difference spectra of photoproducts obtained after UV irradiation of acid aqueous solutions (pH = 2.2) of uracil, adenine, and their equimolar mixture (a) and of uridine, uridine 5'-phosphate, adenosine, and adenosine 5'-phosphate (b). (After Angelov *et al.*, 1980a.)

mixture of uridine and adenosine the selectivity can reach as high as 48 at low intensities. Similarly, the still more complex nucleic acid components, uridine 5′-phosphate (pU) and adenosine 5′-phosphate (pA), were demonstrated to have different photoproduct yields, allowing selectivity to be accomplished.

V. Selectivity in a RNA Component and Multiphoton Experiments with Viruses and Plasmids

The dinucleotide adenylyl-3′-5′-uridine (ApU) can be a component part of single-stranded RNA. Selective action in ApU was also demonstrated at pH 2.2, once again by affecting the uracil portion. Proof of selectivity is demonstrated by several factors: (1) The difference spectrum of ApU is identical to the difference spectrum of pU and to that of a mixture of pU and pA for a small optical density change (Fig. 8); (2) for larger changes in density, however, the difference spectrum remains only identical to that for the mixture of pU and pA. pA photoproducts finally begin to appear with a slight wavelength shift toward the difference spectrum of ApU; (3) after continuous irradiation the UV absorption spectrum of ApU is close to the UV absorption spectrum for pA.

FIG. 8. Difference spectra of photoproducts obtained after UV irradiation of uridine 5′-phosphate, an equimolar mixture of uridine 5′-phosphate and adenosine 5′-phosphate, and the dinucleotide adenylyl ′→ 5′)-uridine (pH = 2.2). (After Angelov et al., 1980a.)

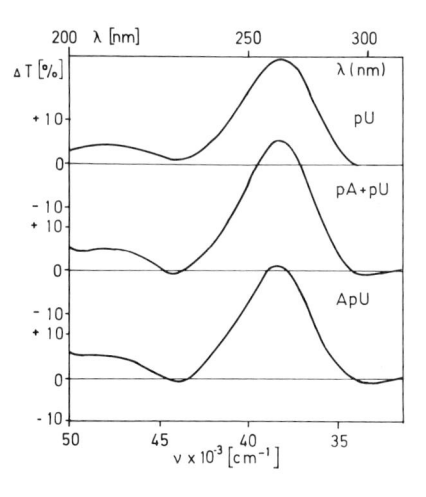

Recently (Angelov et al., 1980) two-photon experiments were conducted with high-power UV radiation on biological objects to test selectivity concepts further. The viruses λ and ϕX174 and the plasmid pBR322 were illuminated with intense picosecond pulses. At intensities between 10^7 and 10^9 W/cm^2 the viruses and plasmids are inactivated, principally because of single-strand breakage in the DNA. At low powers, however, pyrimidine dimers are formed. Selective action on DNA has not yet been proven, but selectivity among the bases provides evidence that selective action on DNA at least should be possible in principle.

It is quite remarkable that effective selective action can be obtained by using nonlinear optical techniques, despite the fact that the UV absorption spectra for acid aqueous solutions of adenine, uracil, and their corresponding nucleotides are almost identical. Although selectivity has been demonstrated, the underlying physical mechanisms are somewhat obscure. Nevertheless, the foundation has been established for more experimental work in this new area.

VI. Theory of Selectivity on Nucleic Acid Components

Because of the complexity of the selection process, present theory must be regarded as tentative. Yet the theory (Kryukov et al., 1979; Angelov et al., 1980a) correlates with several important experimental observations. It allows a prediction of the dependence of the optical density changes on intensity, a determination of the cross sections for optical absorption, in some cases an estimation of the relaxation times, and an estimation of the quantum yields for the photochemical reactions.

A schematic diagram illustrating some elements that a model must include is shown in Fig. 9. A simple level scheme depicted for thymine shows the ground state, first excited singlet state, and higher singlet states. Various relaxation times are indicated for the different levels. Two-photon dissociation processes are indicated for the thymine molecule as well as for the

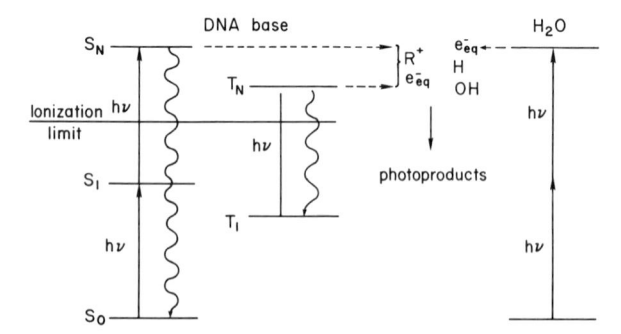

FIG. 9. Model of two-step selection process. Cross section for absorption from S_0 to S_1 is σ_1 and from S_1 to S_n is σ_2. Lifetime of S_1 level is τ_1 and of S_n level τ_2. Triplet states are also shown along with a photodecomposition of the aqueous solution into radicals which can react with molecules in the S_n state. (After Angelov et al., 1980b.)

surrounding aqueous solution. The following simplified set of equations describes the time dependence of the level populations:

$$\frac{dN_0}{dt} = -\sigma_1 I(N_0 - N_1) + \frac{N_1}{\tau_1} + \frac{N_2}{\tau_2}, \tag{2}$$

$$\frac{dN_1}{dt} = \sigma_1 I(N_0 - N_1) - \frac{N_1}{\tau_1} - \sigma_2 N_1 I, \tag{3}$$

$$\frac{dN_2}{dt} = \sigma_2 I N_1 - \frac{N_2}{\tau_2} - \frac{N_2}{\tau_{chem}}. \tag{4}$$

Changes in optical density are considered to take place by absorption within a thin layer of thymine, where I is the UV irradiation intensity, N_0, N_1, and N_2 are the populations of the S_0, S_1, and S_n levels, respectively, τ_1 and τ_2 are the lifetimes of the excited S_1 and S_n states, respectively, and τ_{chem} is the time constant of the chemical reaction from the S_n level. Also indicated is the two-step photodissociation of water into radical fragments including the solvated electron. The time constant τ_{chem} may depend on the concentration of the fragments generated by two-photon dissociation of the aqueous solution. The concentration of these fragments should be highly sensitive to the intensity. However, for simplicity, this dependence will be neglected in order to demonstrate more essential points. Because it is known that the singlet states in thymine deactivate to the ground state in picoseconds (Daniels and Hauswirth, 1971; Callis, 1979), the presence of a triplet population can be neglected, although in general, two-photon absorption through the triplet states cannot be ignored. For this reason also, recovery to the ground state will be complete except for population loss through photochemical reactions. With the realistic assumption based on known quantum efficiencies that τ_1 and $\tau_2 \ll \tau_p = 20$ psec, that $\sigma_2 \leq \sigma_1$, and that $\tau_2 \ll \tau_1$, then $1/(\sigma_1 \tau_1) \ll 1/(\sigma_2 \tau_2)$ and saturation of the $S_1 \to S_n$ transition can be disregarded. Then the change in the number of molecules in the irradiated solution is determined by the photochemical reaction through the excited electronic state S_n. The fractional change in the number of molecules per irradiation pulses will be given by

$$\frac{\Delta N}{N} = \frac{\Delta D'}{D'} = -\left(\frac{1}{N\tau_{chem}}\right) \int_0^\tau N_2 \, dt, \tag{5}$$

where N is the number of molecules in the cell and $\Delta D'/D'$ is the change in the optical density per irradiation pulse. Because the pulse width τ_p is much greater than τ_1 and τ_2, Eqs. (2)–(4) can be solved under steady state conditions, i.e., $dN_1/dt = dN_2/dt = 0$. If the quantum yield for forming photoproducts through the state S_n is defined as $\phi = (1/\tau_{chem})/(1/\tau_2 + 1/\tau_{chem})$, with

ε as the irradiation dose per pulse and M as the total number of pulses, it can be shown that the total change in optical density is

$$\Delta D/D = -(N_0/N)(\phi\sigma_2 M\varepsilon)\{\sigma_1 I/[1/\tau_1 + (\sigma_1 + \sigma_2)I]\}. \tag{6}$$

After introducing the total irradiation dose $E = \varepsilon M$ and correcting by the ratio of the irradiated volume V' to the total volume V of the liquid, the photoproduct yield, plotted experimentally in Fig. 3, is given by

$$\Delta D/DE = -(\phi N_0/N)\{\sigma_1\sigma_2 I/[1/\tau_1 + (\sigma_1 + \sigma_2)I]\}/(V'/V). \tag{7}$$

At low intensities far from saturation, $N_0/N \simeq 1$, $(\sigma_1 + \sigma_2)I \ll 1/\tau_1$, and

$$\Delta D/DE = -\phi\sigma_1\sigma_2 I\tau_1(V'/V). \tag{8}$$

Such a dependence, that is, a linearity with intensity for the photoproduct yield, was in fact observed at low intensities, although the quantum yield may introduce an intensity-dependent factor as well, due to the formation of interacting water radicals.

At high intensities where the $S_0 \to S_1$ transition is saturated, $(\sigma_1 + \sigma_2)I \ll 1/\tau_1$ and the photoproduct yield will be given by

$$\Delta D/DE = -(\phi/2)[\sigma_1\sigma_2/(\sigma_1 + \sigma_2)](V'/V). \tag{9}$$

At a high intensity the photoproduct yield should not be dependent on the intensity, and experimentally this is what is observed (Fig. 3). Once again it is assumed that the quantum yield is constant, as in fact it should be, even with the interaction of water radicals.

After substitution of the experimentally observed values for the change in density, cross sections, and dose, and correction for the volume factor, the formulas above show that the efficiency of the multiphoton photoreactions for adenine and thymine are on the order of several percent, in good agreement with all experimental determinations. It follows from the experimental values of the saturation intensity, about 10^9 W/cm^2 for adenine and thymine, that an upper limit for the excited state lifetime τ_1 can be obtained. The formulas yield $\tau_1 \leq 19$ psec for adenine and ≤ 25 psec for thymine, consistent with the expectation from the quantum yield information that these times should be in the several-picosecond range (Daniels and Hauswirth, 1971; Callis, 1979). More exact measurements for the cross sections and singlet lifetime can be obtained from fitting the complete curve for the photoproduct yield versus intensity. In Fig. 3, for example, a measure of the cross section and lifetime can be obtained from the intensity value at which the curve begins to deviate from linearity (Angelov et al., 1980a), whereas a quantum yield value can be obtained from the plateau of the curves. For adenine and uracil at pH 2.2 the following estimates are obtained: $\tau_{1,\text{Ura}} = 23\beta$ psec,

$\tau_{1,\text{Ade}} = 14\beta$ psec, $\phi_{\text{Ura}} = 3.8\gamma\%$, and $\phi_{\text{Ade}} = 0.2\gamma\%$, where $\beta = 1/(1 + \sigma_1/\sigma_2)$, $\sigma_1 \geq \sigma_2$, and $\gamma = 1/\beta$. Similarly, for uridine and adenosine, which deviate from linearity at much lower intensities, and pU and pA, the lifetime estimates are $\tau_{1,\text{Urd}} = 320\beta$ psec, $\tau_{1,\text{Ado}} = 41\beta$ psec, $\tau_{1,\text{pA}} = 90\beta$ psec, and $\tau_{\text{pU}} = 520\beta$ psec; the corresponding quantum yield estimates are $\phi_{\text{Urd}} = 2.8\gamma\%$, $\phi_{\text{pU}} = 3\gamma\%$, $\phi_{\text{Ado}} = 0.2\gamma\%$, and $\phi_{\text{pA}} = 0.2\gamma\%$. The multiphoton absorption cross sections may also be measured accurately from the photoproduct curves.

To summarize, by irradiating mixtures of bases, molecules of the desired type may be selected. Cross section, lifetime, and quantum yield data are in good agreement with theory.

VII. Selective Photodamage of Dye–Biomolecule Complexes

A much different scheme for selective damage has been proposed and demonstrated by a group in Milan (Andreoni *et al.*, 1980a). Suitable dye molecules are bound to specific sites on the biomolecule. Their scheme then consists of a two-step process. In the first step the bound dye molecules are excited to the first excited singlet state, S_1, by means of a short pulse at frequency v_1. Then in the second step a second pulse, usually at a higher frequency v_2, photoionizes the dye molecule. The damage to the biomolecule then simply arises from the ionized state of the attached dye.

Before attempting selectivity experiments Andreoni *et al.* (1980b) laid the groundwork by studying laser-induced photodamage in the dye quinacrine mustard, both in solution and attached to DNA. Quinacrine mustard, an acridine derivative, is known for its sensitive dependence of the fluoresence on its binding site in DNA. To demonstrate two-step multiphoton damage in quinacrine mustard, a nitrogen laser at 337 nm, 5-nsec pulse width, 200-kW peak power, and a repetition rate of up to 50 Hz is focused onto a 1-mm^2 spot. Simultaneously the dye is illuminated by a weak probe pulse at 420 nm. The 337-nm pulse causes the molecule to photodissociate, whereas the 420-nm 200-psec pulse excites the remaining dye, causing it to emit at 500 nm. As can be seen in Fig. 10, the damage probability for quinacrine mustard in buffer solution, when plotted as a function of intensity on a log–log plot, at first follows a slope of 2 at lower intensities, indicative of a two-step photo-dissociation process. At higher intensities the damage probability appears to saturate. Similarly a wide difference in damage probability is noted for quinacrine mustard bound in poly dA–poly dT than for quinacrine mustard bound to poly dG–poly dC; this is most likely due to the widely different lifetimes of quinacrine mustard in these two compounds.

The Milan group points out two methods of obtaining selectivity at different sites using a two-step scheme (Andreoni *et al.*, 1980a). The first scheme

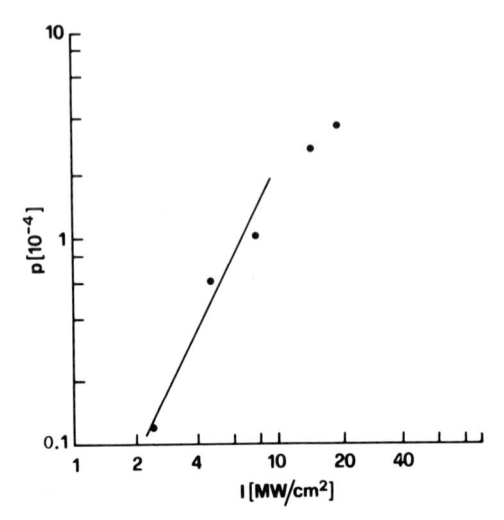

FIG. 10. Damage probability p per laser shot versus laser intensity. Initial slope of 2 indicates a two-step process.

is represented in Fig. 11a. Suppose we have a dye molecule bound to two different sites designated sites 1 and 2. Because the lifetimes of certain dye molecules are sensitive to the binding site environment, the population lifetime at site 1, τ_1, can easily differ by an order of magnitude. Such is the case, for example, for certain acridine dyes bound to different base sites in DNA. First assume the lifetime $\tau_1 \leq \tau_2$. If both complexes are excited with a short pulse of frequency v_1, then the population will disappear from the first excited singlet state of complex 1 much more rapidly. Then a second

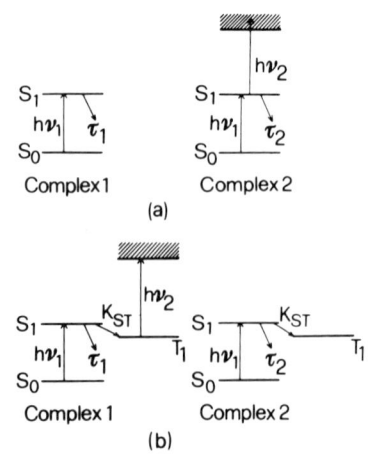

FIG. 11. (a) Selective action on complex 2. Pulse at frequency v_2 is delayed relative to excitation pulse at v_1. Because $\tau_1 \ll \tau_2$, most of the population in S_1, after a suitable delay, favors complex 2, as a delay produces more relative damage at complex 2. (b) Selective action on complex 1. Here deactivation is mainly to a triplet state. For an efficient triplet mechanism, then if $\tau_1 < \tau_2$, at short times, ionization can be accomplished by pumping with a second pulse from the triplet state, which favors complex 1 at early times. (After Andreoni et al., 1980a.)

ionization pulse at v_2, if delayed by a suitable time, will preferentially ionize complex 2 over complex 1 because of the greater population remaining in the first excited singlet state. The destruction of such a complex may be measured in absorption or emission or by any other suitable method such as chromatographic means.

The second method, also using a two-step scheme, is analogous to the aforementioned method, but it allows preferential removal of the shorter-lived complex. This second scheme is depicted schematically in Fig. 11b. As depicted in the figure, the second ionization step takes place from the triplet state instead of the singlet state. In contrast to the previously mentioned scheme, damage will primarily be favored in complex 1, especially at short delay times, at least for a typical case where the short lifetimes reflect energy transfer to a triplet state.

To prove the two-photon damage concept, an acridine derivative, proflavine, was bound to DNA. This dye can have greatly different lifetimes depending on the binding site, and furthermore it is known that the electron released by the photoionized proflavine molecule leads to the formation of a stable radical in the bases surrounding the binding site. As a simple initial test, to provide a standard for comparison, free proflavine in a buffer solution is used as a sample. Two beams are counterpropagated through the same region of the sample. The first beam at 430 nm excites the sample to the first excited singlet state, whereas the second beam at 337 nm photoionizes the dye. After being photoionized the dye no longer fluoresces, so that the amount of undamaged dye as measured by the fluorescence intensity gives a measure of the remaining dye left after a number of shots. No damage is observed with either beam separately. With both beams present the fluorescence intensity decreases with the number of laser shots. If the second beam is delayed relative to the first beam, the probability of damage decreases, presumably because of the relaxation of the singlet state. These results are illustrated in Fig. 12. The pulse repetition rate is 100 Hz, the power per beam is about 100 kW, the beam diameter after focusing on the sample is 1 mm^2, and the pulse widths are 0.25–0.5 nsec. Substantial damage takes on the order of minutes in these 10 MW/cm^2 fluence ranges.

More exciting are the results in two different proflavine–DNA complexes (Andreoni et al., 1980a). For proflavine bound to poly dA–poly dT the probability of damage per laser shot is much higher than for proflavine bound to poly dG–poly dC at all delay times between the two pulses. The selectivity improves from over a factor of about 5 at short delays to a very high value at delays of 8 nsec (Fig. 13). It is suspected that much of the damage to the poly dA–poly dT sample at longer delays arises from the triplet state, because the singlet state probably has mostly decayed away at this lifetime value. Somehow the poly dC–poly dG escapes damage at these longer delays,

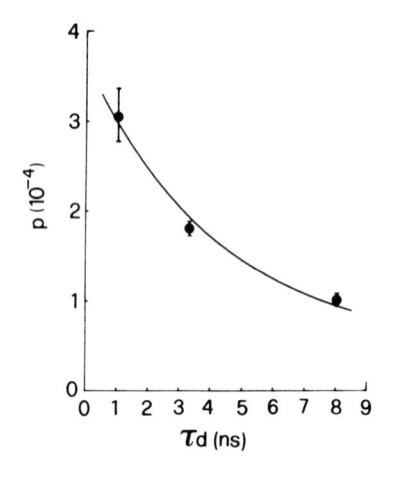

FIG. 12. Damage probability per laser shot p of proflavine in solution at three different delay times τ_D between two laser pulses. Solid line represents theoretical fit. (After Andreoni et al., 1980a.)

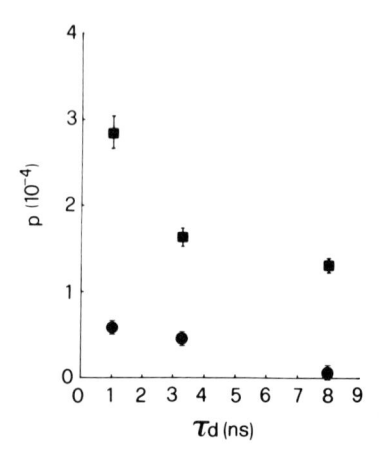

FIG. 13. Damage probability per laser shot p for proflavine bound to poly dA–poly dT (squares) and to poly dG–poly dC (circles) for three different delays between two laser pulses. Selectivity increases dramatically for a delay of 8 nsec. (After Andreoni et al., 1980a.)

possibly as a result of decay to a short-lived state such as might occur in exciplex formation. The validity of the two-step scheme for selectivity has clearly been demonstrated for these two dye–DNA complexes. The main physical mechanisms are also clear, given the damage probability as a function of delay.

VIII. Picosecond Depolarization Measurements and Torsional Motions in DNA

Measurements of the depolarization as a function of time can yield interesting data on molecular dynamics. Although related techniques have been used to measure orientational relaxation times—a nonlinear optical effect induced in a picosecond gating system (Duguay and Hansen, 1969; Chuang and Eisenthal, 1971) and also a transient gating method that observes the disappearance of an induced grating (Phillion et al., 1975)—we confine ourselves to a discussion of emission techniques. Such techniques have been introduced on the picosecond time scale by Chuang and Eisenthal (1971), Porter et al. (1977), and Fleming et al. The latest technique consists mainly of exciting a sample with a polarized picosecond light source and observing the change in polarization of the emission using an analyzer placed before a fast detector, such as a picosecond resolution streak camera. In a recent application by Robbins et al. (1980; Millar et al., 1980) a continuous-wave (cw) mode-locked dye laser was used as an excitation source, and the fluorescence detected by a fast photomultiplier and a photon-counting apparatus.

The time-delay fluorescence measurements allowed Robbins et al. (1980; Millar et al., 1980) to observe fast relaxation processes associated with intercalation of the dye molecule ethidium bromide in calf thymus DNA. In fact, Lerman (1961) originally proposed the intercalation mechanism based on fluorescence and dichroism experiments. As is well known, DNA consists of two helical chains coiled about an axis. The two chains are held together by purine–pyrimidine base pairs hydrogen-bonded together almost perpendicular to the helix axis. The double helix can untwist sufficiently to provide enough space for a molecule such as ethidium bromide to enter at a site which leaves the hydrogen pairing between the bases intact. From Lerman's (1961, 1963) work it can be shown that dye molecules like acridines enter more perpendicularly than parallel to the axis of the double helix. It is these dye molecules that are excited with a picosecond laser, and it is the reorientation of these dye molecules that is measured by the depolarization measurements. That is, a short light pulse preferentially excites dye molecules whose absorption dipoles are aligned along the field. Initially the emission will then be polarized, but the polarization will decay as the excited molecules undergo rotary Brownian motion. Since these dye molecules are probably closely attached to the DNA structure, the observations are of the torsional motions of the double helix itself.

The mechanical properties of DNA such as local stresses are related to the structure and function of the molecule. These include possible biological roles such as the untwisting of the double helix, supercoiling, and packaging. In order to predict theoretically the significance of the mechanical motions, many different models have been devised. In one treatment, for example, DNA can be thought of as a twistable wormlike chain (Barkley and Zimm, 1979). In another representation, DNA is modeled as a set of N coupled coaxial discs (Robinson et al., 1980). Even though there is much information about the motion of large segmental structures, little is known about the local motions in DNA. Picosecond measurements offer a unique method of testing different models for the local motions in DNA.

The main rapid motions the intercalated ethidium bromide molecules will show are the twisting or torsional motions of the DNA chain and wobbling of the intercalated molecule at its site. Because the fluorescence lifetime of the ethidium bromide is 22.6 nsec, all slower motions such as tumbling and rotation of the entire DNA complex will not be observed.

In their first experiments on DNA Robbins et al. (1980) allowed polarized short pulses to excite the sample. The fluorescence intensity as a function of time is displayed for an analyzer perpendicular and parallel to the beam polarization as shown in Fig. 14. Also shown is the anisotropy function defined by

$$r(t) = [I_\parallel(t) - I\perp(t)]/[I_\parallel(t) + 2I\perp(t)],$$

which for DNA–ethidium bromide can be related mainly to the torsional motions of the DNA chain by the relationship given by the elastic model of Barkley and Zimm (1980):

$$r(t) = r_0[\tfrac{1}{4} + \tfrac{3}{4}\exp - \Gamma(t)], \tag{10}$$

where r_0 is the anisotropy function at $t = 0$ and $\Gamma(t)$ is the torsion decay constant. Here the DNA chain is thought of as a flexible cylindrical rod of

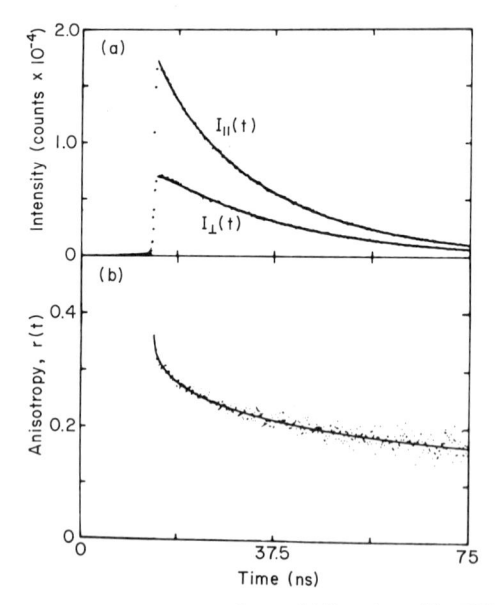

FIG. 14. (a) Fluorescence components of an ethidium bromide–DNA complex parallel $I_\parallel(t)$ and perpendicular $I_\perp(t)$ to the polarization of the incident light. (b) Fluorescence anisotropy obtained from (a). Solid lines are best fit with $\tau = 22.6$ nsec, $b^2\eta C = 2.4 \times 10^{-35}$ erg sec, and $r_0 = 0.36$. (After Robbins et al., 1980.)

length $2L$ and radius b immersed in a liquid medium of viscosity η. The torsion decay function can be approximated at short times by

$$\Gamma(t) = (2K_k T/\pi)\sqrt{t/(b^2\eta C)}, \qquad (11)$$

where C is the torsional rigidity of the helix and k is Boltzmann's constant.

Compared to the experimental data, the torsional rigidity constant can be extracted to be $C = 1.3 \pm 0.2 \times 10^{-9}$ erg cm. This agrees well with a value of C of 1.6×10^{-9} erg cm calculated from classic elasticity theory which gives $C = k\, TP/(1 + \sigma)$, where P is the persistence length of the DNA chain taken as 600 Å and σ is the Poisson ratio taken as 0.5, which relates the bending rigidity to the torsional rigidity.

The agreement of the form of the time function with experiment, the $t^{1/2}$ dependence initially, and the ability to extract the torsional rigidity constant shows that the depolarization method cannot only be used as a powerful technique to probe the internal motions of intercalated molecules but also as a probe for understanding the molecular structure of large macromolecular systems. The initial result for DNA is that it behaves locally like a uniformly elastic rod to the first approximation.

In order to test how the torsional rigidity of DNA and RNA depends on sequence, base pairing, and conformation, Robbins et al. (1980) made further studies on a series of polynucleotides. These compounds included double-stranded DNA and RNA and a DNA–RNA hybrid known to have a triple helical structure. The torsional rigidities C, taken from Table I, display quite interesting variations. The authors note that d(G-C)·d(G-C), rA·rU, and calf thymus DNA have about the same torsional rigidity, showing the insensitivity to the base sequence or helical conformation. On the other hand,

TABLE I

PROPERTIES OF ETHIDIUM–NUCLEIC ACID COMPLEXES

Sample[a]	τ (nsec)	r_0	$b^2\eta C/10^{-35}$ (erg^2 sec)	$C/10^{-19}$ (erg cm)
rA·rU	27.1	0.40	2.4	1.3
d(G-C)·d(G-C)	23.3	0.39	2.6	1.4
CT	22.6	0.36	2.4	1.3
d(A-T)·d(A-T)	24.9	0.39	1.2	0.6
Denatured CT	22.6	0.40	1.2	0.6
dA·dU	22.0	0.38	0.7	0.4
dA·rU·rU	23.3	0.37	6.3	3.1

[a] CT is calf thymus DNA and rA·rU is poly rA·poly rU, where r denotes a ribonucleotide; similarly for the other compounds. After Millar et al. (to be published).

d(A-T)·d(A-T) is as flexible as denatured calf thymus DNA, suggesting local denaturing of the DNA due to weaker adenine–thymine hydrogen bonding.

Even more flexible is dA·dU. Whereas all the other samples can be approximated by uniformly elastic rods, the failure of dA·dU to obey Eq. (10) is taken to mean that the denaturation is even more, with long single-stranded regions resulting in the increased flexibility.

Triple-stranded dA·rU·rU is the most inflexible of the species studied. A large torsional rigidity is expected for a larger cylindrical species, but a simple calculation shows that the triple helical structure is even more inflexible than just its size alone would indicate. The authors concluded from this factor and the other data in the table that the helical structure is the dominant factor determining the structural rigidity as evidenced by the torsional movement. The relationship of the time-dependent depolarization measurements to the underlying structure has therefore been elegantly demonstrated by these sets of measurements. The power of these techniques in examining a wide range of macromolecules such as polymers and proteins is clearly evident.

References

Alfano, R. R., and Shapiro, S. L. (1972). *Phys. Rev. Lett.* **29**, 1655.
Ambartzumian, R. V., and Letokhov, V. S. (1972). *Chem. Phys. Lett.* **13**, 446.
Ambartzumian, R. V., Gorokhov, Y. A., Letokhov, V. S., and Makarov, G. N. (1975). *JETP Lett.* **21**, 171. (*Engl. Transl.*)
Andreoni, A., Cubeddu, R., De Silvestri, S., Laporta, P., and Svelto, O. (1980a). *Phys. Rev. Lett.* **45**, 431.
Andreoni, A., Cubeddu, R., De Silvestri, S., and Laporta, P. (1980b). *Chem. Phys. Lett.* **72**, 448.
Angelov, D. A., Kryukov, P. G., Letokhov, V. S., Nikogosyan, D. N., and Oraevsky, A. A. (1980a). *Appl. Phys.* **21**, 391.
Angelov, D. A., Gruzadyan, G. G., Kryukov, P. G., Letokhov, V. S., Nikogosyan, D. N., and Oraevsky, A. A. (1980b). In "Picosecond Phenomena II" (R. Hochstrasser, W. Kaiser, and C. V. Shank, eds.), Springer Series in Chemical Physics, Vol. 14, p. 338. Springer-Verlag, Berlin and New York.
Barkley, M. D., and Zimm, B. H. (1979). *J. Chem. Phys.* **70**, 2991.
Callis, P. R. (1979). *Chem. Phys. Lett.* **61**, 563.
Chuang, T. J., and Eisenthal, K. B. (1971). *Chem. Phys. Lett.* **11**, 368.
Daniels, M., and Hauswirth, W. (1971). *Science* **171**, 675.
Duguay, M. A., and Hansen, J. W. (1969). *Appl. Phys. Lett.* **15**, 192.
Fleming, G. R., Morris, J. M., and Robinson, G. W. (1976). *Chem. Phys.* **17**, 91.
Kryukov, P. G., Letokhov, V. S., Matveetz, Y. A., Nikogosian, D. N., and Sharkov, A. V. (1978). *Kvantovaya Elektron.* (*Moscow*) **5**, No. 8.
Kryukov, P. G., Letokhov, V. S., Nikogosyan, D. N., Borodavkin, A. V., Budowsky, E. I., and Simukova, N. A. (1979). *Chem. Phys. Lett.* **61**, 375.
Laubereau, A., Seilmeier, A., and Kaiser, W. (1975). *Chem. Phys. Lett.* **36**, 232.
Lerman, L. (1961). *J. Mol. Biol.* **3**, 18.
Lerman, L. (1963). *Proc. Natl. Acad. Sci. U.S.A.* **49**, 94.

Letokhov, V. S. (1975). *J. Photochem.* **4**, 185.
Millar, D. P., Robbins, R. J., and Zewail, A. H. (1980). *Proc. Natl. Acad. Sci. U.S.A.* **77**, 5593.
Millar, D. P., Robbins, R. J., and Zewail, A. H. to be published.
Phillion, D. W., Kuizenga, D. J., and Siegman, A. E. (1975). *Appl. Phys. Lett.* **27**, 85.
Porter, G., Sadowski, P. J., and Tredwell, C. J. (1977). *Chem. Phys. Lett.* **49**, 416.
Robbins, R. J., Millar, D. P., and Zewail, A. H. (1980). *In* "Picosecond Phenomena II" (R. M. Hochstrasser, W. Kaiser, and C. V. Shank, eds.), Springer Series in Chemical Physics, Vol. 14, p. 331. Springer-Verlag, Berlin and New York.
Robinson, B. H., Forgacs, G., Dalton, L. R., and Frisch, H. L. (1980). *J. Chem. Phys.* **73**, 4688.
Shank, C. V., Ippen, E. P., and Teschke, O. (1977). *Chem. Phys. Lett.* **45**, 291.
Shapiro, S. L., ed. (1977). "Ultrashort Light Pulses," Topics in Applied Physics, Vol. 18. Springer-Verlag, Berlin and New York.
Shapiro, S. L., Campillo, A. J., Kollman, V. H., and Goad, W. B. (1975). *Opt. Commun.* **15**, 308.
von der Linde, D., Laubereau, A., and Kaiser, W. (1971). *Phys. Rev. Lett.* **26**, 954.

PART V
Ultrafast Laser Techniques

CHAPTER 17

Picosecond Laser Techniques and Design

A. G. Doukas, J. Buchert, and R. R. Alfano

Picosecond Laser and Spectroscopy Laboratory
Department of Physics
The City College of The City University of New York
New York, New York

I.	Introduction	387
II.	General Principles of Operation of Picosecond Lasers	388
III.	Laser Components and Designs	392
IV.	Single-Pulse Selection and Amplification	397
V.	Obtaining the Right Wavelength	400
VI.	Time Measurements and Applications	405
	A. Fluorescence Kinetics	408
	B. Absorption Kinetics	412
	References	414

I. Introduction

The mode-locked laser has created a new field of spectroscopy for studying the molecular dynamics in the time range from 10^{-9} to 10^{-12} sec (DeMaria et al., 1967; Rentzepis, 1970; Alfano and Shapiro, 1973). Some of the diverse areas of science that have been studied using picosecond laser spectroscopy are energy transfer in chemistry, biology, and solids, primary events in biology, charge transfer reactions, cage effects, rotational and vibrational relaxations, and structural rearrangements of molecules (Alfano and Shapiro, 1970a,b,c, 1972; Shapiro, 1977; Rentzepis, 1978; Holten and Windsor, 1978; Shank et al., 1978; Hochstraser et al., 1980). The time-resolved techniques applied in these studies make use of the physical processes of absorption, fluorescence, and light scattering. The indirect methods which measure the frequency spectrum yield second-hand information and often will not substitute for direct measurements in the time domain. The purpose of this chapter is to introduce the principles of design and operation of the solid state laser and picosecond spectroscopy methods.

This chapter is divided into five sections. The general principles of operation of the mode-locked laser are developed in Section II. The design aspects of the laser cavity and commercially available components are discussed in

Section III. Section IV deals with the selection of a single pulse from a pulse train and its amplification. Section V deals with the nonlinear optical techniques available for changing the laser wavelength. The final section discusses the techniques that have been developed for time measurements and time-resolved spectroscopy on the picosecond time scale.

II. General Principles of Operation of Picosecond Lasers

It has been over 15 years since picosecond laser pulses were first produced by mode-locking a Ruby laser (Mocker and Collins, 1965) and a neodymium–glass laser (DeMaria et al., 1966). During this time the solid state laser has become the workhorse of picosecond laser spectroscopy. A schematic diagram of this laser is shown in Fig. 1. The laser oscillator consists of an appropriately doped glass rod with both ends cut at a Brewster angle and a cell of bleachable dye between two high-reflectivity mirrors. A flash lamp around the glass rod provides the energy for exciting the laser medium. After an atom has been raised to an excited state by absorption, it can decay back to the ground state either spontaneously or, if it is in an electromagnetic field, by stimulated emission (Einstein, 1917). The frequency, the phase, and the direction of the emitted photon are that of the electromagnetic field. It is this process (stimulated emission) that is behind the operation of a laser (light amplification by stimulated emission of radiation).

It is clear that, in order to have laser action, as many atoms as possible have to be excited. The process is called population inversion. In solid state

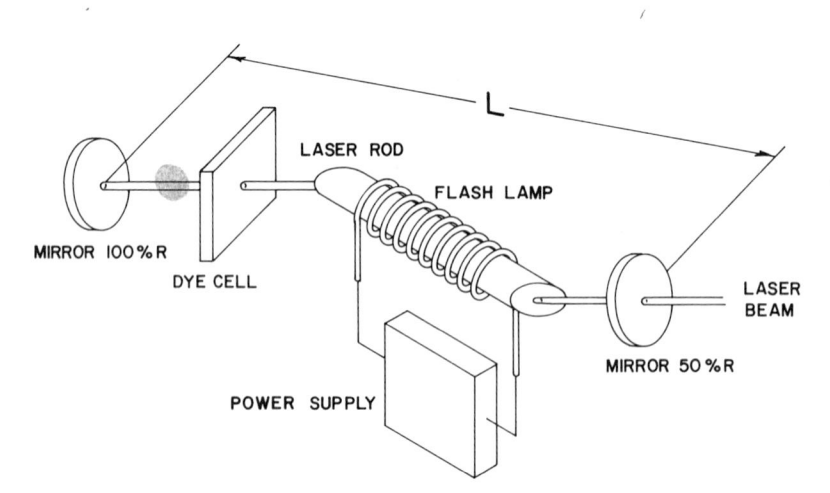

FIG. 1. Schematic diagram of a mode-locked solid state laser. The laser consists of two mirrors, a dye cell, and a laser rod surrounded by a flash lamp.

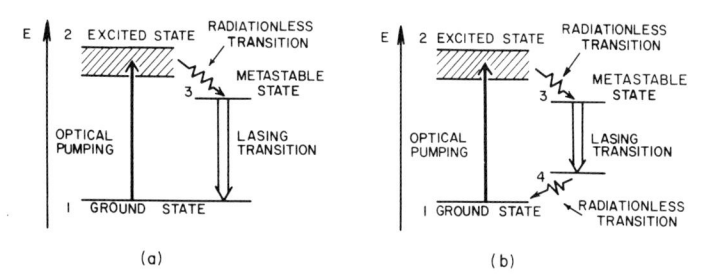

FIG. 2. Energy diagram of a three-level laser (ruby) (a) and a four-level laser (Nd^{3+}) (b).

lasers it is accomplished by the energy provided by the flash lamp. Figure 2 shows a three-level system (ruby) and a four-level system (Nd^{+3}) utilized as a laser medium. In both cases there is a metastable, long-lived state. The metastable state behaves like a trap and creates a population inversion Yariv, 1975). There is always some fluorescence due to the spontaneous emission. These photons bounce back and forth in the laser cavity. As they travel through the laser rod, they stimulate more and more excited atoms to emit photons of the same frequency, phase, and direction. Naturally there is also absorption from the ground state. Since there are more atoms in the excited state, there is a net amplification of the light until the population of the excited state is depleted.

The light emitted by the laser is determined by the fluorescence position bandwidth and quantum yield of the lasing medium, and the configuration of the laser cavity or resonator (Yariv, 1975). From all possible frequencies of the fluorescence spectrum of the lasing material only some are allowed by the laser cavity (Fox and Li, 1961). These frequencies are called modes. It is customary to distinguish two types of resonator modes: longitudinal modes which differ from each other only in frequency, and transverse modes which differ from each other in the distribution of the field as well as the frequency (Yariv, 1975). The longitudinal modes allowed in the resonator are determined by the condition that the field inside the resonator forms a standing wave. These modes satisfy

$$n\lambda_j/2 = L, \qquad (1)$$

where L is the optical length of the cavity, λ_j is the wavelength of an allowed mode, and n is an integer. Figure 3 shows a hypothetical fluorescence profile superimposed on the modes of the resonator. The modes are equally spaced in frequency, with $f = c/2L$. It is clear that only modes that are within the threshold limits $\Delta f'$ of the gain envelope will be amplified. The number N' of longitudinal modes is given by

$$N' = (2L/c)\Delta f'. \qquad (2)$$

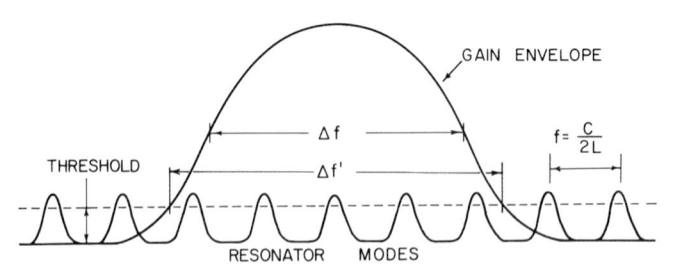

FIG. 3. The gain of the laser line profile superimposed on the allowed modes of the resonator.

The number of the active modes can be very large. For example 10^4 for a neodymium–glass laser of a 1-m cavity. A resonator also may allow many transverse modes (TEM_{mn}, where m, n are integers). For each of the transverse modes many longitudinal modes are determined by Eq. (1). The transverse modes determine beam characteristics such as the energy distribution and beam divergence.

In a free-running laser both longitudinal and transverse modes oscillate simultaneously with random phases. This is not a very useful laser, at least for the study of very fast phenomena. It has a pulse width of a few milliseconds, about the time of the flash lamp, while the intensity has the characteristics of thermal noise. If, however, the different modes are forced to oscillate in phase, a single narrow pulse of high intensity is produced. The process is called mode locking or phase locking (DeMaria et al., 1967). The general relationship between the pulse width (FWHM) and bandwidth (FBHM) is given by

$$\Delta f \, \Delta t \geq k, \tag{3}$$

where k is a constant on the order of unity (Ippen and Shank, 1977). The exact value depends on the pulse shape. For a Gaussian transform-limited pulse $k = 0.441$. Since $\Delta f = (c/2L)N = fN$,

$$\Delta t \geq k/Nf. \tag{4}$$

Equation (4) shows that, the larger the number of modes that oscillate in phase, the shorter the duration of the pulse.

Mocker and Collins (1965) and DeMaria et al. (1966) were the first to produce mode-locked pulses using a saturable absorber. The absorption of a saturable absorber is nonlinear and dependent on the intensity (Yariv, 1975). The onset of nonlinear absorption occurs above 50 MW/cm^2. It is a combination of the lasing medium and the saturable absorber that produces the very short pulses.

The time description of formation of the mode-locked pulse was first introduced by Letokhov (1969a,b). Figure 4 shows a computer simulation of

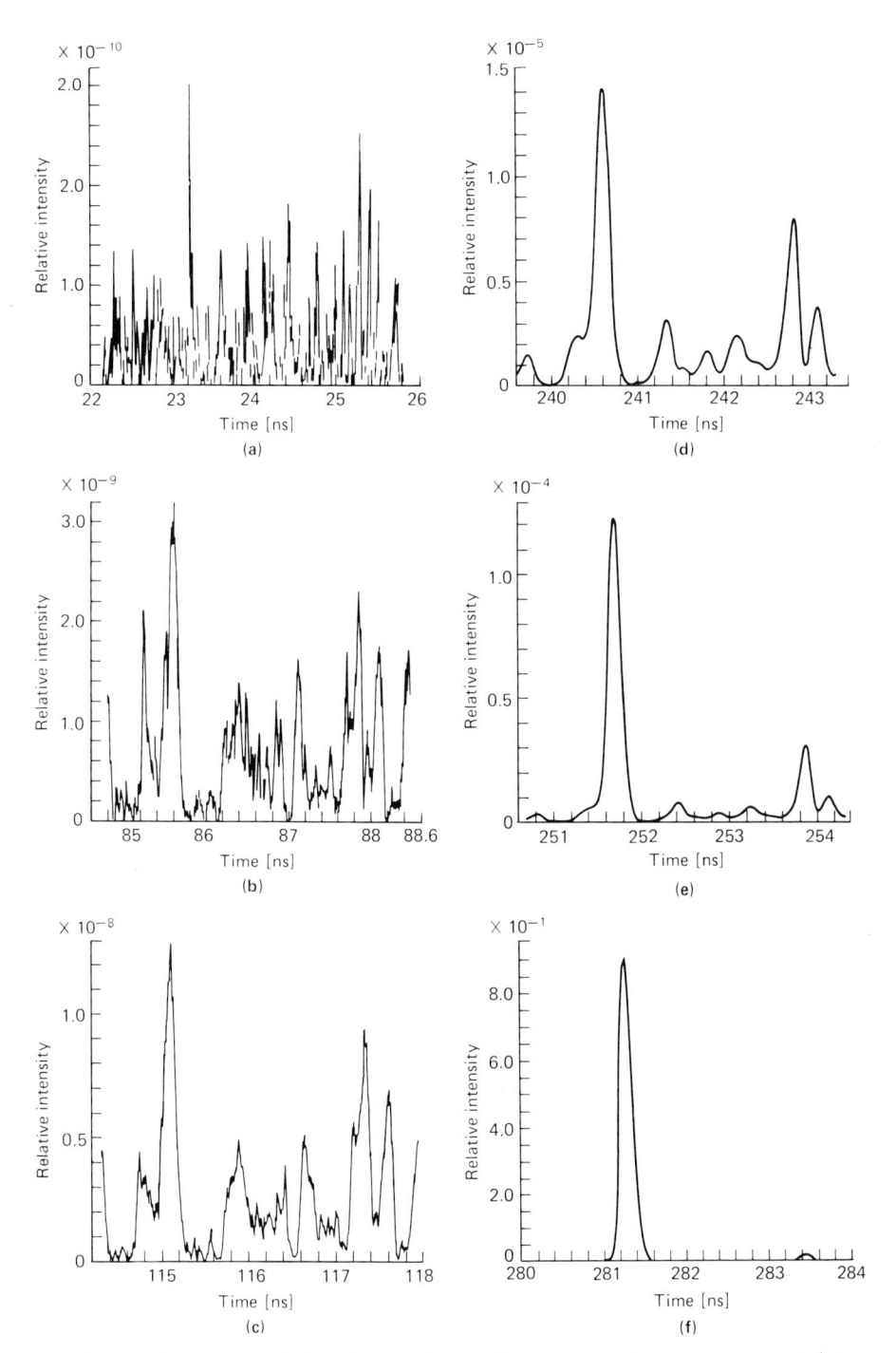

FIG. 4. Computer simulation of the evolution of a mode-locked pulse from noise (Fleck, 1970).

the time evolution of a mode-locked pulse (Fleck, 1970). Initially there is only spontaneous fluorescence inside the cavity. Because of the nonlinear absorbance of the saturable dye, noise pulses above the background will be absorbed less in the dye. As the gain of the laser rod increases, this pulse will be amplified more than the low-intensity noise pulses. The absorber shortens the pulse width and increases the bandwidth. When the intensity of the pulse increases above ~ 50 MW/cm^2, the absorber starts to bleach. The pulse is amplified even more, however, the pulse width increases because of dispersion inside the rod. Finally the population of the excited state is depleted and the pulse decreases. The time evolution of mode-locked pulses (Fig. 4) has been experimentally verified in ruby and glass lasers (Basov et al., 1969; Kriukov and Letokhov, 1972; Von der Linde, 1972, 1973; Chekalin et al., 1974). Alternatively, the saturable absorber can be described as a modulator that locks the modes together (DeMaria et al., 1967; Penzkofer, 1974). The saturable absorber can be considered a shutter of modulating frequency equal to the round-trip time in the cavity.

III. Laser Components and Designs

This section will primarily deal with the design aspects of the laser. We will discuss the components as well as various cavity designs. Our goal is to give the reader enough information so that he or she can evaluate commercially available components. A good source for commercially available lasers and components is the "Laser Focus Buyer's Guide" published by Advanced Technology Publications (Newton, Massachusetts) and "Optical Industry and Systems Purchasing Directory" published by Optical Publishing Company (Pittsfield, Massachusetts).

We will start with the lasing material, since it determines most of the laser characteristics. The lasing material in solid state lasers comes in the form of a glass or crystalline host doped with such ions as Nd^{3+}, Cr^{3+}, etc. There are certain conditions that the laser material must fulfill, namely, sharp fluorescence with a high quantum yield and strong, broad absorption bands in the visible. A number of materials have been investigated and are commercially available. In this section we will limit the discussion to the most commonly used, Nd^{3+} and ruby. However, other materials have been tested for lasing action (Koechner, 1976).

Ruby chemically consists of sapphire (Al_2O_3) in which a small percentage of Al^{3+} has been replaced by Cr^{3+}. The ruby is a three-level system. The lasing line is in the visible at 694.3 nm. The bandwidth, however, is a very narrow 0.5 nm because of the crystalline structure of the material. This sets a lower limit to the pulse width of 10–30 psec depending, of course, on the saturable absorber. The second most commonly used material is Nd^{3+} which can be used either in a crystalline host such as yttrium aluminum garnet (YAG) or glass (silicate or phosphate). All three neodymium-doped materials lase at about 1060 nm. Table I compares some of the physical and optical properties of ruby, neodymium–YAG and neodymium–glass (silicate and phosphate). A new and important material that has not been mode-locked as yet is alexandrite ($BeAl_2O_4:Cr^{3+}$). This material lases in the range from 701 to 818 nm (Walling et al., 1980). It has a wide fluorescent bandwidth which may result in the production of very short pulses.

The laser rod is typically $\frac{1}{2}$ in. in diameter for neodymium–glass, $\frac{1}{8}-\frac{1}{4}$ in. for neodymium–YAG, and $\frac{3}{8}$ in. for ruby. The two ends of the rod are cut at a Brewster angle and polished to $\lambda/10$. This decreases the loss of light traveling in and out of the rod. It also provides a polarized output. In addition, the angle cuts eliminate secondary cavities that can reduce the bandwidth of the laser frequency. This happens because the standing wave inside the cavity has to satisfy Eq. (1) for the secondary cavities as well. The result is that fewer modes are allowed in the cavity. According to Eq. (4) this leads to a longer pulse.

The laser rod and the pumping system constitute the heart of the solid state laser. The unit is also known as the head. It consists of a housing with supports for the laser rod and the flash lamp(s), input and output ports for the cooling liquid, and wire connections to the power supply. Heads are commerically available from companies such as Apollo, Holobeam, and Korad. The same companies also provide power supplies designed to match the particular heads. The cooling system consists of a bath circulator and refrigerator. The coolant is kept at room temperature so that the temperature in the rod is uniform. For ruby, neodymium–YAG, and neodymium–glass silicate lasers, deionized water is the most commonly used coolant. On the other hand, in a neodymium glass phosphate laser a mixture of 50:50 ethylene glycol and deionized water is used.

A number of flash lamps around the laser rod provide the energy for optical pumping of the laser material. The flash lamps can be arranged in different configurations. The type of laser rod, the power of the pulses, and the repetition rate determine selection of the type of flash lamps and the configuration. The main consideration in designing the head is that the laser rod be uniformly pumped. This ensures that the beam will have a uniform profile. The simplest configuration is one or more linear flash lamps around the rod. With small-diameter rods uniformity of pumping can be achieved with linear flash lamps. For this reason they are more often used in neodymium–YAG lasers. For larger-diameter rods a helical flash lamp is chosen, which provides uniform pumping.

The two end mirrors constitute a critical component of the laser cavity. Both mirrors are wedged and have reflectivities of $>99.8\%$ (back mirror) and 30–50% (front mirror), respectively, at the lasing wavelength. The particular

TABLE I

COMPARISON OF OPTICAL AND PHYSICAL OF RUBY, NEODYMIUM–YAG, AND NEODYMIUM–GLASS[a]

Property	Ruby	Neodymium–YAG	Neodymium–glass silicate, ED-2[b]	Neodymium–glass phosphate, LHG-5[c]
Concentration (wt%)	0.05	0.75	3.1	3.31
Doping density (cm^{-3})	1.58×10^{19}	1.38×10^{20}	2.83×10^{20}	3.17×10^{20}
Density (g/cm^3)		4.56	2.547	2.68
Fluorescent linewidth (FWHM) (nm)	0.55	0.4	26	18.6
Spontaneous lifetime at room temperature (μsec)	3000	240	300	290
Laser wavelength (nm)	694.3	1064	1062	1056
Pulse width (psec)	10–30[d]	30[e]	8[e]	6[e]
Cross section for stimulated emission (cm^2)	2.5×10^{-20}	88×10^{-20}	3.03×10^{-20}	3.9×10^{-20}
Brewster angle at lasing wavelength	60°37′	—	57°16′	56°50′
Gain coefficient for 1 J stored energy (cm^{-1})	0.087	4.73	0.160	0.172
Thermal conductivity at room temperature (W cm^{-1} K^{-1})	0.42	0.13	0.0135	0.0119

[a] Koechner (1976).
[b] ED-2 is manufactured by Owens-Illinois.
[c] (LHG-5) is manufactured by Hoya.
[d] DDI in methanol. DDI = 1,1′-diethyl-2,2′-dicarbocyanine iodide.
[e] Eastman Kodak 9860 in dichloroethane.

shape (planar or concave) and the radius of curvature depend on the cavity design. The substrates are usually borosilicate glass (e.g., BK-7A), Pyrex, or quartz. Quartz has a lower thermal expansion coefficient, however, its cost is much higher. The two important specifications for the substrates are flatness and a quality of the surface as defined by "scratch-dig." Flatness is determined as a percentage of the wavelength, and $\lambda/10$ is a reasonable value. It should also be noted that flatness depends on the thermal stability, so for glass substrates it may be hard to obtain a flatness better than $\lambda/20$. The scratch-dig specification refers to the width of scratches (measured in micrometers) and to the diameter of digs (measured in units of 0.01 mm). So a no. 10 scratch refers to 10 μm or 0.01 mm in width and a no. 5 dig refers to 0.05 mm in diameter. The scratch-dig value is very important, as it determines to a large extent the amount of scattered light. A scratch-dig of 10-5 is specified for high-quality laser components. The value 40–20 is specified for elements that are not very critical, such as optical components outside the cavity. It should be noted that most of the time a polished surface of $\lambda/10$ meets the 10-5 scratch-dig specification. The spectral properties of the substrate can be modified by coating. Single or multiple layers of different materials such as MgF_2, ThF_4, ZrO_2, and ZnS can be deposited on a substrate to produce the desired spectral characteristics. A detailed description of coatings is beyond the scope of this chapter. A number of excellent references are available (Heavens, 1955; Baumeister, 1963; Vasicek, 1960; Ritter, 1972). The important parameters that should be specified in coatings are (1) reflectivity and/or transmission at the required wavelength or range of the spectrum, (2) angle of incidence, (3) polarization; "P" (parallel to the plane of incidence), "S" (perpendicular to the plane of incidence), or random, and (4) the power densities that the coatings should withstand. Coatings can be soft or hard. Hard coatings are better because they are more durable and can be cleaned more easily. The specification should be documented by a spectrophotometric trace and a laser radiometric test.

Mirrors should be mounted on orientational devices with good mechanical and thermal stability. The mounts should have two independent degrees of freedom and be blacklash-free. The use of differential micrometers can be very helpful. The oscillator should be placed on a table that isolates vibrations as much as possible. The best choice, of course, is to use a floating table. In our experience, however, a good compromise between cost and isolation is to use an aluminum jig plate about 1 in. in thickness supported by a rigid steel frame. The table rests on rubber isomode pads that isolate it from vibrations. The aluminum plate in addition can be machined to provide a pattern of taped holes used to bolt down the components of the cavity.

As we have seen, in order to produce picosecond pulses a saturable absorber is required inside the cavity. A number of these are commercially

TABLE II

Mode-Locking Dyes

Laser Duration	Wavelength (nm)	Dye	Laser Pulse (psec)
Ruby	694.3	DDI in methanol	10–30
Neodymium–YAG	1064	9860[a], 9740[a] in dichloroethane	30
Neodymium–glass (silicate)	1060	9860[a] in dichloroethane	8–10
Neodymium–glass (phosphate)	1054	9860[a] in dichloroethane	6–7
		5[a,b] in dichloroethane	2–4

[a] Eastman Kodak.
[b] Kopainsky *et al.* (1980), Alfano *et al.* (1981), Kolmeder *et al.* (1981).

available. Table II gives some of the most commonly used dyes. The position of the dye cell in the cavity and the dye concentration parameters are difficult to determine in advance, since there is no well developed theory that can predict the behavior of the laser. These parameters are usually determined experimentally. Figure 5 shows some of the cavity designs used. The position

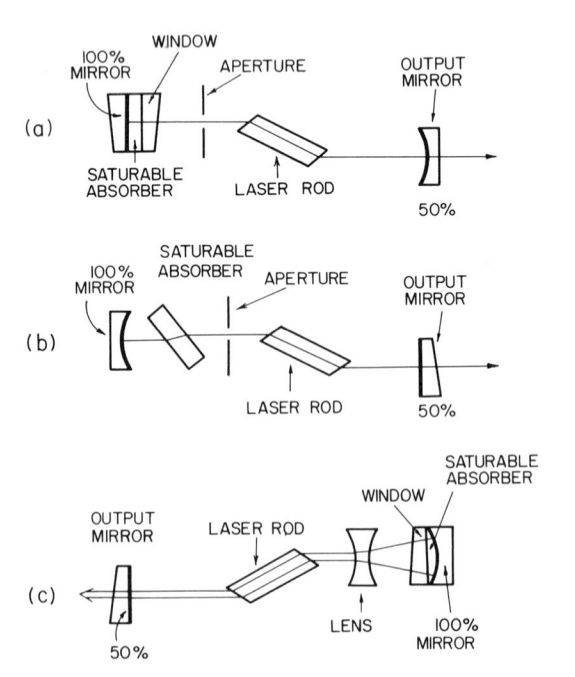

FIG. 5. Three experimental arrangements for passively mode-locked lasers.

of the dye cell is important, as it may cause satellite pulses. A pulse that is not strong enough to bleach the dye can pass through the dye if it coincides with a stronger pulse in the dye cell traveling in the opposite direction. Both pulses will be amplified, causing a double pulse. To avoid this, the dye cell is placed in contact with the rear mirror (Fig. 5a and c). Alternatively, a long dye cell can be used. The cell is usually 1 cm long, as in Fig. 5b. It takes light 50 psec to travel 1 cm in the dye cell. Since the relaxation time of the dye is shorter than the time required to travel through the cell, the dye will be bleached by the strong pulse but still absorb the weak one, eliminating the formation of satellite pulses. The dye 9860 in dichloroethane is usually set at 70% transmission at 1060 nm. The cell itself should be positioned at a Brewster angle close to the rear mirror, though in this case the exact distance from it is not important. A small aperture $\frac{1}{4}$ in. in diameter is placed between the laser head and the saturable absorber which allows only a lower-order transverse mode. The cavity length is set to approximately 1.5 m. The rear mirror is concave and of 99.8% reflectivity at 1060 nm. Typically the radius of curvature is ~ 3 m for high-power applications. The front mirror is a flat-wedged, with a reflectivity of $\sim 50\%$ at 1060 mm. This cavity is quite simple in design and has been found to give a reliable train of pulses.

IV. Single-Pulse Selection and Amplification

The oscillators we have described so far will produce a train of pulses separated by the round-trip time of the cavity. This happens because everytime the pulse in the cavity bounces off the front mirror part of it is transmitted through the mirror.

The first experiments were performed using a train of pulses (Alfano and Shapiro, 1970a; Seibert *et al.*, 1973; Seibert and Alfano, 1974; Yu *et al.*, 1977). Soon, however, it was discovered that this approach could lead to artifacts. Slow processes (e.g., triple states) may build up from successive excitations to the point of introducing significant errors in the measurements. In addition, the use of a train of pulses makes it impossible to study phenomena that are dependent on the power of the pulse, such as fluorescence quenching. Consequently techniques have been developed which allow the selection of a single pulse from the train. Figure 6 shows a schematic of the technique used. A Pockels cells is placed between two dielectric cross-polarizers. The pulses will pass through the first polarizer but not through the second. If, however, a half-wave voltage is applied to the Pockels cell, the plane of polarization will be rotated by 90°. The train can pass through the second polarizer. When the applied voltage lasts for a short time less than the round-trip time in the cavity, only *one* pulse is allowed to pass through the second

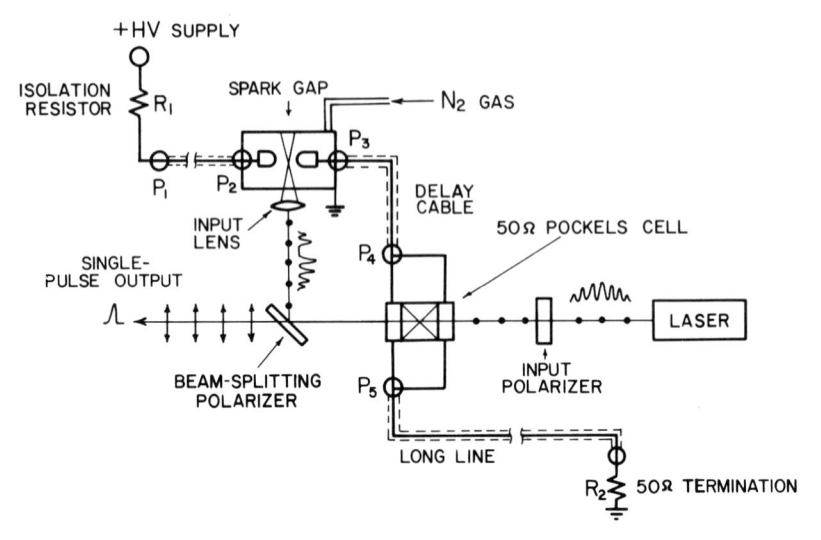

FIG. 6. Typical experimental arrangement for single-pulse selection using a Pockels cell and a spark gap.

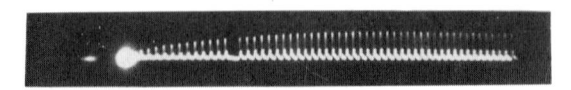

FIG. 7. Train of mode-locked pulses. The time between pulses in 8 nsec, the round-trip time in the laser cavity. The missing pulse is the selected one.

polarizer. Figure 7 shows the output train of pulses. The missing pulse is the one selected.

Pockels cells are made of certain uniaxial crystals. Table III gives some of the most commonly used materials. The crystal is cut perpendicular to the optical axis. When an electric field is applied parallel to the optical axis, the crystal becomes anisotropic, with the new axes (X', Y') making a 45° angle with the X and Y crystallographic axes (Fig. 8). This angle is independent of the intensity of the electric field. The lengths of the eclipse axes are proportional to the values of the indexes of refraction along the two directions. When polarized light along the X axes propagates in the crystal, the two components X' and Y' propagate at different velocities. The phase shift ϕ between the two waves is given by

$$\phi = (2\pi/\lambda)n_0^3 r_{63} V_z, \tag{5}$$

where λ is the wavelength of the light passing through the crystal, n_0 is the ordinary index of refraction of the crystal, r_{63} is the electrooptic constant, and V_z is the voltage applied longitudinally. The particular value of the

17. PICOSECOND LASER TECHNIQUES AND DESIGN 399

TABLE III

CHARACTERISTICS OF LONGITUDINAL MODULATOR CRYSTALS

Material	Electrooptic half-wave (μm/V × 10^{-6})	Typical half-wave voltage at 5461 Å (kV)	Index of refraction
Ammonium dihydrogen phosphate (ADP)	8.5	9.2	1.526
Potassium dihydrogen phosphate (KDP)	10.5	7.5	1.51
Potassium deuterium phosphate (KD*P)	26.4	2.9–3.4[a]	1.52
Potassium dihydrogen arsenate (KDA)	10.9	6.4	1.57
Rubidium dihydrogen phosphate (RDP)	11.0	7.3	–
Ammonium dihydrogen arsenate (ADA)	5.5	13	1.58

[a] Voltage depends on deuterium content, 99% D$_2$ corresponds to 2.9 kV; 7 kV is applied for a 1060-nm pulse.

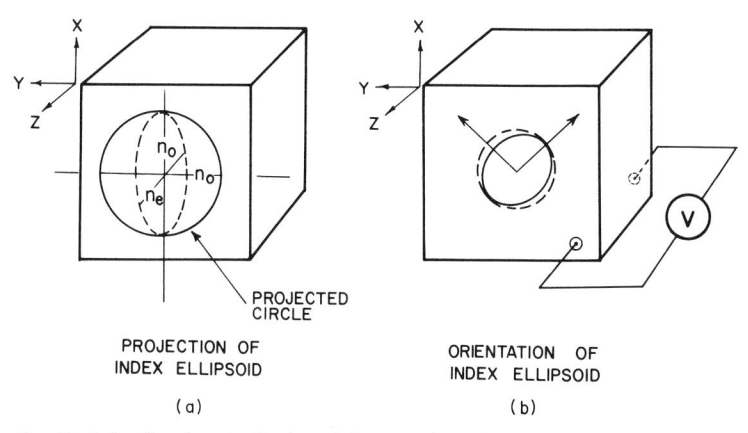

FIG. 8. Pockels effect longitudinal modulator. Orientation of the axes before high voltage is applied (a) and after it has been applied (b).

phase shift that is of interest in the case of an electrooptic shutter is the value which corresponds to a 90° rotation of the polarization plane. The voltage required to produce it is

$$V_{1/2} = \lambda/2n_0^3 r_{63} \quad (6)$$

Pockels cells are commerically available from companies such as Lasermetrics and Inrad. For the high-voltage pulses required to drive the Pockels cell, there are commerically available pulse generators (Lasermetrics, Inrad, Cordin, etc.). They typically have subnanosecond risetimes with an adjustable pulse width and pulse height ($\sim$7 kV and $\sim$8-nsec pulse width for a

1060-nm pulse). Alternatively an inexpensive way is to use a spark gap (Guenther and Bettis, 1971), as shown in Fig. 6. The rejected train is focused between the electrodes of the spark gap. One of the electrodes is held at a high potential. When the intensity of the laser field reaches the breakdown limit of the gas, high voltage is applied to the Pockels cell. This voltage is half the voltage applied to the spark gap. The width of the pulse is $2L/c$, where L is the length of connecting cable $P_1 P_2$ and c is the speed of light. The jitter of the high-voltage pulse is determined by the distance between the electrodes, the pressure of nitrogen gas (~ 10 atm), and the intensity of the laser electric field. These parameters are determined experimentally. The reliability of the spark gap pulse selector depends on the fine-tuning of the system. Under normal conditions, 90% reliability in the selection of a single pulse can be easily achieved. The length of the output cable $P_3 P_4$ determines which pulse in the train will be selected. It is important to select a pulse early in the train where the pulse width is the shortest. For a neodymium–glass laser the pulse width changes from 4 psec at the beginning of the train to 15 psec at the end. The pulse that is selected can be used directly for experiments or amplified if higher power is needed. Higher powers also provide the opportunity to produce other wavelengths by nonlinear optical processes. The extinction ratio of the pulse selector is between $\frac{1}{50}$ and $\frac{1}{300}$; that is, the selected pulse is 50–300 times stronger than the train that has leaked out. To improve the extinction ratio, a saturable absorber is placed after the pulse selector. The saturable absorber preferentially absorbs the weak pulses. Typically, the transmission of the dye is set at 10% at the laser wavelength.

The amplifier consists of one or more stages. Each stage is practically identical with the laser without mirrors we have described in Section III. The amplifier medium is the same as that used in the oscillator with two differences. The diameter of the rods used in the amplification stages should be progressively larger, typically $\frac{1}{2}-\frac{3}{4}$ in. This reduces the chances of surface damage to the rods. The rods are doped at a lower concentration for more uniform pumping. The ends of the rods are cut at a small angle, usually $6°$, to avoid lasing action off the ends and the formation of multiple pulses (Lu et al., 1979). In addition to the linear single amplifier a multiple-pass amplifier is often used. The advantage of course is that a single unit produces more amplification. However, there are two disadvantages: the alignment is more difficult, and it is more likely to damage the surface of the laser rod.

V. Obtaining the Right Wavelength

Solid state neodymium–glass, neodymium–YAG, and ruby lasers can only be operated at a single wavelength. This limitation would greatly reduce the use of solid state lasers in picosecond spectroscopy if there were no means of changing the wavelength of the laser. The extremely high power of the light pulses makes it possible to exploit nonlinear optical effects in order to produce new wavelengths (for review, see Auston, 1977; Kielich and coworkers, 1972, 1977). The laser wavelength can be shifted to nearly any desired wavelength in the near infrared, visible, or near ultraviolet. These techniques include second-, third-, and fourth-harmonic generation, stimulated Raman scattering, stimulated fluorescence, and continuum self-phase modulation.

At a high power density the polarization of materials varies nonlinearly with the electric field of the laser. This gives rise to multiwave interactions. The susceptibility provides the coupling between the light waves. The resulting nonlinear polarization acts as a source term in Maxwell's equations and gives rise to the generation of new frequencies. To describe the nonlinear processes, the polarization of the medium in the electric dipole approximation is expanded in a power series (Bloembergen, 1965):

$$P = X^{(1)}\bar{E} + X^{(2)}\bar{E}\bar{E} + X^{(3)}\bar{E}\bar{E}\bar{E}. \tag{7}$$

The coefficients $X^{(1)}$, $X^{(2)}$, and $X^{(3)}$ denote the second-, third-, and fourth-rank susceptibility tensors, respectively. This approximation is valid when the electric field of the laser is less than the atomic field, 10^9 V/cm. Linear optical phenomena such as absorption are associated with $X^{(1)}$, three-photon interactions such as second-harmonic generation (SHG) are associated with $X^{(2)}$, and four-photon interactions such as third-harmonic generation and self-phase modulation modulation are associated with $X^{(3)}$. A typical value for $X^{(1)}$ is on the order of 10^{-1} esu, for $X^{(2)}$ is 10^{-7}–10^{-9} esu, and for $X^{(3)}$ is 10^{-11}–10^{-14} esu. In media with inversion symmetry (cubic and isotropic solids and liquids) the lowest-order nonlinear effects are associated with $X^{(3)}$, since $X^{(2)} = 0$. SHG in the dipole approximation occurs in material without inversion symmetry. SHG was first observed by Franken et al. (1961). The third-rank tensor X_{ijk} responsible for this process has 27 components, which corresponds to different possible polarizations of P and E. Usually these components are not independent. They are related to each other by the symmetry properties of the crystal.

The intensity of the second harmonic for a plane, monochromatic wave is given by

$$I(2\omega, l) = (X^{2\omega})^2 [\sin(\tfrac{1}{2}l\,\Delta k)/\tfrac{1}{2}l\,\Delta k]^2 I^2(\omega), \tag{8}$$

where

$$\Delta k = |\bar{k}_2 - 2\bar{k}| = (2\omega/c)(n_{2\omega} - n_\omega). \tag{9}$$

Equation (9) is known as the phase mismatch. $I(\omega)$ and $I(2\omega)$ are the intensity of the laser and the generated second harmonic, respectively, and l is the length of the crystal. k and n_ω and k_2 and $n_{2\omega}$ are the wave vector and index of refraction of the fundamental and the second harmonics, respectively. The condition of $\Delta k = 0$ is called phase matching. This can be satisfied by temperature or orientation of the SHG crystal. The efficiency of the harmonic generation depends on the power of the focused beam, the value of the nonlinear susceptibility of the substance used, to the degree that the phase matching condition is satisfied, and for short pulses (picosecond range) the coherence–spatial length of the pulse and its spectral width. For 4-psec pulses the length of the KDP crystal is less than 2.5 cm (Miller, 1968; Shapiro, 1969).

Tables of second-harmonic coefficients of materials have been published by many authors. [See, for example, Becham and Kurtz (1969) and Singh (1971).] There are two types of phase-matching crystals that are commonly used, type I and type II. In type I both beams travel as ordinary waves. In type II one of the beams propagates as an extraordinary wave. In both cases, however, the second harmonic propagates as an extraordinary wave. Recently type II phase matching has become increasingly more popular than type I for SHG. This is because of the increased conversion efficiency and larger phase-matching angle. The laser pulse, however, becomes elliptically polarized in type II crystals.

In a three-photon parametric process a coherent laser pulse passes through a properly oriented nonlinear crystal (Glenn, 1967; Akhmanov et al., 1968). Two new waves are generated called "signal" and "idler" at frequencies ω_s and ω_i. Conservation of energy requires that $\omega_L = \omega_s + \omega_i$. Under proper conditions the signal wave can mix with the laser (or pump) to produce a wave at the signal frequency. The process can continue so that both the signal and idler frequency are amplified at the expense of the initial laser pulse which gradually decreases as a function of the length of the crystal. Laubereau et al. (1974) demonstrated efficient parametric generation in $LiNbO_3$ pumped by 1060-nm picosecond pulses. In order to achieve a high parametric amplification the three polarization waves (pump, signal, and idler) must travel at the same velocity. This condition requires that the indexes of refraction be such that the K vectors satisfy the momentum matching condition $K_L = K_s + K_i$. For colinearly propagating waves the momentum matching condition can be written as

$$n_L/\lambda_L = (n_s/\lambda_s) + (n_i/\lambda_i), \tag{10}$$

where n_L, n_s, and n_i are the indexes of refraction for the laser (pump), signal, and idler waves, respectively. The wavelength of the laser is fixed. A change in the indexes of refraction, however, will cause the signal and idler waves to tune. This can be accomplished by making use of the angular dependence of the birefringence of anisotropic crystals or by variation in the temperature.

TABLE IV

PULSE PROPERTIES OF LIGHT PRODUCED BY PARAMETRIC
GENERATION IN LiNbO$_3$[a,b]

	1060 nm	530 nm
Tuning	2500–7000 cm^{-1}	12,500–17,000 cm^{-1}
	4–1.43 μm	800–588 nm
Frequency width	6.5 cm^{-1}	6 cm^{-1}
Pulse duration	3.5 psec	4 psec
$\Delta f \, \Delta t$	0.7	0.7
Divergence	3 mrad	3 mrad

[a] Laubereau et al. (1978).
[b] 10^9 W/cm intensity.

Table IV gives the pulse properties of the parametric generation in LiNbO$_3$ (Laubereau et al., 1978; Seilmeier and Kaiser, 1980).

Four-photon processes such as third-harmonic generation and four-photon parametric mixing are described by the third-order polarization coefficient. The effect of four-photon interaction can occur in both acentroic and centroic materials, including liquids and gases, according to the symmetry properties of the fourth-order susceptibility tensor $X^{(3)}$. The efficiency of four-photon processes varies from 10^{-1} to 10^{-4} esu.

An important four-photon effect is stimulated Raman scattering (SRS). SRS photons are produced by the interaction of a light wave with vibrational excitation (Carman et al., 1970). An incident light at frequency ω_L interacts with a molecular vibration ω_j to produce a light wave of frequency ω_i, where $\omega_i = \omega_L \pm \omega_j$. The efficiency of the conversion depends on the power of the focused beam and the path length inside the material. Typical power levels used are 10^9–10^{10} W/cm^2 and lengths of 10 cm. Several materials are available for SRS. Table V gives a list of some of the materials used for SRS, the shift and the wavelengths produced, and the conversion efficiencies (Colles, 1969). It should be noted that the conversion efficiencies listed correspond to a power density of the pump of 25 GW/cm^2. Usually the conversion efficiencies are smaller than those given in Table V.

Other useful processes for the generation of different wavelengths are four-photon parametric effects. Four-photon parametric interactions are governed by the $X^{(3)}$ susceptibility tensor and consequently occur in a variety of materials, including isotropic solids, liquids, and gases (Alfano and Shapiro, 1970a). In this process two laser photons are converted to a signal and to an idler photon, $2\omega_L = \omega_s + \omega_i$. A more general interaction can occur when two high-intensity laser pump waves are present: $\omega_1 + \omega_2 = \omega_3 + \omega_4$.

TABLE V

CHARACTERISTICS OF SOME OF THE MATERIALS USED TO PRODUCE NEW LINES BY SRS[a]

Material	Frequency (cm^{-1})	Gain at 530 nm (G cm^{-1})	wavelength produced (nm)	Power conversion (%)
Benzene	992	16.2	559	15
Ethyl alcohol	2928	5.11	627	80
Methyl alcohol	2834	2.27	623	40
	2944	1.76	627	30
Acetone	2925	1.17	627	80
Water	3450	0.14	648	50

[a] Pump, 530 nm at 25 GW/cm^2.

The intensity of the generated wave is simply proportional to the product of the intensities of the incident waves.

Another very important nonlinear process with application in absorption spectroscopy is self-phase modulation (SPM) (Alfano and Shapiro, 1970a,b,c) which can be described by a simple model as follows. An intense optical pulse traveling through a medium can distort the atomic configuration of the material, resulting in a change in the refractive index via the nonlinear coefficients. The electric field of the laser beam in the time domain after traveling a distance z in the material is given by

$$E(t) = \tfrac{1}{2}E_0(t)\exp - i(\omega_L t + n(t)z\omega_L/c) + \text{c.c.}, \qquad (11)$$

where $E_0(t)$ is the envelope of the pulse, ω_L is the laser angular frequency, and n is the total index of refraction. For a medium having inversion symmetry, the first nonlinear coefficient is $X^{(3)}$ and the index of refraction n becomes intensity-dependent:

$$n = n_0 + n_2 E^2. \qquad (12)$$

The intensity-dependent term in the index of refraction modulates the spectral intensity given by the Fourier transform. It does so by modulating the instantaneous phase. This leads to a broadening of the pulse both in Stokes and anit-Stokes frequencies. The broadening $\Delta\omega$ is given by

$$\Delta\omega = \frac{\omega_L z}{c}\frac{\partial n_2(t)}{\partial t}. \qquad (13)$$

The pulse modifies its own spectra through a change in phase and envelope. These processes are SPM and self-steepening, respectively. The effect of envelope shape (supercontinuum) distortion is negligible for sample lengths less than 20 cm. The broadening in liquids is over several thousand wave numbers. Table VI lists typical material used in SPM. The duration of the broadened emission is comparable to the laser pulse. An important application of the four-photon processes and SPM of picosecond pulses is in the use as a "white light" or continuum probe for picosecond absorption spectroscopy. The first measurements of this type were made by Alfano and Shapiro (1970a,b,c).

TABLE VI

Continuum Generation in Liquids and Glasses[a]

Material	Pumping pulse (nm)	Broadening Useful continuum (nm)
CCl_4	1060	440–900
	530	400–900
BK-7A	530	400–800

[a] Numerous other materials have been used, such as H_2O, D_2O, H_3PO_4, and alcohols.

In addition to the described methods for producing new wavelengths there is stimulated fluorescence emission from certain dyes such as rhodamine 6G (Rh6G), rhodamine B3, acridine 13, and several polymethine cyanine dyes that can be used in certain applications (Glenn et al., 1968; Mack, 1969; Malley and Rentzepis, 1970). The wavelength and duration of the pulses emitted by stimulated fluorescence vary with concentration, power of the pump, emission delay time, path length, and solvent. The duration of the stimulated fluorescence is much longer than the pumping pulse, typically on the order of nanoseconds. They provide, however, a large-number wavelength that can be used as a probe beam in an absorption apparatus.

VI. Time Measurements and Applications

Currently available electronic equipment does not have the time response required to measure directly events that take place in picoseconds. The fastest oscilloscope, for example, equipped with a sampling head has a resolution of 300 psec. Indirect methods, however, are available for time measurements. The constant speed of light traveling over a known distance provides the basis for the time-measuring mechanism on a picosecond time scale. It takes light 1 psec to travel 0.3 mm in the air. A length of 0.3 mm can be easily and accurately measured with a micrometer.

This section is divided into two parts. The first part surveys the measuring methods used to measure the pulse duration of the emitted light pulses. The second part surveys techniques that have been developed to measure the dynamic behavior of materials using time-resolved fluorescence and absorption techniques.

The two methods that are described and most often used for pulse width measurements are two-photon fluorescence (TPF) and SHG spatial width (for review, see Ippen and Shank, 1977).

The TPF method was described first by Giordmaine et al. (1967). Figure 9 shows a schematic of the arrangement that is used in TPF. The beam is split into two beams and directed into a cell that contains a dye which does not absorb at the laser wavelength. The dye molecules, however, may absorb two photons simultaneously and be raised to an excited state. Then fluorescence takes place in the normal way. The cross section for two-photon absorption is very small, so that TPF is not ordinarily observed. The probability of a two-photon transition is proportional to the square of the incident light intensity. The size of the TPF is related to the pulse width through the autocorrelation function $G^{(2)}(\tau)$:

$$I_{2f} = C[1 + 2G^{(2)}(\tau)], \tag{14}$$

and

$$G^{(2)}(\tau) = \int_0^{t_0} I(t)I(t+\tau)\,dt \Big/ \int_0^{t_0} I^2(t)\,dt, \tag{15}$$

where C is a constant. From the autocorrelation function and the pulse shape the pulse duration can be estimated. Table VII gives the real values of the pulse width for three different pulse shapes. In addition the ratio of the peak intensity to the background provides information about the mode locking. For ideal single-transverse mode-locked pulses the ratio is 3:1. On the other hand, for ideal multitransverse mode-locked pulses the ratio is 2:1 (Weber and Danielmeyer, 1970). Typical dye solutions used are Rh6G in methanol for 1060 nm and coumarin in methanol or BBOT in cyclohexane for 530 nm.

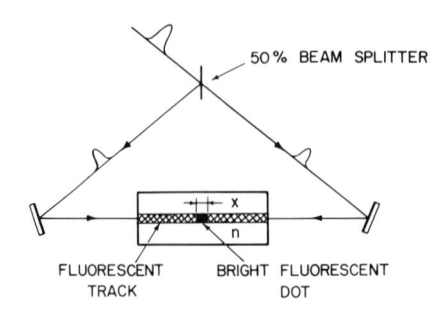

FIG. 9. Experimental arrangement for TPF. The size of the fluorescence spot (x) is related to the convolution through the equation $\Delta t_G = 2xn/c$, where n is the index of refraction and c is the speed of light. The pulse duration depends on the shape of the pulse and is given by the equation $\Delta t = \Delta t_G/\gamma$, where γ is the form factor.

TABLE VII

Relation Between Pulse Shape and Pulse
Width Measured by TPF[a]

Pulse	$\gamma = \Delta t_G/\Delta t$[b]	$\Delta t \, \Delta f$[c]
$\exp(-4ln2\, t^2/\Delta t)$	1.44	0.441
$\text{sech}^2|(1.76t)/\Delta t|$	1.55	0.315
$\exp|-ln2\, t/\Delta t|$	2	0.11

[a] Ippen and Shank (1977).
[b] Δt_G and Δt are (FWHM) of the pulse width as measured by TPF and the true pulse width, respectively. γ is the form factor.
[c] Transform-limited pulses (FWHM).

Figure 10 shows a schematic diagram of the arrangement used for the SHG spatial width (Kolmeder *et al.*, 1979). Two pulses with beam diameters D generated by a beam splitter cross in a nonlinear crystal at an angle ϕ and produce the SHG. The duration of the pulse is projected as the spatial distribution of a single pulse which can be observed with a vidicon system, optical multichannel analyzer, or imaging system with a diode array. The spatial width d (FWHM) and the pulse duration Δt (FWHM) are related by

$$\Delta t = [2d \sin(\phi/2)]/\gamma c, \qquad (16)$$

where c is the group velocity of the incident light pulses with ordinary polarization and γ is a form factor depending upon the shape of the pulse. For Gaussian pulses one has $\gamma = \sqrt{2} = 1.4$. It is important that the geometrical length of the light pulse, $L = c\Delta t$, be smaller than the beam diameter D. For $L < D/3$ this equation holds with a good approximation. For $L \approx D$ a more complex equation involving the temporal as well as the spatial shape of the light pulse has to be used (Jansky *et al.*, 1977). The SHG is more effective

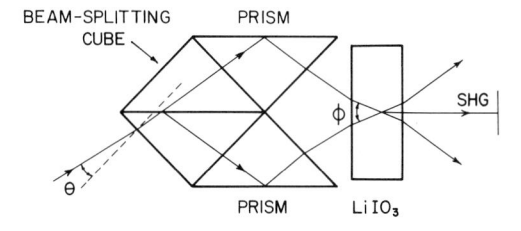

Fig. 10. Experimental arrangement for the SHG spatial width.

than the TPF. The emission of the second harmonic is highly directional, while the TPF radiates into a solid angle of 4π. In addition, the time resolution achieved by the second harmonic is higher. Depending on the nonlinear crystal used ($LiIO_3$, KDP, ADP) and the wavelength region, pulse durations as small as 0.2–0.02 psec can be measured (Kolmeder et al., 1979).

In time-resolved spectroscopy the molecules are excited by a strong laser pulse, and the subsequent relaxation of the system is studied by such spectroscopic techniques such as fluorescence and absorption.

A. Fluorescence Kinetics

The optical Kerr gate (Duguay and Hansen, 1969) consists basically of a Kerr active liquid (e.g., CS_2) situated between two cross-polarizers. Under the intense electric field associated with the laser pulse ($\sim 10^9$ V/cm) the molecules of the Kerr active liquid experience an induced birefringence. As a result, the light passing through the liquid becomes generally elliptically polarized and passes partly through the second polarizer. Light can only pass through the gate while it is coincident with the intense laser pulse that opens the gate. The intense laser pulse can be used to carve out successive portions of the temporal profile of the emitted fluorescence. The resolution of the Kerr gate depends on the pulse duration of the intense laser pulse that opens the gate. Typically it is on the order of 10 psec (Ho and Alfano, 1979).

Figure 11 is a schematic diagram of a fluorescence apparatus which uses an optical Kerr gate. The optical Kerr gate has been the workhorse of fluorescent measurement, but it is time-consuming. Each laser shot provides a single data point of the intensity decay profile. In addition, the signal-to-noise ratio is very low.

The intensity of the fluorescence measured has to be normalized as follows:

$$I_F \propto I_D/I_G^2 I_P, \qquad (17)$$

where I_D is the intensity of the signal detected, I_G is the intensity of the pulse that opens the Kerr gate, and I_P is the intensity of the pump pulse exciting the fluorescence of the sample. Equation (17) holds for small induced phase changes (Ho and Alfano, 1979).

The technique most widely used for fluorescent measurements, despite the high cost, incorporates a streak camera (Fig. 12). The streak camera was first developed by Bradley et al. (1970) and Shelev et al. (1971). Photoelectrons emitted by light striking the photocathode at different times are deflected by an applied voltage ramp which causes the electrons to be transversely streaked across a phosphorescent screen at the same time they are accelerated through the anode. These photoelectrons released at different times from the photocathode strike the phosphorescent screen at different positions,

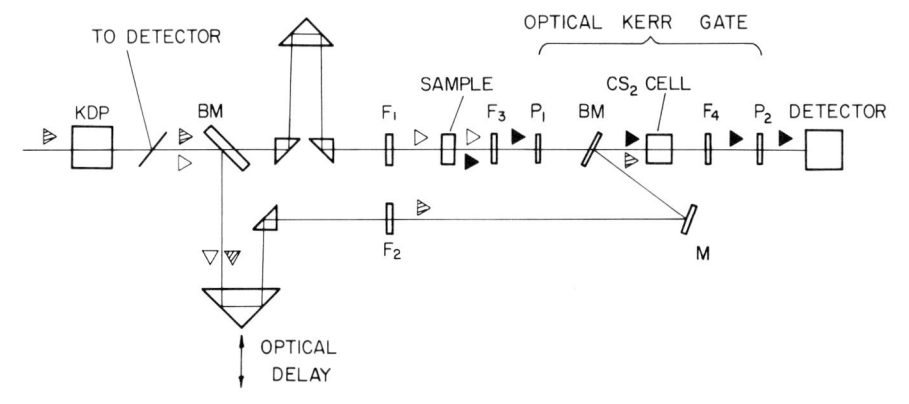

FIG. 11. Experimental arrangement for fluorescence measurements using an optical Kerr gate. A single pulse (infrared) is selected and amplified. The beam passes through a frequency-doubling crystal to produce the second harmonic at 530 nm. A dielectric mirror (BM) separates the two beams. The infrared pulse travels the variable optical path before entering the CS_2 cell. The 530-nm beam excites the sample. The fluorescence is collected by a lens and is passed through the gate. Only the portion of fluorescence that coincides with the infrared beam will pass through the gate. P_1 and P_2 are the two cross-polarizers. F_1, F_2, and F_3 are cutoff filters. ▷, 1060 nm; ▷, 530 nm, ▶, sample fluorescence.

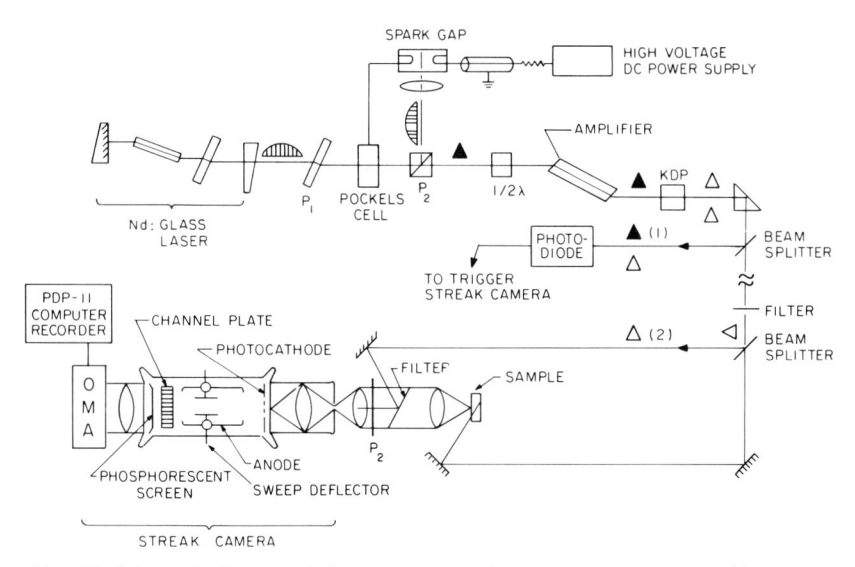

FIG. 12. Schematic diagram of the apparatus used to measure fluorescence kinetics by a streak camera. A single pulse is selected, amplified, and passed through a KDP crystal to produce the second harmonic. A beam splitter provides two side beams. Beam 1 triggers the streak camera. Beam 2 arrives at the streak camera at an earlier time to provide a calibration pulse. The main 530-nm pulse excites the fluorescence from the sample. The streak produced at the phosphorescent screen is recorded by an optical multichannel analyzer (OMA).

TABLE VIII
Comparison of Picosecond Streak Cameras[a,b]

Manufacturer	Hamamatsu	Hamamatsu	Hadland
Model	C979	C1370	675/II (with 50/40)
Time base configuration	Fast plug-in	Ultrafast	Standard
Sweep range full scale	Variable, 1, 2, 5, 10 nsec	Variable, 0.38, 0.5, 1, 2 nsec	Variable, 44 steps, 1.5–500 nsec
Maximum time resolution for S-20 photocathode at 530 nm	10 psec	2 psec	4 psec
Maximum streak velocity	67 psec/mm	25 psec/mm	30 psec/mm
Streak image output format	15 mm diam.	15 mm diam.	15 × 50 mm
Time window (at maximum streak velocity)	1 nsec	375 psec	1.5 nsec
Electronic trigger characteristics			
Minimum delay	11 nsec	11 nsec	12 nsec
Jitter	±50 psec	±50 psec	±25 psec
Electrical pulse volt	2–10 V (50 Ω)	2–10 V (50 Ω)	10 V (50 Ω)
Direct optical input trigger characteristics			
Minimum delay	—	—	1.5 nsec
Jitter	—	—	±25 psec
Optical pulse energy	—	—	10 μJ
Spatial resolution	7 lp/mm	25 lp/mm	7 lp/mm
Photocathode format	4.2 mm diam.	9 mm diam.	7.5 mm diam.
Photocathode	S-1, S-20, multialkali (UV glass)	S-20, multialkali (UV glass)	S-1, S-20, S-20UV
Image intensifier[c]	Built in	Built in	External

causing a track with a spatial intensity profile directly proportional to the incident temporal intensity profile of the fluorescence. The phosphorescent track may be analyzed by photographic techniques or electronically sampled by a video system. In the apparatus shown in Fig. 12 a small fraction of the pulse used to excite the sample is reflected inside the camera to provide a calibration point in time. A modification of this arrangement allows one to measure the temporal profiles of the fluorescence at both polarizations simultaneously (Pellegrino et al., 1980; Seymour and Alfano, 1980). A streak camera can be coupled with a spectrograph and a two-dimensional array in order to provide both the temporal and spatial characteristics of the fluorescence (Robinson et al., 1979). There are limitations, however, imposed by the uncertainty principle. Based on the particular application, one has to choose between temporal resolution or wavelength resolution (Schiller et al., 1980).

TABLE VIII (*continued*)

Hadland 675/II (with 50/40)	Gear PICO-V	Gear PICO-V	Cordin 179-LLL	Thompson TSN-504-04
Ultrafast plug-in	Avalanche krytron module	Ultrafast spark gap module	Individual plug-ins	Plug-in
Fixed, 0.5 nsec	Fixed, 825 psec	Fixed, 260 psec	Fixed, 1, 3, 4, 13 nsec	Variable 1, 2, 5, 10, 20 nsec
2 psec	5 psec	1.5 psec	10 psec	5 psec
10 psec/mm	33 psec/mm	10.5 psec/mm	33 psec/mm	25 psec/mm
15 × 50 mm	25 mm diam.	25 mm diam.	40 mm diam.	25 × 40 mm
500 psec	825 psec	260 psec	1.3 nsec	1 nsec
30 nsec	40 nsec	—	16 nsec	15 nsec
±500 psec	±500 psec	—	±100 psec	≤300 psec
10 V (50 Ω)	10 V (50 Ω)	—	30–50 V (50 Ω)	50 V (50 Ω)
—	—	1 nsec	1.7 nsec	—
—	—	±30 psec	>100 psec	—
—	—	2–15 mJ	1–10 µJ	—
7 lp/mm	8 lp/mm	8 lp/mm	7 lp/mm	12 lp/mm
7.5 mm diam.	1 × 15 mm	1 × 15 mm	13 mm diam.	35 mm diam.
S-1, S-20 S-20UV	S-20, S-25	S-20, S-25	S-1, S-20	S-20, S-20UV
External	External	External	External	Built in

[a] This table was prepared as a supplement by the authors for the article "An ultrafast streak camera system: temporaldisperser and analyzer" by N. H. Schiller *et al.*, which appeared in the June 1980 issue of *Optical Spectra*, p. 55.

[b] Each manufacturer makes a variety of different model streak cameras, plug-ins, and accessories. For the sake of brevity, each manufacturer is represented by no more than two units. The parameters displayed in the table were obtained from information supplied by the manufacturers or compiled from available datasheets. Since there are no standard definitions used to define the various parameters, the authors are not responsible for, or knowledgeable of the testing techniques used by the other manufacturers in obtaining the values used in this table. Equal numbers may not represent an equal comparison in all cases.

[c] For the image intensifier, "Built-in" indicates that the intensifier is built into the steak tube.

The streak camera has, of course, its own problems, namely, the lag of the video tube, the proper triggering of the tube scan, the limited intensity dynamic range of the camera ($<10^3$), and finally the nonlinearity of the streaking rates of the camera. Some of these problems like the last one can be alleviated by data analysis incorporating calibration charts for the streak rates (Schiller *et al.*, 1980). On the other hand, as the technology of the streak camera improves, some of these problems will be solved. A number of different models of streak cameras are commercially available and are summarized in Table VIII.

B. Absorption Kinetics

Figure 13 is a schematic diagram of the picosecond absorption apparatus. The timing mechanism of the apparatus is the speed of light itself. It takes light 3.3 psec to travel 1 mm in the air. If the path length of the pump pulse is decreased by 1 mm, by moving the delay prism, the pump pulse will arrive at the sample site 6.6 psec (twice the change in the path length) before the probe pulse. The kinetics of any absorption changes can be followed point by point by moving the delay prism across the time domain of interest, usually about 300 psec. The probe beam (supercontinuum) is generated by focusing the 1060-nm pulse into a 15-cm-long CCl_4 cell (Alfano and Shapiro 1970a,b,c). The fluorescence emission from a dye can also be used as a probing pulse, provided it has a relatively long lifetime and a high quantum yield (Yoshihara et al., 1979). At the sample site, the probe beam is divided into two pulses of approximately equal intensity designated $I^e(t)$ and $I^r(t)$. $I^e(t)$ is transmitted through the same area of the sample which is excited by the 530-nm pulse. The intensity of the transmitted probe beam depends of course on the changes initiated by the exciting pulse. $I^r(t)$ is the reference beam and is transmitted through a different part of the sample.

The change in optical density for a given delay time is obtained as

$$\Delta OD(t) = -\log\{[I^e(t)/I^r(t)](I_0^r/I_0^e)\}, \tag{18}$$

where the ratio I_0^r/I_0^e is the normalization factor for the two beams before the sample.

An innovative absorption probe technique uses a stack of thin microscope slides or mirrors arranged in step sequences (an echelon) (Busch et al., 1973). Each step introduces a delay which depends on the thickness and the angle of incidence. The probe pulse, as it travels through or reflects from the echelon, is divided into a number of pulses with a step progression of time delays. The probe can then be directed to the sample to interrogate the sample's kinetics on a single shot. A range of 100 psec is easily covered. The probe pulses transmitted through the sample are measured by a spectrograph coupled with either a PM tube, television system, optical multichannel analyzer, or photographic film.

We would like to close this chapter with a word of caution. The laser radiation emitted by the lasers described in the previous sections is extremely dangerous. It is capable of producing permanent eye and skin damage. Even a small percentage reflection of the laser output can cause extensive injury to the retina because the eye will focus the laser beam. Infrared radiation is a particular hazard because one is not always aware of reflections. The common safe procedure for laser personnel is to use safety goggles at all times, block spurious reflections, and keep the laser path covered as much as possible.

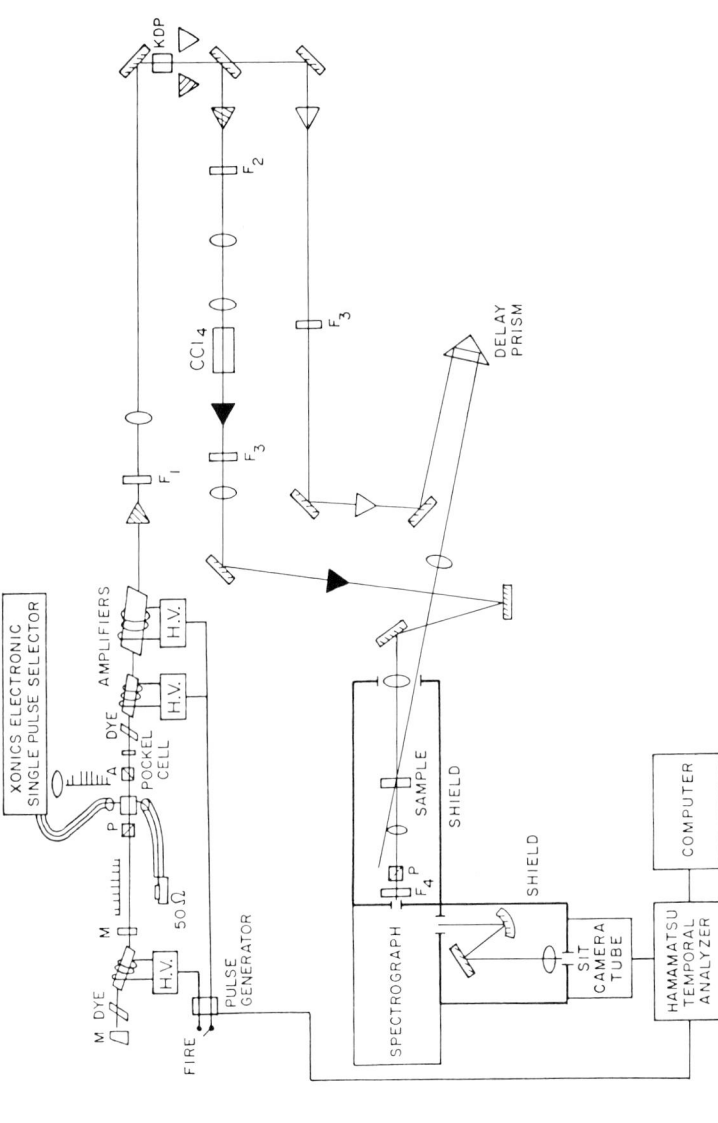

FIG. 13. Schematic diagram of the apparatus used to measure absorption kinetics. A single pulse is selected and amplified. Filter F_1 eliminates the scattered light from the flash lamp. The beam is passed through a KDP crystal to produce 530-nm pulses, and 1060 and 530 nm are separated by a specially coated mirror. Filter F_2 eliminates the 530-nm pulse. A set of lens collimates and focuses the 1060-nm pulse into the CCl_4 cell to produce the "continuum." Filters F_3 eliminate the 1060-nm pulse. The probe beam is analyzed by a spectrograph and projected onto the detector. △, 1.06 μm; △, 0.53 μm (pump); ▲, continuum (probe).

Acknowledgments

This work has been supported by grants from NSF, NIH, AFOSR, NASA, and CUNY FRAP. We thank Mr. Yury Budansky for the excellent drawings and Mrs. M. Weinberg for typing the manuscript. Dr. J. Buchert is on leave from the Non-linear Optics Division, Institute of Physics. A. Mickiewicz University, Poznan, Poland.

References

Akhmanov, S. A., Chirkin, A. S., Drabovich, K. N., Kovrigin, A. I., Khokhlov, R. V., and Sukorukov, A. P. (1968). *IEEE J. Quantum Electron.* **QE-4**, 598.
Alfano, R. R., and Shapiro, S. L. (1970a). *Phys. Rev. Lett.* **24**, 584.
Alfano, R. R., and Shapiro, S. L. (1970b). *Phys. Rev. Lett.* **24**, 592.
Alfano, R. R., and Shapiro, S. L. (1970c). *Phys. Rev. Lett.* **24**, 1217.
Alfano, R. R., and Shapiro, S. L. (1972). *Phys. Rev. Lett.* **29**, 1655.
Alfano, R. R., and Shapiro, S. L. (1973). *Sci. Am.* **228**, 42.
Alfano, R. R., Schiller, N. H., and Reynolds, G. (1981). *IEEE J. Quantum Electron.* **17**, 290.
Auston, D. H. (1977). *In* "Ultrashort Light Pulses" (S. L. Shapiro, ed.), Topics in Applied Physics, Vol. 18, p. 123. Springer-Verlag, Berlin and New York.
Basov, N. G., Drozhbin, Y. A., Kriukov, P. G., Lebedev, V. B., Letokhov, V. S., and Matveetz, Y. A. (1969). *JETP Lett.* **9**, 256.
Baumeister, P. (1963). "Handbook of Optical Design," Sect. 20. U.S. Gov. Print. Off., Washington D.C.
Bechman, R., and Kurtz, S. K. (1969). *In* "Landolt-Böernstein, New Series, Group III, Crystal and Solid State Physics," Vol. 2, Chap. 5. Springer-Verlag, Berlin and New York.
Bloembergen, N. (1965). "Non-linear Optics." Benjamin, New York.
Bradley, D. J., Higgins, J. F., and Key, M. H. (1970). *Appl. Phys. Lett.* **16**, 53.
Busch, G. E., Jones, R. P., and Rentzepis, P. M. (1973). *Chem. Phys. Lett.* **18**, 170.
Carman, R. L., Shimizu, F., Wang, C. S., and Bloembergen, N. (1970). *Phys. Rev. A* **2**, 60.
Chekalin, S. V., Kriukov, P. G., Matveetz, Y. A., and Shatberashvili, O. B. (1974). *Optoelectronics (London)* **6**, 249.
Colles, M. J. (1969). Rep. No. MM 69-1153-15. Bell Telephone Lab. Murray Hill, New Jersey.
DeMaria, A. J., Stetser, D. A., and Heyman, H. (1966). *Appl. Phys. Lett.* **8**, 174
DeMaria, A. J., Stetser, D. A., and Glenn, W. H. (1967). *Science* **156**, 1557.
Duguay, M. A., and Hansen, J. W. (1969). *Appl. Phys. Lett.* **15**, 192.
Einstein, A. (1917). *Phys. Z.* **18**, 21.
Fleck, J. A. (1970). *Phys. Rev. B* **1**, 84.
Fox, A. G., and Li, T. (1961). *Bell Syst. Tech. J.* **40**, 453.
Franken, P., Hill, A., Peters, C., and Weinreich, G. (1961). *Phys. Rev. Lett.* **7**, 118.
Giordmaine, J. A., Rentzepis, P. M., Shapiro, S. L., and Wecht, K. W. (1967). *Appl. Phys. Lett.* **11**, 216.
Glenn, W. H. (1967). *Appl. Phys. Lett.* **11**, 333.
Glenn, W. H., Brienza, M. J., and DeMaria, A. J. (1968). *Appl. Phys. Lett.* **12**, 54.
Guenther, A. H., and Bettis, J. R. (1971). *Proc. IEEE* **59**, 689.
Heavens, D. S. (1955). "Optical Properties of Thin Solid Films." Butterworth, London.
Ho, P. P., and Alfano, R. R. (1979). *Phys. Rev. A* **20**, 2170.
Hochstrasser, R. M., Kaiser, W., and Shank, C. V., eds. (1980). "Picosecond Phenomena II," Springer Series in Chemical Physics, Vol. 14. Springer-Verlag, Berlin and New York.
Holten, D., and Windsor, M. W. (1978). *Annu. Rev. Biophys. Bioeng.* **7**, 189.

Ippen, E. P., and Shank, C. V. (1977). *In* "Ultrashort Laser Pulses" (S. L. Shapiro, ed.), p. 83. Springer-Verlag, Berlin and New York.
Jansky, J., Corradi, G., and Gyuzalian, R. N. (1977). *Opt. Commun.* **23**, 293.
Kielich, S. (1972). *Dielectr. Relat. Mol. Processes* **1**, 192.
Kielich, S. (1977). "Nonlinear Molecular Optics." PWN, Warsaw.
Koechner, W. (1976). "Solid-State Laser Engineering," Springer Series in Optical Sciences, No. 1. Springer-Verlag, Berlin and New York.
Kolmeder, C., and Zinth, W. (1981). *Appl. Phys.* **24**, 341.
Kolmeder, C., Zinth, W., and Kaiser, W. (1979). *Opt. Commun.* **30**, 453.
Kopainsky, B., Kaiser, W., and Drexhage, K. H. (1980). *Opt. Commun.* **32**, 451.
Kriukov, P. G., and Letokhov, V. S. (1972). *IEEE J. Quantum Electron.* **QE-8**, 766.
Laubereau, A., Greiter, L., and Kaiser, W. (1974). *Appl. Phys. Lett.* **25**, 87.
Laubereau, A., Ferat, A., Seilmeier, A., and Kaiser, W. (1978). *In* "Picosecond Phenomena" (C. V. Shank, E. P. Ippen, and S. L. Shapiro, eds.), Springer Series in Chemical Physics, Vol. 4, pp. 89–95. Springer-Verlag, Berlin and New York.
Letokhov, S. V. (1969a). *Sov. Phys.—JETP* **28**, 562.
Letokhov, S. V. (1969b). *Sov. Phys.—JETP* **28**, 1026.
Lu, P. Y., Ho, P. P., and Alfano, R. R. (1979). *IEEE J. Quantum Electron.* **QE-15**, 406.
Mack, M. E. (1969). *Appl. Phys. Lett.* **15**, 166.
Malley, M. M., and Rentzepis, P. M. (1970). *Chem. Phys. Lett.* **7**, 57.
Miller, R. C. (1968). *Phys. Lett. A* **26**, 177.
Mocker, M., and Collins, R. J. (1965). *Appl. Phys. Lett.* **7**, 270.
Pellegrino, F., Sekuler, P., and Alfano, R. R. (1981). *Photochem. Photobiol.* **2**, 15.
Penzkofer, A. (1974). *Opto-electronics (London)* **6**, 87.
Rentzepis, P. M. (1970). *Science* **169**, 239.
Rentzepis, P. M. (1978). *Science* **202**, 174.
Ritter, E. (1972). *In* "Laser Handbook" (F. T. Arrechi and E. O. Schultz-Dubois, eds.), p. 897–921. North-Holland Publ., Amsterdam.
Robinson, C. W., Caughey, T. A., Auerbach, R. A., and Harman, P. J. (1979). *In* "Multichannel Image Detectors" (Y. Talmi, ed.), ACS Symposium, No. 102, Chap. 9. Am. Chem. Soc., Washington, D.C.
Schiller, N. H., Tsuchiya, Y., Inuzuka, E., Suzuki, Y., Kinoshita, K., Kamiya, K., Iida, H., and Alfano, R. R. (1980). *Opt. Spectra* **14**, 55.
Seibert, M., and Alfano, R. R. (1974). *Biophys. J.* **14**, 269.
Seibert, M., Alfano, R. R., and Shapiro, S. L. (1973). *Biochim. Biophys. Acta* **292**, 493.
Seilmeier, A., and Kaiser, W. (1980). *Appl. Phys.* **23**, 113.
Seymour, R. J., and Alfano, R. R. (1980). *Appl. Phys. Lett.* **37**, 231.
Shank, C. V., Ippen, E. P., and Shapiro, S. L., eds. (1978). "Picosecond Phenomena," Springer Series in Chemical Physics, Vol. 4. Springer-Verlag, Berlin and New York.
Shapiro, S. L. (1969). *Appl. Phys. Lett.* **13**, 19.
Shapiro, S. L., ed. (1977). "Ultrashort Light Pulses," Topics in Applied Physics, Vol. 18. Springer-Verlag, Berlin and New York.
Shelev, M. Y., Richardson, M. C., and Alcock, A. J. (1971). *Appl. Phys. Lett.* **18**, 354.
Singh, S. (1971). *In* "Handbook of Lasers" (R. J. Pressley, ed.), Chap. 18. Chem. Rubber Publ. Co., Cleveland, Ohio.
Vasicek, A. (1960). "Optics of Thin Films." North-Holland Publ., Amsterdam.
von de Linde, D. (1972). *IEEE J. Quantum Electron.* **QE-8**, 328.
von de Linde, D. (1973). *Appl. Phys.* **2**, 281.
Walling, J., Peterson, O., and Morri, R. C. (1980). *IEEE J. Quantum Electron.* **QE-16**, 120.
Weber, V. H. P., and Danielmeyer, H. G. (1970). *Phys. Rev. A* **2**, 2074.

Yariv, A. (1975). "Quantum Electronics," 2nd ed. Wiley, New York.
Yoshihara, K., Namiki, A., Sumitani, M. S., and Nakashima, N. (1979). *J. Chem. Phys.* **71**, 2892.
Yu, W., Pellegrino, F., and Alfano, R. R. (1977). *Biochim. Biophys. Acta* **460**, 171.

CHAPTER 18

Subpicosecond Ultrafast Laser Technique—Application and Design

C. V. Shank and B. I. Greene

Bell Telephone Laboratories, Incorporated
Holmdel, New Jersey

I.	Introduction	417
II.	Subpicosecond Pulse Generation	417
III.	Measurement Techniques	420
IV.	Future Directions	426
	References	427

I. Introduction

The last decade has seen dramatic advances in the generation of short laser pulses and their application to the study of picosecond phenomena. New developments that extend this technology into the subpicosecond (10^{-13} sec) regime offer exciting possibilities for accurate studies on previously unresolved processes in biology, chemistry, and physics.

In this chapter we will describe pulse generation methods and measurement techniques applicable for measuring primary events in biological systems. Subpicosecond measurement technology is primarily based on the continuously mode-locked dye laser (Shank *et al.*, 1972; Ippen *et al.*, 1972). Lasers of this type have generated pulses as short as a few tenths of a picosecond (Shank and Ippen, 1974; Ippen and Shank, 1975). In addition to short time resolution, these lasers provide a stable, reproducible train of optical pulses, permitting the use of powerful signal-averaging techniques.

II. Subpicosecond Pulse Generation

Organic dyes (Shank, 1975) provide a nearly ideal medium for generating ultrashort optical pulses. The vibration- and rotation-broadened electronic energy levels result in a gain over a wide range of frequencies. Typically an organic dye laser can be tuned over 100 nm. Since the gain spectrum is primarily homogeneously broadened on a picosecond time scale, the broad range of frequency components can be coherently summed to produce short, tunable pulses.

The process of subpicosecond pulse generation in a dye laser is an interesting phenomenon in itself. There are two favored methods of generating short subpicosecond optical pulses from a dye laser, passive and synchronous mode locking. In the case of passive mode locking a second dye is placed in the cavity, which acts in concert with the gain medium to produce a short pulse. On the other hand, the synchronously mode-locked continuous wave (cw) dye laser (Shank et al., 1972) is pumped by a train of optical pulses from a mode-locked pumping laser. Subpicosecond pulses have also been obtained from this laser system (Heritage and Jain, 1978).

An analysis of the passive mode-locking scheme was first presented by New (1974) who described the conditions under which dramatic pulse shortening occurred. Haus et al. (1975) obtained analytical solutions for the steady state pulse shape and stability in this system.

The process of pulse narrowing as the pulse makes a single pass through the laser may be discussed qualitatively. Two dyes are involved in the process. A slowly recovering saturable absorber acts to steepen only the leading edge of the pulse. Saturation of the gain in combination with the linear loss discriminates against the trailing edge. One condition for short-pulse formation is simply that the absorber saturate more easily than the gain. This provides a net gain at the peak of the pulse and a loss on either side. This process is repeated as the pulse makes multiple passes through the laser. Both dyes substantially recover during the several nanoseconds between passes. Shortening continues until dispersive effects prevent the pulse from becoming narrower.

Passive mode locking of the cw dye laser was first reported in 1972. Subsequent improvements in this system led to the production of pulses (Ippen and Shank, 1975) as short as 0.3 psec. Not only were the pulses short, but the reproducibility from pulse to pulse and the very high repetition rate (greater than 10^5 pulses/sec) obtained with cw systems were perhaps equally important. Powerful signal-averaging techniques could now be applied to picosecond studies. This is especially important for biological studies where it is desirable to make measurements with low-intensity pulses that do not disturb the process under investigation.

Figure 1 is a schematic illustration of a subpicosecond laser system used in our laboratory. The system is pumped by several watts of continuous power from a commercial argon ion laser. The dye gain medium, rhodamine 6G in a free-flowing thin stream of ethylene glycol, is located at a focal point approximately in the center of the resonator. Near one end of the resonator is a second free-flowing ethylene glycol stream containing saturable absorber

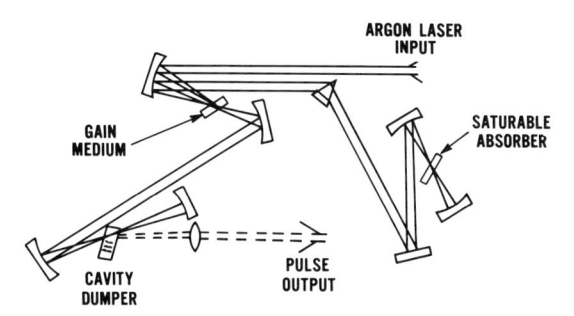

18. SUBPICOSECOND ULTRAFAST LASER TECHNIQUE 419

FIG. 1. A continuously operated subpicosecond dye laser. Pumping (top right) is by a commercial argon ion laser. The two dyes are rhodamine 6G in ethylene glycol (labeled "gain medium") and the saturable absorber, which contains a mixture of DODCI and malachite green. The cavity dumper (lower left) is an acoustooptic deflector positioned to dump single pulses out of the laser at a high repetition rate.

dyes. The laser mode is focused more tightly in the absorber stream than in the gain stream.

Subpicosecond pulses can be obtained with the saturable dye known as DODCI alone in the absorber stream. DODCI has a measured recovery time of 1.2 nsec, so that pulse formation must occur in the manner just described. Experiments show that the addition of a second dye, malachite green, improves the overall stability and allows operation well above laser threshold (Ippen and Shank, 1975). Malachite green alone will not induce mode locking, but it does provide supplementary absorption recovery on a picosecond time scale. With this combination of absorbers the laser in Fig. 1 is a reliable tool for subpicosecond spectroscopy.

An acoustooptic deflector positioned near the other end of the resonator is used to "dump" single pulses from the laser (Shank and Ippen, 1974). This scheme is preferable to using the pulse output from a partially transmitting mirror; the individual pulse energy is greater, and the pulse repetition rate is adjustable. Single pulses can be dumped at rates greater than 10^5 pps to maximize the extent of signal averaging. Alternatively, the repetition rate can be reduced if the system being studied needs more time between pulses for complete recovery.

The subpicosecond pulses produced by this laser each have an energy of approximately 5×10^{-9} J. Beam quality is such that the pulses can easily be focused to energy densities of several millijoules per centimeter squared or photon densities greater than 10^{16} cm^{-2}.

Often it is desirable to generate pulses with high peak power. With intense optical pulses, nonlinear frequency conversion processes such as second-harmonic, stimulated Raman, and continuum generation become possible.

To make these processes efficient it is necessary to produce pulses in the gigawatt range. For example, to amplify the 0.5-psec pulses from the passively mode-locked dye laser to gigawatt peak powers requires a gain of 10^6 (Shank et al., 1979). While such a large amplification and short pulse duration require special considerations, it is possible to realize this goal through the use of a multistage dye amplifier designed specifically for picosecond pulses (Ippen and Shank, 1978).

Several constraints influence amplifier design. The goal is to amplify the pulses efficiently while maintaining the short pulsewidth. The diagram in Fig. 2 shows a three-stage amplifier pumped by a doubled Nd^{3+}–YAG laser at 5300 Å. The dye is rhodamine 6G. The peak repetition rate is 20 Hz. The nanojoule laser pulses from the passively mode-locked laser can be amplified to 1 mJ energy using this system.

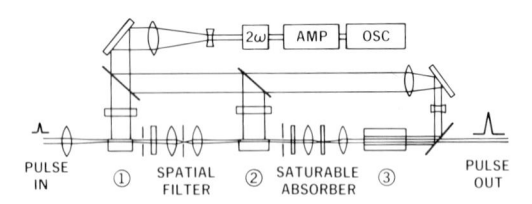

FIG. 2. Subpicosecond pulse amplifier. Optical pulses injected from the left are amplified by a factor of 10^6 by a three-stage amplifier pumped by a synchronized frequency-doubled neodymium–YAG laser. The dyes in the gain region are rhodamine B, and the saturable absorber used to isolate the stages is malachite green.

The technique of picosecond continuum generation first described by Alfano and Shapiro (1970) provides a convenient way of generating a "white" pulse containing a broad frequency spectrum while maintaining the short pulse resolution. A millijoule pulse of subpicosecond duration, when focused into a 2-cm cell containing water, generates a pulse with a useful spectrum that extends from 0.3 to 1.6 μm. The use of this pulse for performing time-resolved absorption spectroscopy will be described in a later section.

III. Measurement Techniques

The continuous train of optical pulses from the passively mode-locked dye laser oscillator can be used to advantage when the experiment requires low photon energy densities to facilitate interpretation. This allows the experiment to be done with low concentrations of excited state species. In this section we will describe how to take advantage of the stable continuous train of optical pulses to perform a subpicosecond time-resolved measurement in a biological system.

An excellent example of using a low photon flux signal-averaging technique is the investigation of the ultrafast dynamics of the photochemistry of bacteriorhodopsin (Ippen *et al.*, 1978). In this example we describe a time-resolved measurement of the primary photochemistry of bacteriorhodopsin, the formation time of the bathobacteriorhodopsin intermediate. The experiment is performed by exciting light-adapted bacteriorhodopsin with a 610-nm subpicosecond optical pulse and observing an induced absorption that corresponds to generation of the batho intermediate.

The experimental apparatus for performing this experiment is shown in Fig. 3. The pulses are obtained from a passively mode-locked cw dye laser system as described in the preceding section. The pulse train is split into two beams at the beam splitter, the probe beam being weaker by at least a factor of 10. A digitally controlled, stepper motor-driven translation stage is used to vary the time delay between the probe pulses and the stronger excitation pulses. The pump (excitation) beam and the probe beam are focused by a lens to the same 10-μm spot in a thin sample cell containing a suspension of membrane fragments with 7.5×10^{-4} M bacteriorhodopsin. The sample is flowed through this thin cell to remove the photochemical products from the laser beam before the next pump and probe pulses reach the sample. A similar flowing sample was used in a previous picosecond experiment to measure the lifetime (Hirsch *et al.*, 1976) of the emission observed at room

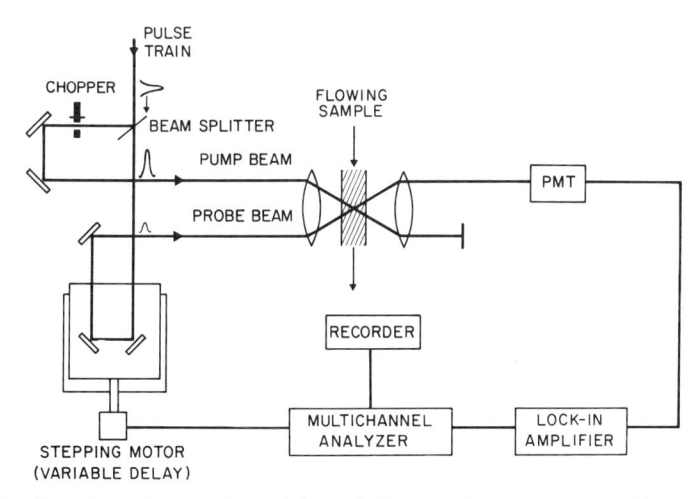

FIG. 3. Experimental setup for making subpicosecond measurements with a passively mode-locked cw dye laser system.

temperature in purple membranes (Lewis et al., 1976). A light source was used to illuminate the sample reservoir to ensure that all bacteriorhodopsin molecules were in the light-adapted bR570 form. At a sample concentration of 7.5 × 10^{-4} M, about 50% of the light is absorbed in a 0.1-mm path length. After passage through the flow cell the pump beam is blocked, and the probe beam passes to a photomultiplier tube. Modulation of the probe beam by the chopped pump is detected with a lock-in amplifier whose output is recorded as a function of delay time in the multichannel analyzer. Averaging is achieved by repetitive scanning of the desired delay interval. In these experiments averaging was performed over periods of 5–10 hr to obtain the time-resolved response of bathobacteriorhodopsin. During this time the laser pulses themselves changed in duration by less than 0.1 psec. Figure 4 shows the induced absorption at 610 nm as a function of relative time delay between the excitation and probing pulses for a suspension of membrane fragments with 7.5 × 10^{-4} M bacteriorhodopsin in doubly distilled water, pH 6.6. An increase in absorption is observed after the excitation pulse, as indicated by the solid experimental curve. The maximum change is less than 1%.

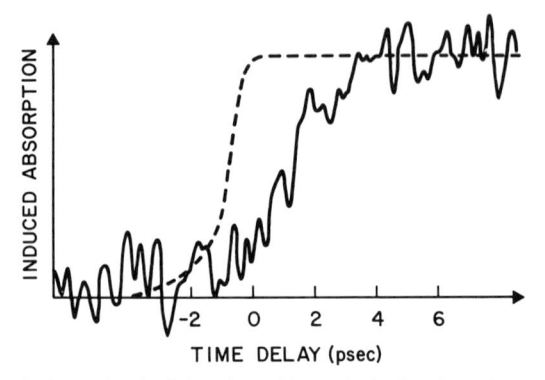

FIG. 4. Induced absorption in light-adapted bacteriorhodopsin at 615 nm plotted against the delay between probe and excitation pulses. The dashed curve is the experimentally determined instantaneous response of the system.

The dashed curve in Fig. 4 is the instantaneous response of the measurement system. It is obtained from autocorrelation measurements of the pulses used in the experiment and includes the appropriate coherence contribution (Shank and Ippen, 1977). Note that the coherence between pump and probe causes the predicted response for an instantaneous process to reach its full value at zero delay. From the data, the time constant with which the system approaches its final level falls in the range 1.0 ± 0.5 psec. We believe that this time corresponds to the formation time of the bathobacteriorhodopsin intermediate.

In the remainder of this section we will discuss the general technique of transient absorption spectroscopy as performed with 10-Hz amplified subpicosecond laser pulses. Experimental results exemplifying the utility of this technique in achieving high-quality transient spectra in biological samples will be presented.

Since the original demonstration of picosecond continuum generation by Alfano and Shapiro (1970) 10 years ago, several researchers have utilized this phenomenon in performing picosecond time-resolved transient absorption measurements. It has only been within the last few years, however, that considerable progress has been made in obtaining data of high spectrophotometric accuracy in experiments performed on this time scale. The reasons for this belated development are largely technical. With the advent of commercially available microprocessor-based multichannel optical detectors, today there remain few, if any, obstacles to achievement of Cary spectrophotometer quality spectra on the picosecond time scale (Weisman and Greene, 1979). Data of this quality have begun to appear in the literature and have served to emphasize repeatedly the absolute necessity for spectral veracity when interpretive speculation on anything other than the grossest effect is to be made (Greene et al., 1979).

The major difficulties remaining in this general experimental technique are those encountered for any complex optical apparatus. Detailed alignment of dozens of mirrors and lens, while conceptually trivial, becomes a time-consuming inevitability when working with low repetition rate (e.g., $\frac{1}{60}$ Hz) lasers. By increasing this repetition rate to 10 Hz, experiments that might have taken several weeks to execute often are accomplished in a single day.

Time resolution in picosecond transient absorption measurements is achieved solely as a consequence of the short temporal duration of the optical pulse. This pulse allows one suddenly to excite a sample optically, and at an arbitrary later time just as suddenly to take a broad-banded optical spectrum of the sample. As a class, these experiments fall into a category often labeled "excite and probe."

For biological samples, optical excitation most often entails direct absorption of laser light to produce electronically excited molecular species. In our laboratory this can readily be accomplished using either the laser fundamental pulse at 610 nm or a frequency-doubled pulse at 305 nm. With the aid of Fig. 5, the mechanics of a transient absorption measurement performed on a solution phase sample will be briefly highlighted.

As described previously, amplified dye laser pulses 0.5 psec in duration, containing 0.5–1.0 mJ pulse at 610 mm, are derived from the laser system at

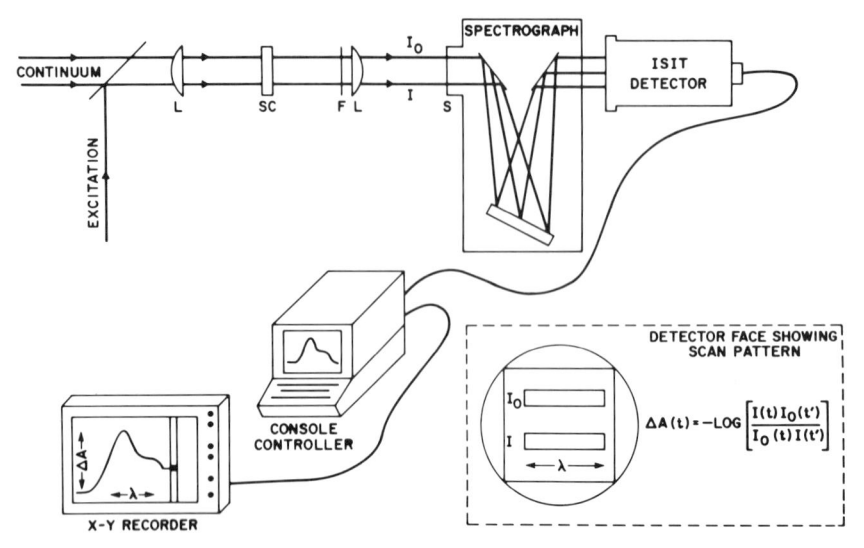

FIG. 5. Schematic diagram of the transient absorption apparatus. L, Lens; SC, flowing sample cell; F, filter; I and I_0, continuum probe beams; S, spectrograph slit.

a repetition rate of 10 Hz. The pulses are divided equally with a beam splitter into two beams, one sent down a variable path length and the other sent down a fixed-length optical path. By adjusting the relative path lengths of the two beams, one can arrange for the two pulses to arrive at a subsequent point of path intersection at exactly the same time. The position of the delay line in the variable-path length arm of the apparatus corresponding to this pulse overlap condition is called the $t = 0$ position. Every 100-μm change in the optical paths of these two pulses from the $t = 0$ position will result in the introduction of a 0.33-psec delay between the arrival of the pulses at the point of path recombination.

For the sample photolysis pulse, the 610-nm light sent down one arm, indicated in the top left-hand corner of Fig. 5 as "excitation," has been frequency-doubled using a 0.5-mm-thick angle-tuned potassium diphosphate (KDP) crystal. Typically 10–20 μJ/pulse is achievable at 305 nm. Broadbanded probing light has been generated by focusing the other 610-nm beam into a 2-cm-long cell of CCl_4. The emergent continuum light is recollimated and divided into two parallel beams, I and I_0, as indicated in Fig. 5. The only relevant timing considerations are between the pulse of light in the I beam and the ultraviolet (UV) excitation pulse.

The excitation beam and the sample probe beam (I) are combined by a dichroic beam splitter and propagate collinearly toward the sample. The reference probe beam, I_0, is directed toward an unexcited region of the sample. All three beams are focused by the same lens; the UV and I beams are directed through a 350-μm-diameter mask placed in the front of the sample, the I_0 beam passing through an unmasked region of the sample. Typically a 2-mm-path length flowing quartz sample cell is used, through which a roughly 5 ml/min flow is sufficient to ensure a totally new sample volume for each laser shot.

After filtration to remove residual excitation light as well as to compensate the transmitted continuum light spectrally, both the I and the I_0 beams are focused onto different heights of a low-dispersion spectrograph. Both probe beams are detected at the exit plane with a Princeton Applied Research Corporation Model 1215/16/54 optical multichannel analyzer (OMA) system (Fig. 5). Difference spectra as a function of delay time are obtained by a comparison of the I/I_0 ratio at time t with the same ratio at a negative delay time. Logarithms of the ratio of these ratios yield the derived ΔA spectra at time t. Typical data derived as the average of six cycles of data accumulation are shown in Fig. 6. Each cycle consists of an accumulation of over approximately 700 laser shots each, at time t, and the negative reference time. During this accumulation, a pulse discriminator selects for UV pulse energies within a preset window. This "good pulse" signal is utilized to blank out unacceptable laser events electronically before they are accumulated in the OMA memory. The result of each cycle is smoothed with an OMA-supplied software Savitzky–Golay routine, choosing a suitable smoothing parameter to coincide with the slit width-determined spectral resolution of 5 nm. The spectra from the six data cycles are then averaged together without further smoothing.

An excellent application of this technique is the investigation of the photochemical dynamics of bilirubin. Bilirubin, a naturally occurring degradation product of hemoglobin in normal adults, is a frequent cause of

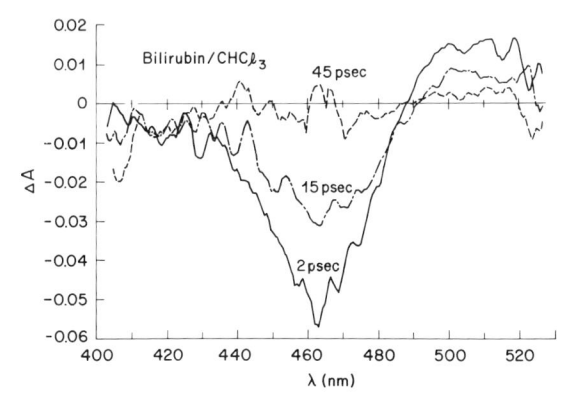

FIG. 6. Induced absorbance changes in bilirubin–$CHCl_3$ at 2 psec (———), 15 psec (·-··-·), and 45 psec (-----) after 305-nm excitation.

jaundice in newborn infants. Since the enzymatic capability for removing bilirubin from the bloodstream is initiated at birth, a transient period of bilirubin toxicity (bilirubinemia) is not uncommon. Many years ago it was demonstrated that, by exposing these infants to light, significant reduction in the bilirubin levels in the blood could be achieved (Cremer et al., 1958). Several classic chemical studies seem to point to a photoinduced structural alteration of the molecule as being responsible for the effect (McDonagh et al., 1980).

A recent picosecond study performed in our laboratory has clarified the details of this process for bilirubin dissolved in $CHCl_3$, as well as for bilirubin bound to human serum albumin (HSA), the state in which the molecule is actually found in the human bloodstream (Greene et al., 1981).

Figure 6 displays difference spectra of bilirubin in $CHCl_3$ taken at three delay times subsequent to pulsed optical excitation. The sample displayed a peak ground state absorption of ~ 2.0 OD. The individual spectra are clearly composites of a ground state bleaching spectrum and an excited state absorption spectrum. The observation of a constant spectral shape as the signal decays rules out the existence of a long-lived intermediate. The ground state is seen to recover roughly simultaneously with the excited state disappearance, characterized by a single exponential lifetime $\tau = 17 \pm 3$ psec.

Very similar results are obtained for bilirubin bound to HSA. This fact is totally unexpected from a simple comparison of fluorescent quantum yields for the two systems. Such data indicate a 25-fold increase in the excited state lifetime for bilirubin–HSA over bilirubin–$CHCl_3$. Arguments involving structural hindrance to rearrangement cocomitant with protein binding had been generally accepted as the explanation for the increased fluorescence of bilirubin–HSA. The direct picosecond measurement points instead to an alteration in radiative rates between bound and unbound molecules. In fact, protein binding is seen not to alter significantly the rate of the nonradiative process responsible for fluorescence quenching.

The picosecond data together with the quantum yield of fluorescence and photoproduct formation as a function of temperature–viscosity strongly suggest that the nonradiative process mentioned above is a configurational photoisomerization about one of the meso ethylenic centers of bilirubin. Recovery to the ground state as either isomer is suggested to occur via a short-lived ($\ll 18$ psec), twisted intermediate form.

IV. Future Directions

A number of new developments are on the horizon which should provide new capabilities for time-resolved measurements on biological systems. Optical pulses less than 0.1 psec will most certainly be obtainable in the near future. This should open the way to measurements with 10-fsec resolution. In addition, subpicosecond streak cameras should come in to use for investigating time-resolved photoemission processes. A low-jitter repetitively triggerable streak camera would allow signal averaging and provide a dramatic increase in dynamic range. Finally, new subpicosecond electronic technologies appear possible that may also have biological applications. Simplification of apparatus and the availability of commercial products should make picosecond measurements a routine laboratory measurement tool in the near future.

Editor's note: Most recently, considerable progress has been achieved in the generation of femtosecond laser pulses at various facilities. An up-to-date review of femtosecond laser technology will appear in a volume now in preparation and will stress ultrafast processes in semiconductors.

REFERENCES

Alfano, R. R., and Shapiro, S. L. (1970). *Phys. Rev. Lett.* **24**, 584–587.
Cremer, R. J., Perryman, P. W., and Richards, D. H. (1958). *Lancet* **i**, 1094.
Greene, B. I., Hochstrasser, R. M., and Weisman, R. B. (1979). *J. Chem. Phys.* **70**, 1247.
Greene, B. I., Lamola, A. A., and Shank, C. V. (1981). *Proc. Natl. Acad. Sci. U.S.A.* **78**, 2008–2012.
Haus, H. A., Shank, C. V., and Ippen, E. P. (1975). *Opt. Commun.* **15**, 29–31.
Heritage, J. P., and Jain, R. K. (1978). *Appl. Phys. Lett.* **32**, 101–104.
Hirsch, M. D., Marcus, M. A., Lewis, A., Mahr, H., and Frigo, N. (1976). *Biophys. J.* **16**, 1399–1406.
Ippen, E. P., and Shank, C. V. (1975). *Appl. Phys. Lett.* **27**, 488–491.
Ippen, E. P., and Shank, C. V. (1977). Chapter in *"Ultrashort Light Pulses"*, edited by S. L. Shapiro, 84–121.
Ippen, E. P., and Shank, C. V. (1978). *In* "Picosecond Phenomena" (C. V. Shank, E. P. Ippen, and S. L. Shapiro, eds.), pp. 103–107. Springer-Verlag, Berlin and New York.
Ippen, E. P., Shank, C. V., and Dienes, A. (1972). *Appl. Phys. Lett.* **21**, 348–351.
Ippen, E. P., Shank, C. V., Lewis, A., and Marcus, M. A. (1978). *Science* **200**, 1279–1281.
Lewis, A., Spoonhower, J. P., and Perrault, G. J. (1976). *Nature (London)* **250**, 675–677.
McDonagh, A. F., Palma, L. A., and Lightner, D. A. (1980). *Science* **208**, 145.
New, G. H. C. (1974). *IEEE J. Quantum Electron* **QE-10**, 115–118.
Shank, C. V. (1975). *Rev. Mod. Phys.* **47**, 649–657.
Shank, C. V., and Ippen, E. P. (1974). *Appl. Phys. Lett.* **24**, 373–375.
Shank, C. V., and Ippen, E. P. (1977). *In* "Dye Lasers" (F. P. Shafer, ed.), pp. 121–143. Springer-Verlag, Berlin and New York.
Shank, C. V., Ippen, E. P., and Dienes, A. (1972). *Dig. Tech. Pap.—Int. Quantum Electron. Conf., 7th, Montreal* p. 78.
Shank, C. V., Fork, R. L., Leheny, R. F., and Shah, J. (1979). *Phys. Rev. Lett.* **42**, 112–115.
Weisman, R. B., and Greene, B. I. (1979). *In* "Multichannel Image Detectors" (Y. Talmi, ed.), p. 227. Am. Chem. Soc., Washington, D.C.

Index

A

Absorption spectroscopy
  of photosynthesis, 55–77, 120
  of polyenes, 283, 284
Accessory pigment complex
  initial light absorption by, 28
  in photosynthesis, 4, 18
Acoustooptic deflector, use in laser spectroscopy, 419
Acridine 13, use in laser spectroscopy, 405
Adenine, photochemical studies on, 366–371, 374
Adenine–thymine pair, hydrogen bonding in, 364
Adenosine, photochemical studies on, 375
Adenosine diphosphate (ADP), in photosynthesis, 56, 97
Adenosine triphosphate (ATP)
  in photosynthesis, 4, 5, 56, 97
  synthesis of, by visual pigments, 240
Adenylyl-3′-5′-uridine (ApU), difference spectrum of, 371
Algae
  energy transfer in accessory complexes of, 46–49
  optical cross section of PSU studies on, 228, 229
  photosynthesis in, optical studies, 120
  photosynthetic pigments of, 7, 8, 197
Allophycocyanin (APC), 46, 211
  as algal pigments, 8, 173
  energy transfer by, 19
Ammonyx LO, as rhodopsin solvent, 260, 263
Amplifiers, for picosecond laser spectroscopy, 400
Anacystis nidulans, fluorescence measurements on, 36, 37
Annihilation (of fluorescence), 157–191, 201–212, 218–224
  kinetics of, 225–227
  statistical aspects of, 218
"Annihilation sink," formation of, 224
Antenna chlorophyll, 4, 121
  as (chlorophyll $a)_n$, 128
  excitation energy of, 197
  fluorescence properties of, 169
  photon capture by, 232
  singlet excited molecules of, 73
Antenna pigments, for energy harvesting, 79, 80, 193, 194
APC fluorescence, from phycobilisomes, 211
Arnon method of chlorophyll estimation, 7
Arrhenius analysis
  of bathorhodopsin formation, 264, 265
  of photodissociated hemoprotein, 331, 334, 335

B

Bacteria
  photosynthetic, 120
    electron transfer reactions in, 79–117
    fluorescence studies on, 181–183
    reaction centers, fluorescence studies on, 187–189
Bacteriochlorophyll (BChl) pigments
  in bacterial reaction centers, 187
  biradical spectroscopy of, 93, 94
  as electron transfer intermediate, 90–92
  fluorescence studies on, 17, 188
  light wavelength absorbed by, 79
  in photosynthetic membranes, 193
  in reaction centers, 81, 97
  triplet states, 87–92
  stimulated emission and lasing for, 161
Bacteriochlorophyll *a*
  coordination interactions in, 123
  EPR signals of, 122
  fluorescence of, 130
  linked pairs of spectral properties, 142

Bacteriochlorophyll *a* (*cont.*)
  spectroscopic and lasing properties of, 132
  structure of, 124
  water adducts of, 125
Bacteriochlorophyll *b*, coordination interactions in, 123
Bacteriochlorophyll *c*
  coordination interactions in, 123
  spectroscopic and lasing properties of, 132
Bacterioopsin, 239
Bacteriophage, picosecond spectroscopy of, 371, 372
Bacteriopheophytin
  in bacterial reaction centers, 81, 187
  fluorescence studies on, 188
Bacteriorhodopsin (bR)
  absorption spectroscopy of, 273–277, 282, 284–286
    low temperature, 273, 274
    ultrafast, 274–276
  components of, 239
  energy transfer by, 240
  fluorescence measurements on, 277–279
    from intermediates, 278
    photochemical applications, 278, 279
    ultrafast measurements, 278
  forms and isomers of, 245
  ground state precursor of, 294
  hypsochromically shifted early intermediates of, 276
  light and dark reactions of, 248–250, 271, 272
  photochemical and thermal reactions of, 247, 272
    kinetics, 276, 277
  in photochemistry of retinal, 240
  photoisomerization in, 249, 281–297
    model for, 289
  photoproduct of 13-*cis* form of, 277
  primary events in, 271–280
  in proton transfer, 295
  resonance Raman spectroscopy of, 245, 246
  structure of, 240
  subpicosecond spectroscopy of, 421, 422
bR548 bacteriorhodopsin, 249
  properties of, 245, 271, 272
bR568 bacteriorhodopsin, 249
  properties of, 245, 271, 272
Bathobacteriorhodopsin, laser spectroscopy of, 422, 423
Bathochromic shift, in photochemistry, 305
Bathorhodopsin
  absorption spectroscopy of, 242
  configuration of, 251
  formation of, 246, 250, 282, 288
    kinetics, 252, 292, 299
  isorhodopsin from, 311
  in photochemical cycle, 250–254, 281
    kinetics, 263, 264
  primary event, 260–268, 286
  photoisomerization of
    quantum yield, 313, 314
    trajectories, 309–316
  potential energy of, comparison with rhodopsin, 306
  resonance Raman spectroscopy of, 251
  in spectral red shift, 253
Beer–Lambert absorption law, 179
p-Benzoquinones, reduction by magnesium-containing porphyrins, 104
Biliproteins, absorbing species of, 8
Bilirubin, laser spectroscopy of, 425, 426
Bimolecular processes, in photosynthetic membranes, 193–214
Bimolecular annihilation constant, expression for, 165
Biradical spectroscopy, of electron transfer in photosynthetic bacteria, 93
Bleaching process, in vision, 259
Born–Oppenheimer local minimum, 308
n-Butylamine, cross-linking of chlorophyll by, 125
n-Butylamine retinal Schiff base, isomerization of, 268

C

C_3 pathway, reactions of, 5
C_4 pathway, reactions of, 5
C705 pigment
  fluorescence properties of, 209
  in PS I, 169
  energy transfer in, 171–173
Calvin cycle, *see* C_3 pathway
Carbohydrate, as photosynthesis product, 3, 12, 55
Carbon dioxide, plant incorporation of, 5

Carbon monoxide, binding and dissociation
    by heme compounds, 340-346
Carboxyhemoglobin
  photodissociation of, 328, 329, 332, 334
  picosecond spectroscopy, 348-357
Carboxymyoglobin
  photodissociation of, 327, 331
  picosecond spectroscopy, 348-354
β-Carotene, as major carotenoid, 7
Carotenoid(s)
  energy transfer by
    efficiency, 19
  in PS II, 35, 40
  initial light absorption by, 28
  properties and occurrence of, 7, 8
  in purple membrane, 277
  triplet formation by, 72, 179, 184, 185
    lifetime, 163
Chlorella
  chloroplast of, 9
  fluorescence yield of, 150, 227, 228
  optical cross section studies on, 228-230
Chlorella pyrenoidosa
  fluorescence measurements on, 30, 35-
    41, 208, 212
    decay time, 167, 168
Chlorophyll(s)
  anion of, in PS II, 22
  coordination interactions and self-
    assembly of, 123-125
  covalently linked pairs, as photoreaction
    center models, 138-144
  determination of, 7
  energy transfer by, 16, 18-20
  EPR properties of, 138, 140
  excimer formations in, 28, 29
  excited, energy dissipation by, 16
  fluorescence of, 16-18, 43, 44, 129-132,
    163-168, 179-183
    self-assembled systems, 149
  laser effects on, 158-162
  lasing properties of, 132-137
  light-harvesting type, 4
  number of in *Chlorella* trap, 231
  optical properties of, 126-129
  photochemistry of, 16
  photophysics of, 135, 158, 159
    exciton-exciton annihilation, 162
  photoreaction center type, *see* P700
    pigment

  in photosynthesis, 3
  photosynthetic unit and, 121-123
  properties and functions of, 5-7
    relevant to models, 123-137
  –protein complexes, 9, 10
  reaction centers of, 10, 11
  spectral properties of, 7, 132, 143
  stimulated emission and lasing for, 161
  structures of, 6, 124
  in thylakoid membranes, 9
  triplets, formation and decay, 181, 182
  water cross-linking of, 138
Chlorophyll *a*
  absorption maximum of, 15, 16
  absorption spectrum of, 30
  as acceptor, 127
  anion radical of, 63, 65
  covalently linked pairs of, 140, 141
  at cryogenic temperatures, 145, 146
  dimers of, 125
    in PS I, 21, 22, 57, 58
    spectral properties, 139
  duplication of properties of, 49
  emission spectrum of, 17
  energy level diagram of, 136
  in energy transfer, 18, 19, 28, 232, 233
  EPR signals of, 122
  fluorescence properties of, 18, 31, 34-
    37, 40, 49, 129, 130, 169, 179-183,
    207
  function of, 18, 19
  initial light absorption by, 28
  laser effects on, 161
    photophysics, 163
  major forms of, 16
  as monomer, 123
  NMR spectroscopy of, 127, 128
  occurrence and structure of, 6
  optical properties of, 126-129
  pheophytin adduct of, 129
  photochemical role of, 9, 10
  photophysics of, 135-137
  in photosynthetic bacteria, 81, *see also*
    bacteriochlorophyll
  in photosynthetic membranes, 193
  radiative lifetime of, 30
  singlet exciton lifetime in, 167, 168
  spectroscopic and lasing properties of,
    132, 134, 135, 142, 145, 146, 161
  structure of, 124

Chlorophyll *a* (*cont.*)
   in thylakoid membranes, 173
   two-photon absorption by, 159
Chlorophyll *b*
   coordination interactions in, 123
   and Emerson enhancement effect, 15
   energy transfer in, 233
   fluorescence properties of, 35–37, 40, 130, 169
   hydration of, 129
   initial light absorption by, 28
   occurrence of, 6
   spectroscopic and lasing properties of, 132, 135
   structure of, 124
Chlorophyll *c*, fluorescence emission from, 46
Chlorophylls c_1 and c_2, occurrence of, 6
Chlorophyll *d*, occurrence of, 6
Chlorophyll dihydrate polycrystal, fluorescence emission of, 50
Chlorophyll special pair (Chl_{sp})
   concept of, 123
      energy transfer by, 150–152
      model, 138
Chloroplast(s), 55
   chemical composition of, 9
   carotenoid triplet formation in, 183
   cooling effects on, 17
   domains in, 167
   fluorescence properties of, 167, 169, 170, 179–183
   laser effect on chlorophyll in, 161, 164
   microsecond pulse excitation of, 183–187
   optical studies on photosynthesis in, 120
   photosynthetic apparatus of, model, 169
   photosynthetic membranes in, 8
   picosecond pulse train excitation of, 186, 187
   structure of, 8, 9
   triplet states in, 73, 184
Chromatic transients, in photosynthesis, 15
Chromatium vinosum
   bacteriochlorophyll of, 81
   reaction centers of, 85, 94
Chromatophore
   antenna–RC complex in, 80
   optical studies on photosynthesis in, 120
   of photosynthetic bacteria, fluorescence emission of, 80

Chromophores, of visual pigments and purple membrane, 239, 240
Circular dichroism
   bacteriochlorophyll studies by, 90–92
   of chlorophyll reaction centers, 10
   of heme–protein binding, 341
   of rhodopsin, 248
Coatings, for laser substrates, 395
Computer
   simulation of mode-locked pulse from noise by, 391
   use in chlorophyll laser studies, 135
   use in photoisomerization calculations, 302
Continuous-wave (CW) mode-locked dye laser, use in DNA studies, 379
Copper-containing porphyrin electron-transfer properties of, 99
Coumarin 6, IR absorption by, 364
Counterion, visual pigment formation of, 244, 245
CP II, fluorescence properties of, 208
Cray computer, use in photoisomerization calculations, 302
Cyanobacteria, 121
   optical studies on photosynthesis in, 120
Cytochromes
   function of, 321
   in photosynthetic bacterial reaction centers, 95
Cytochrome *b*-559, photooxidation of, 15
Cytochrome *c*-553, in bacterial reaction centers, 85, 96
Cytochrome *f*
   photooxidation of, 15
   in photosynthesis, 14
Cytosine, photochemical studies on, 366–371

D

Dark reactions
   occurrence of, 9
   pathways of, 5
   in photosynthesis, 4, 5
Deoxyhemoglobin, resonance Raman spectroscopy of, 334
Depolarization, in DNA, picosecond measurements of, 378–382
Dicarboxylic acid cycle, *see* C_4 pathway

3-(3',4'-Dichlorophenyl)-1,1-dimethylurea (DCMU)
  effect on PS II, 67
  use in fluorescence measurements, 37, 43, 44
Diffusion-limited system, definition of, 198
Digitonin, as rhodopsin solvent, 267
Dihydroretinals
  absorption spectroscopy of, 246
  11,12-, resonance Raman spectroscopy of, 245
Dimer pair P, in bacterial reaction centers, 187
N,N-Dimethylformamide (DMF), effect on electron transfer in porphyrins, 102, 104
Dioxane, effect on, 125
Diporphyrins, energy transfer studies on, 131
DNA
  –dye complexes, selective photodamage in, 375–378
  electronic deactivation of, 363
  picosecond depolarization and torsional motion studies on, 378–382
  synthesis, time-resolved fluorescence spectroscopy of, 31
  ultrafast techniques applied to studies of, 361–383
DODCI, use in laser spectroscopy, 419
Domains
  exciton creation in, 202–205, 231, 232
  role in laser effects on chlorophylls, 166, 167, 171
Dye–biomolecule complexes, selective photodamage of, 375–378
Dye laser
  bacterial pheophytin studies using, 91
  in photosynthesis research, 174
Dyes, mode-locking type, 396

E

Ectothiorhodospira shaposhnikovii, electron transfer reactions in, 90
Electron nuclear double resonance (ENDOR) technique
  of chlorophyll reaction centers, 10
  pheophytin studies using, 71
  photosynthesis studies by, 58, 63

Electron paramagnetic resonance (EPR) spectroscopy
  of bacterial ferroquinone complex, 81
  of chlorophyll, 123
  reaction centers, 10
  doublet development in, 71
  of photosynthesis, 55–77, 120
Electron spin resonance (ESR)
  on bacterial reaction centers, 91, 92
Electron transfer reactions
  from chlorophyll special pair, 152
  in photosynthetic bacteria, 79–97
  intermediate I, 82–92
  in pyrochlorophyll *a* dimers, 105–114
  in pyropheophorbide *a* dimers, 105–114
  in reaction center models, 97–114
Electron transport chain, in photosynthesis, 4
Electron transport chain, primary acceptor in, blockage of, 43
Emerson enhancement effect, in photosynthesis, 15
Energy, storage of, in photochemical cycle, 252
Energy-level diagram, of laser-excited chlorophyll, 160
Energy transfer
  in photosynthesis, 28, 29, 162, 193
  in phycobilisomes, 8, 173, 174
  picosecond spectroscopy of, 387
  by visual pigments, 240
Escarole chloroplasts, fluorescence emission studies on, 35
Ethanethiol, cross-linking of chlorophyll by, 125
Ethanol, cross-linking of chlorophyll by, 125
Ethidium bromide–DNA complex, forsional movement studies on, 380, 381
Ethyl chlorophyllide *a*, self-assembled special pairs of, 146
Ethyl chlorophyllide *a*, water adduct of, 128
Ethylene glycol, bridge of, for chlorophyll *a* dimers, 140
EXAFS, as source of myoglobin spatial structure, 322, 323
Excited molecules, kinetic equations for, 174, 175

Excited-state absorption, of chlorophylls, 160
Exciton
  in antenna–RC complex, 80
  percolation, application to photosynthesis, 29
  theory, application to chlorophyll spectra, 126
Exciton–exciton annihilation, 157–191, 201–212, 218–224
  in chlorophyll photochemistry, 162
  kinetics of, 225–227
  in phycobiliproteins, 174
  in time-resolved fluorescence spectroscopy, 38–43

F

Ferredoxin-NADP reductase (FNR), in photosynthesis, 14
Ferroquinone complex, in reaction bacteria centers, 81–85, 95, 97
Ferrous ion, EPR doublet restoration and, 71
Flash photolysis, of dark-adapted bacteriorhodopsin, 272
Fluorescence
  of excited chlorophyll, 16–18, 129–132
  lifetimes, phase measurement of, 35
  plant emission of, 119
  quenching, exciton annihilation and, 41, 42
  yield of, trap effects on, 233, 234
Fluorescence decay kinetics, in photosynthetic membranes, 193–214
Fluorescence spectroscopy
  of bacteriorhodopsin, 277–279
  of chloroplasts, 29, 30
  in photosynthesis studies, 120
  picosecond type, *see* Picosecond laser spectroscopy
  time resolved, *see* Time-resolved spectroscopy
Forster critical distance, 209
Forster dipole–dipole approximation, 194, 195
Förster's inductive resonance mechanism
  in excitation transfer in photosynthesis, 20, 131
  in hemoprotein photodissociation, 327

Forster pairwise transfer, in energy transfer in accessory complexes, 47
Forster transfer rate, 210
Frank–Condon factors, 314
  in electron transfer reactions, 103
Fucoxanthol, energy transfer by, efficiency, 19

G

Grana, in chloroplasts, 8, 9
Ground state depletion, in phycobiliproteins, 159
Guanine, photochemical studies on, 366–371
Guanine–cytosine pair, hydrogen bonding in, 364

H

Halobacterium halobium, bacteriorhodopsin of, 240, 271, 282
Hatch–Slack pathway, *see* C_4 pathway
Hemoglobin (Hb)
  function of, 321
  oxygen and carbon monoxide binding by, 339–355
  structure and states of, 322–325
  x-ray diffraction of, 323
Hemoproteins, 319–357
  conformational changes in, 333, 334
  fast elementary processes of, 326, 327
  functions of, 321
  ligand binding to, potential energy diagram of, 329, 330, 344
  occurrence of, 321
  photodissociation of, 326, 327
    intermediate states, 328, 329
    light absorption and energy transfer, 327, 328
    picosecond spectroscopy, 346–355
    quantum yields, 330
  picosecond spectroscopy of photolysis of, 339–357
  structure and states of, 322–325
Heptofluorodimethyl octane dionato europium [Eu(fod)$_3$], as electrophile, 127
Hubbard–Kropf isomerization hypothesis, of vision primary event, 259

INDEX 435

Human serum albumin, bilirubin binding to, laser spectroscopy of, 426
Hypsorhodopsin
 formation of, 250, 267
 in vision primary event, 263
 kinetics, 267

I

IND-CISD semiempirical MO formalism, application to photoisomerization, 302, 303
Infrared (IR) spectroscopy, of chlorophyll, 123
Infrared-ultraviolet method, as vibrational method, 363, 364
Ionone, binding site for, in visual pigments, 243, 285, 303
β-Ionylidene ring, in retinal, 241
Iron porphyrin, binding studies on, 341
Iron protoporphyrin, structure of, 322
Isomerization hypothesis
 in photochemical cycle, 250, 251, 253, 260, 261, 266, 268
  charge separation and, 288–290
  dynamics, 292–294
Isorhodopsin as artificial visual pigment, 243
 photochemistry of, 266, 311
 resonance Raman spectroscopy of, 244
Isotopes, separation of, using lasers, 362

J

Jaundice, in newborns, bilirubin spectroscopy studies on, 425, 426

K

Kerr gate, fluorescence emission detection by, 33–36, 408
Kinetics
 of annihilation, 225–227
 of excited molecules, 174, 175

L

"Lake" model, of energy migration, 42, 81, 195, 203, 207, 211, 219

Lasers
 comparison among, 394
 components and designs of, 392–397
 modes of, 389, 390
 picosecond type
  energy diagram, 389
  general principles of, 388–392
  effects on chlorophyll systems, 158–162
  photophysical processes, 162, 163
 mode-locked gas type, 162, 417
  fluorescence studies by, 35, 167, 168
 ultrafast techniques using, 362, 385–427
Lasing properties
 of chlorophylls, 132–135
 of triester models, 151, 152
Light-harvesting complex (LHC)
 domain size in, 208
 fluorescence emission from, 35, 44, 46
Light-harvesting protein complex (LHCP), 10, 57
 triplet-fluorescence quenching in, 40
Light reactions, in photosynthesis, 4
Linear dichroism, bacteriochlorophyll studies by, 90, 92, 96
Lithium chlorate, in studies of EPR doublet, 71
Lithium niobate, parametric generation in, 402, 403
Lutein, as photosynthesis pigment, 7
Lysine
 ε-amino group of, binding with chromophore, 243, 282, 303
 conformational energy of in visual pigments, 305

M

Mackinney method of chlorophyll estimation, 7
Magnesium
 in chlorophyll, 6, 123, 124
 ion effect on energy transfer from PS II to PS I, 44, 195
 EPR doublet restoration and, 71
Magnesium-containing porphyrin (MgP)
 electron transfer properties of, 98–105, 131
 molecular properties of, 100
Magnesium hexafluoroacetylacetonate as electrophile, 127

INDEX

Magnesium octaethylporphyrin, in reaction center studies, 84, 85
Magnesium tris(pyrochlorophyllide a)-1,1,1-tris(hydroxymethylethane) triester, structure and spectral properties of, 151, 152
Magnetic circular dichroism, of heme–protein binding, 341
Magnetic interactions, in bacterial reaction centers, 95
Malachite green, use in laser spectroscopy, 419
Manganese ion, EPR doublet restoration and, 71
Mauzerell limit, application to fluorescence decay kinetics, 207
Metalloporphyrins, photodissociation of, 342
Metarhodopsins, 272
  resonance Raman spectroscopy of, 248
Methylpheophorbide, spectroscopic and lasing properties of, 132, 133
14-Methylretinal, resonance Raman spectroscopy of, 244
Microalgae, optical studies on photosynthesis in, 120
Monte Carlo techniques, in fluorescence decay kinetics, 200
Microsecond pulse excitation, of chloroplasts, 183–187
Mode-locked lasers
  development of, 27, 387
  diagram of, 388
  pulse formation by, 390, 391
Modulator crystals, characteristics of, 399
Molecular dynamics
  formalism for, 306–309
  of rhodopsin photochemistry, 299–317
Multiphoton dissociation methods, in DNA studies, 362, 363
Myoglobin (Mb)
  absorption spectroscopy of, 324, 325
  active center of, schematic cross section through, 324
  compounds
    photodissociation studies on, 331, 341–355
    picosecond spectroscopy, 346–357
  conformational changes in, 324, 334
  function of, 321
  structure and states of, 322–325

N

NADP, in photosynthesis, 56
NADPH, in photosynthesis, 5
Nanosecond pulse excitation, 215
  exciton annihilation effects on, 177, 178
  criteria for, 178, 179
  in photosynthesis studies, 174–183
    exciton annihilation effects, 177, 178
    intensity-dependent transmittance, 176
    kinetic equations, 174, 175
    steady state approximation, 175–177
Neodymium glass lasers
  design aspects of, 393
  photochemical studies using, 260, 261
  properties of, 394
  for time-resolved fluorescence spectroscopy, 31, 33, 35, 36
  use of picosecond spectroscopy, 347
  wavelength of, 400
Neodymium lasers
  design aspects of, 392–397
  energy diagram of, 389
  mode-locked pulses in, 392
  properties of, 394
  wavelength of, 400
Neodymium-YAG lasers
  for time-resolved fluorescence spectroscopy, 31, 91
  of DNA, 363
Neoxanthin, as photosynthesis pigment, 7
Newton's equations of motion, 306, 309
Newton's second law, 301
Nitrogen lasers
  effect on RGG dye, 176
  in photosynthesis research, 174
Nostoc sp.
  excitation energy distribution in, 173, 174
  fluorescence studies on, 42, 43, 48, 211
Nuclear magnetic resonance (NMR) spectroscopy
  of bacteriochlorophyll a linked pairs, 142
  of chlorophyll, 123, 127, 128
Nucleic acids
  photochemical reactions in induced, by visible and UV pulses, 365–379
  theory of selectivity in, 372–375

INDEX

O

Octaethylporphyrin, in reaction center studies, 84
Opsin, 239, 291, 303
　chromophore binding to, 285
　retinal dissociation from, 246
Optical dyes, for subpicosecond ultrafast laser techniques, 417, 418
Optical effects, nonlinear, 363–365
Optical multichannel analyzer (OMA)
　fluorescence detection by, 184
　use in laser spectroscopy, 425
Oxygen, binding and dissociation of, by heme compounds, 340–346
Oxyhemoglobin
　photodissociation of, 332, 334
　　picosecond spectroscopy, 348–357
Oxymyoglobin
　photodissociation of, 332
　　picosecond spectroscopy, 348–357

P

P^F and P^R triplet states, in photosynthetic bacteria, 87–92
π-electron, delocalization of, role in absorption spectroscopy, 283
π systems, of chlorophyll, 145
P430 pigment, as iron–sulfur protein, 66
P680 pigment
　photooxidation of, in PS II, 122
　as PS II primary electron donor, 58, 59
　　EPR spectroscopy, 66–69, 75
　separation of, 15
　spectral properties of, 11
P700 pigment, 121
　model for, optical and EPR properties, 140, 143
　photo-EPR signal of, 122
　photooxidation of, 15, 122
　as PS I primary electron donor, 58
　　EPR spectroscopy, 61–65
　redox properties of, 141, 142
　separation of, 15
　spectral properties of, 10
P865 pigment, 121
　photo-EPR signal of, 122
　photooxidation of, 122
　redox properties of, 141

Paillotin theory, application to exciton–exciton annihilation, 200, 202, 203, 207, 208, 211, 212
Pauli form, use in exciton annihilation estimates, 202, 203
Pentose phosphate pathway, see C_3 pathway
Phase fluorimetry, in time-resolved fluorescence spectroscopy, 32, 34
4,5-Phenanthroline, effect on bacterial reaction centers, 87
Pheophytin
　in photosynthetic bacteria, 82, 97
　in electron transfer, 89, 91, 92
　phototrapping of, in TSF-IIa fragments, 74
　reduction of, spectroscopic studies of, 68, 69
Pheophytin a
　chlorophyll a aggregate with, 129
　fluorescence studies on, 129, 130
　spectroscopic and lasing properties of, 132, 133
Phosphoglyceraldehyde, in C_3 pathway, 5
Phospholipids, in thylakoid membrane, 9
Phosphorescence, plant emission of, 119
Photochemical cycle
　chromophore isomerization in, 249
　energy storage in, 252
　isomerization in, 250, 253, 260, 261, 266, 268
　kinetics of, 252
　low-temperature studies of, 265
　models for primary event in, 253, 254
　primary event of, 250–254
　　ground state energetics, 286, 287
　　molecular dynamics, 299–317
　quantum yields of, 253
　spectral red shift in, 253
Photodissociation, of hemoproteins, 326–335
Photoisomerization, 250, 251, 253, 260, 261, 266, 268
　charge separation and, 288–290
　dynamics, 292–294, 302–316
　potential surfaces of, 303–306
　quantum yield from, 291, 292, 302, 309–315
　trajectories, 309–315
Photomultiplier detectors, in time-resolved fluorescence spectroscopy, 32

Photon counting, in time-resolved fluorescence spectroscopy, 32
Photoreaction centers, synthetic, spectral properties of, 139
Photosynthetic bacteria, electron transfer reactions in, 79–117
Photosynthetic systems, multiple excitation in, statistical theory of, 215–235
Photosynthetic membranes
  CP complexes in, 195
  fluorescence decay kinetics of, 193–214
  structure and function of, 8, 9
Photosynthetic unit (PSU)
  chlorophyll and, 121–123
  concept of, 12, 195
  connectivity between, 166, 167
  in domains, 204, 205, 232
  energy transfer in, 20, 197
  excitation energy distribution in, 168–173
  fluorescence studies on, 28, 39, 42, 43, 210, 211
    annihilation in, 218–223
    multitrapped units, 219–223
  laser effects on, 164, 165
  models for, 20
  optical cross sections of, 215, 228–230
  traps in, 228, 229
Photosystem I (PS I), 4, 56, 121, 195
  antenna pigment of, 173
  C705 pigment in, 169
  chlorophyll–protein complexes in, 10
  fluorescence emission studies on, 17, 35, 36, 40–46, 172, 173
  fluorescence quenching studies on, 185
  primary electron donor of, 57–59
  primary photochemistry in, 21, 22
Photosystem II (PS II), 4, 56, 121, 122, 195
  chlorophyll antenna molecules in, 169
  chlorophyll–protein complexes in, 10, 150
  closing of reaction center traps in, 37, 43, 44
  early photochemical events in, 59–75
  electron acceptors in reaction centers of, 71
  fluorescence emission studies on, 17, 35, 36, 40–46, 69
  primary electron donor of, 58–60
  primary photochemistry of, 22

reaction center of, fluorescence properties, 179, 180
reduced reaction centers of, 72
Photosynthesis
  absorption and EPR spectroscopy of, 55–77
  apparatus of, 5–11
  in chlorophyll model systems, 119–155
  Emerson enhancement effect in, 15
  early photochemical events in, 55–77
  energy transfer in, 18–20, 28, 29
    mechanism, 19, 20
  fluorescence in, 17, 18
  fundamental concepts of, 11–15
  history of, 4
  light absorption in, 15, 16
  outside living cell, 119, 120
  oxygen yield of, 11, 12
  partial reactions of, 15
  photosynthetic unit of, 11, 12
  pigments of, 5–8
  primary photochemistry of, 20–22
  primary processes of, 3–25
  reaction centers of, 10, 11
  red drop phenomena in, 14
  time-resolved fluorescence emission in, 29–31
    environmental effects, 43–46
    interpretation, 37, 38
  two light reactions in, 13–15
  Z scheme of, 13
Phycobilins
  Emerson enhancement effect and, 15
  fluorescence of, 17
Phycobiliproteins
  as algal photosynthesis pigment, 7, 8
  arrangement of, 46
  chromophores in, 47
  energy transfer in, 48, 49, 233
  fluorescence quantum yield of, 42, 43
  ground state depletion in, 159
  initial light absorption by, 28
  types and occurrence of, 8
Phycobilisomes
  energy transfer in, 8, 46–49, 195, 211, 233
  fluorescence decay kinetics of, 200
  fluorescence quantum yield in, 42, 43
  pigments in, 173, 211

Phycocyanin (PC), 46, 211
  as algal pigment, 8, 173
  energy transfer by, 19
Phycoerythrin (PE), 46, 211
  as algal pigments, 8, 173
  energy transfer by, 19
Picosecond laser spectroscopy
  of bacteriorhodopsin, 274–276
  of chlorophyll a, 159
  electron-transfer studies by, 79, 80
  of hemoprotein photolysis, 339–357
  of hypsorhodopsin photochemistry, 267
  of isorhopsin photochemistry, 266
  lasers for, see Lasers
  photosynthesis studies by, 34–46
  of phycobiliproteins, 174
  single-pulse selection and amplification in, 397–400
  techniques and designs in, 387–415
  time measurements and applications of, 405–413
  of vision primary event, 259, 261–268, 283
  wavelength attainment in, 400–405
Picosecond pulse excitation, 215, 229
  of bacterial reaction centers, 187–189
  of chlorophylls, 163–174, 202–203
  of chloroplasts, 186, 187
Pigments, of photosynthesis, 5–8
Plasmid pBR322, picosecond spectroscopy of, 371, 372
Plastocyanin
  in photosynthesis, 14
  in PSI, 21
Plastoquinone (PQ)
  P680 pigment as, 66, 67
  photooxidation of, 15
  in photosynthesis, 14
Pockels cell, 397, 400
  components of, 398, 399
Poisson distribution, of multiple excitations, 221, 223, 225
Poisson factor, 226
Poisson statistics
  in analysis of fluorescence-quenching data, 167
  application to fluorescence quantum yield, 38, 39
  of multiple excitations, 216, 217

Polyenes, absorption spectroscopy of, 283, 284
Polyene–lysine system, vibrational energy studies on, 314
Polymethine cyanine dyes, use in laser spectroscopy, 405
Porphyridium cruentum
  energy transfer in phycobilisomes of, 47
  excitation energy distribution in, 173, 174, 211
  fluorescence measurements on, 37
  pigments in, 173
Porphyrins
  in reaction center models, 97–105
  triplet states of, 218
Potassium diphosphate (KDP)
  use in laser spectroscopy, 35, 424
  length, for picosecond pulses, 402
Potassium isocyanate, effect on bacterial reaction centers, 87
Prelumirhodopsin, see Bathorhodopsin
Primary electron donors, of photosystems I and II, 57–59
Proflavine–DNA complex, laser-induced photo damage to, 377, 378
Protein–chlorophyll complexes, 9
Proton pumping and transfer, in photochemical cycle, 249, 253, 254, 294, 295
Proton transfer time, in photoisomerization, 316
"Puddle" model, of energy migration, 81, 195, 219
Purple membranes, 239–258
Purple photosynthetic bacteria, bacteriochlorophyll a reconstruction of, 125
Pyrochlorophyll a
  linked pairs of, 140, 141
  electron transfer studies on, 152
  optical properties, 142
  as reaction center model, electron transfer in, 105–114
  spectroscopic and lasing properties of, 132
  structure of, 124
  trimer of, structural drawing, 110
Pyropheophorbide a, as reaction center model, electron transfer in, 105–114
Pyropheophytin a, fluorescence studies on, 129, 130

Q

Q acceptor molecule
  concept of, 14
  as a fluorescence quencher, 67, 68
Q bands, of chlorophyll spectrum, 7
Q-switched ruby lasers, in photosynthesis research, 174
Quantasomes, concept of, 12
Quantum yield(s)
  of hemoprotein dissociation, 330
  of photochemical process, 253, 307
  from photoisomerization, 291, 292
  from PSU excitation, 217
Quencher
  definition of, 198
  effect on annihilation process, 224, 225
Quinacrine–DNA complex, laser-induced photodamage to, 375–377

R

Reaction centers (RC)
  of chlorophyll, 10, 11
  energy trapped in, 122
  models of
    electron transfer reactions in, 97–114
    porphyrins, 98–105
    pyrochlorophyll and pyropheophorbide, 105–114
  of photosynthetic bacteria, 79
    components, 81, 82
    distance estimates, 94–96
    drawing of, 86
    fluorescence studies on, 187–189
    thermodynamics, 96, 97
Reaction path, prediction of, in molecular dynamics, 300–302
Red drop phenomena, in photosynthesis, 14
Resonance Raman spectroscopy
  of hemoprotein photodissociation, 331, 332
  of myoglobin, 324, 325
  of visual pigments, 244, 248, 251, 265, 268, 282
Retinal(s)
  absorption spectroscopy of, 242, 243, 283
    "cis peak," 242
  artificial pigment formation by, 243
  as chromophore, 239
    in bacteriorhodopsin, 271
  isomers of, 9
  photochemistry of, 243
  resonance Raman spectroscopy of, 244, 283, 284
  Schiff base of, protanation, 283, 284
11-*cis*-Retinal, 241
  conformation of, 242
11-*cis*-Retinal-*n*-butylamine, absorption spectroscopy of, 242
trans-Retinal
  formation of, 246
  in vision primary event, 262, 263
Rhodamine B, use in laser spectroscopy, 405
Rhodamine 6G
  laser effects on, 161, 177, 405
    fluorescence lifetime, 178
    triplet yield from, 176
  as optical dye, 418, 420
  relaxation time in, 362
Rhodopseudomonas sphaeroides
  bacteriochlorophylls of, 80–83, 89
  chromatophores of, 199
  electron transfer reactions in, 92, 104
    biradical spectroscopy, 93, 94
    thermodynamics, 96, 97
  fluorescence studies on, 188, 207, 212
  mutants, 43
  laser effects on, 166
  reaction centers of, 86–88, 94, 95
Rhodopseudomonas viridis
  bacteriochlorophyll of, 81, 82, 90
  reaction centers of, 97
Rhodopsin
  absorption spectroscopy of, 242, 276, 285, 286
  active site model of, 304
  binding and color of, 243–245
  bleaching-sequence diagram of, 247
  cis–trans-isomerization of, 289, 299
    molecular dynamics, 293–295, 299–317
    quantum yield, 307, 308, 312, 313, 315
    time, 315
    trajectories, 309–316
  components of, 239
  light and dark reactions of, 246–248

in photochemical cycle, 250–254, 267–268, 281
  kinetics, 252, 295
  primary events, 259–265, 299–317
  potential energy of
    compared to bathorhodopsin, 306
    curve for, 287, 291
  in primary photochemical event, 286
  resonance Raman spectroscopy of, 244, 251
  in retinal photochemistry, 240
Rhodospirillum rubrum, electron transfer reactions in, 89–91
Ribulose diphosphate, in C_3 pathway, 5
Rieske protein, in photosynthesis, 14
RNA, selectivity in component of, 371, 372
Ruby lasers
  design aspects of, 392–397
  energy diagram of, 389
  mode-locked pulses in, 392
  properties of, 394
  wavelength of, 400

S

Schiff bases of visual pigments
  absorption spectroscopy of, 243
  protonation of, 243, 253, 265, 271, 275, 276, 295, 296
  singlet state energy of, 309
  UV absorption of, 244
Scratch-dig values, for laser components, 395
Second-harmonic generation (SHG), in linear optic phenomena, 401, 402
Self-phase modulation, in absorption spectroscopy, 403, 405
Semiannihilation, excitation lifetime and, 233, 234
SHG spatial width spectroscopy, principles of, 401–407
Singlet excitons, 174
  lifetime of, 168
Singlet–singlet exciton annihilation
  fluorescence and, 165
  on nanosecond pulse excitation, 177, 178
Singlet–triplet annihilation effects, on nanosecond pulse excitation, 177, 178

Soret bands, of chlorophyll spectrum, 7
Spectral red shift, in photochemical cycle, 253
Spinach chloroplasts
  fluorescence studies on, 35–37, 40, 42, 185, 205, 206, 211, 212
  at low temperatures, 45
  laser effects on, 161, 165
Squid, rhodopsin studies on, 246
Statistical theory, of multiple excitation in photosynthetic systems, 215–235
Steady state approximation
  of nanosecond pulse excitation, 175–177
  three-level system in, 176, 177
Stern–Volmer kinetics, equation for singlet–singlet exciton annihilation based on, 165
Stimulated Raman scattering (SRS), as four-photon effect, 403
Stokes frequencies, of laser pulses, 404
Streak cameras
  comparison of types of, 410, 411
  fluorescence emission detection by, 33, 34, 36, 37, 50, 408–410
  subpicosecond type, 427
Stokes shifts, laser behavior and, 134
Stroma lamellae, in chloroplasts, 9
Subchloroplast particles, spectra properties of, 11
Subnanosecond fluorescence kinetics measurement by streak camera, 33
Subpicosecond ultrafast laser techniques, 417–427
  for bacteriorhodopsin, 275
  future developments in, 426, 427
  measurement techniques in, 420–427
  pulse generation in, 417–420

T

Tetracene, exciton interactions in, 42
Tetrahydrofuran diporphyrins in, absorbance spectroscopy of, 101
m-Tetraphenylporphyrin in electron transfer reactions, 105
Tetrapropylammonium perchlorate in reaction center studies, 84
Thiocapsa roseoperscina electron transfer reactions in, 90

Thylakoid(s)
  chlorophyll in, 8
  fluorescence and, 129
  energy transfer in, 18
  membrane
    chlorophyll states in, 16
    conformational changes in, 44
    fluid mosaic model of, 9
    as photosynthesis site, 55
    pigments, 173
  optical studies on photosynthesis in, 120
Thymine
  IR spectrum of, 364
  photochemical studies on, 366–374
Time-resolved spectroscopy, 27–53, 406
  absorption kinetics of, 412, 413
  apparatus for, diagram, 33
  of Chl a–water aggregate systems, 50
  energy transfer studies by, 46–49
  excitation sources for, 31, 32
  fluorescence kinetics of, 408–411
  fluorescence quenching and exciton annihilation effects in, 38
  at low temperatures, 45, 46
  of photosynthesis, 27, 29–31, 120
    environmental effects, 43, 44
  techniques in, 31–34
Torsional kinetic energy, sources of, restraints by molecular dynamics formalism, 306, 307, 310
Torsional motion, in DNA, picosecond measurements of, 378–382
Transient absorption apparatus, schematic diagram of, 424
Traps
  for excitations, 218–227
    effect on excitation yield, 233, 234
Tribonema aequale, fluorescence of, 147–150
Triplets
  formation, by chlorophyll pairs, 143, 144
  from mode-locked pulse trains, 167, 168
  role in fluorescence quenching, 42
  yield of, from carotenoids, 184
Triplet excited states, 174
  quantum efficiency of formation of, 173
Tryptophan
  fluorescence property change related to protein stability by, 248
  in hemoprotein dissociation, 327

TSF-IIa fragments, absorbance changes in, 69, 70, 73, 74
Two-photon absorption, for chlorophyll a, 159
Two-photon fluorescence spectroscopy development and principles of, 406
Tyrosine, in hemoprotein dissociation, 327

U

Ultrafast techniques, applied to DNA studies, 361–383
Uracil, photochemical studies on, 366–367, 374
Uridine, photochemical studies on, 375

V

Violaxanthin, as photosynthesis pigment, 7
Vision
  laser spectroscopy of events in, 237–317
  primary events in, 259–268
Visual pigments, 239–258
  absorption spectroscopy of, 282, 284–286
  chromophore binding and color, 243–246
  energy transfer by, 240
  excited state processes of, 290–296
  formation of, 243
  light and dark reactions of, 246–250
  photoisomerization of, 281–297
    charge separation and, 288–290
    conformation changes, 288, 289
    in primary photochemical event, 286–290
    models, 289
  resonance Raman spectroscopy of, 244, 282
  thermal isomerization of, 287

W

Water, hydrogen fuel from dissociation of, 50

X

X-ray scattering, in bacterial reaction center studies, 95
Xanthophylls, occurrence of, 7

Z

Z electron donor, in PS II, 22
"Zero-mode" approximation, in fluorescence decay kinetics, 198
Zinc tris(pyrochlorophyllide *a*)-1,1,1-tris(hydroxymethyl ethane) triester
structure and spectral properties of, 151, 152
Z scheme
of photosynthesis, 13
diagram, 56